W9-BDI-245

Advanced Calculus of Several Variables

Advanced Calculus of Several Variables

C. H. EDWARDS, JR.
The University of Georgia

DOVER PUBLICATIONS, INC.
New York

Bibliographical Note

This Dover edition, first published in 1994, is an unabridged, corrected republication of the work first published by Academic Press, New York, 1973.

Library of Congress Cataloging-in-Publication Data

Edwards, C. H. (Charles Henry), 1937–
Advanced calculus of several variables / C.H. Edwards, Jr.
p. cm.
Originally published: New York : Academic Press, 1973.
Includes bibliographical references and index.
ISBN 0-486-68336-2 (pbk.)
1. Calculus. I. Title.
QA303.E22 1994
515′.84—dc20 94-24204
CIP

Manufactured in the United States of America
Dover Publications, Inc., 31 East 2nd Street, Mineola, N.Y. 11501

To My Parents

CONTENTS

IV Multiple Integrals

V Line and Surface Integrals; Differential Forms and Stokes' Theorem

VI The Calculus of Variations

PREFACE

This book has developed from junior–senior level advanced calculus courses that I have taught during the past several years. It was motivated by a desire to provide a modern conceptual treatment of multivariable calculus, emphasizing the interplay of geometry and analysis via linear algebra and the approximation of nonlinear mappings by linear ones, while at the same time giving equal attention to the classical applications and computational methods that are responsible for much of the interest and importance of this subject.

In addition to a satisfactory treatment of the theory of functions of several variables, the reader will (hopefully) find evidence of a healthy devotion to matters of exposition as such—for example, the extensive inclusion of motivational and illustrative material and applications that is intended to make the subject attractive and accessible to a wide range of "typical" science and mathematics students. The many hundreds of carefully chosen examples, problems, and figures are one result of this expository effort.

This book is intended for students who have completed a standard introductory calculus sequence. A slightly faster pace is possible if the students' first course included some elementary multivariable calculus (partial derivatives and multiple integrals). However this is not essential, since the treatment here of multivariable calculus is fully self-contained. We do not review single-variable calculus; with the exception of Taylor's formula in Section II.6 (Section 6 of Chapter II) and the fundamental theorem of calculus in Section IV.1.

Chapter I deals mainly with the linear algebra and geometry of Euclidean n-space $\mathscr{R}^n$. With students who have taken a typical first course in elementary linear algebra, the first six sections of Chapter I can be omitted; the last two sections of Chapter I deal with limits and continuity for mappings of Euclidean spaces, and with the elementary topology of $\mathscr{R}^n$ that is needed in calculus. The only linear algebra that is actually needed to start Chapter II is a knowledge of the correspondence between linear mappings and matrices. With students having this minimal knowledge of linear algebra, Chapter I might (depending upon the taste of the instructor) best be used as a source for reference as needed.

Chapters II through V are the heart of the book. Chapters II and III treat multivariable differential calculus, while Chapters IV and V treat multivariable integral calculus.

In Chapter II the basic ingredients of single-variable differential calculus are generalized to higher dimensions. We place a slightly greater emphasis than usual on maximum–minimum problems and Lagrange multipliers—experience has shown that this is pedagogically sound from the standpoint of student motivation. In Chapter III we treat the fundamental existence theorems of multivariable calculus by the method of successive approximations. This approach is equally adaptable to theoretical applications and numerical computations.

Chapter IV centers around Sections 4 and 5 which deal with iterated integrals and change of variables, respectively. Section IV.6 is a discussion of improper multiple integrals. Chapter V builds upon the preceding chapters to give a comprehensive treatment, from the viewpoint of differential forms, of the classical material associated with line and surface integrals, Stokes' theorem, and vector analysis. Here, as throughout the book, we are not concerned solely with the development of the theory, but with the development of conceptual understanding and computational facility as well.

Chapter VI presents a modern treatment of some venerable problems of the calculus of variations. The first part of the Chapter generalizes (to normed vector spaces) the differential calculus of Chapter II. The remainder of the Chapter treats variational problems by the basic method of " ordinary calculus "—equate the first derivative to zero, and then solve for the unknown (now a function). The method of Lagrange multipliers is generalized so as to deal in this context with the classical isoperimetric problems.

There is a sense in which the exercise sections may constitute the most important part of this book. Although the mathematician may, in a rapid reading, concentrate mainly on the sequence of definitions, theorems and proofs, this is not the way that a textbook is read by students (nor is it the way a course should be taught). The student's actual course of study may be more nearly defined by the problems than by the textual material. Consequently, those ideas and concepts that are not dealt with by the problems may well remain unlearned by the students. For this reason, a substantial portion of my effort has gone into the approximately 430 problems in the book. These are mainly concrete computational problems, although not all routine ones, and many deal with physical applications. A proper emphasis on these problems, and on the illustrative examples and applications in the text, will give a course taught from this book the appropriate intuitive and conceptual flavor.

I wish to thank the successive classes of students who have responded so enthusiastically to the class notes that have evolved into this book, and who have contributed to it more than they are aware. In addition, I appreciate the excellent typing of Janis Burke, Frances Chung, and Theodora Schultz.

Advanced Calculus of Several Variables

I

Euclidean Space and Linear Mappings

Introductory calculus deals mainly with real-valued functions of a single variable, that is, with functions from the real line $\mathscr{R}$ to itself. Multivariable calculus deals in general, and in a somewhat similar way, with mappings from one Euclidean space to another. However a number of new and interesting phenomena appear, resulting from the rich geometric structure of n-dimensional Euclidean space $\mathscr{R}^n$.

In this chapter we discuss $\mathscr{R}^n$ in some detail, as preparation for the development in subsequent chapters of the calculus of functions of an arbitrary number of variables. This generality will provide more clear-cut formulations of theoretical results, and is also of practical importance for applications. For example, an economist may wish to study a problem in which the variables are the prices, production costs, and demands for a large number of different commodities; a physicist may study a problem in which the variables are the coordinates of a large number of different particles. Thus a "real-life" problem may lead to a high-dimensional mathematical model. Fortunately, modern techniques of automatic computation render feasible the numerical solution of many high-dimensional problems, whose manual solution would require an inordinate amount of tedious computation.

1 THE VECTOR SPACE $\mathscr{R}^n$

As a set, $\mathscr{R}^n$ is simply the collection of all ordered n-tuples of real numbers. That is,

$$\mathscr{R}^n = \{(x_1, x_2, \dots, x_n) : \text{each } x_i \in \mathscr{R}\}.$$

Recalling that the Cartesian product $A \times B$ of the sets A and B is by definition the set of all pairs (a, b) such that $a \in A$ and $b \in B$, we see that $\mathscr{R}^n$ can be regarded as the Cartesian product set $\mathscr{R} \times \cdots \times \mathscr{R}$ (n times), and this is of course the reason for the symbol $\mathscr{R}^n$.

The geometric representation of $\mathscr{R}^3$, obtained by identifying the triple (x_1, x_2, x_3) of numbers with that point in space whose coordinates with respect to three fixed, mutually perpendicular "coordinate axes" are x_1, x_2, x_3 respectively, is familiar to the reader (although we frequently write (x, y, z) instead of (x_1, x_2, x_3) in three dimensions). By analogy one can imagine a similar geometric representation of $\mathscr{R}^n$ in terms of n mutually perpendicular coordinate axes in higher dimensions (however there is a valid question as to what "perpendicular" means in this general context; we will deal with this in Section 3).

The elements of $\mathscr{R}^n$ are frequently referred to as *vectors*. Thus a vector is simply an n-tuple of real numbers, and *not* a directed line segment, or equivalence class of them (as sometimes defined in introductory texts).

The set $\mathscr{R}^n$ is endowed with two algebraic operations, called *vector addition* and *scalar multiplication* (numbers are sometimes called scalars for emphasis). Given two vectors $\mathbf{x} = (x_1, \ldots, x_n)$ and $\mathbf{y} = (y_1, \ldots, y_n)$ in $\mathscr{R}^n$, their *sum* $\mathbf{x} + \mathbf{y}$ is defined by

$$\mathbf{x} + \mathbf{y} = (x_1 + y_1, \ldots, x_n + y_n),$$

that is, by coordinatewise addition. Given $a \in \mathscr{R}$, the *scalar multiple* $a\mathbf{x}$ is defined by

$$a\mathbf{x} = (ax_1, \ldots, ax_n).$$

For example, if $\mathbf{x} = (1, 0, -2, 3)$ and $\mathbf{y} = (-2, 1, 4, -5)$ then $\mathbf{x} + \mathbf{y} = (-1, 1, 2, -2)$ and $2\mathbf{x} = (2, 0, -4, 6)$. Finally we write $\mathbf{0} = (0, \ldots, 0)$ and $-\mathbf{x} = (-1)\mathbf{x}$, and use $\mathbf{x} - \mathbf{y}$ as an abbreviation for $\mathbf{x} + (-\mathbf{y})$.

The familiar associative, commutative, and distributive laws for the real numbers imply the following basic properties of vector addition and scalar multiplication:

V1 $\mathbf{x} + (\mathbf{y} + \mathbf{z}) = (\mathbf{x} + \mathbf{y}) + \mathbf{z}$
V2 $\mathbf{x} + \mathbf{y} = \mathbf{y} + \mathbf{x}$
V3 $\mathbf{x} + \mathbf{0} = \mathbf{x}$
V4 $\mathbf{x} + (-\mathbf{x}) = \mathbf{0}$
V5 $(ab)\mathbf{x} = a(b\mathbf{x})$
V6 $(a + b)\mathbf{x} = a\mathbf{x} + b\mathbf{x}$
V7 $a(\mathbf{x} + \mathbf{y}) = a\mathbf{x} + a\mathbf{y}$
V8 $1\mathbf{x} = \mathbf{x}$

(Here $\mathbf{x}, \mathbf{y}, \mathbf{z}$ are arbitrary vectors in $\mathscr{R}^n$, and a and b are real numbers.) V1–V8 are all immediate consequences of our definitions and the properties of $\mathscr{R}$. For example, to prove V6, let $\mathbf{x} = (x_1, \ldots, x_n)$. Then

$$\begin{aligned}(a+b)\mathbf{x} &= ((a+b)x_1, \ldots, (a+b)x_n)\\ &= (ax_1 + bx_1, \ldots, ax_n + bx_n)\\ &= (ax_1, \ldots, ax_n) + (bx_1, \ldots, bx_n)\\ &= a\mathbf{x} + b\mathbf{x}.\end{aligned}$$

The remaining verifications are left as exercises for the student.

A *vector space* is a set V together with two mappings $V \times V \to V$ and $\mathscr{R} \times V \to V$, called vector addition and scalar multiplication respectively, such that V1–V8 above hold for all $\mathbf{x}, \mathbf{y}, \mathbf{z} \in V$ and $a, b \in \mathscr{R}$ (V3 asserts that there exists $\mathbf{0} \in V$ such that $\mathbf{x} + \mathbf{0} = \mathbf{x}$ for all $\mathbf{x} \in V$, and V4 that, given $\mathbf{x} \in V$, there exists $-\mathbf{x} \in V$ such that $\mathbf{x} + (-\mathbf{x}) = \mathbf{0}$). Thus V1–V8 may be summarized by saying that $\mathscr{R}^n$ is a vector space. For the most part, all vector spaces that we consider will be either Euclidean spaces, or subspaces of Euclidean spaces.

By a *subspace* of the vector space V is meant a subset W of V that is itself a vector space (with the same operations). It is clear that the subset W of V is a subspace if and only if it is "closed" under the operations of vector addition and scalar multiplication (that is, the sum of any two vectors in W is again in W, as is any scalar multiple of an element of W)—properties V1–V8 are then inherited by W from V. Equivalently, W is a subspace of V if and only if any linear combination of two vectors in W is also in W (why?). Recall that a *linear combination* of the vectors $\mathbf{v}_1, \ldots, \mathbf{v}_k$ is a vector of the form $a_1\mathbf{v}_1 + \cdots + a_k\mathbf{v}_k$, where the $a_i \in \mathscr{R}$. The *span* of the vectors $\mathbf{v}_1, \ldots, \mathbf{v}_k \in \mathscr{R}^n$ is the set S of all linear combinations of them, and it is said that S is *generated* by the vectors $\mathbf{v}_1, \ldots, \mathbf{v}_k$.

Example 1 $\mathscr{R}^n$ is a subspace of itself, and is generated by the *standard basis vectors*

$$\begin{aligned}\mathbf{e}_1 &= (1, 0, 0, \ldots, 0),\\ \mathbf{e}_2 &= (0, 1, 0, \ldots, 0),\\ &\vdots\\ \mathbf{e}_n &= (0, 0, 0, \ldots, 0, 1),\end{aligned}$$

since $(x_1, x_2, \ldots, x_n) = x_1\mathbf{e}_1 + x_2\mathbf{e}_2 + \cdots + x_n\mathbf{e}_n$. Also the subset of $\mathscr{R}^n$ consisting of the zero vector alone is a subspace, called the *trivial* subspace of $\mathscr{R}^n$.

Example 2 The set of all points in $\mathscr{R}^n$ with last coordinate zero, that is, the set of all $(x_1, \ldots, x_{n-1}, 0) \in \mathscr{R}^n$, is a subspace of $\mathscr{R}^n$ which may be identified with $\mathscr{R}^{n-1}$.

Example 3 Given $(a_1, a_2, \ldots, a_n) \in \mathscr{R}^n$, the set of all $(x_1, x_2, \ldots, x_n) \in \mathscr{R}^n$ such that $a_1x_1 + \cdots + a_nx_n = 0$ is a subspace of $\mathscr{R}^n$ (see Exercise 1.1).

Example 4 The span S of the vectors $\mathbf{v}_1, \ldots, \mathbf{v}_k \in \mathscr{R}^n$ is a subspace of $\mathscr{R}^n$ because, given elements $\mathbf{a} = \sum_1^k a_i \mathbf{v}_i$ and $\mathbf{b} = \sum_1^k b_i \mathbf{v}_i$ of S, and real numbers r and s, we have $r\mathbf{a} + s\mathbf{b} = \sum_1^k (ra_i + sb_i)\mathbf{v}_i \in S$.

Lines through the origin in $\mathscr{R}^3$ are (essentially by definition) those subspaces of $\mathscr{R}^3$ that are generated by a single nonzero vector, while planes through the origin in $\mathscr{R}^3$ are those subspaces of $\mathscr{R}^3$ that are generated by a pair of non-collinear vectors. We will see in the next section that every subspace V of $\mathscr{R}^n$ is generated by some finite number, at most n, of vectors; the dimension of the subspace V will be defined to be the minimal number of vectors required to generate V. Subspaces of $\mathscr{R}^n$ of all dimensions between 0 and n will then generalize lines and planes through the origin in $\mathscr{R}^3$.

Example 5 If V and W are subspaces of $\mathscr{R}^n$, then so is their intersection $V \cap W$ (the set of all vectors that lie in both V and W). See Exercise 1.2.

Although most of our attention will be confined to subspaces of Euclidean spaces, it is instructive to consider some vector spaces that are not subspaces of Euclidean spaces.

Example 6 Let $\mathscr{F}$ denote the set of all real-valued functions on $\mathscr{R}$. If $f + g$ and af are defined by $(f + g)(x) = f(x) + g(x)$ and $(af)(x) = af(x)$, then $\mathscr{F}$ is a vector space (why?), with the zero vector being the function which is zero for all $x \in \mathscr{R}$. If $\mathscr{C}$ is the set of all continuous functions and $\mathscr{P}$ is the set of all polynomials, then $\mathscr{P}$ is a subspace of $\mathscr{C}$, and $\mathscr{C}$ in turn is a subspace of $\mathscr{F}$. If $\mathscr{P}_n$ is the set of all polynomials of degree at most n, then $\mathscr{P}_n$ is a subspace of $\mathscr{P}$ which is generated by the polynomials $1, x, x^2, \ldots, x^n$.

Exercises

1.1 Verify Example 3.

1.2 Prove that the intersection of two subspaces of $\mathscr{R}^n$ is also a subspace.

1.3 Given subspaces V and W of $\mathscr{R}^n$, denote by $V + W$ the set of all vectors $v + w$ with $v \in V$ and $w \in W$. Show that $V + W$ is a subspace of $\mathscr{R}^n$.

1.4 If V is the set of all $(x, y, z) \in \mathscr{R}^3$ such that $x + 2y = 0$ and $x + y = 3z$, show that V is a subspace of $\mathscr{R}^3$.

1.5 Let $\mathscr{D}_0$ denote the set of all differentiable real-valued functions on $[0, 1]$ such that $f(0) = f(1) = 0$. Show that $\mathscr{D}_0$ is a vector space, with addition and multiplication defined as in Example 6. Would this be true if the condition $f(0) = f(1) = 0$ were replaced by $f(0) = 0, f(1) = 1$?

1.6 Given a set S, denote by $\mathscr{F}(S, \mathscr{R})$ the set of all real-valued functions on S, that is, all maps $S \to R$. Show that $\mathscr{F}(S, \mathscr{R})$ is a vector space with the operations defined in Example 6. Note that $\mathscr{F}(\{1, \ldots, n\}, \mathscr{R})$ can be interpreted as $\mathscr{R}^n$ since the function $\varphi \in \mathscr{F}(\{1, \ldots, n\}, \mathscr{R})$ may be regarded as the n-tuple $(\varphi(1), \varphi(2), \ldots, \varphi(n))$.

2 SUBSPACES OF $\mathscr{R}^n$

In this section we will define the dimension of a vector space, and then show that $\mathscr{R}^n$ has precisely $n-1$ types of *proper* subspaces (that is, subspaces other than $\mathbf{0}$ and $\mathscr{R}^n$ itself)—namely, one of each dimension 1 through $n-1$.

In order to define dimension, we need the concept of linear independence. The vectors $\mathbf{v}_1, \mathbf{v}_2, \ldots, \mathbf{v}_k$ are said to be *linearly independent* provided that no one of them is a linear combination of the others; otherwise they are *linearly dependent.* The following proposition asserts that the vectors $\mathbf{v}_1, \ldots, \mathbf{v}_k$ are linearly independent if and only if $x_1\mathbf{v}_1 + x_2\mathbf{v}_2 + \cdots + x_k\mathbf{v}_k = \mathbf{0}$ implies that $x_1 = x_2 = \cdots = x_k = 0$. For example, the fact that $x_1\mathbf{e}_1 + x_2\mathbf{e}_2 + \cdots + x_n\mathbf{e}_n = (x_1, x_2, \ldots, x_n)$ then implies immediately that the standard basis vectors $\mathbf{e}_1, \mathbf{e}_2, \ldots, \mathbf{e}_n$ in $\mathscr{R}^n$ are linearly independent.

Proposition 2.1 The vectors $\mathbf{v}_1, \mathbf{v}_2, \ldots, \mathbf{v}_k$ are linearly dependent if and only if there exist numbers $x_1, x_2, \ldots, x_k$, not all zero, such that $x_1\mathbf{v}_1 + x_2\mathbf{v}_2 + \cdots + x_k\mathbf{v}_k = \mathbf{0}$.

PROOF If there exist such numbers, suppose, for example, that $x_1 \neq 0$. Then

$$\mathbf{v}_1 = -\frac{x_2}{x_1}\mathbf{v}_2 - \cdots - \frac{x_k}{x_1}\mathbf{v}_k,$$

so $\mathbf{v}_1, \mathbf{v}_2, \ldots, \mathbf{v}_k$ are linearly dependent. If, conversely, $\mathbf{v}_1 = a_2\mathbf{v}_2 + \cdots + a_k\mathbf{v}_k$, then we have $x_1\mathbf{v}_1 + x_2\mathbf{v}_2 + \cdots + x_k\mathbf{v}_k = \mathbf{0}$ with $x_1 = -1 \neq 0$ and $x_i = a_i$ for $i > 1$. ∎

Example 1 To show that the vectors $\mathbf{x} = (1, 1, 0)$, $\mathbf{y} = (1, 1, 1)$, $\mathbf{z} = (0, 1, 1)$ are linearly independent, suppose that $a\mathbf{x} + b\mathbf{y} + c\mathbf{z} = \mathbf{0}$. By taking components of this vector equation we obtain the three scalar equations

$$\begin{aligned} a + b \phantom{{}+c} &= 0, \\ a + b + c &= 0, \\ b + c &= 0. \end{aligned}$$

Subtracting the first from the second, we obtain $c = 0$. The last equation then gives $b = 0$, and finally the first one gives $a = 0$.

Example 2 The vectors $\mathbf{x} = (1, 1, 0)$, $\mathbf{y} = (1, 2, 1)$, $\mathbf{z} = (0, 1, 1)$ are linearly dependent, because $\mathbf{x} - \mathbf{y} + \mathbf{z} = \mathbf{0}$.

It is easily verified (Exercise 2.7) that any two collinear vectors, and any three coplanar vectors, are linearly dependent. This motivates the following definition of the dimension of a vector space. The vector space V has *dimension n*, $\dim V = n$, provided that V contains a set of n linearly independent vectors, while any $n+1$ vectors in V are linearly dependent; if there is no integer n for which this is true, then V is said to be *infinite-dimensional*. Thus the dimension of a finite-dimensional vector space is the largest number of linearly independent vectors which it contains; an infinite-dimensional vector space is one that contains n linearly independent vectors for every positive integer n.

Example 3 Consider the vector space $\mathscr{F}$ of all real-valued functions on $\mathscr{R}$. The functions $1, x, x^2, \ldots, x^n$ are linearly independent because a polynomial $a_0 + a_1 x + \cdots + a_n x^n$ can vanish identically only if all of its coefficients are zero. Therefore $\mathscr{F}$ is infinite-dimensional.

One certainly expects the above definition of dimension to imply that Euclidean n-space $\mathscr{R}^n$ does indeed have dimension n. We see immediately that its dimension is at least n, since it contains the n linearly independent vectors $\mathbf{e}_1, \ldots, \mathbf{e}_n$. To show that the dimension of $\mathscr{R}^n$ is precisely n, we must prove that any $n+1$ vectors in $\mathscr{R}^n$ are linearly dependent.

Suppose that $\mathbf{v}_1, \ldots, \mathbf{v}_k$ are $k > n$ vectors in $\mathscr{R}^n$, and write

$$\mathbf{v}_j = (a_{1j}, a_{2j}, \ldots, a_{nj}), \qquad j = 1, \ldots, k.$$

We want to find real numbers $x_1, \ldots, x_k$, not all zero, such that

$$\begin{aligned}\mathbf{0} &= x_1 \mathbf{v}_1 + x_2 \mathbf{v}_2 + \cdots + x_k \mathbf{v}_k \\ &= \sum_{j=1}^{k} x_j (a_{1j}, a_{2j}, \ldots, a_{nj}).\end{aligned}$$

This will be the case if $\sum_{j=1}^{k} a_{ij} x_j = 0$, $i = 1, \ldots, n$. Thus we need to find a nontrivial solution of the homogeneous linear equations

$$\begin{aligned} a_{11}x_1 + a_{12}x_2 + \cdots + a_{1k}x_k &= 0, \\ a_{21}x_1 + a_{22}x_2 + \cdots + a_{2k}x_k &= 0, \\ \vdots \qquad\qquad & \quad \vdots \\ a_{n1}x_1 + a_{n2}x_2 + \cdots + a_{nk}x_k &= 0. \end{aligned} \tag{1}$$

By a *nontrivial* solution $(x_1, x_2, \ldots, x_k)$ of the system (1) is meant one for which not all of the x_i are zero. But $k > n$, and (1) is a system of n homogeneous linear equations in the k unknowns $x_1, \ldots, x_k$. (Homogeneous meaning that the right-hand side constants are all zero.)

It is a basic fact of linear algebra that any system of homogeneous linear equations, with more unknowns than equations, has a nontrivial solution. The proof of this fact is an application of the elementary algebraic technique of elimination of variables. Before stating and proving the general theorem, we consider a special case.

Example 4 Consider the following three equations in four unknowns:

$$\begin{aligned} x_1 + 2x_2 - x_3 + 2x_4 &= 0, \\ x_1 - x_2 + 2x_3 + x_4 &= 0, \\ 2x_1 + x_2 - x_3 - x_4 &= 0. \end{aligned} \tag{2}$$

We can eliminate x_1 from the last two equations of (2) by subtracting the first equation from the second one, and twice the first equation from the third one. This gives two equations in three unknowns:

$$\begin{aligned} -3x_2 + 3x_3 - x_4 &= 0, \\ -3x_2 + x_3 - 5x_4 &= 0. \end{aligned} \tag{3}$$

Subtraction of the first equation of (3) from the second one gives the single equation

$$-2x_3 - 4x_4 = 0 \tag{4}$$

in two unknowns. We can now choose x_4 arbitrarily. For instance, if $x_4 = 1$, then $x_3 = -2$. The first equation of (3) then gives $x_2 = -\frac{3}{7}$, and finally the first equation of (2) gives $x_1 = -\frac{22}{7}$. So we have found the nontrivial solution $(-\frac{22}{7}, -\frac{3}{7}, -2, 1)$ of the system (2).

The procedure illustrated in this example can be applied to the general case of n equations in the unknowns $x_1, \ldots, x_k$, $k > n$. First we order the n equations so that the first equation contains x_1, and then eliminate x_1 from the remaining equations by subtracting the appropriate multiple of the first equation from each of them. This gives a system of $n - 1$ homogeneous linear equations in the $k - 1$ variables $x_2, \ldots, x_k$. Similarly we eliminate x_2 from the last $n - 2$ of these $n - 1$ equations by subtracting multiples of the first one, obtaining $n - 2$ equations in the $k - 2$ variables $x_3, x_4, \ldots, x_k$. After $n - 2$ steps of this sort, we end up with a single homogeneous linear equation in the $k - n + 1$ unknowns x_n, $x_{n+1}, \ldots, x_k$. We can then choose arbitrary nontrivial values for the "extra" variables $x_{n+1}, x_{n+2}, \ldots, x_k$ (such as $x_{n+1} = 1$, $x_{n+2} = \cdots = x_k = 0$), solve the final equation for x_n, and finally proceed backward to solve successively for each of the eliminated variables $x_{n-1}, x_{n-2}, \ldots, x_1$. The reader may (if he likes) formalize this procedure to give a proof, by induction on the number n of equations, of the following result.

Theorem 2.2 If $k > n$, then any system of n homogeneous linear equations in k unknowns has a nontrivial solution.

By the discussion preceding Eqs. (1) we now have the desired result that $\dim \mathscr{R}^n = n$.

Corollary 2.3 Any $n + 1$ vectors in $\mathscr{R}^n$ are linearly dependent.

We have seen that the linearly independent vectors $\mathbf{e}_1, \mathbf{e}_2, \ldots, \mathbf{e}_n$ generate $\mathscr{R}^n$. A set of linearly independent vectors that generates the vector space V is called a *basis* for V. Since $\mathbf{x} = (x_1, x_2, \ldots, x_n) = x_1\mathbf{e}_1 + x_2\mathbf{e}_2 + \cdots + x_n\mathbf{e}_n$, it is clear that the basis vectors $\mathbf{e}_1, \ldots, \mathbf{e}_n$ generate V *uniquely*; that is, if $\mathbf{x} = y_1\mathbf{e}_1 + y_2\mathbf{e}_2 + \cdots + y_n\mathbf{e}_n$ also, then $x_i = y_i$ for each i. Thus each vector in $\mathscr{R}^n$ can be expressed in one and only one way as a linear combination of $\mathbf{e}_1, \ldots, \mathbf{e}_n$. Any set of n linearly independent vectors in an n-dimensional vector space has this property.

Theorem 2.4 If the vectors $\mathbf{v}_1, \ldots, \mathbf{v}_n$ in the n-dimensional vector space V are linearly independent, then they constitute a basis for V, and furthermore generate V uniquely.

PROOF Given $\mathbf{v} \in V$, the vectors $\mathbf{v}, \mathbf{v}_1, \ldots, \mathbf{v}_n$ are linearly dependent, so by Proposition 2.1 there exist numbers $x, x_1, \ldots, x_n$, not all zero, such that

$$x\mathbf{v} + x_1\mathbf{v}_1 + \cdots + x_n\mathbf{v}_n = \mathbf{0}.$$

If $x = 0$, then the fact that $\mathbf{v}_1, \ldots, \mathbf{v}_n$ are linearly independent implies that $x_1 = \cdots = x_n = 0$. Therefore $x \neq 0$, so we solve for $\mathbf{v}$:

$$\mathbf{v} = -\frac{x_1}{x}\mathbf{v}_1 - \frac{x_2}{x}\mathbf{v}_2 + \cdots - \frac{x_n}{x}\mathbf{v}_n.$$

Thus the vectors $\mathbf{v}_1, \ldots, \mathbf{v}_n$ generate V, and therefore constitute a basis for V. To show that they generate V uniquely, suppose that

$$a_1\mathbf{v}_1 + \cdots + a_n\mathbf{v}_n = a_1'\mathbf{v}_1 + \cdots + a_n'\mathbf{v}_n.$$

Then

$$(a_1 - a_1')\mathbf{v}_1 + \cdots + (a_n - a_n')\mathbf{v}_n = \mathbf{0}.$$

So, since $\mathbf{v}_1, \ldots, \mathbf{v}_n$ are linearly independent, it follows that $a_i - a_i' = 0$, or $a_i = a_i'$, for each i. ∎

There remains the possibility that $\mathscr{R}^n$ has a basis which contains fewer than n elements. But the following theorem shows that this cannot happen.

Theorem 2.5 If $\dim V = n$, then each basis for V consists of exactly n vectors.

PROOF Let $\mathbf{w}_1, \mathbf{w}_2, \ldots, \mathbf{w}_n$ be n linearly independent vectors in V. If there were a basis $\mathbf{v}_1, \mathbf{v}_2, \ldots, \mathbf{v}_m$ for V with $m < n$, then there would exist numbers $\{a_{ij}\}$ such that

$$\begin{aligned} \mathbf{w}_1 &= a_{11}\mathbf{v}_1 + \cdots + a_{m1}\mathbf{v}_m, \\ &\ \vdots \\ \mathbf{w}_n &= a_{1n}\mathbf{v}_1 + \cdots + a_{mn}\mathbf{v}_m. \end{aligned}$$

Since $m < n$, Theorem 2.2 supplies numbers $x_1, \ldots, x_n$ *not all zero*, such that

$$a_{11}x_1 + \cdots + a_{1n}x_n = 0,$$
$$\vdots$$
$$a_{m1}x_1 + \cdots + a_{mn}x_n = 0.$$

But this implies that

$$\begin{aligned} x_1\mathbf{w}_1 + \cdots + x_n\mathbf{w}_n &= \sum_{j=1}^{n} x_j(a_{1j}\mathbf{v}_1 + \cdots + a_{mj}\mathbf{v}_m) \\ &= \sum_{i=1}^{m} (a_{i1}x_1 + \cdots + a_{in}x_n)\mathbf{v}_i \\ &= \mathbf{0}, \end{aligned}$$

which contradicts the fact that $\mathbf{w}_1, \ldots, \mathbf{w}_n$ are linearly independent. Consequently no basis for V can have $m < n$ elements. ∎

We can now completely describe the general situation as regards subspaces of $\mathscr{R}^n$. If V is a subspace of $\mathscr{R}^n$, then $k = \dim V \leqq n$ by Corollary 2.3, and if $k = n$, then $V = \mathscr{R}^n$ by Theorem 2.4. If $k > 0$, then any k linearly independent vectors in V generate V, and no basis for V contains fewer than k vectors (Theorem 2.5).

Exercises

2.1 Why is it true that the vectors $\mathbf{v}_1, \ldots, \mathbf{v}_k$ are linearly dependent if any one of them is zero? If any subset of them is linearly dependent?

2.2 Which of the following sets of vectors are bases for the appropriate space $\mathscr{R}^n$?
(a) (1, 0) and (1, 1).
(b) (1, 0, 0), (1, 1, 0), and (0, 0, 1).
(c) (1, 1, 1), (1, 1, 0), and (1, 0, 0).
(d) (1, 1, 1, 0), (1, 0, 0, 0), (0, 1, 0, 0), and (0, 0, 1, 0).
(e) (1, 1, 1, 1), (1, 1, 1, 0), (1, 1, 0, 0), and (1, 0, 0, 0).

2.3 Find the dimension of the subspace V of $\mathscr{R}^4$ that is generated by the vectors (0, 1, 0, 1), (1, 0, 1, 0), and (1, 1, 1, 1).

2.4 Show that the vectors (1, 0, 0, 1), (0, 1, 0, 1), (0, 0, 1, 1) form a basis for the subspace V of $\mathscr{R}^4$ which is defined by the equation $x_1 + x_2 + x_3 - x_4 = 0$.

2.5 Show that any set $\mathbf{v}_1, \ldots, \mathbf{v}_k$, of linearly independent vectors in a vector space V can be extended to a basis for V. That is, if $k < n = \dim V$, then there exist vectors $\mathbf{v}_{k+1}, \ldots, \mathbf{v}_n$ in V such that $\mathbf{v}_1, \ldots \mathbf{v}_n$ is a basis for V.

2.6 Show that Theorem 2.5 is equivalent to the following theorem: Suppose that the equations

$$a_{11}x_1 + \cdots + a_{1n}x_n = 0,$$
$$\vdots$$
$$a_{n1}x_1 + \cdots + a_{nn}x_n = 0$$

have only the trivial solution $x_1 = \cdots = x_n = 0$. Then, for each $\mathbf{b} = (b_1, \ldots, b_n)$, the equations

$$\begin{aligned} a_{11}x_1 + \cdots + a_{1n}x_n &= b_1, \\ &\vdots \\ a_{n1}x_1 + \cdots + a_{nn}x_n &= b_n \end{aligned}$$

have a *unique* solution. *Hint:* Consider the vectors $\mathbf{a}_j = (a_{1j}, a_{2j}, \ldots, a_{nj}), j = 1, \ldots, n$.

2.7 Verify that any two collinear vectors, and any three coplanar vectors, are linearly dependent.

3 INNER PRODUCTS AND ORTHOGONALITY

In order to obtain the full geometric structure of $\mathscr{R}^n$ (including the concepts of distance, angles, and orthogonality), we must supply $\mathscr{R}^n$ with an inner product. An *inner* (scalar) *product* on the vector space V is a function $V \times V \to \mathscr{R}$, which associates with each pair $(\mathbf{x}, \mathbf{y})$ of vectors in V a real number $\langle \mathbf{x}, \mathbf{y} \rangle$, and satisfies the following three conditions:

SP1 $\langle \mathbf{x}, \mathbf{x} \rangle > 0$ if $\mathbf{x} \neq \mathbf{0}$ (positivity).

SP2 $\langle \mathbf{x}, \mathbf{y} \rangle = \langle \mathbf{y}, \mathbf{x} \rangle$ (symmetry).

SP3 $\langle a\mathbf{x} + b\mathbf{y}, \mathbf{z} \rangle = a\langle \mathbf{x}, \mathbf{z} \rangle + b\langle \mathbf{y}, \mathbf{z} \rangle$.

The third of these conditions is linearity in the first variable; symmetry then gives linearity in the second variable also. Thus an inner product on V is simply a positive, symmetric, bilinear function on $V \times V$. Note that SP3 implies that $\langle \mathbf{0}, \mathbf{0} \rangle = 0$ (see Exercise 3.1).

The *usual inner product* on $\mathscr{R}^n$ is denoted by $\mathbf{x} \cdot \mathbf{y}$ and is defined by

$$\mathbf{x} \cdot \mathbf{y} = x_1 y_1 + \cdots + x_n y_n, \tag{1}$$

where $\mathbf{x} = (x_1, \ldots, x_n)$, $\mathbf{y} = (y_1, \ldots, y_n)$. It should be clear that this definition satisfies conditions SP1, SP2, SP3 above. There are many inner products on $\mathscr{R}^n$ (see Example 2 below), but we shall use only the usual one.

Example 1 Denote by $\mathscr{C}[a, b]$ the vector space of all continuous functions on the interval $[a, b]$, and define

$$\langle f, g \rangle = \int_a^b f(t)g(t)\, dt$$

for any pair of functions $f, g \in \mathscr{C}[a, b]$. It is obvious that this definition satisfies conditions SP2 and SP3. It also satisfies SP1, because if $f(t_0) \neq 0$, then by continuity $(f(t))^2 > 0$ for all t in some neighborhood of t_0, so

$$\langle f, f \rangle = \int_a^b f(t)^2\, dt > 0.$$

Therefore we have an inner product on $\mathscr{C}[a, b]$.

Example 2 Let a, b, c be real numbers with $a > 0$, $ac - b^2 > 0$, so that the quadratic form $q(\mathbf{x}) = ax_1^2 + 2bx_1x_2 + cx_2^2$ is positive-definite (see Section II.4). Then $\langle \mathbf{x}, \mathbf{y} \rangle = ax_1y_1 + bx_1y_2 + bx_2y_1 + cx_2y_2$ defines an inner product on $\mathscr{R}^2$ (why?). With $a = c = 1$, $b = 0$ we obtain the usual inner product on $\mathscr{R}^2$.

An inner product on the vector space V yields a notion of the length or "size" of a vector $\mathbf{x} \in V$, called its *norm* $|\mathbf{x}|$. In general, a *norm* on the vector space V is a real-valued function $\mathbf{x} \to |\mathbf{x}|$ on V satisfying the following conditions:

N1 $|\mathbf{x}| > 0$ if $\mathbf{x} \neq 0$ (positivity),
N2 $|a\mathbf{x}| = |a|\,|\mathbf{x}|$ (homogeneity),
N3 $|\mathbf{x} + \mathbf{y}| \leqq |\mathbf{x}| + |\mathbf{y}|$ (triangle inequality),

for all $\mathbf{x}, \mathbf{y} \in V$ and $a \in \mathscr{R}$. Note that N2 implies that $|\mathbf{0}| = 0$.

The norm associated with the inner product $\langle\ ,\ \rangle$ on V is defined by

$$|\mathbf{x}| = \sqrt{\langle \mathbf{x}, \mathbf{x} \rangle} \tag{2}$$

It is clear that SP1–SP3 and this definition imply conditions N1 and N2, but the triangle inequality is not so obvious; it will be verified below.

The most commonly used norm on $\mathscr{R}^n$ is the *Euclidean norm*

$$|\mathbf{x}| = (x_1^2 + \cdots + x_n^2)^{1/2},$$

which comes in the above way from the usual inner product on $\mathscr{R}^n$. Other norms on $\mathscr{R}^n$, not necessarily associated with inner products, are occasionally employed, but henceforth $|\mathbf{x}|$ will denote the Euclidean norm unless otherwise specified.

Example 3 $\|\mathbf{x}\| = \max\{|x_1|, \ldots, |x_n|\}$, the maximum of the absolute values of the coordinates of $\mathbf{x}$, defines a norm on $\mathscr{R}^n$ (see Exercise 3.2).

Example 4 $|\mathbf{x}|_1 = |x_1| + |x_2| + \cdots + |x_n|$ defines still another norm on $\mathscr{R}^n$ (again see Exercise 3.2).

A norm on V provides a definition of the *distance* $d(\mathbf{x}, \mathbf{y})$ between any two points $\mathbf{x}$ and $\mathbf{y}$ of V:

$$d(\mathbf{x}, \mathbf{y}) = |\mathbf{x} - \mathbf{y}|.$$

Note that a distance function d defined in this way satisfies the following three conditions:

D1 $d(\mathbf{x}, \mathbf{y}) > 0$ unless $\mathbf{x} = \mathbf{y}$ (positivity),
D2 $d(\mathbf{x}, \mathbf{y}) = d(\mathbf{y}, \mathbf{x})$ (symmetry),
D3 $d(\mathbf{x}, \mathbf{z}) \leqq d(\mathbf{x}, \mathbf{y}) + d(\mathbf{y}, \mathbf{z})$ (triangle inequality),

for any three points **x**, **y**, **z**. Conditions D1 and D2 follow immediately from N1 and N2, respectively, while

$$\begin{aligned} d(\mathbf{x}, \mathbf{z}) = |\mathbf{x} - \mathbf{z}| &= |(\mathbf{x} - \mathbf{y}) + (\mathbf{y} - \mathbf{z})| \\ &\leqq |\mathbf{x} - \mathbf{y}| + |\mathbf{y} - \mathbf{z}| \\ &= d(\mathbf{x}, \mathbf{y}) + d(\mathbf{y}, \mathbf{z}) \end{aligned}$$

by N3. Figure 1.1 indicates why N3 (or D3) is referred to as the *triangle inequality*.

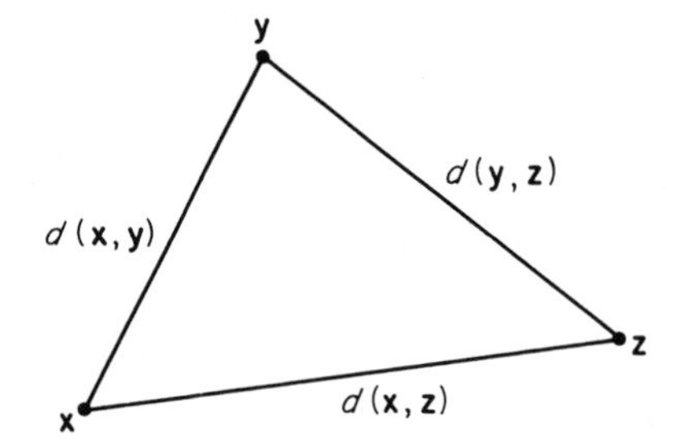

Figure 1.1

The distance function that comes in this way from the Euclidean norm is the familiar Euclidean distance function

$$d(\mathbf{x}, \mathbf{y}) = [(x_1 - y_1)^2 + \cdots + (x_n - y_n)^2]^{1/2}.$$

Thus far we have seen that an inner product on the vector space V yields a norm on V, which in turn yields a distance function on V, except that we have not yet verified that the norm associated with a given inner product does indeed satisfy the triangle inequality. The triangle inequality will follow from the *Cauchy–Schwarz inequality* of the following theorem.

Theorem 3.1 If $\langle\ ,\ \rangle$ is an inner product on a vector space V, then

$$|\langle \mathbf{x}, \mathbf{y} \rangle| \leqq |\mathbf{x}|\ |\mathbf{y}|$$

for all $\mathbf{x}, \mathbf{y} \in V$ [where the norm is the one defined by (2)].

PROOF The inequality is trivial if either **x** or **y** is zero, so assume neither is. If $\mathbf{u} = \mathbf{x}/|\mathbf{x}|$ and $\mathbf{v} = \mathbf{y}/|\mathbf{y}|$, then $|\mathbf{u}| = |\mathbf{v}| = 1$. Hence

$$\begin{aligned} 0 \leqq |\mathbf{u} - \mathbf{v}|^2 &= \langle \mathbf{u} - \mathbf{v}, \mathbf{u} - \mathbf{v} \rangle \\ &= |\mathbf{u}|^2 - 2\langle \mathbf{u}, \mathbf{v} \rangle + |\mathbf{v}|^2 \\ &= 2 - 2\langle \mathbf{u}, \mathbf{v} \rangle. \end{aligned}$$

So $\langle \mathbf{u}, \mathbf{v} \rangle \leqq 1$, that is $\langle \mathbf{x}/|\mathbf{x}|, \mathbf{y}/|\mathbf{y}| \rangle \leqq 1$, or

$$\langle \mathbf{x}, \mathbf{y} \rangle \leqq |\mathbf{x}|\ |\mathbf{y}|.$$

Replacing $\mathbf{x}$ by $-\mathbf{x}$, we obtain

$$-\langle \mathbf{x}, \mathbf{y} \rangle \leqq |\mathbf{x}|\, |\mathbf{y}|$$

also, so the inequality follows. ∎

The Cauchy–Schwarz inequality is of fundamental importance. With the usual inner product in $\mathscr{R}^n$, it takes the form

$$\left(\sum_{i=1}^{n} x_i y_i\right)^2 \leqq \left(\sum_{i=1}^{n} x_i^2\right)\left(\sum_{i=1}^{n} y_i^2\right),$$

while in $\mathscr{C}[a, b]$, with the inner product of Example 1, it becomes

$$\left(\int_a^b fg\right)^2 \leqq \left(\int_a^b f^2\right)\left(\int_a^b g^2\right).$$

PROOF OF THE TRIANGLE INEQUALITY Given $\mathbf{x}, \mathbf{y} \in V$ note that

$$\begin{aligned} |\mathbf{x}+\mathbf{y}|^2 &= \langle \mathbf{x}+\mathbf{y}, \mathbf{x}+\mathbf{y} \rangle \\ &= |\mathbf{x}|^2 + 2\langle \mathbf{x}, \mathbf{y} \rangle + |\mathbf{y}|^2 \\ &\leqq |\mathbf{x}|^2 + 2|\mathbf{x}|\, |\mathbf{y}| + |\mathbf{y}|^2 \qquad \text{(Cauchy–Schwarz)} \\ &= (|\mathbf{x}| + |\mathbf{y}|)^2, \end{aligned}$$

which implies that $|\mathbf{x}+\mathbf{y}| \leqq |\mathbf{x}| + |\mathbf{y}|$. ∎

Notice that, if $\langle \mathbf{x}, \mathbf{y} \rangle = 0$, in which case $\mathbf{x}$ and $\mathbf{y}$ are perpendicular (see the definition below), then the second equality in the above proof gives

$$|\mathbf{x}+\mathbf{y}|^2 = |\mathbf{x}|^2 + |\mathbf{y}|^2.$$

This is the famous theorem associated with the name of Pythagoras (Fig. 1.2).

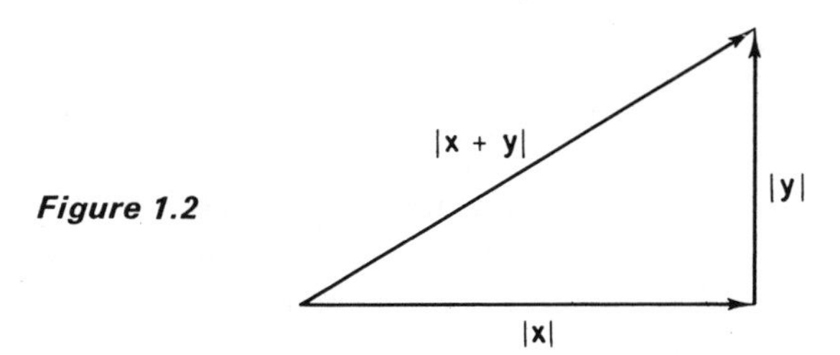

Figure 1.2

Recalling the formula $\mathbf{x} \cdot \mathbf{y} = |\mathbf{x}|\, |\mathbf{y}| \cos \theta$ for the usual inner product in $\mathscr{R}^2$, we are motivated to *define* the *angle* $\angle(\mathbf{x}, \mathbf{y})$ between the vectors $\mathbf{x}, \mathbf{y} \in V$ by

$$\angle(\mathbf{x}, \mathbf{y}) = \arccos \frac{\langle \mathbf{x}, \mathbf{y} \rangle}{|\mathbf{x}|\, |\mathbf{y}|} \in [0, \pi].$$

Notice that this makes sense because $\langle \mathbf{x}, \mathbf{y} \rangle / |\mathbf{x}|\,|\mathbf{y}| \in [-1, 1]$ by the Cauchy–Schwarz inequality. In particular we say that $\mathbf{x}$ and $\mathbf{y}$ are *orthogonal* (or perpendicular) if and only if $\mathbf{x} \cdot \mathbf{y} = 0$, because then $\angle(\mathbf{x}, \mathbf{y}) = \arccos \pi/2 = 0$.

A set of nonzero vectors $\mathbf{v}_1, \mathbf{v}_2, \ldots$ in V is said to be an *orthogonal set* if

$$\langle \mathbf{v}_i, \mathbf{v}_j \rangle = 0$$

whenever $i \neq j$. If in addition each $\mathbf{v}_i$ is a unit vector, $\langle \mathbf{v}_i, \mathbf{v}_i \rangle = 1$, then the set is said to be *orthonormal.*

Example 5 The standard basis vectors $\mathbf{e}_1, \ldots, \mathbf{e}_n$ form an orthonormal set in $\mathscr{R}^n$.

Example 6 The (infinite) set of functions

$$1, \ \cos x, \ \sin x, \ \ldots, \ \cos nx, \ \sin nx, \ \ldots$$

is orthogonal in $\mathscr{C}[-\pi, \pi]$ (see Example 1 and Exercise 3.11). This fact is the basis for the theory of Fourier series.

The most important property of orthogonal sets is given by the following result.

Theorem 3.2 Every finite orthogonal set of nonzero vectors is linearly independent.

PROOF Suppose that

$$a_1 \mathbf{v}_1 + \cdots + a_k \mathbf{v}_k = \mathbf{0}. \tag{3}$$

Taking the inner product with $\mathbf{v}_i$, we obtain

$$a_i \langle \mathbf{v}_i, \mathbf{v}_i \rangle = 0$$

because $\langle \mathbf{v}_i, \mathbf{v}_j \rangle = 0$ for $i \neq j$ if the vectors $\mathbf{v}_1, \ldots, \mathbf{v}_k$ are orthogonal. But $\langle \mathbf{v}_i, \mathbf{v}_i \rangle \neq 0$, so $a_i = 0$. Thus (3) implies $a_1 = \cdots = a_k = 0$, so the orthogonal vectors $\mathbf{v}_1, \ldots, \mathbf{v}_k$ are linearly independent. ∎

We now describe the important *Gram–Schmidt orthogonalization process* for constructing orthogonal bases. It is motivated by the following elementary construction. Given two linearly independent vectors $\mathbf{v}$ and $\mathbf{w}_1$, we want to find a nonzero vector $\mathbf{w}_2$ that lies in the subspace spanned by $\mathbf{v}$ and $\mathbf{w}_1$, and is orthogonal to $\mathbf{w}_1$. Figure 1.3 suggests that such a vector $\mathbf{w}_2$ can be obtained by subtracting from $\mathbf{v}$ an appropriate multiple $c\mathbf{w}_1$ of $\mathbf{w}_1$. To determine c,

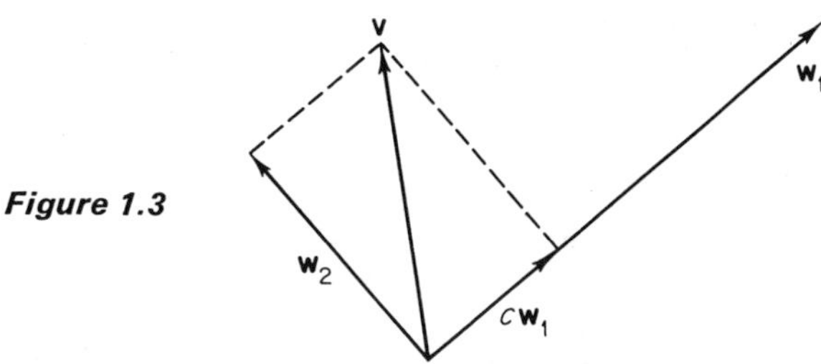

Figure 1.3

we simply solve the equation $\langle \mathbf{w}_1, \mathbf{v} - c\mathbf{w}_1 \rangle = 0$ for $c = \langle \mathbf{v}, \mathbf{w}_1 \rangle / \langle \mathbf{w}_1, \mathbf{w}_1 \rangle$. The desired vector is therefore

$$\mathbf{w}_2 = \mathbf{v} - \frac{\langle \mathbf{v}, \mathbf{w}_1 \rangle}{\langle \mathbf{w}_1, \mathbf{w}_1 \rangle} \mathbf{w}_1,$$

obtained by subtracting from $\mathbf{v}$ the "component of $\mathbf{v}$ parallel to $\mathbf{w}_1$." We immediately verify that $\langle \mathbf{w}_2, \mathbf{w}_1 \rangle = 0$, while $\mathbf{w}_2 \neq 0$ because $\mathbf{v}$ and $\mathbf{w}_1$ are linearly independent.

Theorem 3.3 If V is a finite-dimensional vector space with an inner product, then V has an orthogonal basis.

In particular, every subspace of $\mathscr{R}^n$ has an orthogonal basis.

PROOF We start with an arbitrary basis $\mathbf{v}_1, \ldots, \mathbf{v}_n$ for V. Let $\mathbf{w}_1 = \mathbf{v}_1$. Then, by the preceding construction, the nonzero vector

$$\mathbf{w}_2 = \mathbf{v}_2 - \frac{\langle \mathbf{v}_2, \mathbf{w}_1 \rangle}{\langle \mathbf{w}_1, \mathbf{w}_1 \rangle} \mathbf{w}_1$$

is orthogonal to $\mathbf{w}_1$ and lies in the subspace generated by $\mathbf{v}_1$ and $\mathbf{v}_2$.

Suppose inductively that we have found an orthogonal basis $\mathbf{w}_1, \ldots, \mathbf{w}_k$ for the subspace of V that is generated by $\mathbf{v}_1, \ldots, \mathbf{v}_k$. The idea is then to obtain $\mathbf{w}_{k+1}$ by subtracting from $\mathbf{v}_{k+1}$ its components parallel to each of the vectors $\mathbf{w}_1, \ldots, \mathbf{w}_k$. That is, define

$$\mathbf{w}_{k+1} = \mathbf{v}_{k+1} - c_1\mathbf{w}_1 - c_2\mathbf{w}_2 - \cdots - c_k\mathbf{w}_k,$$

where $c_i = \langle \mathbf{v}_{k+1}, \mathbf{w}_i \rangle / \langle \mathbf{w}_i, \mathbf{w}_i \rangle$. Then $\langle \mathbf{w}_{k+1}, \mathbf{w}_i \rangle = \langle \mathbf{v}_{k+1}, \mathbf{w}_i \rangle - c_i \langle \mathbf{w}_i, \mathbf{w}_i \rangle = 0$ for $i \leqq k$, and $\mathbf{w}_{k+1} \neq 0$, because otherwise $\mathbf{v}_{k+1}$ would be a linear combination of the vectors $\mathbf{w}_1, \ldots, \mathbf{w}_k$, and therefore of the vectors $\mathbf{v}_1, \ldots, \mathbf{v}_k$. It follows that the vectors $\mathbf{w}_1, \ldots, \mathbf{w}_{k+1}$ form an orthogonal basis for the subspace of V that is generated by $\mathbf{v}_1, \ldots, \mathbf{v}_{k+1}$.

After a finite number of such steps we obtain the desired orthogonal basis $\mathbf{w}_1, \ldots, \mathbf{w}_n$ for V. ∎

It is the method of proof of Theorem 3.3 that is known as the Gram–Schmidt orthogonalization process, summarized by the equations

$$\begin{aligned}
\mathbf{w}_1 &= \mathbf{v}_1,\\
\mathbf{w}_2 &= \mathbf{v}_2 - \frac{\langle \mathbf{v}_2, \mathbf{w}_1\rangle}{\langle \mathbf{w}_1, \mathbf{w}_1\rangle}\mathbf{w}_1,\\
\mathbf{w}_3 &= \mathbf{v}_3 - \frac{\langle \mathbf{v}_3, \mathbf{w}_1\rangle}{\langle \mathbf{w}_1, \mathbf{w}_1\rangle}\mathbf{w}_1 - \frac{\langle \mathbf{v}_3, \mathbf{w}_2\rangle}{\langle \mathbf{w}_2, \mathbf{w}_2\rangle}\mathbf{w}_2,\\
&\vdots\\
\mathbf{w}_n &= \mathbf{v}_n - \frac{\langle \mathbf{v}_n, \mathbf{w}_1\rangle}{\langle \mathbf{w}_1, \mathbf{w}_1\rangle}\mathbf{w}_1 - \cdots - \frac{\langle \mathbf{v}_n, \mathbf{w}_{n-1}\rangle}{\langle \mathbf{w}_{n-1}, \mathbf{w}_{n-1}\rangle}\mathbf{w}_{n-1},
\end{aligned}$$

defining the orthogonal basis $\mathbf{w}_1, \ldots, \mathbf{w}_n$ in terms of the original basis $\mathbf{v}_1, \ldots, \mathbf{v}_n$.

Example 7 To find an orthogonal basis for the subspace V of $\mathscr{R}^4$ spanned by the vectors $\mathbf{v}_1 = (1, 1, 0, 0)$, $\mathbf{v}_2 = (1, 0, 1, 0)$, $\mathbf{v}_3 = (0, 1, 0, 1)$, we write

$$\begin{aligned}
\mathbf{w}_1 &= \mathbf{v}_1 = (1, 1, 0, 0),\\
\mathbf{w}_2 &= \mathbf{v}_2 - \frac{\mathbf{v}_2 \cdot \mathbf{w}_1}{\mathbf{w}_1 \cdot \mathbf{w}_1}\mathbf{w}_1\\
&= (1, 0, 1, 0) - \tfrac{1}{2}(1, 1, 0, 0) = (\tfrac{1}{2}, -\tfrac{1}{2}, 1, 0),\\
\mathbf{w}_3 &= \mathbf{v}_3 - \frac{\mathbf{v}_3 \cdot \mathbf{w}_1}{\mathbf{w}_1 \cdot \mathbf{w}_1}\mathbf{w}_1 - \frac{\mathbf{v}_3 \cdot \mathbf{w}_2}{\mathbf{w}_2 \cdot \mathbf{w}_2}\mathbf{w}_2\\
&= (0, 1, 0, 1) - \tfrac{1}{2}(1, 1, 0, 0) + \tfrac{1}{3}(\tfrac{1}{2}, -\tfrac{1}{2}, 1, 0)\\
&= (-\tfrac{1}{3}, \tfrac{1}{3}, \tfrac{1}{3}, 1).
\end{aligned}$$

Example 8 Let $\mathscr{P}$ denote the vector space of polynomials in x, with inner product defined by

$$\langle p, q\rangle = \int_{-1}^{1} p(x)q(x)\,dx.$$

By applying the Gram–Schmidt orthogonalization process to the linearly independent elements $1, x, x^2, \ldots, x^n, \ldots$, one obtains an infinite sequence $\{p_n(x)\}_{n=0}^{\infty}$, the first five elements of which are $p_0(x) = 1$, $p_1(x) = x$, $p_2(x) = x^2 - \frac{1}{3}$, $p_3(x) = x^3 - \frac{3}{5}x$, $p_4(x) = x^4 - \frac{6}{7}x^2 + \frac{3}{35}$ (see Exercise 3.12). Upon multiplying the polynomials $\{p_n(x)\}$ by appropriate constants, one obtains the famous *Legendre polynomials* $P_0(x) = p_0(x)$, $P_1(x) = p_1(x)$, $P_2(x) = \frac{3}{2}p_2(x)$, $P_3(x) = \frac{5}{2}p_3(x)$, $P_4(x) = \frac{35}{8}p_4(x)$, etc.

One reason for the importance of orthogonal bases is the ease with which a vector $\mathbf{v} \in V$ can be expressed as a linear combination of orthogonal basis vectors $\mathbf{w}_1, \ldots, \mathbf{w}_n$ for V. Writing

$$\mathbf{v} = a_1 \mathbf{w}_1 + \cdots + a_n \mathbf{w}_n,$$

and taking the inner product with $\mathbf{w}_i$, we immediately obtain

$$a_i = \frac{\mathbf{v} \cdot \mathbf{w}_i}{\mathbf{w}_i \cdot \mathbf{w}_i},$$

so

$$\mathbf{v} = \frac{\mathbf{v} \cdot \mathbf{w}_1}{\mathbf{w}_1 \cdot \mathbf{w}_1} \mathbf{w}_1 + \cdots + \frac{\mathbf{v} \cdot \mathbf{w}_n}{\mathbf{w}_n \cdot \mathbf{w}_n} \mathbf{w}_n. \tag{4}$$

This is especially simple if $\mathbf{w}_1, \ldots, \mathbf{w}_n$ is an orthonormal basis for V:

$$\mathbf{v} = (\mathbf{v} \cdot \mathbf{w}_1)\mathbf{w}_1 + (\mathbf{v} \cdot \mathbf{w}_2)\mathbf{w}_2 + \cdots + (\mathbf{v} \cdot \mathbf{w}_n)\mathbf{w}_n. \tag{5}$$

Of course orthonormal basis vectors are easily obtained from orthogonal ones, simply by dividing by their lengths. In this case the coefficient $\mathbf{v} \cdot \mathbf{w}_i$ of $\mathbf{w}_i$ in (5) is sometimes called the *Fourier coefficient* of $\mathbf{v}$ with respect to $\mathbf{w}_i$. This terminology is motivated by an analogy with Fourier series. The orthonormal functions in $\mathscr{C}[-\pi, \pi]$ corresponding to the orthogonal functions of Example 6 are

$$\frac{1}{\sqrt{(2\pi)}}, \frac{\cos x}{\sqrt{\pi}}, \frac{\sin x}{\sqrt{\pi}}, \ldots, \frac{\cos nx}{\sqrt{\pi}}, \frac{\sin nx}{\sqrt{\pi}}, \ldots.$$

Writing

$$\varphi_n(x) = \frac{\cos nx}{\sqrt{\pi}} \quad \text{and} \quad \psi_n(x) = \frac{\sin nx}{\sqrt{\pi}},$$

one defines the Fourier coefficients of $f \in \mathscr{C}[-\pi, \pi]$ by

$$a_n = \langle f, \varphi_n \rangle = \frac{1}{\sqrt{\pi}} \int_{-\pi}^{\pi} f(x) \cos nx \, dx$$

and

$$b_n = \langle f, \psi_n \rangle = \frac{1}{\sqrt{\pi}} \int_{-\pi}^{\pi} f(x) \sin nx \, dx.$$

It can then be established, under appropriate conditions on f, that the infinite series

$$a_0 + \sum_{n=1}^{\infty} (a_n \cos nx + b_n \sin nx)$$

converges to $f(x)$. This infinite series may be regarded as an infinite-dimensional analog of (5).

Given a subspace V of $\mathscr{R}^n$, denote by $V^\perp$ the set of all those vectors in $\mathscr{R}^n$, each of which is orthogonal to every vector in V. Then it is easy to show that $V^\perp$ is a subspace of $\mathscr{R}^n$, called the *orthogonal complement* of V (Exercise 3.3). The significant fact about this situation is that the dimensions add up as they should.

Theorem 3.4 If V is a subspace of $\mathscr{R}^n$, then

$$\dim V + \dim V^\perp = n. \tag{6}$$

PROOF By Theorem 3.3, there exists an orthonormal basis $\mathbf{v}_1, \ldots, \mathbf{v}_r$ for V, and an orthonormal basis $\mathbf{w}_1, \ldots, \mathbf{w}_s$ for $V^\perp$. Then the vectors $\mathbf{v}_1, \ldots, \mathbf{v}_r$, $\mathbf{w}_1, \ldots, \mathbf{w}_s$ are orthornormal, and therefore linearly independent. So in order to conclude from Theorem 2.5 that $r + s = n$ as desired, it suffices to show that these vectors generate $\mathscr{R}^n$. Given $\mathbf{x} \in \mathscr{R}^n$, define

$$\mathbf{y} = \mathbf{x} - \sum_{i=1}^{r} (\mathbf{x} \cdot \mathbf{v}_i)\mathbf{v}_i. \tag{7}$$

Then $\mathbf{y} \cdot \mathbf{v}_i = \mathbf{x} \cdot \mathbf{v}_i - (\mathbf{x} \cdot \mathbf{v}_i)(\mathbf{v}_i \cdot \mathbf{v}_i) = 0$ for each $i = 1, \ldots, r$. Since $\mathbf{y}$ is orthogonal to each element of a basis for V, it follows easily that $\mathbf{y} \in V^\perp$ (Exercise 3.4). Therefore Eq. (5) above gives

$$\mathbf{y} = \sum_{i=1}^{s} (\mathbf{y} \cdot \mathbf{w}_i)\mathbf{w}_i.$$

This and (7) then yield

$$\mathbf{x} = \sum_{i=1}^{r} (\mathbf{x} \cdot \mathbf{v}_i)\mathbf{v}_i + \sum_{i=1}^{s} (\mathbf{y} \cdot \mathbf{w}_i)\mathbf{w}_i,$$

so the vectors $\mathbf{v}_1, \ldots, \mathbf{v}_r, \mathbf{w}_1, \ldots, \mathbf{w}_s$ constitute a basis for $\mathscr{R}^n$. ∎

Example 9 Consider the system

$$\begin{aligned} a_{11}x_1 + a_{12}x_2 + \cdots + a_{1n}x_n &= 0, \\ a_{21}x_1 + a_{22}x_2 + \cdots + a_{2n}x_n &= 0, \\ &\vdots \\ a_{k1}x_1 + a_{k2}x_2 + \cdots + a_{kn}x_n &= 0, \end{aligned} \tag{8}$$

of $k \leqq n$ homogeneous linear equations in $x_1, \ldots, x_n$. If $\mathbf{a}_i = (a_{i1}, \ldots, a_{in})$, $i = 1, \ldots, k$, then these equations can be rewritten as

$$\begin{aligned} \mathbf{a}_1 \cdot \mathbf{x} &= 0, \\ \mathbf{a}_2 \cdot \mathbf{x} &= 0, \\ &\vdots \\ \mathbf{a}_k \cdot \mathbf{x} &= 0. \end{aligned}$$

Therefore the set S of all solutions of (8) is simply the set of all those vectors $\mathbf{x} \in \mathscr{R}^n$ that are orthogonal to the vectors $\mathbf{a}_1, \ldots, \mathbf{a}_k$. If V is the subspace of $\mathscr{R}^n$ generated by $\mathbf{a}_1, \ldots, \mathbf{a}_k$, it follows that $S = V^\perp$ (Exercise 3.4). If the vectors $\mathbf{a}_1, \ldots, \mathbf{a}_k$ are linearly independent, we can then conclude from Theorem 3.4 that $\dim S = n - k$.

Exercises

3.1 Conclude from condition SP3 that $\langle \mathbf{0}, \mathbf{0} \rangle = 0$.

3.2 Verify that the functions defined in Examples 3 and 4 are norms on $\mathscr{R}^n$.

3.3 If V is a subspace of $\mathscr{R}^n$, prove that $V^\perp$ is also a subspace.

3.4 If the vectors $\mathbf{a}_1, \ldots, \mathbf{a}_k$ generate the subspace V of $\mathscr{R}^n$, and $x \in \mathscr{R}^n$ is orthogonal to each of these vectors, show that $\mathbf{x} \in V^\perp$.

3.5 Verify the "polarization identity" $\mathbf{x} \cdot \mathbf{y} = \frac{1}{4}(|\mathbf{x} + \mathbf{y}|^2 - |\mathbf{x} - \mathbf{y}|^2)$.

3.6 Let $\mathbf{a}_1, \mathbf{a}_2, \ldots, \mathbf{a}_n$ be an orthonormal basis for $\mathscr{R}^n$. If $\mathbf{x} = s_1\mathbf{a}_1 + \cdots + s_n\mathbf{a}_n$ and $\mathbf{y} = t_1\mathbf{a}_1 + \cdots + t_n\mathbf{a}_n$, show that $\mathbf{x} \cdot \mathbf{y} = s_1t_1 + \cdots + s_nt_n$. That is, in computing $\mathbf{x} \cdot \mathbf{y}$, one may replace the coordinates of $\mathbf{x}$ and $\mathbf{y}$ by their components relative to any orthonormal basis for $\mathscr{R}^n$.

3.7 Orthogonalize the basis $(1, 0, 0, 1)$, $(-1, 0, 2, 1)$, $(0, 1, 2, 0)$, $(0, 0, -1, 1)$ in $\mathscr{R}^4$.

3.8 Orthogonalize the basis

$$\mathbf{e}_1' = (1, 0, \ldots, 0), \qquad \mathbf{e}_2' = (1, 1, \ldots, 0), \qquad \ldots, \qquad \mathbf{e}_n' = (1, 1, \ldots, 1)$$

in $\mathscr{R}^n$.

3.9 Find an orthogonal basis for the 3-dimensional subspace V of $\mathscr{R}^4$ that consists of all solutions of the equation $x_1 + x_2 + x_3 - x_4 = 0$. *Hint:* Orthogonalize the vectors $\mathbf{v}_1 = (1, 0, 0, 1)$, $\mathbf{v}_2 = (0, 1, 0, 1)$, $\mathbf{v}_3 = (0, 0, 1, 1)$.

3.10 Consider the two equations

$$x_1 + 2x_2 - x_3 + x_4 = 0, \qquad (*)$$
$$2x_1 + x_2 + x_3 - x_4 = 0. \qquad (**)$$

Let V be the set of all solutions of $(*)$ and W the set of all solutions of both equations. Then W is a 2-dimensional subspace of the 3-dimensional subspace V of $\mathscr{R}^4$ (why?).

(a) Solve $(*)$ and $(**)$ to find a basis $\mathbf{v}_1, \mathbf{v}_2$ for W.

(b) Find by inspection a vector $\mathbf{v}_3$ which is in V but not in W. Why is $\mathbf{v}_1, \mathbf{v}_2, \mathbf{v}_3$ then a basis for V?

(c) Orthogonalize $\mathbf{v}_1, \mathbf{v}_2, \mathbf{v}_3$ to obtain an orthogonal basis $\mathbf{w}_1, \mathbf{w}_2, \mathbf{w}_3$ for V, with $\mathbf{w}_1$ and $\mathbf{w}_2$ in W.

(d) Normalize $\mathbf{w}_1, \mathbf{w}_2, \mathbf{w}_3$ to obtain an orthonormal basis $\mathbf{u}_1, \mathbf{u}_2, \mathbf{u}_3$ for V. Express $\mathbf{v} = (11, 3, 6, -11)$ as a linear combination of $\mathbf{u}_1, \mathbf{u}_2, \mathbf{u}_3$.

(e) Find vectors $\mathbf{x} \in W$ and $\mathbf{y} \in W^\perp$ such that $\mathbf{v} = \mathbf{x} + \mathbf{y}$.

3.11 Show that the functions

$$\frac{1}{\sqrt{(2\pi)}}, \frac{\cos x}{\sqrt{\pi}}, \frac{\sin x}{\sqrt{\pi}}, \ldots, \frac{\cos nx}{\sqrt{\pi}}, \frac{\sin nx}{\sqrt{\pi}}, \ldots$$

are orthogonal in the inner product space $\mathscr{C}[-\pi, \pi]$ of Example 1.

3.12 Orthogonalize in $\mathscr{C}[-1, 1]$ the functions $1, x, x^2, x^3, x^4$ to obtain the polynomials $p_0(x), \ldots, p_4(x)$ listed in Example 8.

4 LINEAR MAPPINGS AND MATRICES

In this section we introduce an important special class of mappings of Euclidean spaces—those which are linear (see definition below). One of the central ideas of multivariable calculus is that of approximating nonlinear mappings by linear ones.

Given a mapping $f: \mathscr{R}^n \to \mathscr{R}^m$, let $f_1, \ldots, f_m$ be the *component functions* of f. That is, $f_1, \ldots, f_m$ are the real-valued functions on $\mathscr{R}^n$ defined by writing $f(\mathbf{x}) = (f_1(\mathbf{x}), \ldots, f_m(\mathbf{x})) \in \mathscr{R}^m$. In Chapter II we will see that, if the component functions of f are continuously differentiable at $\mathbf{a} \in \mathscr{R}^n$, then there exists a linear mapping $L : \mathscr{R}^n \to \mathscr{R}^m$ such that

$$f(\mathbf{a} + \mathbf{h}) = f(\mathbf{a}) + L(\mathbf{h}) + R(\mathbf{h})$$

with $\lim_{\mathbf{h}\to\mathbf{0}} R(\mathbf{h})/|\mathbf{h}| = \mathbf{0}$. This fact will be the basis of much of our study of the differential calculus of functions of several variables.

Given vector spaces V and W, the mapping $L : V \to W$ is called *linear* if and and only if

$$L(a\mathbf{x} + b\mathbf{y}) = aL(\mathbf{x}) + bL(\mathbf{y}) \tag{1}$$

for all $\mathbf{x}, \mathbf{y} \in V$ and $a, b \in \mathscr{R}$. It is easily seen (Exercise 4.1) that the mapping L satisfies (1) if and only if

$$L(a\mathbf{x}) = aL(\mathbf{x}) \qquad \text{(homogeneity)} \tag{2}$$

and

$$L(\mathbf{x} + \mathbf{y}) = L(\mathbf{x}) + L(\mathbf{y}) \qquad \text{(additivity)} \tag{3}$$

for all $\mathbf{x}, \mathbf{y} \in V$ and $a \in \mathscr{R}$.

Example 1 Given $b \in \mathscr{R}$, the mapping $f: \mathscr{R} \to \mathscr{R}$ defined by $f(x) = bx$ obviously satisfies conditions (2) and (3), and is therefore linear. Conversely, if f is linear, then $f(x) = f(x \cdot 1) = xf(1)$, so f is of the form $f(x) = bx$ with $b = f(1)$.

Example 2 The identity mapping $I : V \to V$, defined by $I(\mathbf{x}) = \mathbf{x}$ for all $\mathbf{x} \in V$, is linear, as is the zero mapping $\mathbf{x} \to \mathbf{0}$. However the constant mapping $\mathbf{x} \to \mathbf{c}$ is not linear if $\mathbf{c} \neq \mathbf{0}$ (why?).

Example 3 Let $f: \mathscr{R}^3 \to \mathscr{R}^2$ be the vertical projection mapping defined by $f(x, y, z) = (x, y)$. Then f is linear.

Example 4 Let $\mathscr{D}$ denote the vector space of all infinitely differentiable functions from $\mathscr{R}$ to $\mathscr{R}$. If $D(f)$ denotes the derivative of $f \in \mathscr{D}$, then $D : \mathscr{D} \to \mathscr{D}$ is linear.

Example 5 Let $\mathscr{C}$ denote the vector space of all continuous functions on $[a, b]$. If $\mathscr{I}(f) = \int_a^b f$, then $\mathscr{I}: \mathscr{C} \to \mathscr{R}$ is linear.

Example 6 Given $\mathbf{a} = (a_1, a_2, a_3) \in \mathscr{R}^3$, the mapping $f: \mathscr{R}^3 \to \mathscr{R}$ defined by $f(\mathbf{x}) = \mathbf{a} \cdot \mathbf{x} = a_1x_1 + a_2x_2 + a_3x_3$ is linear, because $\mathbf{a} \cdot (\mathbf{x} + \mathbf{y}) = \mathbf{a} \cdot \mathbf{x} + \mathbf{a} \cdot \mathbf{y}$ and $\mathbf{a} \cdot (c\mathbf{x}) = c(\mathbf{a} \cdot \mathbf{x})$.

Conversely, the approach of Example 1 can be used to show that a function $f: \mathscr{R}^3 \to \mathscr{R}$ is linear only if it is of the form $f(x_1, x_2, x_3) = a_1x_1 + a_2x_2 + a_3x_3$. In general, we will prove in Theorem 4.1 that the mapping $f: \mathscr{R}^n \to \mathscr{R}^m$ is linear if and only if there exist numbers a_{ij}, $i = 1, \ldots, m$, $j = 1, \ldots, n$, such that the coordinate functions $f_1, \ldots, f_m$ of f are given by

$$
\begin{aligned}
f_1(\mathbf{x}) &= a_{11}x_1 + a_{12}x_2 + \cdots + a_{1n}x_n, \\
f_2(\mathbf{x}) &= a_{21}x_1 + a_{22}x_2 + \cdots + a_{2n}x_n, \\
&\vdots \\
f_m(\mathbf{x}) &= a_{m1}x_1 + a_{m2}x_2 + \cdots + a_{mn}x_n.
\end{aligned} \tag{4}
$$

for all $\mathbf{x} = (x_1, \ldots, x_n) \in \mathscr{R}^n$. Thus the linear mapping f is completely determined by the rectangular array

$$
A = \begin{pmatrix} a_{11} & a_{12} & \cdots & a_{1n} \\ a_{21} & a_{22} & \cdots & a_{2n} \\ \vdots & & & \vdots \\ a_{m1} & a_{m2} & \cdots & a_{mn} \end{pmatrix}
$$

of numbers. Such a rectangular array of real numbers is called a *matrix*. The horizontal lines of numbers in a matrix are called *rows*; the vertical ones are called *columns*. A matrix having m rows and n columns is called an $m \times n$ matrix. Rows are numbered from top to bottom, and columns from left to right. Thus the *element* a_{ij} of the matrix A above is the one which is in the ith row and the jth column of A. This type of notation for the elements of a matrix is standard—the first subscript gives the row and the second the column. We frequently write $A = (a_{ij})$ for brevity.

The set of all $m \times n$ matrices can be made into a vector space as follows. Given two $m \times n$ matrices $A = (a_{ij})$ and $B = (b_{ij})$, and a number r, we define

$$
A + B = (a_{ij} + b_{ij}) \qquad \text{and} \qquad rA = (ra_{ij}).
$$

That is, the ijth element of $A + B$ is the sum of the ijth elements of A and B, and the ijth element of rA is r times that of A. It is a simple matter to check that these two definitions satisfy conditions V1–V8 of Section 1.

Indeed these operations for matrices are simply an extension of those for vectors. A $1 \times n$ matrix is often called a *row vector*, and an $m \times 1$ matrix is similarly called a *column vector*. For example, the ith row

$$
A_i = (a_{i1} \;\; a_{i2} \;\; \cdots \;\; a_{in})
$$

of the matrix A is a row vector, and the jth column

$$A^j = \begin{pmatrix} a_{1j} \\ a_{2j} \\ \vdots \\ a_{mj} \end{pmatrix}$$

of A is a column vector. In terms of the rows and columns of a matrix A, we will sometimes write

$$A = \begin{pmatrix} A_1 \\ \vdots \\ A_m \end{pmatrix} = (A^1 \ \cdots \ A^n),$$

using subscripts for rows and superscripts for columns.

Next we define an operation of multiplication of matrices which generalizes the inner product for vectors. We define the product AB first in the special case when A is a *row* vector and B is a *column* vector of the same dimension. If

$$A = (a_1 \ a_2 \ \cdots \ a_n) \qquad \text{and} \qquad B = \begin{pmatrix} b_1 \\ \vdots \\ b_n \end{pmatrix},$$

we define $AB = \sum_{i=1}^n a_i b_i$. Thus in this case AB is just the scalar product of A and B as vectors, and is therefore a real number (which we may regard as a 1×1 matrix).

The product AB in general is defined only when the number of columns of A is equal to the number of rows of B. So let $A = (a_{ij})$ be an $m \times n$ matrix, and $B = (b_{ij})$ an $n \times p$ matrix. Then the *product* AB of A and B is by definition the $m \times p$ matrix

$$AB = (A_i B^j) \tag{5}$$

whose ijth element is the product of the ith row of A and the jth column of B (note that it is the fact that the number of columns of A equals the number of rows of B which allows the row vector A_i and the column vector B^j to be multiplied). The product matrix AB then has the same number of rows as A, and the same number of columns as B. If we write $AB = (c_{ij})$, then this product is given in terms of matrix elements by

$$c_{ij} = \sum_{k=1}^n a_{ik} b_{kj}. \tag{6}$$

So as to familiarize himself with the definition completely, the student should multiply together several suitable pairs of matrices.

Let us note now that Eqs. (4) can be written very simply in matrix notation. In terms of the ith row vector A_i of A and the column vector

$$\mathbf{x} = \begin{pmatrix} x_1 \\ x_2 \\ \vdots \\ x_n \end{pmatrix},$$

the ith equation of (4) becomes

$$f_i(\mathbf{x}) = A_i \mathbf{x}.$$

Consequently, by the definition of matrix multiplication, the m scalar equations (4) are equivalent to the single matrix equation

$$f(\mathbf{x}) = A\mathbf{x}, \tag{7}$$

so that our linear mapping from $\mathscr{R}^n$ to $\mathscr{R}^m$ takes a form which in notation is precisely the same as that for a linear real-valued function of one variable [$f(x) = ax$]. Of course $f(\mathbf{x})$ on the left-hand side of (7) is a column vector, like $\mathbf{x}$ on the right-hand side. In order to take advantage of matrix notation, we shall hereafter regard points of $\mathscr{R}^n$ interchangeably as n-tuples or n-dimensional column vectors; in all matrix contexts they will be the latter. The fact that Eqs. (4) take the simple form (7) in terms of matrices, together with the fact that multiplication as defined by (5) turns out to be associative (Theorem 4.3 below), is the main motivation for the definition of matrix multiplication.

Now let an $m \times n$ matrix A be given, and define a function $f: \mathscr{R}^n \to \mathscr{R}^m$ by $f(\mathbf{x}) = A\mathbf{x}$. Then f is linear, because

$$f_i(\mathbf{x} + \mathbf{y}) = A_i(\mathbf{x} + \mathbf{y}) = A_i \mathbf{x} + A_i \mathbf{y} = f_i(\mathbf{x}) + f_i(\mathbf{y})$$

by the distributivity of the scalar product of vectors, so $f(\mathbf{x} + \mathbf{y}) = f(\mathbf{x}) + f(\mathbf{y})$, and $f(r\mathbf{x}) = rf(\mathbf{x})$ similarly. The following theorem asserts not only that every mapping of the form $f(\mathbf{x}) = A\mathbf{x}$ is linear, but conversely that every linear mapping from $\mathscr{R}^n$ to $\mathscr{R}^m$ is of this form.

Theorem 4.1 The mapping $f: \mathscr{R}^n \to \mathscr{R}^m$ is linear if and only if there exists a matrix A such that $f(\mathbf{x}) = A\mathbf{x}$ for all $\mathbf{x} \in \mathscr{R}^n$. Then A is that $m \times n$ matrix whose jth column is the column vector $f(\mathbf{e}_j)$, where $\mathbf{e}_j = (0, \ldots, 1, \ldots, 0)$ is the jth unit vector in $\mathscr{R}^n$.

PROOF Given the linear mapping $f: \mathscr{R}^n \to \mathscr{R}^m$, write

$$f(\mathbf{e}_j) = \begin{pmatrix} a_{1j} \\ \vdots \\ a_{mj} \end{pmatrix} \quad \text{and} \quad A = (a_{ij}).$$

Then, given $\mathbf{x} = (x_1, \ldots, x_n) = x_1 \mathbf{e}_1 + \cdots + x_n \mathbf{e}_n$ in $\mathscr{R}^n$, we have

$$\begin{aligned} f(\mathbf{x}) &= f(x_1\mathbf{e}_1 + \cdots + x_n \mathbf{e}_n) \\ &= x_1 f(\mathbf{e}_1) + \cdots + x_n f(\mathbf{e}_n) \qquad \text{(linearity)} \\ &= x_1 \begin{pmatrix} a_{11} \\ \vdots \\ a_{m1} \end{pmatrix} + \cdots + x_n \begin{pmatrix} a_{1n} \\ \vdots \\ a_{mn} \end{pmatrix} \\ &= \begin{pmatrix} a_{11}x_1 + \cdots + a_{1n} x_n \\ a_{21}x_1 + \cdots + a_{2n}x_n \\ \vdots \qquad\qquad \vdots \\ a_{m1}x_1 + \cdots + a_{mn} x_n \end{pmatrix} \\ &= A\mathbf{x} \end{aligned}$$

as desired. ∎

Example 7 If $f: \mathscr{R}^n \to \mathscr{R}^1$ is a linear function on $\mathscr{R}^n$, then the matrix A provided by the theorem has the form

$$A = (a_{11} \ a_{12} \ \cdots \ a_{1n}).$$

Hence, deleting the first subscript, we have

$$f(\mathbf{x}) = a_1 x_1 + \cdots + a_n x_n.$$

Thus the linear mapping $f: \mathscr{R}^n \to \mathscr{R}^1$ can be written

$$f(\mathbf{x}) = \mathbf{a} \cdot \mathbf{x},$$

where $\mathbf{a} = (a_1, \ldots, a_n) \in \mathscr{R}^n$.

Example 8 If $f: \mathscr{R}^1 \to \mathscr{R}^m$ is a linear mapping, then the matrix A has the form

$$A = \begin{pmatrix} a_{11} \\ \vdots \\ a_{m1} \end{pmatrix},$$

Writing $\mathbf{a} = (a_1, \ldots, a_m) \in \mathscr{R}^m$ (second subscripts deleted), we then have

$$f(t) = t\mathbf{a}$$

for all $t \in \mathscr{R}^1$. The image under f of $\mathscr{R}^1$ in $\mathscr{R}^m$ is thus the line through $\mathbf{0}$ in $\mathscr{R}^m$ determined by $\mathbf{a}$.

Example 9 The matrix which Theorem 4.1 associates with the identity transformation

$$f(\mathbf{x}) = \mathbf{x}$$

of $\mathscr{R}^n$ is the $n \times n$ matrix

$$I = \begin{pmatrix} 1 & & & 0 \\ & 1 & & \\ & & \ddots & \\ 0 & & & 1 \end{pmatrix}$$

having every element on the *principal diagonal* (all the elements a_{ij} with $i = j$) equal to 1, and every other element zero. I is called the $n \times n$ *identity matrix*. Note that $AI = IA = A$ for every $n \times n$ matrix A.

Example 10 Let $R(\alpha) : \mathscr{R}^2 \to \mathscr{R}^2$ be a counterclockwise rotation of $\mathscr{R}^2$ about $\mathbf{0}$ through the angle α (Fig. 1.4). We show that $R(\alpha)$ is linear by computing its matrix explicitly. If (r, θ) are the polar coordinates of $\mathbf{x} = (x_1, x_2)$, so

$$x_1 = r \cos \theta, \qquad x_2 = r \sin \theta,$$

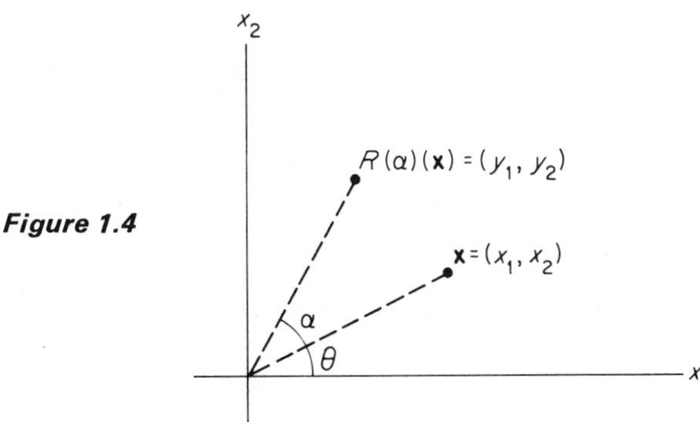

Figure 1.4

then the polar coordinates of $(y_1, y_2) = R(\alpha)(\mathbf{x})$ are $(r, \theta + \alpha)$, so

$$\begin{aligned} y_1 &= r \cos (\theta + \alpha) \\ &= r \cos \theta \cos \alpha - r \sin \theta \sin \alpha \\ &= x_1 \cos \alpha - x_2 \sin \alpha \end{aligned}$$

and

$$\begin{aligned} y_2 &= r \sin (\theta + \alpha) \\ &= r \cos \theta \sin \alpha + r \sin \theta \cos \alpha \\ &= x_1 \sin \alpha + x_2 \cos \alpha. \end{aligned}$$

Therefore

$$\begin{pmatrix} y_1 \\ y_2 \end{pmatrix} = \begin{pmatrix} \cos \alpha & -\sin \alpha \\ \sin \alpha & \cos \alpha \end{pmatrix} \begin{pmatrix} x_1 \\ x_2 \end{pmatrix}.$$

Theorem 4.1 sets up a one-to-one correspondence between the set of all *linear* maps $f : \mathscr{R}^n \to \mathscr{R}^m$ and the set of all $m \times n$ matrices. Let us denote by M_f the matrix of the linear map f.

The following theorem asserts that the problem of finding the composition of two linear mappings is a purely computational matter—we need only multiply their matrices.

Theorem 4.2 If the mappings $f: \mathscr{R}^n \to \mathscr{R}^m$ and $g: \mathscr{R}^m \to \mathscr{R}^p$ are linear, then so is $g \circ f: \mathscr{R}^n \to \mathscr{R}^p$, and

$$M_{g \circ f} = M_g M_f .$$

PROOF Let $M_g = (a_{ij})$ and $M_f = (b_{ij})$. Then $M_g M_f = (A_i B^j)$, where A_i is the ith row of M_g and B^j is the jth column of M_f. What we want to prove is that

$$g \circ f(\mathbf{x}) = (M_g M_f)\mathbf{x},$$

that is, that

$$z_i = \sum_{j=1}^{n} (A_i B^j) x_j, \tag{8}$$

where $\mathbf{x} = (x_1, \ldots, x_n)$ and $g \circ f(\mathbf{x}) = (z_1, \ldots, z_p)$. Now

$$f(\mathbf{x}) = M_f \mathbf{x} = \begin{pmatrix} B_1 \mathbf{x} \\ B_2 \mathbf{x} \\ \vdots \\ B_m \mathbf{x} \end{pmatrix}, \tag{9}$$

where $B_k = (b_{k1} \cdots b_{kn})$ is the kth row of M_f, so

$$B_k \mathbf{x} = \sum_{j=1}^{n} b_{kj} x_j . \tag{10}$$

Also

$$g \circ f(\mathbf{x}) = g(f(\mathbf{x})) = M_g(M_f \mathbf{x}) = \begin{pmatrix} a_{11} & \cdots & a_{1m} \\ \vdots & & \\ a_{p1} & \cdots & a_{pm} \end{pmatrix} \begin{pmatrix} B_1 \mathbf{x} \\ \vdots \\ B_m \mathbf{x} \end{pmatrix}$$

using (9). Therefore

$$\begin{aligned} z_i &= \sum_{k=1}^{m} a_{ik}(B_k \mathbf{x}) \\ &= \sum_{k=1}^{m} a_{ik}\left(\sum_{j=1}^{n} b_{kj} x_j\right) \qquad \text{using (10)} \\ &= \sum_{j=1}^{n}\left(\sum_{k=1}^{m} a_{ik} b_{kj}\right) x_j \\ &= \sum_{j=1}^{n} (A_i B^j) x_j \end{aligned}$$

as desired [recall (8)]. In particular, since we have shown that $g \circ f(\mathbf{x}) = (M_g M_f)\mathbf{x}$, it follows from Theorem 4.1 that $g \circ f$ is linear. ∎

Finally, we list the standard algebraic properties of matrix addition and multiplication.

Theorem 4.3 Addition and multiplication of matrices obey the following rules:

(a) $A(BC) = (AB)C$ (associativity).

(b) $A(B + C) = AB + AC$ (distributivity).

(c) $(A + B)C = AC + BC$

(d) $(rA)B = r(AB) = A(rB)$.

PROOF We prove (a) and (b), leaving (c) and (d) as exercises for the reader.

Let the matrices A, B, C be of dimensions $k \times l$, $l \times m$, and $m \times n$ respectively. Then let $f : \mathscr{R}^l \to \mathscr{R}^k$, $g : \mathscr{R}^m \to \mathscr{R}^l$, $h : \mathscr{R}^n \to \mathscr{R}^m$ be the linear maps such that $M_f = A$, $M_g = B$, $M_h = C$. Then

$$\begin{aligned}(f \circ (g \circ h))(\mathbf{x}) &= f(g \circ h(\mathbf{x})) \\ &= f(g(h)\mathbf{x}))) = (f \circ g)(h(\mathbf{x})) \\ &= ((f \circ g) \circ h)(\mathbf{x})\end{aligned}$$

for all $\mathbf{x} \in \mathscr{R}^n$, so $f \circ (g \circ h) = (f \circ g) \circ h$. Theorem 4.2 therefore implies that

$$\begin{aligned}A(BC) &= M_f M_{g \circ h} = M_{f \circ (g \circ h)} \\ &= M_{(f \circ g) \circ h} = M_{f \circ g} M_h \\ &= (M_f M_g) M_h \\ &= (AB)C,\end{aligned}$$

thereby verifying associativity.

To prove (b), let A be an $l \times m$ matrix, and B, C $m \times n$ matrices. Then let $f : \mathscr{R}^m \to \mathscr{R}^l$ and $g, h : \mathscr{R}^n \to \mathscr{R}^m$ be the linear maps such that $M_f = A$, $M_g = B$, and $M_h = C$. Then $f \circ (g + h) = f \circ g + f \circ h$, so Theorem 4.2 and Exercise 4.9 give

$$\begin{aligned}A(B + C) &= M_f(M_g + M_h) \\ &= M_f M_{g+h} \\ &= M_{f \circ (g+h)} = M_{f \circ g} + M_{f \circ h} \\ &= M_f M_g + M_f M_h \\ &= AB + AC,\end{aligned}$$

thereby verifying distributivity. ∎

The student should not leap from Theorem 4.3 to the conclusion that the algebra of matrices enjoys *all* of the familiar properties of the algebra of real numbers. For example, there exist $n \times n$ matrices A and B such that $AB \neq BA$, so the multiplication of matrices is, in general, *not* commutative (see Exercise 4.12). Also there exist matrices A and B such that $AB = 0$ but neither A nor B is the *zero matrix* whose elements are all 0 (see Exercise 4.13). Finally not every non-zero matrix has an inverse (see Exercise 4.14). The $n \times n$ matrices A and B are called *inverses* of each other if $AB = BA = I$.

Exercises

4.1 Show that the mapping $f: V \to W$ is linear if and only if it satisfies conditions (2) and (3).

4.2 Tell whether or not $f: \mathscr{R}^3 \to \mathscr{R}^2$ is linear, if f is defined by

(a) $f(x, y, z) = (z, x)$,
(b) $f(x, y, z) = (xy, yz)$,
(c) $f(x, y, z) = (x + y, y + z)$,
(d) $f(x, y, z) = (x + y, z + 1)$,
(e) $f(x, y, z) = (2x - y - z, x + 3y + z)$.

For each of these mappings that *is* linear, write down its matrix.

4.3 Show that, if $b \neq 0$, then the function $f(x) = ax + b$ is *not* linear. Although such functions are sometimes loosely referred to as linear ones, they should be called *affine*—an affine function is the sum of a linear function and a constant function.

4.4 Show directly from the definition of linearity that the composition $g \circ f$ is linear if both f and g are linear.

4.5 Prove that the mapping $f: \mathscr{R}^n \to \mathscr{R}^m$ is linear if and only if its coordinate functions $f_1, \ldots, f_m$ are all linear.

4.6 The linear mapping $L: \mathscr{R}^n \to \mathscr{R}^n$ is called *norm preserving* if $|L(\mathbf{x})| = |\mathbf{x}|$, and *inner product preserving* if $L(\mathbf{x}) \cdot L(\mathbf{y}) = \mathbf{x} \cdot \mathbf{y}$. Use Exercise 3.5 to show that L is norm preserving if and only if it is inner product preserving.

4.7 Let $R(\alpha)$ be the counterclockwise rotation of $\mathscr{R}^2$ through an angle α. Then, as shown in Example 10, the matrix of $R(\alpha)$ is

$$M_{R(\alpha)} = \begin{pmatrix} \cos\alpha & -\sin\alpha \\ \sin\alpha & \cos\alpha \end{pmatrix}.$$

It is geometrically clear that $R(\alpha) \circ R(\beta) = R(\alpha + \beta)$, so Theorem 4.2 gives $M_{R(\alpha)} M_{R(\beta)} = M_{R(\alpha+\beta)}$. Verify this by matrix multiplication.

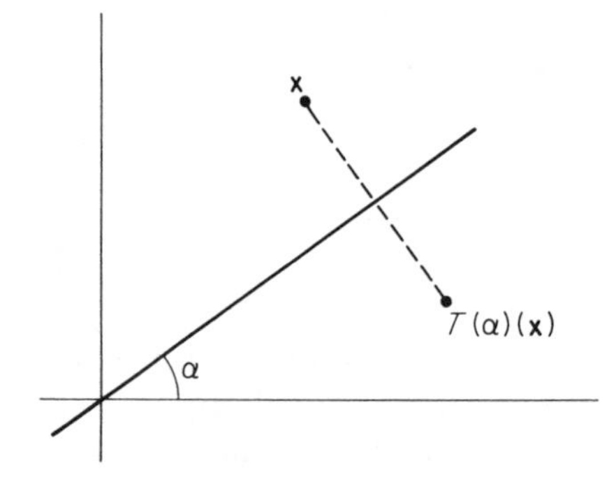

Figure 1.5

4.8 Let $T(\alpha): \mathscr{R}^2 \to \mathscr{R}^2$ be the reflection in $\mathscr{R}^2$ through the line through $\mathbf{0}$ at an angle α from the horizontal (Fig. 1.5). Note that $T(0)$ is simply reflection in the x_1-axis, so

$$M_{T(0)} = \begin{pmatrix} 1 & 0 \\ 0 & -1 \end{pmatrix}.$$

Using the geometrically obvious fact that $T(\alpha) = R(\alpha) \circ T(0) \circ R(-\alpha)$, apply Theorem 4.2 to compute $M_{T(\alpha)}$ by matrix multiplication.

4.9 Show that the composition of two reflections in $\mathscr{R}^2$ is a rotation by computing the matrix product $M_{T(\alpha)}M_{T(\beta)}$. In particular, show that $M_{T(\alpha)}M_{T(\beta)} = M_{R(\gamma)}$ for some γ, identifying γ in terms of α and β.

4.10 If f and g are linear mappings from $\mathscr{R}^n$ to $\mathscr{R}^m$, show that $f + g$ is also linear, with $M_{f+g} = M_f + M_g$.

4.11 Show that $(A + B)C = AC + BC$ by a proof similar to that of part (b) of Theorem 4.3.

4.12 If $f = R(\pi/2)$, the rotation of $\mathscr{R}^2$ through the angle $\pi/2$, and $g = T(0)$, reflection of $\mathscr{R}^2$ through the x_1-axis, then $g(f(1, 0)) = (0, -1))$, while $f(g(1, 0)) = (0, 1)$. Hence it follows from Theorem 4.2 that $M_g M_f \neq M_f M_g$. Consulting Exercises 4.7 and 4.8 for M_f and M_g, verify this by matrix multiplication.

4.13 Find two linear maps $f, g: \mathscr{R}^2 \to \mathscr{R}^2$, neither identically zero, such that the image of f and the kernel of g are both the x_1-axis. Then M_f and M_g will be nonzero matrices such that $M_g M_f = 0$. Verify this by matrix multiplication.

4.14 Show that, if $ad = bc$, then the matrix $\left(\begin{smallmatrix} a & b \\ c & d \end{smallmatrix}\right)$ has no inverse.

4.15 If $A = \left(\begin{smallmatrix} a & b \\ c & d \end{smallmatrix}\right)$ and $B = \left(\begin{smallmatrix} d & -b \\ -c & a \end{smallmatrix}\right)$, compute AB and BA. Conclude that, if $ad - bc \neq 0$, then A has an inverse.

4.16 Let $P(\alpha)$, $Q(\alpha)$, and $R(\alpha)$ denote the rotations of $\mathscr{R}^3$ through an angle α about the x_1-, x_2-, and x_3-axes respectively. Using the facts that

$$M_{P(\alpha)} = \begin{pmatrix} 1 & 0 & 0 \\ 0 & \cos\alpha & -\sin\alpha \\ 0 & \sin\alpha & \cos\alpha \end{pmatrix},$$

$$M_{Q(\alpha)} = \begin{pmatrix} \cos\alpha & 0 & \sin\alpha \\ 0 & 1 & 0 \\ -\sin\alpha & 0 & \cos\alpha \end{pmatrix},$$

$$M_{R(\alpha)} = \begin{pmatrix} \cos\alpha & -\sin\alpha & 0 \\ \sin\alpha & \cos\alpha & 0 \\ 0 & 0 & 1 \end{pmatrix},$$

show by matrix multiplication that

$$P\left(\frac{\pi}{2}\right) \circ Q\left(\frac{\pi}{2}\right) \circ P\left(-\frac{\pi}{2}\right) = R\left(\frac{\pi}{2}\right).$$

5 THE KERNEL AND IMAGE OF A LINEAR MAPPING

Let $L: V \to W$ be a linear mapping of vector spaces. By the *kernel* of L, denoted by Ker L, is meant the set of all those vectors $\mathbf{v} \in V$ such that $L(\mathbf{v}) = \mathbf{0} \in W$,

$$\text{Ker } L = \{\mathbf{v} \in V : L(\mathbf{v}) = \mathbf{0}\}.$$

By the *image* of L, denoted by Im L or $f(L)$, is meant the set of all those vectors $\mathbf{w} \in W$ such that $\mathbf{w} = L(\mathbf{v})$ for some vector $\mathbf{v} \in V$,

$$\operatorname{Im} L = \{\mathbf{w} \in W : \text{there exists } \mathbf{v} \in V \text{ such that } L(\mathbf{v}) = \mathbf{w}\}.$$

It follows easily from these definitions, and from the linearity of L, that the sets Ker L and Im L are subspaces of V and W respectively (Exercises 5.1 and 5.2). We are concerned in this section with the dimensions of these subspaces.

Example 1 If $\mathbf{a}$ is a nonzero vector in $\mathscr{R}^n$, and $L : \mathscr{R}^n \to \mathscr{R}$ is defined by $L(\mathbf{x}) = \mathbf{a} \cdot \mathbf{x}$, then Ker L is the $(n-1)$-dimensional subspace of $\mathscr{R}^n$ that is orthogonal to the vector $\mathbf{a}$, and Im $L = \mathscr{R}$.

Example 2 If $P : \mathscr{R}^3 \to \mathscr{R}^2$ is the projection $P(x_1, x_2, x_3) = (x_1, x_2)$, then Ker P is the x_3-axis and Im $P = \mathscr{R}^2$.

The assumption that the kernel of $L : V \to W$ is the zero vector alone, Ker $L = \mathbf{0}$, has the important consequence that L is *one-to-one*, meaning that $L(\mathbf{v}_1) = L(\mathbf{v}_2)$ implies that $\mathbf{v}_1 = \mathbf{v}_2$ (that is, L is one-to-one if no two vectors of V have the same image under L).

Theorem 5.1 Let $L : V \to W$ be linear, with V being n-dimensional. If Ker $L = \mathbf{0}$, then L is one-to-one, and Im L is an n-dimensional subspace of W.

PROOF To show that L is one-to-one, suppose $L(\mathbf{v}_1) = L(\mathbf{v}_2)$. Then $L(\mathbf{v}_1 - \mathbf{v}_2) = \mathbf{0}$, so $\mathbf{v}_1 - \mathbf{v}_2 = \mathbf{0}$ since Ker $L = \mathbf{0}$.

To show that the subspace Im L is n-dimensional, start with a basis $\mathbf{v}_1, \ldots, \mathbf{v}_n$ for V. Since it is clear (by linearity of L) that the vectors $L(\mathbf{v}_1), \ldots, L(\mathbf{v}_n)$ generate Im L, it suffices to prove that they are linearly independent. Suppose

$$t_1 L(\mathbf{v}_1) + \cdots + t_n L(\mathbf{v}_n) = \mathbf{0}.$$

Then

$$L(t_1 \mathbf{v}_1 + \cdots + t_n \mathbf{v}_n) = \mathbf{0},$$

so $t_1 \mathbf{v}_1 + \cdots + t_n \mathbf{v}_n = \mathbf{0}$ because Ker $L = \mathbf{0}$. But then $t_1 = \cdots = t_n = 0$ because the vectors $\mathbf{v}_1, \ldots, \mathbf{v}_n$ are linearly independent. ∎

An important special case of Theorem 5.1 is that in which W is also n-dimensional; it then follows that Im $L = W$ (see Exercise 5.3).

Theorem 5.2 Let $L : \mathscr{R}^n \to \mathscr{R}^m$ be defined by $L(\mathbf{x}) = A\mathbf{x}$, where $A = (a_{ij})$ is an $m \times n$ matrix. Then

(a) Ker L is the orthogonal complement of that subspace of $\mathscr{R}^n$ that is generated by the row vectors $A_1, \ldots, A_m$ of A, and

(b) $\operatorname{Im} L$ is the subspace of $\mathscr{R}^m$ that is generated by the column vectors $A^1, \ldots, A^n$ of A.

PROOF (a) follows immediately from the fact that L is described by the scalar equations

$$\begin{aligned} L_1(\mathbf{x}) &= A_1\mathbf{x}, \\ L_2(\mathbf{x}) &= A_2\mathbf{x}, \\ &\vdots \\ L_m(\mathbf{x}) &= A_m\mathbf{x}, \end{aligned}$$

so that the ith coordinate $L_i(\mathbf{x})$ is zero if and only if $\mathbf{x}$ is orthogonal to the row vector A_i.

(b) follows immediately from the fact that $\operatorname{Im} L$ is generated by the images $L(\mathbf{e}_1), \ldots, L(\mathbf{e}_n)$ of the standard basis vectors in $\mathscr{R}^n$, whereas $L(\mathbf{e}_i) = A^i$, $i = 1, \ldots, n$, by the definition of matrix multiplication. ∎

Example 3 Suppose that the matrix of $L : \mathscr{R}^3 \to \mathscr{R}^3$ is

$$A = \begin{pmatrix} 2 & -1 & -2 \\ 1 & 2 & 1 \\ 3 & 1 & -1 \end{pmatrix}.$$

Then $A_3 = A_1 + A_2$, but A_1 and A_2 are not collinear, so it follows from 5.2(*a*) that $\operatorname{Ker} L$ is 1-dimensional, since it is the orthogonal complement of the 2-dimensional subspace of $\mathscr{R}^3$ that is spanned by A_1 and A_2. Since the column vectors of A are linearly dependent, $3A^1 = 4A^2 - 5A^3$, but not collinear, it follows from 5.2(b) that $\operatorname{Im} L$ is 2-dimensional.

Note that, in this example, $\dim \operatorname{Ker} L + \dim \operatorname{Im} L = 3$. This is an illustration of the following theorem.

Theorem 5.3 If $L : V \to W$ is a linear mapping of vector spaces, with $\dim V = n$, then

$$\dim \operatorname{Ker} L + \dim \operatorname{Im} L = n.$$

PROOF Let $\mathbf{w}_1, \ldots, \mathbf{w}_p$ be a basis for $\operatorname{Im} L$, and choose vectors $\mathbf{v}_1, \ldots, \mathbf{v}_p \in V$ such that $L(\mathbf{v}_i) = \mathbf{w}_i$ for $i = 1, \ldots, p$. Also let $\mathbf{u}_1, \ldots, \mathbf{u}_q$ be a basis for $\operatorname{Ker} L$. It will then suffice to prove that the vectors $\mathbf{v}_1, \ldots, \mathbf{v}_p, \mathbf{u}_1, \ldots, \mathbf{u}_q$ constitute a basis for V.

To show that these vectors generate V, consider $\mathbf{v} \in V$. Then there exist numbers $a_1, \ldots, a_p$ such that

$$L(\mathbf{v}) = a_1\mathbf{w}_1 + \cdots + a_p\mathbf{w}_p,$$

because $\mathbf{w}_1, \ldots, \mathbf{w}_p$ is a basis for Im L. Since $\mathbf{w}_i = L(\mathbf{v}_i)$ for each i, by linearity we have

$$L(\mathbf{v}) = L(a_1\mathbf{v}_1 + \cdots + a_p\mathbf{v}_p),$$

or

$$L(\mathbf{v} - a_1\mathbf{v}_1 - \cdots - a_p\mathbf{v}_p) = \mathbf{0},$$

so $\mathbf{v} - a_1\mathbf{v}_1 - \cdots - a_p\mathbf{v}_p \in \operatorname{Ker} L$. Hence there exist numbers $b_1, \ldots, b_q$ such that

$$\mathbf{v} - a_1\mathbf{v}_1 - \cdots - a_p\mathbf{v}_p = b_1\mathbf{u}_1 + \cdots + b_q\mathbf{u}_q,$$

or

$$\mathbf{v} = a_1\mathbf{v}_1 + \cdots + a_p\mathbf{v}_p + b_1\mathbf{u}_1 + \cdots + b_q\mathbf{u}_q,$$

as desired.

To show that the vectors $\mathbf{v}_1, \ldots, \mathbf{v}_p, \mathbf{u}_1, \ldots, \mathbf{u}_q$ are linearly independent, suppose that

$$s_1\mathbf{v}_1 + \cdots + s_p\mathbf{v}_p + t_1\mathbf{u}_1 + \cdots + t_q\mathbf{u}_q = \mathbf{0}.$$

Then

$$s_1\mathbf{w}_1 + \cdots + s_p\mathbf{w}_p = \mathbf{0}$$

because $L(\mathbf{v}_i) = \mathbf{w}_i$ and $L(\mathbf{u}_j) = \mathbf{0}$. Since $\mathbf{w}_1, \ldots, \mathbf{w}_p$ are linearly independent, it follows that $s_1 = \cdots = s_p = 0$. But then $t_1\mathbf{u}_1 + \cdots + t_q\mathbf{u}_q = \mathbf{0}$ implies that $t_1 = \cdots = t_q = 0$ also, because the vectors $\mathbf{u}_1, \ldots, \mathbf{u}_q$ are linearly independent. By Proposition 2.1 this concludes the proof. ∎

We give an application of Theorem 5.3 to the theory of linear equations. Consider the system

$$\begin{aligned} a_{11}x_1 + \cdots + a_{1n}x_n &= 0, \\ a_{21}x_1 + \cdots + a_{2n}x_n &= 0, \\ &\vdots \\ a_{m1}x_1 + \cdots + a_{mn}x_n &= 0 \end{aligned} \tag{1}$$

of homogeneous linear equations in $x_1, \ldots, x_n$. As we have observed in Example 9 of Section 3, the space S of solutions $(x_1, \ldots, x_n)$ of (1) is the orthogonal complement of the subspace of $\mathscr{R}^n$ that is generated by the row vectors of the $m \times n$ matrix $A = (a_{ij})$. That is,

$$S = \operatorname{Ker} L,$$

where $L : \mathscr{R}^n \to \mathscr{R}^m$ is defined by $L(\mathbf{x}) = A\mathbf{x}$ (see Theorem 5.2).

Now the *row rank* of the $m \times n$ matrix A is by definition the dimension of the subspace of $\mathscr{R}^n$ generated by the row vectors of A, while the *column rank* of A is the dimension of the subspace of $\mathscr{R}^m$ generated by the column vectors of A.

Theorem 5.4 The row rank of the $m \times n$ matrix $A = (a_{ij})$ and the column rank of A are equal to the same number r. Furthermore $\dim S = n - r$, where S is the space of solutions of the system (1) above.

PROOF We have observed that S is the orthogonal complement to the subspace of $\mathscr{R}^n$ generated by the row vectors of A, so

$$(\text{row rank of } A) + \dim S = n \tag{2}$$

by Theorem 3.4. Since $S = \text{Ker } L$, and by Theorem 5.2, Im L is the subspace of $\mathscr{R}^m$ generated by the column vectors of A, we have

$$(\text{column rank of } A) + \dim S = n \tag{3}$$

by Theorem 5.3. But Eqs. (2) and (3) immediately give the desired results. ∎

Recall that if U and V are subspaces of $\mathscr{R}^n$, then

$$U \cap V = \{\mathbf{x} \in \mathscr{R}^n : \text{both } \mathbf{x} \in U \text{ and } \mathbf{x} \in V\}$$

and

$$U + V = \{\mathbf{x} \in \mathscr{R}^n : \mathbf{x} = \mathbf{u} + \mathbf{v} \text{ with } \mathbf{u} \in U \text{ and } \mathbf{v} \in V\}$$

are both subspaces of $\mathscr{R}^n$ (Exercises 1.2 and 1.3). Let

$$U \times V = \{(\mathbf{x}, \mathbf{y}) \in \mathscr{R}^{2n} : \mathbf{x} \in U \text{ and } \mathbf{y} \in V\}.$$

Then $U \times V$ is a subspace of $\mathscr{R}^{2n}$ with $\dim(U \times V) = \dim U + \dim V$ (Exercise 5.4).

Theorem 5.5 If U and V are subspaces of $\mathscr{R}^n$, then

$$\dim(U + V) + \dim(U \cap V) = \dim U + \dim V. \tag{4}$$

In particular, if $U + V = \mathscr{R}^n$, then

$$\dim U + \dim V - \dim(U \cap V) = n.$$

PROOF Let $L : U \times V \to \mathscr{R}^n$ be the linear mapping defined by

$$L(\mathbf{u}, \mathbf{v}) = \mathbf{u} - \mathbf{v}.$$

Then $\text{Im } L = U + V$ and $\text{Ker } L = \{(\mathbf{x}, \mathbf{x}) \in \mathscr{R}^{2n} : \mathbf{x} \in U \cap V\}$, so $\dim \text{Im } L = \dim(U + V)$ and $\dim \text{Ker } L = \dim(U \cap V)$. Since $\dim U \times V = \dim U + \dim V$ by the preceding remark, Eq. (4) now follows immediately from Theorem 5.3. ∎

Theorem 5.5 is a generalization of the familiar fact that two planes in $\mathscr{R}^3$ "generally" intersect in a line ("generally" meaning that this is the case if the two planes together contain enough linearly independent vectors to span $\mathscr{R}^3$). Similarly a 3-dimensional subspace and a 4-dimensional subspace of $\mathscr{R}^7$ generally intersect in a point (the origin); two 7-dimensional subspaces of $\mathscr{R}^{10}$ generally intersect in a 4-dimensional subspace.

Exercises

5.1 If $L: V \to W$ is linear, show that Ker L is a subspace of V.

5.2 If $L: V \to W$ is linear, show that Im L is a subspace of W.

5.3 Suppose that V and W are n-dimensional vector spaces, and that $F: V \to W$ is linear, with Ker $F = 0$. Then F is one-to-one by Theorem 5.1. Deduce that Im $F = W$, so that the inverse mapping $G = F^{-1}: W \to V$ is defined. Prove that G is also linear.

5.4 If U and V are subspaces of $\mathscr{R}^n$, prove that $U \times V \subset \mathscr{R}^{2n}$ is a subspace of $\mathscr{R}^{2n}$, and that $\dim(U \times V) = \dim U + \dim V$. *Hint:* Consider bases for U and V.

5.5 Let V and W be n-dimensional vector spaces. If $L: V \to W$ is a linear mapping with Im $L = W$, show that Ker $L = \mathbf{0}$.

5.6 Two vector spaces V and W are called *isomorphic* if and only if there exist linear mappings $S: V \to W$ and $T: W \to V$ such that $S \circ T$ and $T \circ S$ are the identity mappings of W and V respectively. Prove that two finite-dimensional vector spaces are isomorphic if and only if they have the same dimension.

5.7 Let V be a finite-dimensional vector space with an inner product $\langle\ ,\ \rangle$. The *dual space* V^* of V is the vector space of all linear functions $V \to \mathscr{R}$. Prove that V and V^* are isomorphic. *Hint:* Let $\mathbf{v}_1, \ldots, \mathbf{v}_n$ be an orthonormal basis for V, and define $\theta_j \in V^*$ by $\theta_j(\mathbf{v}_i) = 0$ unless $i = j$, $\theta_j(\mathbf{v}_j) = 1$. Then prove that $\theta_1, \ldots, \theta_n$ constitute a basis for V^*.

6 DETERMINANTS

It is clear by now that a method is needed for deciding whether a given n-tuple of vectors $\mathbf{a}_1, \ldots, \mathbf{a}_n$ in $\mathscr{R}^n$ are linearly independent (and therefore constitute a basis for $\mathscr{R}^n$). We discuss in this section the method of determinants. The *determinant* of an $n \times n$ matrix A is a real number denoted by det A or $|A|$.

The student is no doubt familiar with the definition of the determinant of a 2×2 or 3×3 matrix. If A is 2×2, then

$$\det\begin{pmatrix} a & b \\ c & d \end{pmatrix} = \begin{vmatrix} a & b \\ c & d \end{vmatrix} = ad - bc.$$

For 3×3 matrices we have expansions by rows and columns. For example, the formula for expansion by the first row is

$$\det\begin{pmatrix} a_{11} & a_{12} & a_{13} \\ a_{21} & a_{22} & a_{23} \\ a_{31} & a_{32} & a_{33} \end{pmatrix} = a_{11}\begin{vmatrix} a_{22} & a_{23} \\ a_{32} & a_{33} \end{vmatrix} - a_{12}\begin{vmatrix} a_{21} & a_{23} \\ a_{31} & a_{33} \end{vmatrix} + a_{13}\begin{vmatrix} a_{21} & a_{22} \\ a_{31} & a_{32} \end{vmatrix}.$$

Formulas for expansions by rows or columns are greatly simplified by the following notation. If A is an $n \times n$ matrix, let A_{ij} denote the $(n-1) \times (n-1)$ submatrix obtained from A by deletion of the ith row and the jth column of A. Then the above formula can be written

$$\det A = a_{11} \det A_{11} - a_{12} \det A_{12} + a_{13} \det A_{13}.$$

The formula for expansion of the $n \times n$ matrix A by the ith row is

$$\det A = \sum_{j=1}^{n} (-1)^{i+j} a_{ij} \det A_{ij}, \tag{1}$$

while the formula for expansion by the jth column is

$$\det A = \sum_{i=1}^{n} (-1)^{i+j} a_{ij} \det A_{ij}. \tag{2}$$

For example, with $n = 3$ and $j = 2$, (2) gives

$$\det A = -a_{12} \det A_{12} + a_{22} \det A_{22} - a_{32} \det A_{32}$$

as the expansion of a 3×3 matrix by its second column.

One approach to the problem of defining determinants of matrices is to define the determinant of an $n \times n$ matrix by means of formulas (1) and (2), assuming inductively that determinants of $(n-1) \times (n-1)$ matrices have been previously defined. Of course it must be verified that expansions along different rows and/or columns give the same result. Instead of carrying through this program, we shall state without proof the basic properties of determinants (I–IV below), and then proceed to derive from them the specific facts that will be needed in subsequent chapters. For a development of the theory of determinants, including proofs of these basic properties, the student may consult the chapter on determinants in any standard linear algebra textbook.

In the statement of Property I, we are thinking of a matrix A as being a function of the column vectors of A, $\det A = D(A^1, \ldots, A^n)$.

(I) There exists a unique (that is, one and only one) alternating, multilinear function D, from n-tuples of vectors in $\mathscr{R}^n$ to real numbers, such that $D(\mathbf{e}_1, \ldots, \mathbf{e}_n) = 1$.

The assertion that D is *multilinear* means that it is linear in each variable separately. That is, for each $i = 1, \ldots, n$,

$$\begin{aligned} D(\mathbf{a}_1, \ldots, x\mathbf{a}_i + y\mathbf{b}_i, \ldots, \mathbf{a}_n) = {} & xD(\mathbf{a}_1, \ldots, \mathbf{a}_i, \ldots, \mathbf{a}_n) \\ & + yD(\mathbf{a}_1, \ldots, \mathbf{b}_i, \ldots, \mathbf{a}_n). \end{aligned} \tag{3}$$

The assertion that D is *alternating* means that $D(\mathbf{a}_1, \ldots, \mathbf{a}_n) = 0$ if $\mathbf{a}_i = \mathbf{a}_j$ for some $i \neq j$. In Exercises 6.1 and 6.2, we ask the student to derive from the alternating multilinearity of D that

$$D(\mathbf{a}_1, \ldots, r\mathbf{a}_i, \ldots, \mathbf{a}_n) = rD(\mathbf{a}_1, \ldots, \mathbf{a}_i, \ldots, \mathbf{a}_n), \tag{4}$$

$$D(\mathbf{a}_1, \ldots, \mathbf{a}_i + r\mathbf{a}_j, \ldots, \mathbf{a}_n) = D(\mathbf{a}_1, \ldots, \mathbf{a}_i, \ldots, \mathbf{a}_n), \tag{5}$$

and

$$D(\mathbf{a}_1, \ldots, \mathbf{a}_j, \ldots, \mathbf{a}_i, \ldots, \mathbf{a}_n) = -D(\mathbf{a}_1, \ldots, \mathbf{a}_i, \ldots, \mathbf{a}_j, \ldots, \mathbf{a}_n) \quad \text{if} \quad i \neq j. \tag{6}$$

Given the alternating multilinear function provided by (I), the determinant of the $n \times n$ matrix A can then be defined by

$$\det A = D(A^1, \ldots, A^n), \tag{7}$$

where $A^1, \ldots, A^n$ are as usual the column vectors of A. Then (4) above says that the determinant of A is multiplied by r if some column of A is multiplied by r, (5) that the determinant of A is unchanged if a multiple of one column is added to another column, while (6) says that the sign of det A is changed by an interchange of any two columns of A. By virtue of the following fact, the word "column" in each of these three statements may be replaced throughout by the word "row."

(II) The determinant of the matrix A is equal to that of its transpose A^t.

The *transpose* A^t of the matrix $A = (a_{ij})$ is obtained from A by interchanging the elements a_{ij} and a_{ji}, for each i and j. Another way of saying this is that the matrix A is reflected through its principal diagonal. We therefore write $A^t = (a_{ji})$ to state the fact that the element in the ith row and jth column of A^t is equal to the one in the jth row and ith column of A. For example, if

$$A = \begin{pmatrix} 1 & 2 & 3 \\ 4 & 5 & 6 \\ 7 & 8 & 9 \end{pmatrix}, \qquad \text{then} \qquad A^t = \begin{pmatrix} 1 & 4 & 7 \\ 2 & 5 & 8 \\ 3 & 6 & 9 \end{pmatrix}.$$

Still another way of saying this is that A^t is obtained from A by changing the rows of A to columns, and the columns to rows.

(III) The determinant of a matrix can be calculated by expansions along rows and columns, that is, by formulas (1) and (2) above.

In a systematic development, it would be proved that formulas (1) and (2) give definitions of det A that satisfy the conditions of Property I and therefore, by the uniqueness of the function D, each must agree with the definition in (7) above.

The fourth basic property of determinants is the fact that the determinant of the product of two matrices is equal to the product of their determinants.

(IV) $\det AB = (\det A)(\det B)$.

As an application, recall that the $n \times n$ matrix B is said to be an *inverse* of the $n \times n$ matrix A if and only if $AB = BA = I$, where I denotes the $n \times n$

identity matrix. In this case we write $B = A^{-1}$ (the matrix A^{-1} is unique if it exists at all—see Exercise 6.3), and say A is *invertible*. Since the fact that $D(\mathbf{e}_1, \ldots, \mathbf{e}_n) = 1$ means that $\det I = 1$, (IV) gives $(\det A)(\det A^{-1}) = 1 \neq 0$. So a necessary condition for the existence of A^{-1} is that $\det A \neq 0$. We prove in Theorem 6.3 that this condition is also sufficient. The $n \times n$ matrix A is called *nonsingular* if $\det A \neq 0$, *singular* if $\det A = 0$.

We can now give the determinant criterion for the linear independence of n vectors in $\mathscr{R}^n$.

Theorem 6.1 The n vectors $\mathbf{a}_1, \ldots, \mathbf{a}_n$ in $\mathscr{R}^n$ are linearly independent if and only if

$$D(\mathbf{a}_1, \ldots, \mathbf{a}_n) \neq 0.$$

PROOF Suppose first that they are linearly dependent; we then want to show that $D(\mathbf{a}_1, \ldots, \mathbf{a}_n) = 0$. Some one of them is then a linear combination of the others; suppose, for instance, that,

$$\mathbf{a}_1 = t_2 \mathbf{a}_2 + \cdots + t_n \mathbf{a}_n .$$

Then

$$\begin{aligned} D(\mathbf{a}_1, \ldots, \mathbf{a}_n) &= D(t_2 \mathbf{a}_2 + \cdots + t_n \mathbf{a}_n, \mathbf{a}_2, \ldots, \mathbf{a}_n) \\ &= \sum_{i=2}^{n} t_i D(\mathbf{a}_i, \mathbf{a}_2, \ldots, \mathbf{a}_n) \qquad \text{(multilinearity)} \\ &= 0 \end{aligned}$$

because each $D(\mathbf{a}_i, \mathbf{a}_2, \ldots, \mathbf{a}_n) = 0$, $i = 2, \ldots, n$, since D is alternating.

Conversely, suppose that the vectors $\mathbf{a}_1, \ldots, \mathbf{a}_n$ are linearly independent. Let A be the $n \times n$ matrix whose column vectors are $\mathbf{a}_1, \ldots, \mathbf{a}_n$, and define the linear mapping $L : \mathscr{R}^n \to \mathscr{R}^n$ by $L(\mathbf{x}) = A\mathbf{x}$ for each (column) vector $\mathbf{x} \in \mathscr{R}^n$. Since $L(\mathbf{e}_i) = \mathbf{a}_i$ for each $i = 1, \ldots, n$, $\operatorname{Im} L = \mathscr{R}^n$ and L is one-to-one by Theorem 5.1. It therefore has a linear inverse mapping $L^{-1} : \mathscr{R}^n \to \mathscr{R}^n$ (Exercise 5.3); denote by B the matrix of L^{-1}. Then $AB = BA = I$ by Theorem 4.2, so it follows from the remarks preceding the statement of the theorem that $\det A \neq 0$, as desired. ∎

Determinants also have important applications to the solution of linear systems of equations. Consider the system

$$\begin{aligned} a_{11}x_1 + \cdots + a_{1n}x_n &= b_1, \\ a_{21}x_1 + \cdots + a_{2n}x_n &= b_2, \\ &\vdots \\ a_{n1}x_2 + \cdots + a_{nn}x_n &= b_n \end{aligned} \tag{8}$$

of n equations in n unknowns. In terms of the column vectors of the coefficient matrix $A = (a_{ij})$, (8) can be rewritten

$$x_1 A^1 + x_2 A^2 + \cdots + x_n A^n = B = \begin{pmatrix} b_1 \\ \vdots \\ b_n \end{pmatrix}. \tag{9}$$

The situation then depends upon whether A is singular or nonsingular. If A is singular then, by Theorem 6.1, the vectors $A^1, \ldots, A^n$ are linearly dependent, and therefore generate a proper subspace V of $\mathscr{R}^n$. If $B \notin V$, then (9) clearly has *no* solution, while if $B \in V$, it is easily seen that (9) infinitely many solutions (Exercise 6.5).

If the matrix A is nonsingular then, by Theorem 6.1, the vectors $A^1, \ldots, A^n$ constitute a basis for $\mathscr{R}^n$, so Eq. (9) has exactly one solution. The formula given in the following theorem, for this unique solution of (8) or (9), is known as *Cramer's Rule.*

Theorem 6.2 Let A be a nonsingular $n \times n$ matrix and let B be a column vector. If $(x_1, \ldots, x_n)$ is the unique solution of (9), then, for each $j = 1, \ldots, n$,

$$x_j = \frac{D(A^1, \ldots, B, \ldots, A^n)}{D(A^1, \ldots, A^n)}, \tag{10}$$

where B occurs in the jth place instead of A^j. That is,

$$x_j = \frac{\begin{vmatrix} a_{11} & \cdots & b_1 & \cdots & a_{1n} \\ \vdots & & \vdots & & \vdots \\ a_{n1} & \cdots & b_n & \cdots & a_{nn} \end{vmatrix}}{\det A}.$$

PROOF If $x_1 A^1 + \cdots + x_n A^n = B$, then

$$\begin{aligned} D(A^1, \ldots, B, \ldots, A^n) &= D(A^1, \ldots, x_1 A^1 + \cdots + x_n A^n, \ldots, A^n) \\ &= x_1 D(A^1, \ldots, A^1, \ldots, A^n) + \cdots \\ &\quad + x_j D(A^1, \ldots, A^j, \ldots, A^n) + \cdots \\ &\quad + x_n D(A^1, \ldots, A^n, \ldots, A^n) \end{aligned}$$

by the multilinearity of D. Then each term of this sum except the jth one vanishes by the alternating property of D, so

$$D(A^1, \ldots, B, \ldots, A^n) = x_j D(A^1, \ldots, A^j, \ldots, A^n).$$

But, since $\det A \neq 0$, this is Eq. (10). ∎

Example 1 Consider the system

$$\begin{aligned} x + 2y + z &= 1, \\ 2x + y - 2z &= 0, \\ x + 2y - 3z &= 1. \end{aligned}$$

Then $\det A = 12 \neq 0$, so (10) gives the solution

$$x = \frac{1}{12}\begin{vmatrix} 1 & 2 & 1 \\ 0 & 1 & -2 \\ 1 & 2 & -3 \end{vmatrix}, \qquad y = \frac{1}{12}\begin{vmatrix} 1 & 1 & 1 \\ 2 & 0 & -2 \\ 1 & 1 & -3 \end{vmatrix}, \qquad z = \frac{1}{12}\begin{vmatrix} 1 & 2 & 1 \\ 2 & 1 & 0 \\ 1 & 2 & 1 \end{vmatrix}.$$

We have noted above that an invertible $n \times n$ matrix A must be nonsingular. We now prove the converse, and give an explicit formula for A^{-1}.

Theorem 6.3 Let $A = (a_{ij})$ be a nonsingular $n \times n$ matrix. Then A is invertible, with its inverse matrix $B = (b_{ij})$ given by

$$b_{ij} = \frac{D(A^1, \ldots, E^j, \ldots, A^n)}{\det A}, \tag{11}$$

where the jth unit column vector occurs in the ith place.

PROOF Let $X = (x_{ij})$ denote an unknown $n \times n$ matrix. Then, from the definition of matrix products, we find that $AX = I$ if and only if

$$x_{1j}A^1 + x_{2j}A^2 + \cdots + x_{nj}A^n = E^j.$$

for each $j = 1, \ldots, n$. For each fixed j, this is a system of n linear equations in the n unknowns $x_{1j}, \ldots, x_{nj}$, with coefficient matrix A. Since A is nonsingular, Cramer's rule gives the solution

$$x_{ij} = \frac{D(A^1, \ldots, E^j, \ldots, A^n)}{\det A}.$$

This is the formula of the theorem, so the matrix B defined by (11) satisfies $AB = I$.

It remains only to prove that $BA = I$ also. Since $\det A^t = \det A \neq 0$, the method of the preceding paragraph gives a matrix C such that $A^tC = I$. Taking transposes, we obtain $C^tA = I$ (see Exercise 6.4). Therefore

$$\begin{aligned} BA &= I(BA) = (C^tA)(BA) \\ &= C^t(AB)A = C^tIA = C^tA = I \end{aligned}$$

as desired. ∎

Formula (11) can be written

$$b_{ij} = \frac{\begin{vmatrix} a_{11} & \cdots & 0 & \cdots & a_{1n} \\ \vdots & & \vdots & & \vdots \\ a_{j1} & \cdots & 1 & \cdots & a_{jn} \\ \vdots & & \vdots & & \vdots \\ a_{n1} & \cdots & 0 & \cdots & a_{nn} \end{vmatrix}}{\det A}.$$

Expanding the numerator along the ith column in which E^j appears, and noting the reversal of subscripts which occurs because the 1 is in the jth row and ith column, we obtain

$$b_{ij} = \frac{(-1)^{i+j} \det A_{ji}}{\det A}.$$

This gives finally the formula

$$A^{-1} = \left(\frac{(-1)^{i+j} \det A_{ij}}{\det A} \right)^{t} \tag{12}$$

for the inverse of the nonsingular matrix A.

Exercises

6.1 Deduce formulas (4) and (5) from Property I.

6.2 Deduce formula (6) from Property I. *Hint:* Compute $D(\mathbf{a}_1, \ldots, \mathbf{a}_i + \mathbf{a}_j, \ldots, \mathbf{a}_i + \mathbf{a}_j, \ldots, \mathbf{a}_n)$, where $\mathbf{a}_i + \mathbf{a}_j$ appears in both the ith place and the jth place.

6.3 Prove that the inverse of an $n \times n$ matrix is unique. That is, if B and C are both inverses of the $n \times n$ matrix A, show that $B = C$. *Hint:* Look at the product CAB.

6.4 If A and B are $n \times n$ matrices, show that $(AB)^t = B^t A^t$.

6.5 If the linearly dependent vectors $\mathbf{a}_1, \ldots, \mathbf{a}_n$ generate the subspace V of $\mathscr{R}^n$, and $\mathbf{b} \in V$, show that $\mathbf{b}$ can be expressed in infinitely many ways as a linear combination of $\mathbf{a}_1, \ldots, \mathbf{a}_n$.

6.6 Suppose that $A = (\mathbf{a}_{ij})$ is an $n \times n$ triangular matrix in which all elements below the principal diagonal are zero; that is, $\mathbf{a}_{ij} = 0$ if $i > j$. Show that $\det A = a_{11}a_{22}\cdots a_{nn}$. In particular, this is true if A is a *diagonal* matrix in which all elements off the principal diagonal are zero.

6.7 Compute using formula (12) the inverse A^{-1} of the coefficient matrix

$$A = \begin{pmatrix} 1 & 2 & 1 \\ 2 & 1 & -2 \\ 1 & 2 & -3 \end{pmatrix}$$

of the system of equations in Example 1. Then show that the solution

$$\begin{pmatrix} x \\ y \\ z \end{pmatrix} = A^{-1} \begin{pmatrix} 1 \\ 0 \\ 1 \end{pmatrix}$$

agrees with that found using Cramer's rule.

6.8 Let $\mathbf{a}_i = (a_{i1}, a_{i2}, \ldots, a_{in})$, $i = 1, \ldots, k < n$, be k linearly dependent vectors in $\mathscr{R}^n$. Then show that every $k \times k$ submatrix of the matrix

$$\begin{pmatrix} a_{11} & \cdots & a_{1n} \\ \vdots & & \vdots \\ a_{k1} & \cdots & a_{kn} \end{pmatrix}$$

has zero determinant.

6.9 If A is an $n \times n$ matrix, and $\mathbf{x}$ and $\mathbf{y}$ are (column) vectors in $\mathscr{R}^n$, show that $(A\mathbf{x})\cdot\mathbf{y} = \mathbf{x}\cdot(A^t\mathbf{y})$, and then that $(A\mathbf{x})\cdot(A\mathbf{y}) = \mathbf{x}\cdot(A^tA\mathbf{y})$.

6.10 The $n \times n$ matrix A is said to *orthogonal* if and only if $AA^t = I$, so A is invertible with $A^{-1} = A^t$. The linear transformation $L: \mathscr{R}^n \to \mathscr{R}^n$ is said to be *orthogonal* if and only if its matrix is orthogonal. Use the identity of the previous exercise to show that the linear transformation L is orthogonal if and only if it is inner product preserving (see Exercise 4.6)

6.11 (a) Show that the $n \times n$ matrix A is orthogonal if and only if its column vectors are othonormal. (b) Show that the $n \times n$ matrix A is orthogonal if and only if its row vectors are orthonormal.

6.12 If $\mathbf{a}_1, \ldots, \mathbf{a}_n$ and $\mathbf{b}_1, \ldots, \mathbf{b}_n$ are two different orthonormal bases for $\mathscr{R}^n$, show that there is an orthogonal transformation $L: \mathscr{R}^n \to \mathscr{R}^n$ with $L(\mathbf{a}_i) = \mathbf{b}_i$ for each $i = 1, \ldots, n$. *Hint:* If A and B are the $n \times n$ matrices whose column vectors are $\mathbf{a}_1, \ldots, \mathbf{a}_n$ and $\mathbf{b}_1, \ldots, \mathbf{b}_n$, respectively, why is the matrix BA^{-1} orthogonal?

7 LIMITS AND CONTINUITY

We now generalize to higher dimensions the familiar single-variable definitions of limits and continuity. This is largely a matter of straightforward repetition, involving merely the use of the norm of a vector in place of the absolute value of a number.

Let D be a subset of $\mathscr{R}^n$, and f a *mapping* of D into $\mathscr{R}^m$ (that is, a rule that associates with each point $\mathbf{x} \in D$ a point $f(\mathbf{x}) \in \mathscr{R}^m$). We write $f: D \to \mathscr{R}^m$, and call D the *domain* (of definition) of f.

In order to define $\lim_{\mathbf{x}\to\mathbf{a}} f(\mathbf{x})$, the *limit* of f at $\mathbf{a}$, it will be necessary that f be defined at points arbitrarily close to $\mathbf{a}$, that is, that D contains points arbitrarily close to $\mathbf{a}$. However we do not want to insist that $\mathbf{a} \in D$, that is, that f be defined at $\mathbf{a}$. For example, when we define the derivative $f'(a)$ of a real-valued single-variable function as the limit of its difference quotient at a,

$$f'(a) = \lim_{x\to a} \frac{f(x) - f(a)}{x - a},$$

this difference quotient is not defined at a.

This consideration motivates the following definition. The point $\mathbf{a}$ is a *limit point* of the set D if and only if every open ball centered at $\mathbf{a}$ contains points of D other than $\mathbf{a}$ (this is what is meant by the statement that D contains points arbitrarily close to $\mathbf{a}$). By the *open ball* of radius r centered at $\mathbf{a}$ is meant the set

$$B_r(\mathbf{a}) = \{\mathbf{x} \in \mathscr{R}^n : |\mathbf{x} - \mathbf{a}| < r\}.$$

Note that $\mathbf{a}$ may, or may not, be itself a point of D. *Examples*: (a) A finite set of points has no limit points; (b) every point of $\mathscr{R}^n$ is a limit point of $\mathscr{R}^n$ (c) *the origin* $\mathbf{0}$ is a limit point of the set $\mathscr{R}^n - \mathbf{0}$; (d) every point of $\mathscr{R}^n$ is a limit point of the set Q of all those points of $\mathscr{R}^n$ having rational coordinates; (e) the *closed ball*

$$\bar{B}_r(\mathbf{a}) = \{\mathbf{x} \in \mathscr{R}^n : |\mathbf{x} - \mathbf{a}| \leqq r\}$$

is the set of all limit points of the open ball $B_r(\mathbf{a})$.

Given a mapping $f: D \to \mathscr{R}^m$, a limit point $\mathbf{a}$ of D, and a point $\mathbf{b} \in \mathscr{R}^m$, we say that $\mathbf{b}$ is the limit of f at $\mathbf{a}$, written

$$\lim_{\mathbf{x} \to \mathbf{a}} f(\mathbf{x}) = \mathbf{b},$$

if and only if, given $\varepsilon > 0$, there exists $\delta > 0$ such that $\mathbf{x} \in D$ and $0 < |\mathbf{x} - \mathbf{a}| < \delta$ imply $|f(\mathbf{x}) - \mathbf{b}| < \varepsilon$.

The idea is of course that $f(\mathbf{x})$ can be made arbitrarily close to $\mathbf{b}$ by choosing $\mathbf{x}$ sufficiently close to $\mathbf{a}$, but not equal to $\mathbf{a}$. In geometrical language (Fig. 1.6),

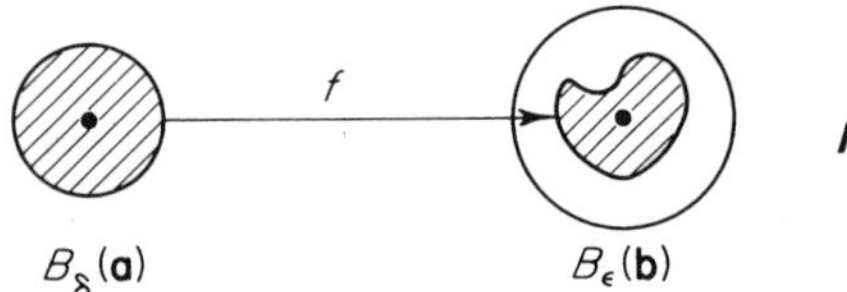

Figure 1.6

the condition of the definition is that, given any open ball $B_\varepsilon(\mathbf{b})$ centered at $\mathbf{b}$, there exists an open ball $B_\delta(\mathbf{a})$ centered at $\mathbf{a}$, whose intersection with $D - \mathbf{a}$ is sent by f into $B_\varepsilon(\mathbf{b})$.

Example 1 Consider the function $f: \mathscr{R}^2 \to \mathscr{R}$ defined by

$$f(x, y) = x^2 + xy + y.$$

In order to prove that $\lim_{\mathbf{x} \to (1,1)} f(x, y) = 3$, we first write

$$\begin{aligned} |f(x, y) - 3| &= |x^2 + xy + y - 3| \\ &\leqq |x^2 - 1| + |y - 1| + |xy - 1| \\ &= |x + 1|\,|x - 1| + |y - 1| + |xy - y + y - 1|, \end{aligned}$$

$$|f(x, y) - 3| \leqq |x + 1|\,|x - 1| + 2|y - 1| + |y|\,|x - 1|. \tag{1}$$

Given $\varepsilon > 0$, we want to find $\delta > 0$ such that $|(x, y) - (1, 1)| = [(x - 1)^2 + (y - 1)^2]^{1/2} < \delta$ implies that the right-hand side of (1) is $< \varepsilon$. Clearly we need bounds for the coefficients $|x+1|$ and $|y|$ of $|x-1|$ in (1). So let us first agree to choose $\delta \leqq 1$, so

$$\begin{aligned}|(x, y)-(1, 1)| < \delta &\Rightarrow |x-1| < 1 \quad \text{and} \quad |y-1| < 1\\ &\Rightarrow 0 < x < 2 \quad \text{and} \quad 0 < y < 2\\ &\Rightarrow |x+1| < 3 \quad \text{and} \quad |y| < 2,\end{aligned}$$

which then implies that

$$|f(x, y)-3| \leqq 5|x-1| + 2|y-1|$$

by (1). It is now clear that, if we take

$$\delta = \min\left(1, \frac{\varepsilon}{7}\right),$$

then

$$|(x, y)-(1, 1)| < \delta \Rightarrow |f(x, y)-3| < 7\,\delta = \varepsilon$$

as desired.

Example 2 Consider the function $f: \mathscr{R}^2 \to \mathscr{R}$ defined by

$$f(x, y) = \begin{cases} \dfrac{xy}{x^2-y^2} & \text{if} \quad x \neq \pm y,\\ 0 & \text{if} \quad x = \pm y.\end{cases}$$

To investigate $\lim_{(x,y)\to\mathbf{0}} f(x, y)$, let us consider the value of $f(x, y)$ as (x, y) approaches $\mathbf{0}$ along the straight line $y = \alpha x$. The lines $y = \pm x$, along which $f(x, y) = 0$, are given by $\alpha = \pm 1$. If $\alpha \neq \pm 1$, then

$$f(x, \alpha x) = \frac{\alpha x^2}{x^2-\alpha^2 x^2} = \frac{\alpha}{1-\alpha^2}.$$

For instance,

$$f(x, 2x) = -\tfrac{2}{3} \qquad \text{and} \qquad f(x, -2x) = +\tfrac{2}{3}$$

for all $x \neq 0$. Thus $f(x, y)$ has different constant values on different straight lines through $\mathbf{0}$, so it is clear that $\lim_{(x,y)\to\mathbf{0}} f(x, y)$ does not exist (because, given any proposed limit b, the values $-\frac{2}{3}$ and $+\frac{2}{3}$ of f cannot both be within ε of b if $\varepsilon < \frac{2}{3}$).

Example 3 Consider the function $f: \mathscr{R} \to \mathscr{R}^2$ defined by $f(t) = (\cos t, \sin t)$, the familiar parametrization of the unit circle. We want to prove that

$$\lim_{t\to 0} f(t) = f(0) = (1, 0).$$

Given $\varepsilon > 0$, we must find $\delta > 0$ such that

$$|t| < \delta \Rightarrow [(\cos t - 1)^2 + (\sin t)^2]^{1/2} < \varepsilon.$$

In order to simplify the square root, write $a = \cos t - 1$ and $b = \sin t$. Then

$$\begin{aligned}[(\cos t - 1)^2 + (\sin t)^2]^{1/2} &= (a^2 + b^2)^{1/2} \\ &\leqq (|a|^2 + 2|a|\,|b| + |b|^2)^{1/2} \\ &= |a| + |b| \\ &= |\cos t - 1| + |\sin t|,\end{aligned}$$

so we see that it suffices to find $\delta > 0$ such that

$$|t| < \delta \Rightarrow |\cos t - 1| < \frac{\varepsilon}{2} \qquad \text{and} \qquad |\sin t| < \frac{\varepsilon}{2}.$$

But we can do this by the fact (from introductory calculus) that the functions $\cos t$ and $\sin t$ are continuous at 0 (where $\cos 0 = 1$, $\sin 0 = 0$).

Example 3 illustrates the fact that limits can be evaluated coordinatewise. To state this result precisely, consider $f: D \to \mathscr{R}^m$, and write $f(\mathbf{x}) = (f_1(\mathbf{x}), \ldots, f_m(\mathbf{x})) \in \mathscr{R}^m$ for each $\mathbf{x} \in D$. Then $f_1, \ldots, f_m$ are real-valued functions on D, called as usual the *coordinate functions* of f, and we write $f = (f_1, \ldots, f_m)$. For the function f of Example 3 we have $f = (f_1, f_2)$ where $f_1(t) = \cos t$, $f_2(t) = \sin t$, and we found that

$$\lim_{t \to 0} f(t) = \left(\lim_{t \to 0} f_1(t), \lim_{t \to 0} f_2(t)\right) = (1, 0).$$

Theorem 7.1 Suppose $f = (f_1, \ldots, f_m) : D \to \mathscr{R}^m$, that $\mathbf{a}$ is a limit point of D, and $\mathbf{b} = (b_1, \ldots, b_m) \in \mathscr{R}^m$. Then

$$\lim_{\mathbf{x} \to \mathbf{a}} f(\mathbf{x}) = \mathbf{b} \tag{2}$$

if and only if

$$\lim_{\mathbf{x} \to \mathbf{a}} f_i(\mathbf{x}) = b_i, \qquad i = 1, \ldots, m. \tag{3}$$

PROOF First assume (2). Then, given $\varepsilon > 0$, there exists $\delta > 0$ such that $\mathbf{x} \in D$ and $0 < |\mathbf{x} - \mathbf{a}| < \delta$ imply that $|f(\mathbf{x}) - \mathbf{b}| < \varepsilon$. But then

$$0 < |\mathbf{x} - \mathbf{a}| < \delta \Rightarrow |f_i(\mathbf{x}) - b_i| \leqq |f(\mathbf{x}) - \mathbf{b}| < \varepsilon,$$

so (3) holds for each $i = 1, \ldots, m$.

Conversely, assume (3). Then, given $\varepsilon > 0$, for each $i = 1, \ldots, m$ there exists a $\delta_i > 0$ such that

$$\mathbf{x} \in D \qquad \text{and} \qquad 0 < |\mathbf{x} - \mathbf{a}| < \delta_i \Rightarrow |f_i(\mathbf{x}) - b_i| < \frac{\varepsilon}{\sqrt{m}}. \tag{4}$$

If we now choose $\delta = \min(\delta_1, \ldots, \delta_m)$, then $\mathbf{x} \in D$ and

$$0 < |\mathbf{x} - \mathbf{a}| < \delta \Rightarrow |f(\mathbf{x}) - \mathbf{b}| = \left[\sum_{i=1}^{m} |f_i(\mathbf{x}) - b_i|^2\right]^{1/2}$$
$$< \left(m \cdot \frac{\varepsilon^2}{m}\right)^{1/2} = \varepsilon$$

by (4), so we have shown that (3) implies (2). ∎

The student should recall the concept of continuity introduced in single-variable calculus. Roughly speaking, a continuous function is one which has nearby values at nearby points, and thus does not change values abruptly. Precisely, the function $f\colon D \to \mathscr{R}^m$ is said to be *continuous at* $\mathbf{a} \in D$ if and only if

$$\lim_{\mathbf{x} \to \mathbf{a}} f(\mathbf{x}) = f(\mathbf{a}). \tag{5}$$

f is said to be *continuous on* D (or, simply, *continuous*) if it is continuous at every point of D.

Actually we cannot insist upon condition (5) if $\mathbf{a} \in D$ is not a limit point of D, for in this case the limit of f at $\mathbf{a}$ cannot be discussed. Such a point, which belongs to D but is not a limit point of D, is called an *isolated point* of D, and we remedy this situation by including in the definition the stipulation that f is automatically continuous at every isolated point of D.

Example 4 If D is the open ball $B_1(\mathbf{0})$ together with the point $(2, 0)$, then any function f on D is continuous at $(2, 0)$, while f is continuous at $\mathbf{a} \in B_1(\mathbf{0})$ if and only if condition (5) is satisfied.

Example 5 If D is the set of all those points (x, y) of $\mathscr{R}^2$ such that both x and y are integers, then every point of D is an isolated point, so every function on D is continuous (at every point of D).

The following result is an immediate corollary to Theorem 7.1.

Theorem 7.2 The mapping $f\colon D \to \mathscr{R}^m$ is continuous at $\mathbf{a} \in D$ if and only if each coordinate function of f is continuous at $\mathbf{a}$.

Example 6 The identity mapping $\pi : \mathscr{R}^n \to \mathscr{R}^n$, defined by $\pi(\mathbf{x}) = \mathbf{x}$, is obviously continuous. Its ith coordinate function, $\pi_i(x_1, \ldots, x_n) = x_i$, is called the *ith projection function*, and is continuous by Theorem 7.2.

Example 7 The real-valued functions s and p on $\mathscr{R}^2$, defined by $s(x, y) = x + y$ and $p(x, y) = xy$, are continuous. The proofs are left as exercises.

The continuity of many mappings can be established without direct recourse to the definition of continuity—instead we apply the known continuity of the elementary single-variable functions, elementary facts such as Theorem 7.2 and Examples 6 and 7, and the fact that a composition of continuous functions is continuous. Given $f: D_1 \to \mathscr{R}^m$ and $g: D_2 \to \mathscr{R}^k$, where $D_1 \subset \mathscr{R}^n$ and $D_2 \subset \mathscr{R}^m$, the *composition*

$$g \circ f: D \to \mathscr{R}^k$$

of f and g is defined as usual by $g \circ f(\mathbf{x}) = g(f(\mathbf{x}))$ for all $\mathbf{x} \in \mathscr{R}^n$ such that $\mathbf{x} \in D_1$ and $f(\mathbf{x}) \in D_2$. That is, the domain of $g \circ f$ is

$$D = \{\mathbf{x} \in \mathscr{R}^n : \mathbf{x} \in D_1 \text{ and } f(\mathbf{x}) \in D_2\}.$$

(This is simply the set of all $\mathbf{x}$ such that $g(f(\mathbf{x}))$ is meaningful.)

Theorem 7.3 If f is continuous at $\mathbf{a}$ and g is continuous at $f(\mathbf{a})$, then $g \circ f$ is continuous at $\mathbf{a}$.

This follows immediately from the following lemma [upon setting $\mathbf{b} = f(\mathbf{a})$].

Lemma 7.4 Given $f: D_1 \to \mathscr{R}^m$ and $g: D_2 \to \mathscr{R}^k$ where $D_1 \subset \mathscr{R}^n$ and $D_2 \subset \mathscr{R}^m$, suppose that

$$\lim_{\mathbf{x} \to \mathbf{a}} f(\mathbf{x}) = \mathbf{b}, \tag{6}$$

and that

$$g \text{ is continuous at } \mathbf{b}. \tag{7}$$

Then

$$\lim_{\mathbf{x} \to \mathbf{a}} g \circ f(\mathbf{x}) = g(\mathbf{b}).$$

PROOF Given $\varepsilon > 0$, we must find $\delta > 0$ such that $|g(f(\mathbf{x})) - g(\mathbf{b})| < \varepsilon$ if $0 < |\mathbf{x} - \mathbf{a}| < \delta$ and $\mathbf{x} \in D$, the domain of $g \circ f$. By (7) there exists $\eta > 0$ such that

$$\mathbf{y} \in D_2 \quad \text{and} \quad |\mathbf{y} - \mathbf{b}| < \eta \Rightarrow |g(\mathbf{y}) - g(\mathbf{b})| < \varepsilon. \tag{8}$$

Then by (6) there exists $\delta > 0$ such that

$$\mathbf{x} \in D_1 \quad \text{and} \quad 0 < |\mathbf{x} - \mathbf{a}| < \delta \Rightarrow |f(\mathbf{x}) - \mathbf{b}| < \eta.$$

But then, upon substituting $\mathbf{y} = f(\mathbf{x})$ in (8), we obtain

$$\mathbf{x} \in D \quad \text{and} \quad 0 < |\mathbf{x} - \mathbf{a}| < \delta \Rightarrow |g(f(\mathbf{x})) - g(\mathbf{b})| < \varepsilon$$

as desired. ∎

It may be instructive for the student to consider also the following geometric formulation of the proof of Theorem 7.3 (see Fig. 1.7).

Given $\varepsilon > 0$, we want $\delta > 0$ so that

$$g \circ f(B_\delta(\mathbf{a})) \subset B_\varepsilon(g(f(\mathbf{a}))).$$

Since g is continuous at $f(\mathbf{a})$, there exists $\eta > 0$ such that

$$g(B_\eta(f(\mathbf{a}))) \subset B_\varepsilon(g(f(\mathbf{a}))).$$

Then, since f is continuous at $\mathbf{a}$, there exists $\delta > 0$ such that

$$f(B_\delta(\mathbf{a})) \subset B_\eta(f(\mathbf{a})).$$

Then

$$g(f(B_\delta(\mathbf{a}))) \subset g(B_\eta(f(\mathbf{a}))) \subset B_\varepsilon(g(f(\mathbf{a})))$$

as desired.

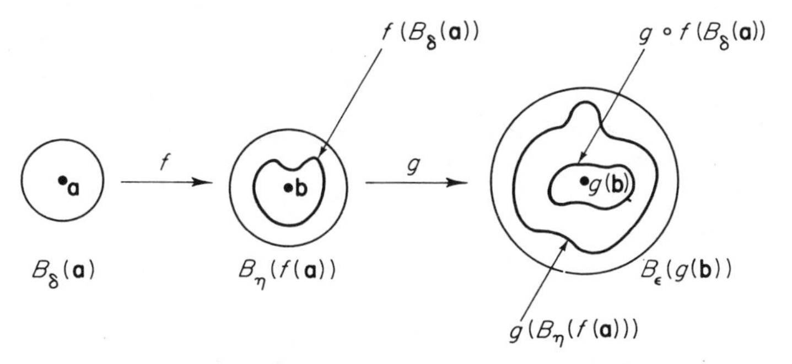

Figure 1.7

As an application of the above results, we now prove the usual theorem on limits of sums and products without mentioning ε and δ.

Theorem 7.5 Let f and g be real-valued functions on $\mathscr{R}^n$. Then

$$\lim_{\mathbf{x}\to\mathbf{a}}(f(\mathbf{x}) + g(\mathbf{x})) = \lim_{\mathbf{x}\to\mathbf{a}} f(\mathbf{x}) + \lim_{\mathbf{x}\to\mathbf{a}} g(\mathbf{x}) \tag{9}$$

and

$$\lim_{\mathbf{x}\to\mathbf{a}} f(\mathbf{x})g(\mathbf{x}) = \left(\lim_{\mathbf{x}\to\mathbf{a}} f(\mathbf{x})\right)\left(\lim_{\mathbf{x}\to\mathbf{a}} g(\mathbf{x})\right), \tag{10}$$

provided that $\lim_{\mathbf{x}\to\mathbf{a}} f(\mathbf{x})$ and $\lim_{\mathbf{x}\to\mathbf{a}} g(\mathbf{x})$ exist.

PROOF We prove (9), and leave the similar proof of (10) to the exercises. Note first that

$$f + g = s \circ (f, g),$$

where $s(x, y) = x + y$ is the sum function on $\mathscr{R}^2$ of Example 7. If

$$b_1 = \lim_{\mathbf{x}\to\mathbf{a}} f(\mathbf{x}) \qquad \text{and} \qquad b_2 = \lim_{\mathbf{x}\to\mathbf{a}} g(\mathbf{x}),$$

then $\lim_{\mathbf{x}\to\mathbf{a}} (f(\mathbf{x}), g(\mathbf{x})) = (b_1, b_2)$ by Theorem 7.1, so

$$\begin{aligned}\lim_{\mathbf{x}\to\mathbf{a}}(f(\mathbf{x}) + g(\mathbf{x})) &= \lim_{\mathbf{x}\to\mathbf{a}} s(f(\mathbf{x}), g(\mathbf{x})) \\ &= s(b_1, b_2) = b_1 + b_2 \\ &= \lim_{\mathbf{x}\to\mathbf{a}} f(\mathbf{x}) + \lim_{\mathbf{x}\to\mathbf{a}} g(\mathbf{x})\end{aligned}$$

by Lemma 7.4. ∎

Example 8 It follows by mathematical induction from Theorem 7.5 that a sum of products of continuous functions is continuous. For instance, any linear real-valued function

$$f(x_1, \ldots, x_n) = \sum_{i=1}^{n} a_i x_i,$$

or polynomial in $x_1, \ldots, x_n$, is continuous. It then follows from Theorem 7.1 that any linear mapping $L : \mathscr{R}^n \to \mathscr{R}^m$ is continuous.

Example 9 To see that $f: \mathscr{R}^3 \to \mathscr{R}$, defined by $f(x, y, z) = \sin(x + \cos yz)$, is continuous, note that

$$f = \sin \circ (s \circ (\pi_1, \cos \circ p \circ (\pi_2, \pi_3))),$$

where π_1, π_2, π_3 are the projection functions on $\mathscr{R}^3$, and s and p are the sum and product functions on $\mathscr{R}^2$.

Exercises

7.1 Verify that the functions s and p of Example 7 are continuous. *Hint:* $xy - x_0y_0 = (xy - xy_0) + (xy_0 - x_0y_0)$.

7.2 Give an $\varepsilon - \delta$ proof that the function $f: \mathscr{R}^3 \to \mathscr{R}$ defined by $f(x, y, z) = x^2y + 2xz^2$ is continuous at $(1, 1, 1)$. *Hint:* $x^2y + 2xz^2 - 3 = (x^2y - y) + (y - 1) + (2xz^2 - 2x) + (2x - 2)$.

7.3 If $f(x, y) = (x^2 - y^2)/(x^2 + y^2)$ unless $x = y = 0$, and $f(0, 0) = 0$, show that $f: \mathscr{R}^2 \to \mathscr{R}$ is not continuous at $(0, 0)$. *Hint:* Consider the behavior of f on straight lines through the origin.

7.4 Let $f(x, y) = 2x^2y/(x^4 + y^2)$ unless $x = y = 0$, and $f(0, 0) = 0$. Define $\varphi(t) = (t, at)$ and $\psi(t) = (t, t^2)$.
(a) Show that $\lim_{t\to 0} f(\varphi(t)) = 0$. Thus f is continuous on any straight line through $(0, 0)$.
(b) Show that $\lim_{t\to 0} f(\psi(t)) = 1$. Conclude that f is *not* continuous at $(0, 0)$.

7.5 Prove the second part of Theorem 7.5.

7.6 The point $\mathbf{a}$ is called a *boundary point* of the set $D \subset \mathscr{R}^n$ if and only if every open ball centered at $\mathbf{a}$ contains both a point of D and a point of the complementary set $\mathscr{R}^n - D$.

For example, the set of all boundary points of the open ball $B_r(\mathbf{p})$ is the sphere $S_r(\mathbf{p}) = \{\mathbf{x} \in \mathscr{R}^n : |\mathbf{x} - \mathbf{p}| = r\}$. Show that every boundary point of D is either a point of D or a limit point of D.

7.7 Let D^* denote the set of all limit points of the set D^*. Then prove that the set $D \cup D^*$ contains all of its limit points.

7.8 Let $f: \mathscr{R}^n \to \mathscr{R}^m$ be continuous at the point $\mathbf{a}$. If $\{\mathbf{a}_n\}_1^\infty$ is a sequence of points of $\mathscr{R}^n$ which converges to $\mathbf{a}$, prove that the sequence $\{f(\mathbf{a}_n)\}_1^\infty$ converges to the point $f(\mathbf{a})$.

8 ELEMENTARY TOPOLOGY OF $\mathscr{R}^n$

In addition to its linear structure as a vector space, and the metric structure provided by the usual inner product, Euclidean n-space $\mathscr{R}^n$ possesses a *topological* structure (defined below). Among other things, this topological structure enables us to define and study a certain class of subsets of $\mathscr{R}^n$, called compact sets, that play an important role in maximum–minimum problems. Once we have defined compact sets, the following two statements will be established.

(A) If D is a compact set in $\mathscr{R}^n$, and $f: D \to \mathscr{R}^m$ is continuous, then its image $f(D)$ is a compact set in $\mathscr{R}^m$ (Theorem 8.7).

(B) If C is a compact set on the real line $\mathscr{R}$, then C contains a maximal element b, that is, a number $b \in C$ such that $x \leqq b$ for all $x \in C$.

It follows immediately from (A) and (B) that, *if $f: D \to \mathscr{R}$ is a continuous real-valued function on the compact set $D \subset \mathscr{R}^n$, then $f(\mathbf{x})$ attains an absolute maximum value at some point $\mathbf{a} \in D$.* For if b is the maximal element of the compact set $f(D) \subset \mathscr{R}$, and $\mathbf{a}$ is a point of D such that $f(\mathbf{a}) = b$, then it is clear that $f(\mathbf{a}) = b$ is the maximum value attained by $f(\mathbf{x})$ on D. The existence of maximum (and, similarly, minimum) values for continuous functions on compact sets, together with the fact that compact sets turn out to be easily recognizable as such (Theorem 8.6), enable compact sets to play the same role in multivariable maximum–minimum problems as do closed intervals in single-variable ones.

By a *topology* (or topological structure) for the set S is meant a collection $\mathscr{T}$ of subsets, called *open* subsets of S, such that $\mathscr{T}$ satisfies the following three conditions:

(i) The empty set $\varnothing$ and the set S itself are open.
(ii) The union of any collection of open sets is an open set.
(iii) The intersection of a *finite* number of open sets is an open set.

The subset A of $\mathscr{R}^n$ is called *open* if and only if, given any point $\mathbf{a} \in A$, there exists an open ball $B_r(\mathbf{a})$ (with $r > 0$) which is centered at $\mathbf{a}$ and is wholly contained in A. Put the other way around, A is open if there does *not* exist a point $\mathbf{a} \in A$ such that every open ball $B_r(\mathbf{a})$ contains points that are not in A. It is easily verified that, with this definition, the collection of all open subsets of $\mathscr{R}^n$ satisfies conditions (i)–(iii) above (Exercise 8.1).

Examples (a) An open interval is an open subset of $\mathscr{R}$, but a closed interval is not. (b) More generally, an open ball in $\mathscr{R}^n$ is an open subset of $\mathscr{R}^n$ (Exercise 8.3) but a closed ball is not (points on the boundary violate the definition). (c) If F is a finite set of points in $\mathscr{R}^n$, then $\mathscr{R}^n - F$ is an open set. (d) Although $\mathscr{R}$ is an open subset of itself, it is not an open subset of the plane $\mathscr{R}^2$.

The subset B of $\mathscr{R}^n$ is called *closed* if and only if its complement $\mathscr{R}^n - B$ is open. It is easily verified (Exercise 8.2) that conditions (i)–(iii) above imply that the collection of all closed subsets of $\mathscr{R}^n$ satisfies the following three analogous conditions:

(i′) $\varnothing$ and $\mathscr{R}^n$ are closed.
(ii′) The intersection of any collection of closed sets is a closed set.
(iii′) The union of a *finite* number of closed sets is a closed set.

Examples: (a) A closed interval is a closed subset of $\mathscr{R}$. (b) More generally, a closed ball in $\mathscr{R}^n$ is a closed subset of $\mathscr{R}^n$ (Exercise 8.3). (c) A finite set F of points is a closed set. (d) The real line $\mathscr{R}$ is a closed subset of $\mathscr{R}^2$. (e) If A is the set of points of the sequence $\{1/n\}_1^\infty$, together with the limit point 0, then A is a closed set (why?)

The last example illustrates the following useful alternative characterization of closed sets.

Proposition 8.1 The subset A of $\mathscr{R}^n$ is closed if and only if it contains all of its limit points.

PROOF Suppose A is closed, and that $\mathbf{a}$ is a limit point of A. Since every open ball centered at $\mathbf{a}$ contains points of A, and $\mathscr{R}^n - A$ is open, $\mathbf{a}$ cannot be a point of $\mathscr{R}^n - A$. Thus $\mathbf{a} \in A$.

Conversely, suppose that A contains all of its limit points. If $\mathbf{b} \in \mathscr{R}^n - A$, then $\mathbf{b}$ is not a limit point of A, so there exists an open ball $B_r(\mathbf{b})$ which contains no points of A. Thus $\mathscr{R}^n - A$ is open, so A is closed. ∎

If, given $A \subset \mathscr{R}^n$, we denote by $\bar{A}$ the union of A and the set of all limit points of A, then Proposition 8.1 implies that A is closed if and only if $A = \bar{A}$.

The empty set $\varnothing$ and $\mathscr{R}^n$ itself are the only subsets of $\mathscr{R}^n$ that are both open and closed (this is not supposed to be obvious—see Exercise 8.6). However there are many subsets of $\mathscr{R}^n$ that are neither open nor closed. For example, the set Q of all rational numbers is such a subset of $\mathscr{R}$.

The following theorem is often useful in verifying that a set is open or closed (as the case may be).

Theorem 8.2 The mapping $f: \mathscr{R}^n \to \mathscr{R}^m$ is continuous if and only if, given any open set $U \subset \mathscr{R}^m$, the inverse image $f^{-1}(U)$ is open in $\mathscr{R}^n$. Also, f is continuous if and only if, given any closed set $C \subset \mathscr{R}^m$, $f^{-1}(C)$ is closed in $\mathscr{R}^n$.

PROOF The inverse image $f^{-1}(U)$ is the set of points in $\mathscr{R}^n$ that map under f into U, that is,

$$f^{-1}(U) = \{\mathbf{x} \in \mathscr{R}^n : f(\mathbf{x}) \in U\}.$$

We prove the "only if" part of the Theorem, and leave the converse as Exercise 8.4.

Suppose f is continuous. If $U \subset \mathscr{R}^m$ is open, and $\mathbf{a} \in f^{-1}(U)$, then there exists an open ball $B_r(f(\mathbf{a})) \subset U$. Since f is continuous, there exists an open ball $B_\delta(\mathbf{a})$ such that $f(B_\delta(\mathbf{a})) \subset B_r(f(\mathbf{a})) \subset U$. Hence $B_\delta(\mathbf{a}) \subset f^{-1}(U)$; this shows that $f^{-1}(U)$ is open.

If $C \subset \mathscr{R}^m$ is closed, then $\mathscr{R}^m - C$ is open, so $f^{-1}(\mathscr{R}^m - C)$ is open by what has just been proved. But $f^{-1}(\mathscr{R}^m - C) = \mathscr{R}^n - f^{-1}(C)$, so it follows that $f^{-1}(C)$ is closed. ∎

As an application of Theorem 8.2, let $f: \mathscr{R}^n \to \mathscr{R}$ be the continuous mapping defined by $f(\mathbf{x}) = |\mathbf{x} - \mathbf{a}|$, where $\mathbf{a} \in \mathscr{R}^n$ is a fixed point. Then $f^{-1}((-r, r))$ is the open ball $B_r(\mathbf{a})$, so it follows that this open ball is indeed an open set. Also $f^{-1}([0, r]) = \bar{B}_r(\mathbf{a})$, so the closed ball is indeed closed. Finally,

$$f^{-1}(r) = S_r(\mathbf{a}) = \{x \in \mathscr{R}^n : |\mathbf{x} - \mathbf{a}| = r\},$$

so the $(n-1)$-sphere of radius r, centered at $\mathbf{a}$, is a closed set.

The subset A of $\mathscr{R}^n$ is said to be *compact* if and only if every infinite subset of A has a limit point which lies in A. This is equivalent to the statement that every sequence of points of A has a subsequence $\{\mathbf{a}_n\}_1^\infty$ which converges to a point $\mathbf{a} \in A$. (This means the same thing in $\mathscr{R}^n$ as on the real line: Given $\varepsilon > 0$, there exists N such that $n \geqq N \Rightarrow |\mathbf{a}_n - \mathbf{a}| < \varepsilon$.) The equivalence of this statement and the definition is just a matter of language (Exercise 8.7).

Examples: (a) $\mathscr{R}$ is not compact, because the set of all integers is an infinite subset of $\mathscr{R}$ that has no limit point at all. Similarly, $\mathscr{R}^n$ is not compact. (b) The open interval $(0, 1)$ is not compact, because the sequence $\{1/n\}_1^\infty$ is an infinite subset of $(0, 1)$ whose limit point 0 is not in the interval. Similarly, open balls fail to be compact. (c) If the set F is finite, then it is automatically compact because it has no infinite subsets which could cause problems.

Closed intervals do not appear to share the problems (in regard to compactness) of open intervals. Indeed the Bolzano–Weierstrass theorem says precisely that every closed interval *is* compact (see the Appendix). We will see presently that every closed ball is compact. Note that a closed ball is both closed and bounded, meaning that it lies inside some ball $B_r(\mathbf{0})$ centered at the origin.

Lemma 8.3 Every compact set is both closed and bounded.

PROOF Suppose that $A \subset \mathscr{R}^n$ is compact. If $\mathbf{a}$ is a limit point of A then, for each integer n, there is a point $\mathbf{a}_n$ such that $|\mathbf{a}_n - \mathbf{a}| < 1/n$. Then the point $\mathbf{a}$ is the only limit point of the sequence $\{\mathbf{a}_n\}_1^\infty$. But, since A is compact, the infinite set $\{\mathbf{a}_n\}_1^\infty$ must have a limit point in A. Therefore $\mathbf{a} \in A$, so it follows from Proposition 8.1 that A is closed.

If A were not bounded then, for each positive integer n, there would exist a point $\mathbf{b}_n \in A$ with $|\mathbf{b}_n| > n$. But then $\{\mathbf{b}_n\}_1^\infty$ would be an infinite subset of A having no limit point (Exercise 8.8), thereby contradicting the compactness of A. ∎

Lemma 8.4 A closed subset of a compact set is compact.

PROOF Suppose that A is closed, B is compact, and $A \subset B$. If S is an infinite subset of A, then S has a limit point $\mathbf{b} \in B$, because B is compact. But $\mathbf{b} \in A$ also, because $\mathbf{b}$ is a limit point of A, and A is closed. Thus every infinite subset of A has a limit point in A, so A is compact. ∎

In the next theorem and its proof we use the following notation. Given $\mathbf{x} = (x_1, \ldots, x_m) \in \mathscr{R}^m$ and $\mathbf{y} = (y_1, \ldots, y_n) \in \mathscr{R}^n$, write $(\mathbf{x}, \mathbf{y}) = (x_1, \ldots, x_m, y_1, \ldots, y_n) \in \mathscr{R}^{m+n}$. If $A \subset \mathscr{R}^m$ and $B \subset \mathscr{R}^n$, then the Cartesian product

$$A \times B = \{(\mathbf{a}, \mathbf{b}) \in \mathscr{R}^{m+n} : \mathbf{a} \in A \text{ and } \mathbf{b} \in B\}$$

is a subset of $\mathscr{R}^{m+n}$.

Theorem 8.5 If A is a compact subset of $\mathscr{R}^m$ and B is a compact subset of $\mathscr{R}^n$, then $A \times B$ is a compact subset of $\mathscr{R}^{m+n}$.

PROOF Given a sequence $\{\mathbf{c}_i\}_1^\infty = \{(\mathbf{a}_i, \mathbf{b}_i)\}_1^\infty$ of points of $A \times B$, we want to show that it has a subsequence converging to a point of $A \times B$. Since A is compact, the sequence $\{\mathbf{a}_i\}_{i=1}^\infty$ has a subsequence $\{\mathbf{a}_{i_j}\}_{j=1}^\infty$ which converges to a point $\mathbf{a} \in A$. Since B is compact, the sequence $\{\mathbf{b}_{i_j}\}_{j=1}^\infty$ has a subsequence $\{\mathbf{b}(i_{j_k})\}_{k=1}^\infty$ which converges to a point $\mathbf{b} \in B$. Then $\{(\mathbf{a}(i_{j_k}), \mathbf{b}(i_{j_k}))\}_{k=1}^\infty$ is a subsequence of the original sequence $\{(\mathbf{a}_i, \mathbf{b}_i)_{i=1}^\infty\}$ which converges to the point $(\mathbf{a}, \mathbf{b}) \in A \times B$. ∎

We are now ready for the criterion that will serve as our recognition test for compact sets.

Theorem 8.6 A subset of $\mathscr{R}^n$ is compact if and only if it is both closed and bounded.

PROOF We have already proved in Lemma 8.3 that every compact set is closed and bounded, so now suppose that A is a closed and bounded subset of $\mathscr{R}^n$.

Figure 1.8

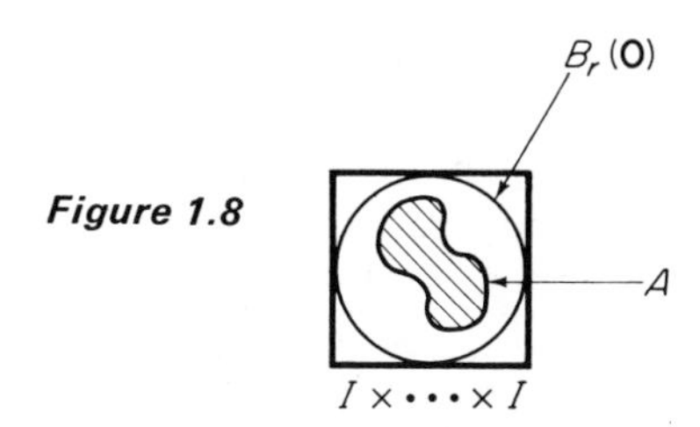

Choose $r > 0$ so large that $A \subset B_r(\mathbf{0})$. (See Fig. 1.8.) If $I = [-r, r]$, then A is a closed subset of the product $I \times I \times \cdots \times I$ (n factors), which is compact by repeated application of Theorem 8.5. It therefore follows from Lemma 8.4 that A is compact. ∎

For example, since spheres and closed balls are closed and bounded, it follows from the theorem that they are compact.

We now prove statement (A) at the beginning of the section.

Theorem 8.7 If A is a compact subset of $\mathscr{R}^n$, and $f\colon A \to \mathscr{R}^m$ is continuous, then $f(A)$ is a compact subset of $\mathscr{R}^m$.

PROOF Given an infinite set T of points of $f(A)$, we want to prove that T contains a sequence of points which converges to a point of $f(A)$. If $S = f^{-1}(T)$, then S is an infinite set of points of A. Since A is compact, S contains a sequence $\{\mathbf{a}_n\}_1^\infty$ of points that converges to a point $\mathbf{a} \in A$. Since f is continuous, $\{f(\mathbf{a}_n)\}_1^\infty$ is then a sequence of points of T that converges to the point $f(\mathbf{a}) \in f(A)$ (see Exercise 7.8). Therefore $f(A)$ is compact. ∎

Statement (B) will be absorbed into the proof of the maximum–minimum value theorem.

Theorem 8.8 If D is a compact set in $\mathscr{R}^n$, and $f\colon D \to \mathscr{R}$ is a continuous function, then f attains maximum and minimum values at points of D. That is, there exist points $\mathbf{a}$ and $\mathbf{b}$ of D such that $f(\mathbf{a}) \leqq f(\mathbf{x}) \leqq f(\mathbf{b})$ for all $\mathbf{x} \in D$.

PROOF We deal only with the maximum value; the treatment of the minimum value is similar. By the previous theorem, $f(D)$ is a compact subset of $\mathscr{R}$, and is therefore closed and bounded by Theorem 8.6. Then $f(D)$ has a least upper bound b, the least number such that $t \leqq b$ for all $t \in f(D)$ (see the Appendix). We want to show that $b \in f(D)$. Since b is the *least* upper bound for $f(D)$, either $b \in f(D)$ or, for each positive integer n, there exists a point $t_n \in f(D)$ with $b - 1/n < t_n < b$. But then the sequence $\{t_n\}_1^\infty$ of points of $f(D)$ converges to b, so b is a limit point of $f(D)$. Since $f(D)$ is closed, it follows that $b \in f(D)$ as desired. If now $\mathbf{b} \in D$ is a point such that $f(\mathbf{b}) = b$, it is clear that $f(\mathbf{x}) \leqq f(\mathbf{b})$ for all $\mathbf{x} \in D$. ∎

For example, we now know that every continuous function on a sphere or closed ball attains maximum and minimum values.

Frequently, in applied maximum–minimum problems, one wants to find a maximum or minimum value of a continuous function $f\colon D \to \mathscr{R}$ where D is *not* compact. Often Theorem 8.8 can still be utilized. For example, suppose we can find a compact subset C of D and a number c such that $f(\mathbf{x}) \geqq c$ for all $\mathbf{x} \in D - C$, whereas f attains values less than c at various points of C. Then it is clear that the minimum of f on C, which exists by Theorem 8.8, is also its minimum on all of D.

Two additional applications of compactness will be needed later. Theorem 8.9 below gives an important property of continuous functions defined on compact sets, while the Heine–Borel theorem deals with coverings of compact sets by open sets.

First recall the familiar definition of continuity: The mapping $f\colon D \to \mathscr{R}$ is continuous if, given $\mathbf{a} \in D$ and $\varepsilon > 0$, there exists $\delta > 0$ such that

$$\mathbf{x} \in D, \qquad |\mathbf{x} - \mathbf{a}| < \delta \Rightarrow |f(\mathbf{x}) - f(\mathbf{a})| < \varepsilon.$$

In general, δ will depend upon the point $\mathbf{a} \in D$. If this is *not* the case, then f is called *uniformly* continuous on D. That is, $f\colon D \to \mathscr{R}$ is *uniformly continuous* if, given $\varepsilon > 0$, there exists $\delta > 0$ such that

$$\mathbf{x}, \mathbf{y} \in D, \qquad |\mathbf{x} - \mathbf{y}| < \delta \Rightarrow |f(\mathbf{x}) - f(\mathbf{y})| < \varepsilon.$$

Not every continuous mapping is uniformly continuous. For example, the function $f\colon (0, 1) \to \mathscr{R}$ defined by $f(x) = 1/x$ is continuous but not uniformly continuous on $(0, 1)$ (why?).

Theorem 8.9 If $f\colon C \to \mathscr{R}$ is continuous, and $C \subset \mathscr{R}^n$ is compact, then f is uniformly continuous on C.

PROOF Suppose, to the contrary, that there exists a number $\varepsilon > 0$ such that, for every positive integer n, there exist two points $\mathbf{x}_n$ and $\mathbf{y}_n$ of C such that

$$|\mathbf{x}_n - \mathbf{y}_n| < 1/n \qquad \text{while} \qquad |f(\mathbf{x}_n) - f(\mathbf{y}_n)| \geqq \varepsilon.$$

Since C is compact we may assume, taking subsequences if necessary, that the sequences $\{\mathbf{x}_n\}_1^\infty$ and $\{\mathbf{y}_n\}_1^\infty$ both converge to the point $\mathbf{a} \in C$. But then we have an easy contradiction to the continuity of f at $\mathbf{a}$. ∎

The Heine–Borel theorem states that, if C is a compact set in $\mathscr{R}^n$ and $\{U_\alpha\}_{\alpha \in A}$ is a collection of open sets whose union contains C, then there is a *finite* subcollection $U_{\alpha_1}, \ldots, U_{\alpha_k}$ of these open sets whose union contains C. We will assume that the collection $\{U_\alpha\}_{\alpha \in A}$ is countable (although it turns out that this assumption involves no loss of generality). So let $\{U_k\}_{k=1}^\infty$ be a sequence of open sets such that $C \subset \bigcup_{k=1}^\infty U_k$. If $V_k = \bigcup_{i=1}^k U_i$ for each $k > 1$, then $\{V_k\}_1^\infty$ is an *increasing* sequence of sets—that is, $V_k \subset V_{k+1}$ for each $k \geqq 1$—and it suffices to prove that $C \subset V_k$ for some integer k.

Theorem 8.10 Let C be a compact subset of $\mathscr{R}^n$, and let $\{V_k\}_1^\infty$ be an increasing sequence of open subsets of $\mathscr{R}^n$ such that $C \subset \bigcup_{k=1}^\infty V_k$. Then there exists a positive integer k such that $C \subset V_k$.

PROOF To the contrary suppose that, for each $k \geqq 1$, there exists a point $\mathbf{x}_k$ of C that is not in V_k. Then no one of the sets $\{V_k\}_1^\infty$ contains infinitely many of the points $\{\mathbf{x}_k\}_1^\infty$ (why?).

Since C is compact, we may assume (taking a subsequence if necessary) that the sequence $\{\mathbf{x}_k\}_1^\infty$ converges to a point $\mathbf{x}_0 \in C$. But then $\mathbf{x}_0 \in V_k$ for some k, and since the set V_k is open, it must contain infinitely many elements of the sequence $\{\mathbf{x}_k\}_1^\infty$. This contradiction proves the theorem. ∎

Exercises

8.1 Verify that the collection of all open subsets of $\mathscr{R}^n$ satisfies conditions (i)—(iii).

8.2 Verify that the collection of all closed subsets of $\mathscr{R}^n$ satisfies conditions (i′)—(iii′). *Hint:* If $\{A_\alpha\}$ is a collection of subsets of $\mathscr{R}^n$, then $\mathscr{R}^n - \bigcup A_\alpha = \bigcap (\mathscr{R}^n - A_\alpha)$ and $\mathscr{R}^n - \bigcap A_\alpha = \bigcup (\mathscr{R}^n - A_\alpha)$.

8.3 Show, directly from the definitions of open and closed sets, that open and closed balls are respectively open and closed sets.

8.4 Complete the proof of Theorem 8.2.

8.5 The point $\mathbf{a}$ is called a *boundary point* of the set A if and only if every open ball centered at $\mathbf{a}$ intersects both A and $\mathscr{R}^n - A$. The *boundary* of the set A is the set of all of its boundary points. Show that the boundary of A is a closed set. Noting that the sphere $S_r(\mathbf{p})$ is the boundary of the ball $B_r(\mathbf{p})$, this gives another proof that spheres are closed sets.

8.6 Show that $\mathscr{R}^n$ is the only nonempty subset of itself that is both open and closed. *Hint:* Use the fact that this is true in the case $n = 1$ (see the Appendix), and the fact that $\mathscr{R}^n$ is a union of straight lines through the origin.

8.7 Show that A is compact if and only if every sequence of points of A has a subsequence that converges to a point of A.

8.8 If $|\mathbf{b}_n| > n$ for each n, show that the sequence $\{\mathbf{b}_n\}_1^\infty$ has no limit.

8.9 Prove that the union or intersection of a finite number of compact sets is compact.

8.10 Let $\{A_n\}_1^\infty$ be a decreasing sequence of compact sets (that is, $A_{n+1} \subset A_n$ for all n). Prove that the intersection $\bigcap_{n=1}^\infty A_n$ is compact and nonempty. Give an example of a decreasing sequence of closed sets whose intersection is empty.

8.11 Given two sets C and D in $\mathscr{R}^n$, define the *distance* $d(C, D)$ between them to be the greatest lower bound of the numbers $|\mathbf{a} - \mathbf{b}|$ for $\mathbf{a} \in C$ and $\mathbf{b} \in D$. If $\mathbf{a}$ is a point of $\mathscr{R}^n$ and D is a closed set, show that there exists $\mathbf{d} \in D$ such that $d(\mathbf{a}, D) = |\mathbf{a} - \mathbf{d}|$. *Hint:* Let B be an appropriate closed ball centered at $\mathbf{a}$, and consider the continuous function $f: B \cap D \to \mathscr{R}$ defined by $f(\mathbf{x}) = |\mathbf{x} - \mathbf{a}|$.

8.12 If C is compact and D is closed, prove that there exist points $\mathbf{c} \in C$ and $\mathbf{d} \in D$ such that $d(C, D) = |\mathbf{c} - \mathbf{d}|$. *Hint:* Consider the continuous function $f: C \to \mathscr{R}^n$ defined by $f(\mathbf{x}) = d(\mathbf{x}, D)$.

II

Multivariable Differential Calculus

Our study in Chapter I of the geometry and topology of $\mathscr{R}^n$ provides an adequate foundation for the study in this chapter of the differential calculus of mappings from one Euclidean space to another. We will find that the basic idea of multivariable differential calculus is the approximation of nonlinear mappings by linear ones.

This idea is implicit in the familiar single-variable differential calculus. If the function $f: \mathscr{R} \to \mathscr{R}$ is differentiable at a, then the tangent line at $(a, f(a))$ to the graph $y = f(x)$ in $\mathscr{R}^2$ is the straight line whose equation is

$$y - f(a) = f'(a)(x - a).$$

The right-hand side of this equation is a linear function of $x - a$; we may regard

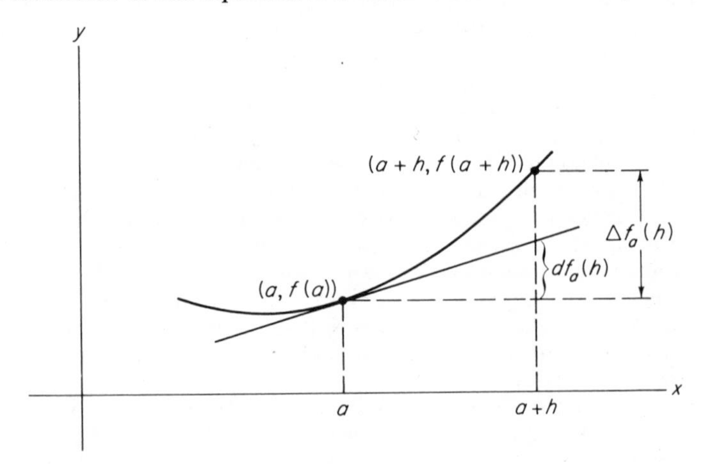

Figure 2.1

it as a linear approximation to the actual change $f(x) - f(a)$ in the value of f between a and x. To make this more precise, let us write $h = x - a$, $\Delta f_a(h) = f(a + h) - f(a)$, and $df_a(h) = f'(a)h$ (see Fig. 2.1). The *linear* mapping $df_a: \mathscr{R} \to \mathscr{R}$, defined by $df_a(h) = f'(a)h$, is called the *differential* of f at a; it is simply that linear mapping $\mathscr{R} \to \mathscr{R}$ whose matrix is the *derivative* $f'(a)$ of f at a (the matrix of a linear mapping $\mathscr{R} \to \mathscr{R}$ being just a real number). With this terminology, we find that when h is small, the linear change $df_a(h)$ is a good approximation to the actual change $\Delta f_a(h)$, in the sense that

$$\lim_{h \to 0} \frac{\Delta f_a(h) - df_a(h)}{h} = \lim_{h \to 0} \frac{f(a + h) - f(a) - f'(a)h}{h} = 0.$$

Roughly speaking, our point of view in this chapter will be that a mapping $f: \mathscr{R}^n \to \mathscr{R}^m$ is (by definition) differentiable at $\mathbf{a}$ if and only if it has near $\mathbf{a}$ an appropriate linear approximation $df_{\mathbf{a}}: \mathscr{R}^n \to \mathscr{R}^m$. In this case $df_{\mathbf{a}}$ will be called the differential of f at $\mathbf{a}$; its $(m \times n)$ matrix will be called the derivative of f at $\mathbf{a}$, thus preserving the above relationship between the differential (a linear mapping) and the derivative (its matrix). We will see that this approach is geometrically well motivated, and permits the basic ingredients of differential calculus (for example, the chain rule, etc.) to be developed and utilized in a multivariable setting.

1 CURVES IN $\mathscr{R}^m$

We consider first the special case of a mapping $f: \mathscr{R} \to \mathscr{R}^m$. Motivated by curves in $\mathscr{R}^2$ and $\mathscr{R}^3$, one may think of a curve in $\mathscr{R}^m$, traced out by a moving point whose position at time t is the point $f(t) \in \mathscr{R}^m$, and attempt to define its velocity at time t. Just as in the single-variable case, $m = 1$, this problem leads to the definition of the *derivative* f' of f. The change in position of the particle from time a to time $a + h$ is described by the vector $f(a + h) - f(a)$, so the average velocity of the particle during this time interval is the familiar-looking difference quotient

$$\frac{f(a + h) - f(a)}{h}$$

whose limit (if it exists) as $h \to 0$ should (by definition) be the instantaneous velocity at time a. So we define

$$f'(a) = \lim_{h \to 0} \frac{f(a + h) - f(a)}{h} \tag{1}$$

if this limit exists, in which case we say that f is *differentiable* at $a \in \mathscr{R}$. The *derivative* vector $f'(a)$ of f at a may be visualized as a tangent vector to the image curve of f at the point $f(a)$ (see Fig. 2.2); its length $|f'(a)|$ is the *speed* at time $t = a$ of the moving point $f(t)$, so $f'(a)$ is often called the *velocity vector* at time a.

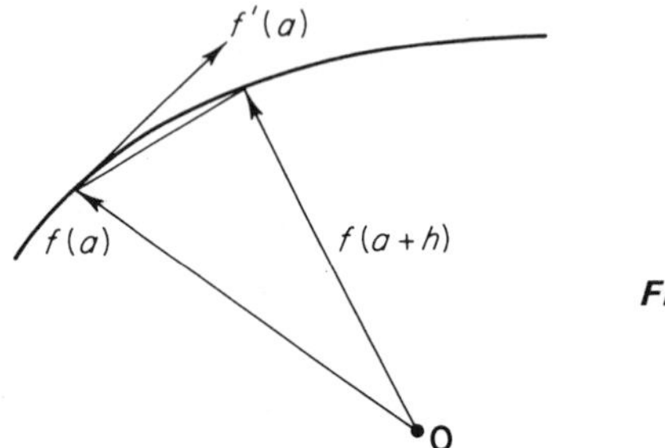

Figure 2.2

If the derivative mapping $f' : \mathscr{R} \to \mathscr{R}^m$ is itself differentiable at a, its derivative at a is the *second derivative* $f''(a)$ of f at a. Still thinking of f in terms of the motion of a moving point (or particle) in $\mathscr{R}^m$, $f''(a)$ is often called the *acceleration vector* at time a. Exercises 1.3 and 1.4 illustrate the usefulness of the concepts of velocity and acceleration for points moving in higher-dimensional Euclidean spaces.

By Theorem 7.1 of Chapter I (limits in $\mathscr{R}^m$ may be taken coordinatewise), we see that $f : \mathscr{R} \to \mathscr{R}^m$ is differentiable at a if and only if each of its coordinate functions $f_1, \ldots, f_m$ is differentiable at a, in which case

$$f' = (f_1', \ldots, f_m').$$

That is, the differentiable function $f : \mathscr{R} \to \mathscr{R}^m$ may be differentiated coordinatewise. Applying coordinatewise the familiar facts about derivatives of real-valued functions, we therefore obtain the results listed in the following theorem.

Theorem 1.1 Let f and g be mappings from $\mathscr{R}$ to $\mathscr{R}^m$, and $\phi : \mathscr{R} \to \mathscr{R}$, all differentiable. Then

$$(f + g)' = f' + g', \tag{2}$$

$$(\varphi f)' = \varphi' f + \varphi f', \tag{3}$$

$$(f \cdot g)' = f' \cdot g + f \cdot g', \tag{4}$$

and

$$(f \circ \varphi)'(t) = \varphi'(t) f'(\varphi(t)). \tag{5}$$

Formula (5) is the *chain rule* for the composition $\mathscr{R} \xrightarrow{\varphi} \mathscr{R} \xrightarrow{f} \mathscr{R}^m$.

Notice the familiar pattern for the differentiation of a product in formulas (3) and (4). The proofs of these formulas are all the same—simply apply componentwise the corresponding formula for real-valued functions. For example, to prove (5), we write

$$\begin{aligned}(f \circ \varphi)'(t) &= ((f_1 \circ \varphi)'(t), (f_2 \circ \varphi)'(t), (f_3 \circ \varphi)'(t)) \\ &= (f_1'(\varphi(t))\varphi'(t), f_2'(\varphi(t))\varphi'(t), f_3'(\varphi(t))\varphi'(t)) \\ &= \varphi'(t)(f_1'(\varphi(t)), f_2'(\varphi(t)), f_3'(\varphi(t))) \\ &= \varphi'(t)f'(\varphi(t)),\end{aligned}$$

applying componentwise the single-variable chain rule, which asserts that

$$(f \circ g)'(t) = f'(g(t))g'(t)$$

if the functions $f, g : \mathscr{R} \to \mathscr{R}$ are differentiable at $g(t)$ and t respectively.

We see below (Exercise 1.12) that the mean value theorem does not hold for vector-valued functions. However it *is* true that two vector-valued functions differ only by a constant (vector) if they have the same derivative; we see this by componentwise application of this fact for real-valued functions.

The *tangent line* at $f(a)$ to the image curve of the differentiable mapping $f: \mathscr{R} \to \mathscr{R}^m$ is, by definition, that straight line which passes through $f(a)$ and is parallel to the tangent vector $f'(a)$. We now inquire as to how well this tangent line approximates the curve close to $f(a)$. That is, how closely does the mapping $h \to f(a) + hf'(a)$ of $\mathscr{R}$ into $\mathscr{R}^m$ (whose image is the tangent line) approximate the mapping $h \to f(a+h)$? Let us write

$$\Delta f_a(h) = f(a+h) - f(a)$$

for the actual change in f from a to $a + h$, and

$$df_a(h) = hf'(a)$$

for the linear (as a function of h) change along the tangent line. Then Fig. 2.3 makes it clear that we are simply asking how small the difference vector

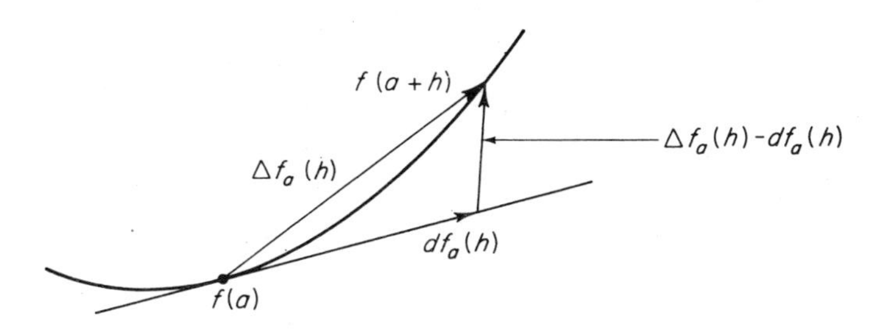

Figure 2.3

$\Delta f_a(h) - df_a(h)$ is when h is small. The answer is that it goes to zero even faster than h does. That is,

$$\begin{aligned}\lim_{h\to 0} \frac{\Delta f_a(h) - df_a(h)}{h} &= \lim_{h\to 0} \frac{f(a+h) - f(a) - hf'(a)}{h} \\ &= \left(\lim_{h\to 0} \frac{f(a+h) - f(a)}{h}\right) - f'(a) \\ &= \mathbf{0}\end{aligned}$$

by the definition of $f'(a)$. Noting that $df_{\mathbf{a}} : \mathscr{R} \to \mathscr{R}^m$ is a linear mapping, we have proved the "only if" part of the following theorem.

Theorem 1.2 The mapping $f: \mathscr{R} \to \mathscr{R}^m$ is differentiable at $a \in \mathscr{R}$ if and only if there exists a linear mapping $L: \mathscr{R} \to \mathscr{R}^m$ such that

$$\lim_{h\to 0} \frac{f(a+h) - f(a) - L(h)}{h} = \mathbf{0}, \tag{6}$$

in which case L is defined by $L(h) = df_a(h) = hf'(a)$.

To prove the "if" part, suppose that there exists a linear mapping satisfying (6). Then there exists $\mathbf{b} \in \mathscr{R}^m$ such that L is defined by $L(h) = h\mathbf{b}$; we must show that $f'(a)$ exists and equals $\mathbf{b}$. But

$$\lim_{h\to 0} \frac{f(a+h) - f(a)}{h} = \left(\lim_{h\to 0} \frac{f(a+h) - f(a) - h\mathbf{b}}{h}\right) + \mathbf{b} = \mathbf{b}$$

by (6). ∎

If $f: \mathscr{R} \to \mathscr{R}^m$ is differentiable at a, then the linear mapping $df_{\mathbf{a}} : \mathscr{R} \to \mathscr{R}^m$, defined by $df_a(h) = hf'(a)$, is called the *differential* of f at a. Notice that the derivative vector $f'(a)$ is, as a column vector, the matrix of the linear mapping df_a, since

$$df_a(h) = hf'(a) = \begin{pmatrix} f_1'(a) \\ \vdots \\ f_m'(a) \end{pmatrix} (h).$$

When in the next section we define derivatives and differentials of mappings from $\mathscr{R}^n$ to $\mathscr{R}^m$, this relationship between the two will be preserved—the differential will be a linear mapping whose matrix is the derivative.

The following discussion provides some motivation for the notation df_a for the differential of f at a. Let us consider the identity function $\mathscr{R} \to \mathscr{R}$, and write x for its name as well as its value at x. Since its derivative is 1 everywhere, its differential at a is defined by

$$dx_a(h) = 1 \cdot h = h.$$

If f is real-valued, and we substitute $h = dx_a(h)$ into the definition of $df_a : \mathscr{R} \to \mathscr{R}$, we obtain

$$df_a(h) = f'(a)h = f'(a)\, dx_a(h),$$

so the two linear mappings df_a and $f'(a)\, dx_a$ are equal,

$$df_a = f'(a)\, dx_a.$$

If we now use the Leibniz notation $f'(a) = df/dx$ and drop the subscript a, we obtain the famous formula

$$df = \frac{df}{dx}\, dx,$$

which now not only makes sense, but is true! It is an actual equality of linear mappings of the real line into itself.

Now let f and g be two differentiable functions from $\mathscr{R}$ to $\mathscr{R}$, and write $h = g \circ f$ for the composition. Then the chain rule gives

$$\begin{aligned} dh_a(t) &= h'(a)t \\ &= g'(f(a))[f'(a)t] \\ &= g'(f(a))[df_a(t)] \\ &= dg_{f(a)}(df_a(t)), \end{aligned}$$

so we see that the single-variable chain rule takes the form

$$dh_a = dg_{f(a)} \circ df_a.$$

In brief, the differential of the composition $h = g \circ f$ is the composition of the differentials of g and f. It is this elegant formulation of the chain rule that we will generalize in Section 3 to the multivariable case.

Exercises

1.1 Let $f: \mathscr{R} \to \mathscr{R}^n$ be a differentiable mapping with $f'(t) \neq \mathbf{0}$ for all $t \in \mathscr{R}$. Let $\mathbf{p}$ be a fixed point not on the image curve of f as in Fig. 2.4. If $\mathbf{q} = f(t_0)$ is the point of the curve closest to $\mathbf{p}$, that is, if $|\mathbf{p} - \mathbf{q}| \leqq |\mathbf{p} - f(t)|$ for all $t \in \mathscr{R}$, show that the vector $\mathbf{p} - \mathbf{q}$ is orthogonal to the curve at $\mathbf{q}$. *Hint:* Differentiate the function $\varphi(t) = |\mathbf{p} - f(t)|^2$.

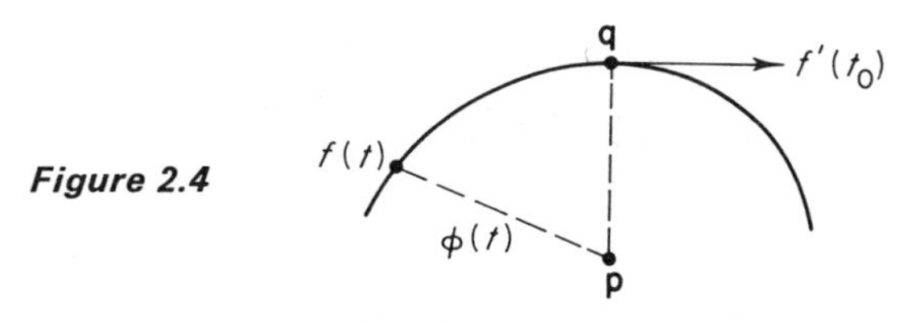

Figure 2.4

1.2 (a) Let $f: \mathscr{R} \to \mathscr{R}^n$ and $g: \mathscr{R} \to \mathscr{R}^n$ be two differentiable curves, with $f'(t) \neq \mathbf{0}$ and $g'(t) \neq \mathbf{0}$ for all $t \in \mathscr{R}$. Suppose the two points $\mathbf{p} = f(s_0)$ and $\mathbf{q} = g(t_0)$ are closer than any other pair of points on the two curves. Then prove that the vector $\mathbf{p} - \mathbf{q}$ is orthogonal to both velocity vectors $f'(s_0)$ and $g'(t_0)$. *Hint:* The point (s_0, t_0) must be a critical point for the function $\rho: \mathscr{R}^2 \to \mathscr{R}$ defined by $\rho(s, t) = |f(s) - g(t)|^2$.
(b) Apply the result of (a) to find the closest pair of points on the "skew" straight lines in $\mathscr{R}^3$ defined by $f(s) = (s, 2s, -s)$ and $g(t) = (t+1, t-2, 2t+3)$.

1.3 Let $F: \mathscr{R}^n \to \mathscr{R}^n$ be a *conservative* force field on $\mathscr{R}^n$, meaning that there exists a continuously differentiable *potential function* $V: \mathscr{R}^n \to \mathscr{R}$ such that $F(\mathbf{x}) = -\nabla V(\mathbf{x})$ for all $\mathbf{x} \in \mathscr{R}^n$ [recall that $\nabla V = (\partial V/\partial x_1, \ldots, \partial V/\partial x_n)$]. Call the curve $\varphi: \mathscr{R} \to \mathscr{R}^n$ a "quasi-Newtonian particle" if and only if there exist constants $m_1, m_2, \ldots, m_n$, called its "mass components," such that

$$F_i(\varphi(t)) = m_i\, \varphi_i''(t) \qquad (F = ma)$$

for each $i = 1, \ldots, n$. Thus, with respect to the x_i-direction, it behaves as though it has mass m_i. Define its *kinetic energy* $K(t)$ and *potential energy* $P(t)$ at time t by

$$K(t) = \tfrac{1}{2} \sum_{i=1}^{n} m_i [\varphi_i'(t)]^2, \qquad P(t) = V(\varphi(t)).$$

Now prove that the law of the *conservation of energy* holds for quasi-Newtonian particles, that is, $K + P =$ constant. *Hint:* Differentiate $K(t) + P(t)$, using the chain rule in the form $P'(t) = \nabla V(\varphi(t)) \cdot \varphi'(t)$, which will be verified in Section 3.

1.4 (*n-body problem*) Deduce from Exercise 1.3 the law of the conservation of energy for a system of n particles moving in $\mathscr{R}^3$ (without colliding) under the influence of their mutual gravitational attractions. You may take $n = 2$ for brevity, although the method is general. *Hint:* Denote by m_1 and m_2 the masses of the two particles, and by $\mathbf{r}_1 = (x_1, x_2, x_3)$ and $\mathbf{r}_2 = (x_4, x_5, x_6)$ their positions at time t. Let $r_{12} = |\mathbf{r}_1 - \mathbf{r}_2|$ be the distance between them. We then have a quasi-Newtonian particle in $\mathscr{R}^6$ with mass components $m_1, m_1, m_1, m_2, m_2, m_2$ and force field F defined by

$$F(\mathbf{r}_1, \mathbf{r}_2) = \frac{Gm_1 m_2}{r_{12}^3} (\mathbf{r}_2 - \mathbf{r}_1,\ \mathbf{r}_1 - \mathbf{r}_2)$$

for $\mathbf{r}_1 \neq \mathbf{r}_2 \in \mathscr{R}^3$. If

$$V(\mathbf{r}_1, \mathbf{r}_2) = -\frac{Gm_1\, m_2}{r_{12}},$$

verify that $F = -\nabla V$. Then apply Exercise 1.3 to conclude that

$$\tfrac{1}{2} m_1 |\mathbf{r}_1'(t)|^2 + \tfrac{1}{2} m_2 |\mathbf{r}_2'(t)|^2 + V(\mathbf{r}_1(t), \mathbf{r}_2(t)) = \text{constant}.$$

Remark: In the general case of a system of n particles, the potential function would be

$$V(\mathbf{r}_1, \ldots, \mathbf{r}_n) = -\sum_{i \leq i < j \leq n} \frac{Gm_i m_j}{r_{ij}}$$

where $r_{ij} = |\mathbf{r}_j - \mathbf{r}_j|$.

1.5 If $f: \mathscr{R} \to \mathscr{R}^m$ is linear, prove that $f'(a)$ exists for all $a \in \mathscr{R}$, with $df_a = f$.

1.6 If L_1 and L_2 are two linear mappings from $\mathscr{R}$ ro $\mathscr{R}^n$ satisfying formula (6), prove that $L_1 = L_2$. *Hint:* Show first that

$$\lim_{h \to 0} \frac{L_1(h) - L_2(h)}{h} = 0.$$

1.7 Let $f, g: \mathscr{R} \to \mathscr{R}$ both be differentiable at a.
(a) Show that $d(fg)_a = g(a)\, df_a + f(a)\, dg_a$.

(*b*) Show that

$$d\left(\frac{f}{g}\right)_a = \frac{g(a)\,df_a - f(a)\,dg_a}{(g(a))^2} \qquad \text{if } g(a) \neq 0.$$

1.8 Let $\gamma(t)$ be the position vector of a particle moving with constant acceleration vector $\gamma''(t) = \mathbf{a}$. Then show that $\gamma(t) = \frac{1}{2}t^2\mathbf{a} + t\mathbf{v}_0 + \mathbf{p}_0$, where $\mathbf{p}_0 = \gamma(0)$ and $\mathbf{v}_0 = \gamma'(0)$. If $\mathbf{a} = \mathbf{0}$, conclude that the particle moves along a straight line through $\mathbf{p}_0$ with velocity vector $\mathbf{v}_0$ (*the law of inertia*).

1.9 Let $\gamma: \mathscr{R} \to \mathscr{R}^n$ be a differentiable curve. Show that $|\gamma(t)|$ is constant if and only if $\gamma(t)$ and $\gamma'(t)$ are orthogonal for all t.

1.10 Suppose that a particle moves around a circle in the plane $\mathscr{R}^2$, of radius r centered at $\mathbf{0}$, with constant speed v. Deduce from the previous exercise that $\gamma(t)$ and $\gamma''(t)$ are both orthogonal to $\gamma'(t)$, so it follows that $\gamma''(t) = k(t)\gamma(t)$. Substitute this result into the equation obtained by differentiating $\gamma(t)\cdot\gamma'(t) = \mathbf{0}$ to obtain $k = -v^2/r^2$. Thus the acceleration vector always points towards the origin and has constant length v^2/r.

1.11 Given a particle in $\mathscr{R}^3$ with mass m and position vector $\gamma(t)$, its *angular momentum* vector is $\mathbf{L}(t) = \gamma(t) \times m\gamma'(t)$, and its *torque* is $\mathbf{T}(t) = \gamma(t) \times m\gamma''(t)$.

(a) Show that $\mathbf{L}'(t) = \mathbf{T}(t)$, so the angular momentum is constant if the torque is zero (this is the law of the conservation of angular momentum).

(b) If the particle is moving in a central force field, that is, $\gamma(t)$ and $\gamma''(t)$ are always collinear, conclude from (a) that it remains in some fixed plane through the origin.

1.12 Consider a particle which moves on a circular helix in $\mathscr{R}^3$ with position vector

$$\gamma(t) = (a \cos \omega t, a \sin \omega t, b\omega t).$$

(a) Show that the speed of the particle is constant.

(b) Show that its velocity vector makes a constant nonzero angle with the z-axis.

(c) If $t_1 = 0$ and $t_2 = 2\pi/\omega$, notice that $\gamma(t_1) = (a, 0, 0)$ and $\gamma(t_2) = (a, 0, 2\pi b)$, so the vector $\gamma(t_2) - \gamma(t_1)$ is vertical. Conclude that the equation

$$\gamma(t_2) - \gamma(t_1) = (t_2 - t_1)\gamma'(\tau)$$

cannot hold for any $\tau \in (t_1, t_2)$. Thus the mean value theorem does not hold for vector-valued functions.

2 DIRECTIONAL DERIVATIVES AND THE DIFFERENTIAL

We have seen that the definition of the derivative of a function of a single variable is motivated by the problem of defining tangent lines to curves. In a similar way the concept of differentiability for functions of several variables is motivated by the problem of defining tangent planes to surfaces.

It is customary to describe the graph in $\mathscr{R}^3$ of a function $f: \mathscr{R}^2 \to \mathscr{R}$ as a "surface" lying "over" the xy-plane $\mathscr{R}^2$. This graph may be regarded as the image of the mapping $F: \mathscr{R}^2 \to \mathscr{R}^3$ defined by $F(x, y) = (x, y, f(x, y))$. Generalizing this geometric interpretation, we will (at least in the case $m > n$) think of the image of a mapping $F: \mathscr{R}^n \to \mathscr{R}^m$ as an n-dimensional surface in $\mathscr{R}^m$. So here we are using the word "surface" only in an intuitive way; we defer its precise definition to Section 5.

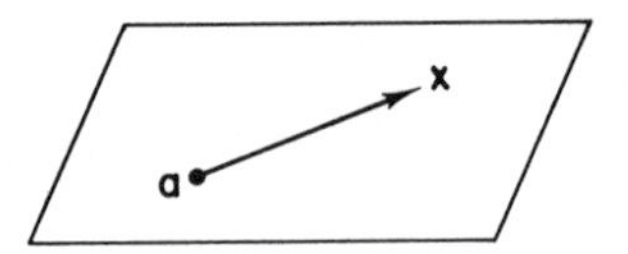

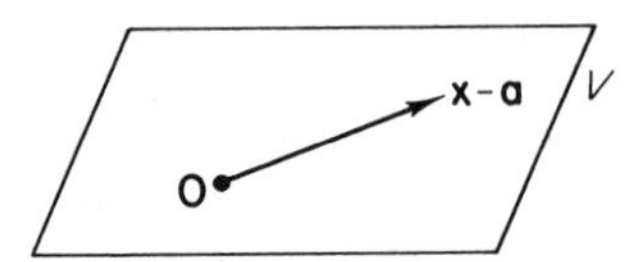

Figure 2.5

One would naturally expect an n-dimensional surface in $\mathscr{R}^m$ $(m > n)$ to have at each point an n-dimensional tangent plane. By an *n-dimensional plane* in $\mathscr{R}^m$ will be meant a parallel translate of an n-dimensional subspace (through the origin) of $\mathscr{R}^m$. If V is a subspace of $\mathscr{R}^m$, by the *parallel translate* of V through the point $\mathbf{a} \in \mathscr{R}^m$ (or the parallel translate of V *to* $\mathbf{a}$) is meant the set of all points $\mathbf{x} \in \mathscr{R}^m$ such that $\mathbf{x} - \mathbf{a} \in V$ (Fig. 2.5). If V is the solution set of the linear equation

$$A\mathbf{x} = \mathbf{0},$$

where A is a matrix and $\mathbf{x}$ a column vector, then this parallel translate of V is the solution set of the equation

$$A(\mathbf{x} - \mathbf{a}) = \mathbf{0}.$$

Given a mapping $F : \mathscr{R}^n \to \mathscr{R}^m$ and a point $\mathbf{a} \in \mathscr{R}^n$, let us try to define the plane (if any) in $\mathscr{R}^m$ that is tangent to the image surface S of F at $F(\mathbf{a})$. The basic idea is that this tangent plane should consist of all straight lines through $F(\mathbf{a})$ which are tangent to curves in the surface S (see Fig. 2.6).

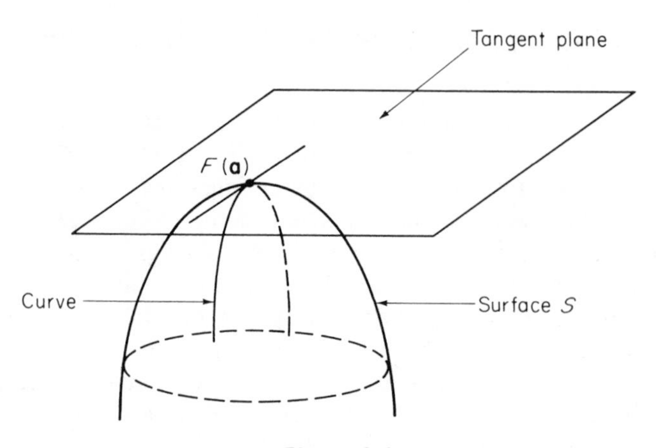

Figure 2.6

Given $\mathbf{v} \in \mathscr{R}^n$, we consider, as a fairly typical such curve, the image under F of the straight line in $\mathscr{R}^n$ which passes through the point $\mathbf{a}$ and is parallel to the vector $\mathbf{v}$. So we define $\gamma_{\mathbf{v}} : \mathscr{R} \to \mathscr{R}^m$ by

$$\gamma_{\mathbf{v}}(t) = F(\mathbf{a} + t\mathbf{v})$$

for each $t \in \mathscr{R}$.

We then define the *directional derivative with respect to* $\mathbf{v}$ *of* F *at* $\mathbf{a}$ to be the velocity vector $\gamma_{\mathbf{v}}'(0)$, that is,

$$D_{\mathbf{v}} F(\mathbf{a}) = \lim_{h \to 0} \frac{F(\mathbf{a} + h\mathbf{v}) - F(\mathbf{a})}{h} \tag{1}$$

provided that the limit exists. The vector $D_{\mathbf{v}} F(\mathbf{a})$, translated to $F(\mathbf{a})$, is then a tangent vector to S at $F(\mathbf{a})$ (see Fig. 2.7).

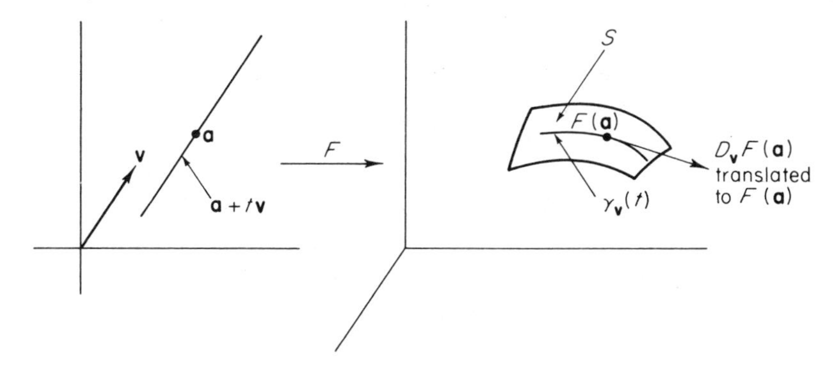

Figure 2.7

For an intuitive interpretation of the directional derivative, consider the following physical example. Suppose $f(\mathbf{p})$ denotes the temperature at the point $\mathbf{p} \in \mathscr{R}^n$. If a particle travels along a straight line through $\mathbf{p}$ with constant velocity vector $\mathbf{v}$, then $D_{\mathbf{v}} f(\mathbf{p})$ is the rate of change of temperature which the particle is experiencing as it passes through the point $\mathbf{p}$ (why?).

For another interpretation, consider the special case $f : \mathscr{R}^2 \to \mathscr{R}$, and let $\mathbf{v} \in \mathscr{R}^2$ be a unit vector. Then $D_{\mathbf{v}} f(\mathbf{p})$ is the slope at $(\mathbf{p}, f(\mathbf{p})) \in \mathscr{R}^3$ of the curve in which the surface $z = f(x, y)$ intersects the vertical plane which contains the point $\mathbf{p} \in \mathscr{R}^2$ and is parallel to the vector $\mathbf{v}$ (why?).

Of special interest are the directional derivatives of F with respect to the standard unit basis vectors $\mathbf{e}_1, \ldots, \mathbf{e}_n$. These are called the *partial derivatives* of F. The ith partial derivative of F at $\mathbf{a}$, denoted by

$$D_i F(\mathbf{a}) \qquad \text{or} \qquad \frac{\partial F}{\partial x_i}(\mathbf{a}),$$

is defined by

$$D_i F(\mathbf{a}) = \frac{\partial F}{\partial x_i}(\mathbf{a}) = D_{\mathbf{e}_i} F(\mathbf{a}). \tag{2}$$

If $\mathbf{a} = (a_1, \ldots, a_n)$, we see that

$$\begin{aligned} D_i F(\mathbf{a}) &= \lim_{h \to 0} \frac{F(\mathbf{a} + h\mathbf{e}_i) - F(\mathbf{a})}{h} \\ &= \lim_{h \to 0} \frac{F(a_1, \ldots, a_i + h, \ldots, a_n) - F(a_1, \ldots, a_i, \ldots, a_n)}{h}, \end{aligned}$$

so $D_i F(\mathbf{a})$ is simply the result of differentiating F as a function of the single variable x_i, holding the remaining variables fixed.

Example 1 If $f(x, y) = xy$, then $D_1 f(x, y) = y$ and $D_2 f(x, y) = x$. If $g(x, y) = e^x \sin y$, then $D_1 g(x, y) = e^x \sin y$ and $D_2 g(x, y) = e^x \cos y$.

To return to the matter of the tangent plane to S at $F(\mathbf{a})$, note that

$$\begin{aligned} D_{c\mathbf{v}} F(\mathbf{a}) &= \lim_{h \to 0} \frac{F(\mathbf{a} + hc\mathbf{v}) - F(\mathbf{a})}{h} \\ &= c \lim_{h \to 0} \frac{F(\mathbf{a} + hc\mathbf{v}) - F(\mathbf{a})}{hc} \\ &= c \lim_{k \to 0} \frac{F(\mathbf{a} + k\mathbf{v}) - F(\mathbf{a})}{k} \\ &= c\, D_{\mathbf{v}} F(\mathbf{a}), \end{aligned}$$

so $D_{\mathbf{v}} F(\mathbf{a})$ and $D_{\mathbf{w}} F(\mathbf{a})$ are collinear vectors in $\mathscr{R}^m$ if $\mathbf{v}$ and $\mathbf{w}$ are collinear in $\mathscr{R}^n$. Thus every straight line through the origin in $\mathbf{R}^n$ determines a straight line through the origin in $\mathscr{R}^m$. Obviously we would like the union

$$\mathscr{L}_{\mathbf{a}} = \{D_{\mathbf{v}} F(\mathbf{a}) \in \mathscr{R}^m : \text{for all } \mathbf{v} \in \mathscr{R}^n\},$$

of all straight lines in $\mathscr{R}^m$ obtained in this manner, to be a *subspace* of $\mathbf{R}^m$. If this were the case, then the parallel translate of $\mathscr{L}_{\mathbf{a}}$ to $F(\mathbf{a})$ would be a likely candidate for the tangent plane to S at $F(\mathbf{a})$.

The set $\mathscr{L}_{\mathbf{a}} \subset \mathscr{R}^m$ is simply the image of $\mathscr{R}^n$ under the mapping $L : \mathscr{R}^n \to \mathscr{R}^m$ defined by

$$L(\mathbf{v}) = D_{\mathbf{v}} F(\mathbf{a}) \tag{3}$$

for all $\mathbf{v} \in \mathscr{R}^n$. Since the image of a linear mapping is always a subspace, we can therefore ensure that $\mathscr{L}_{\mathbf{a}}$ is a subspace of $\mathscr{R}^m$ by requiring that L be a *linear* mapping.

We would also like our tangent plane to "closely fit" the surface S near $F(\mathbf{a})$. This means that we want $L(\mathbf{v})$ to be a good approximation to $F(\mathbf{a}+\mathbf{v})-F(\mathbf{a})$ when $|\mathbf{v}|$ is small. But we have seen this sort of condition before, namely in Theorem 1.2. The necessary and sufficient condition for differentiability in the case $n=1$ now becomes our definition for the general case.

The mapping F, from an open subset D of $\mathscr{R}^n$ to $\mathscr{R}^m$, is *differentiable* at the point $\mathbf{a} \in D$ if and only if there exists a *linear* mapping $L : \mathscr{R}^n \to \mathscr{R}^m$ such that

$$\lim_{\mathbf{h}\to\mathbf{0}} \frac{F(\mathbf{a}+\mathbf{h})-F(\mathbf{a})-L(\mathbf{h})}{|\mathbf{h}|} = \mathbf{0}. \tag{4}$$

The linear mapping L is then denoted by $dF_{\mathbf{a}}$, and is called the *differential* of F at $\mathbf{a}$. Its matrix $F'(\mathbf{a})$ is called the *derivative* of F at $\mathbf{a}$. Thus $F'(\mathbf{a})$ is the (unique) $m \times n$ matrix, provided by Theorem 4.1 of Chapter I, such that

$$dF_{\mathbf{a}}(\mathbf{x}) = F'(\mathbf{a})\mathbf{x} \qquad \text{(matrix multiplication)} \tag{5}$$

for all $\mathbf{x} \in \mathscr{R}^n$. In Theorem 2.4 below we shall prove that the differential of F at $\mathbf{a}$ is well defined by proving that

$$F'(\mathbf{a}) = \left(\frac{\partial F_i}{\partial x_j}(\mathbf{a})\right) = (D_j F_i(\mathbf{a})) \tag{6}$$

in terms of the partial derivatives of the coordinate functions $F_1, \ldots, F_m$ of F. For then if L_1 and L_2 were two linear mappings both satisfying (4) above, then each would have the same matrix given by (6), so they would in fact be the same linear mapping.

To reiterate, the relationship between the differential $dF_{\mathbf{a}}$ and the derivative $F'(\mathbf{a})$ is the same as in Section 1—the differential $dF_{\mathbf{a}} : \mathscr{R}^n \to \mathscr{R}^m$ is a linear mapping represented by the $m \times n$ matrix $F'(\mathbf{a})$.

Note that, if we write $\Delta F_{\mathbf{a}}(\mathbf{h}) = F(\mathbf{a}+\mathbf{h}) - F(\mathbf{a})$, then (4) takes the form

$$\lim_{\mathbf{h}\to\mathbf{0}} \frac{\Delta F_{\mathbf{a}}(\mathbf{h}) - dF_{\mathbf{a}}(\mathbf{h})}{|\mathbf{h}|} = \mathbf{0},$$

which says (just as in the case $n=1$ of Section 1) that the difference, between the actual change in the value of F from $\mathbf{a}$ to $\mathbf{a}+\mathbf{h}$ and the approximate change $dF_{\mathbf{a}}(\mathbf{h})$, goes to zero faster than $\mathbf{h}$ as $\mathbf{h} \to \mathbf{0}$. We indicate this by writing $\Delta F_{\mathbf{a}}(\mathbf{h}) \approx dF_{\mathbf{a}}(\mathbf{h})$, or $F(\mathbf{a}+\mathbf{h}) \approx F(\mathbf{a}) + dF_{\mathbf{a}}(\mathbf{h})$. We will see presently that $dF_{\mathbf{a}}(\mathbf{h})$ is quite easy to compute (if we know the partial derivatives of F at $\mathbf{a}$), so this gives an approximation to the actual value $F(\mathbf{a}+\mathbf{h})$ if $|\mathbf{h}|$ is small. However we will not be able to say *how* small $|\mathbf{h}|$ need be, or to estimate the "error" $\Delta F_{\mathbf{a}}(\mathbf{h}) - dF_{\mathbf{a}}(\mathbf{h})$ made in replacing the actual value $F(\mathbf{a}+\mathbf{h})$ by the approximation $F(\mathbf{a}) + dF_{\mathbf{a}}(\mathbf{h})$, until the multivariable Taylor's formula is available (Section 7). The picture of the *graph* of F when $n=2$ and $m=1$ is instructive (Fig. 2.8).

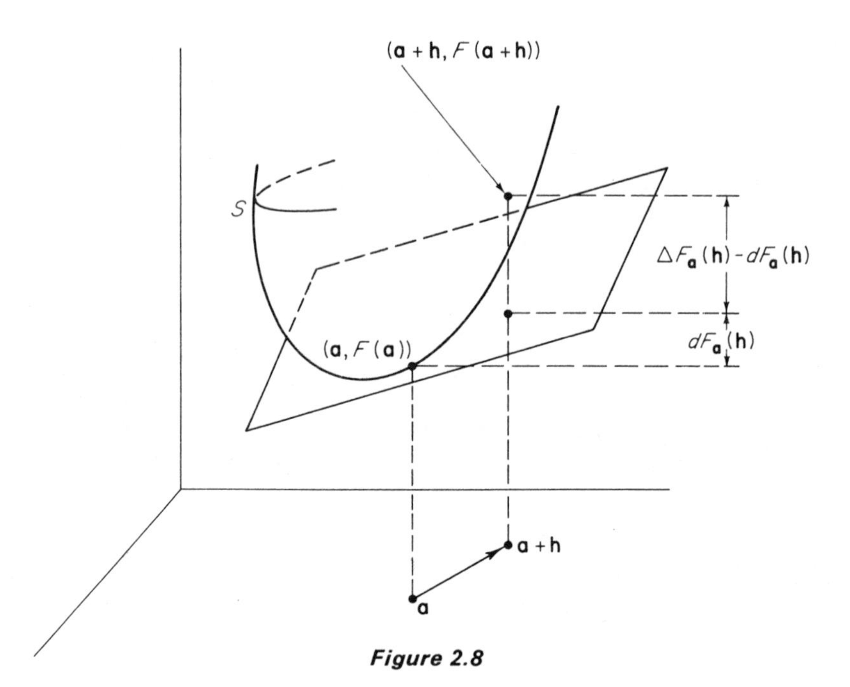

Figure 2.8

Example 2 If $F : \mathscr{R}^n \to \mathscr{R}^m$ is constant, that is, there exists $\mathbf{b} \in \mathscr{R}^m$ such that $F(\mathbf{x}) = \mathbf{b}$ for all $\mathbf{x} \in \mathscr{R}^n$, then F is differentiable everywhere, with $dF_{\mathbf{a}} = \mathbf{0}$ (so the derivative of a constant is zero as expected), because

$$\lim_{\mathbf{h} \to \mathbf{0}} \frac{F(\mathbf{a} + \mathbf{h}) - F(\mathbf{a}) - \mathbf{0}}{|\mathbf{h}|} = \lim_{\mathbf{h} \to \mathbf{0}} \frac{\mathbf{b} - \mathbf{b}}{|\mathbf{h}|} = \mathbf{0}.$$

Example 3 If $F : \mathscr{R}^n \to \mathscr{R}^m$ is *linear*, then F is differentiable everywhere, and

$$dF_{\mathbf{a}} = F \qquad \text{for all} \quad \mathbf{a} \in \mathscr{R}^n.$$

In short, a linear mapping is its own differential, because

$$\lim_{\mathbf{h} \to \mathbf{0}} \frac{F(\mathbf{a} + \mathbf{h}) - F(\mathbf{a}) - F(\mathbf{h})}{|\mathbf{h}|} = \lim_{\mathbf{h} \to \mathbf{0}} \frac{\mathbf{0}}{|\mathbf{h}|} = \mathbf{0}$$

by linearity of F.

For instance, if $s : \mathscr{R}^2 \to \mathscr{R}$ is defined by $s(x, y) = x + y$, then $ds_{\mathbf{a}} = s$ for all $\mathbf{a} \in \mathscr{R}^2$.

The following theorem relates the differential to the directional derivatives which motivated its definition.

Theorem 2.1 If $F : \mathscr{R}^n \to \mathscr{R}^m$ is differentiable at $\mathbf{a}$, then the directional derivative $D_{\mathbf{v}} F(\mathbf{a})$ exists for all $\mathbf{v} \in \mathscr{R}^n$, and

$$D_{\mathbf{v}} F(\mathbf{a}) = dF_{\mathbf{a}}(\mathbf{v}). \tag{7}$$

PROOF We substitute $\mathbf{h} = t\mathbf{v}$ into (4) and let $t \to 0$. Then

$$\mathbf{0} = \lim_{t\to 0} \frac{F(\mathbf{a} + t\mathbf{v}) - F(\mathbf{a}) - dF_{\mathbf{a}}(t\mathbf{v})}{|t\mathbf{v}|}$$
$$= \frac{1}{|\mathbf{v}|}\left[\lim_{t\to 0} \frac{F(\mathbf{a} + t\mathbf{v}) - F(\mathbf{a})}{t} - dF_{\mathbf{a}}(\mathbf{v})\right],$$

so it is clear that

$$D_{\mathbf{v}} F(\mathbf{a}) = \lim_{t\to 0} \frac{F(\mathbf{a} + t\mathbf{v}) - F(\mathbf{a})}{t}$$

exists and equals $dF_{\mathbf{a}}(\mathbf{v})$. ∎

However the converse of Theorem 2.1 is false. That is, a function may possess directional derivatives in all directions, yet still fail to be differentiable.

Example 4 Let $f : \mathscr{R}^2 \to \mathscr{R}$ be defined by

$$f(x, y) = \frac{2x^2y}{x^4 + y^2}$$

unless $x = y = 0$, and $f(0, 0) = 0$. In Exercise 7.4 of Chapter I it was shown that f is not continuous at (0, 0). By Exercise 2.1 below it follows that f is *not* differentiable at (0, 0). However, if $\mathbf{v} = (a, b)$ with $b \neq 0$, then

$$D_{\mathbf{v}} f(0, 0) = \lim_{h\to 0} \frac{f(ah, bh) - f(0, 0)}{h}$$
$$= \lim_{h\to 0} \frac{2h^2a^2b}{a^4h^4 + b^2h^2} = \frac{2a^2}{b}$$

exists, while clearly $D_{\mathbf{v}} f(0, 0) = 0$ if $b = 0$. Other examples of nondifferentiable functions that nevertheless possess directional derivatives are given in Exercises 2.3 and 2.4.

The next theorem proceeds a step further, expressing directional derivatives in terms of partial derivatives (which presumably are relatively easy to compute).

Theorem 2.2 If $F : \mathscr{R}^n \to \mathscr{R}^m$ is differentiable at $\mathbf{a}$, and $\mathbf{v} = (v_1, \ldots, v_n)$, then

$$D_{\mathbf{v}} F(\mathbf{a}) = \sum_{j=1}^{n} v_j\, D_j F(\mathbf{a}). \tag{8}$$

PROOF

$$\begin{aligned} D_{\mathbf{v}} F(\mathbf{a}) &= dF_{\mathbf{a}}(\mathbf{v}) \qquad \text{(by Theorem 2.1)} \\ &= dF_{\mathbf{a}}(v_1 \mathbf{e}_1 + \cdots + v_n \mathbf{e}_n) \\ &= \sum_{j=1}^{n} v_j \, dF_{\mathbf{a}}(\mathbf{e}_j) \qquad \text{(linearity),} \end{aligned}$$

so $D_{\mathbf{v}} F(\mathbf{a}) = \sum_{j=1}^{n} v_j D_{\mathbf{e}_j} F(\mathbf{a}) = \sum_{j=1}^{n} v_j D_j F(\mathbf{a})$, applying Theorem 2.1 again. ∎

In the case $m = 1$ of a differentiable real-valued function $f: \mathscr{R}^n \to \mathscr{R}$, the vector

$$\nabla f(\mathbf{a}) = (D_1 f(\mathbf{a}), \ldots, D_n f(\mathbf{a})) \in \mathscr{R}^n, \tag{9}$$

whose components are the partial derivatives of f, is called the *gradient vector* of f at $\mathbf{a}$. In terms of $\nabla f(\mathbf{a})$, Eq. (8) becomes

$$D_{\mathbf{v}} f(\mathbf{a}) = \nabla f(\mathbf{a}) \cdot \mathbf{v}, \tag{10}$$

which is a strikingly simple expression for the directional derivative in terms of partial derivatives.

Example 5 We use Eq. (10) and the approximation $\Delta f_{\mathbf{a}}(\mathbf{h}) \approx df_{\mathbf{a}}(\mathbf{h})$ to estimate $[(13.1)^2 - (4.9)^2]^{1/2}$. Let $f(x, y) = (x^2 - y^2)^{1/2}$, $\mathbf{a} = (13, 5)$, $\mathbf{h} = (\frac{1}{10}, -\frac{1}{10})$. Then $f(\mathbf{a}) = 12$, $D_1 f(\mathbf{a}) = \frac{13}{12}$, $D_2 f(\mathbf{a}) = -\frac{5}{12}$, so

$$\begin{aligned} [13.1)^2 - (4.9)^2]^{1/2} &= f(13.1, 4.9) \\ &\approx f(13, 5) + \nabla f(13, 5) \cdot (\tfrac{1}{10}, -\tfrac{1}{10}) \\ &= 12 + (\tfrac{13}{12})(\tfrac{1}{10}) + (-\tfrac{5}{12})(-\tfrac{1}{10}) \\ &= 12.15. \end{aligned}$$

To investigate the significance of the gradient vector, let us consider a differentiable function $f: \mathscr{R}^n \to \mathscr{R}$ and a point $\mathbf{a} \in \mathscr{R}^n$, where $\nabla f(\mathbf{a}) \neq \mathbf{0}$. Suppose that we want to determine the direction in which f increases most rapidly at $\mathbf{a}$. By a "direction" here we mean a unit vector $\mathbf{u}$. Let $\theta_{\mathbf{u}}$ denote the angle between $\mathbf{u}$ and $\nabla f(\mathbf{a})$. Then (10) gives

$$D_{\mathbf{u}} f(\mathbf{a}) = \nabla f(\mathbf{a}) \cdot \mathbf{u} = |\nabla f(\mathbf{a})| \cos \theta_{\mathbf{u}}.$$

But $\cos \theta_{\mathbf{u}}$ attains its maximum value of $+1$ when $\theta_{\mathbf{u}} = 0$, that is, when $\mathbf{u}$ and $\nabla f(\mathbf{a})$ are collinear and point in the same direction. We conclude that $|\nabla f(\mathbf{a})|$ is the maximum value of $D_{\mathbf{u}} f(\mathbf{a})$ for $\mathbf{u}$ a unit vector, and that this maximum value is attained with $\mathbf{u} = \nabla f(\mathbf{a})/|\nabla f(\mathbf{a})|$.

For example, suppose that $f(\mathbf{a})$ denotes the temperature at the point $\mathbf{a}$. It is a common physical assumption that heat flows in a direction opposite to that of greatest increase of temperature (heat seeks cold). This principle and the above remarks imply that the direction of heat flow at $\mathbf{a}$ is given by the vector $-\nabla f(\mathbf{a})$.

If $\nabla f(\mathbf{a}) = \mathbf{0}$, then $\mathbf{a}$ is called a *critical point* of f. If f is a differentiable real-valued function defined on an open set D in $\mathscr{R}^n$, and f attains a local maximum (or local minimum) at the point $\mathbf{a} \in D$, then it follows that $\mathbf{a}$ must be a critical point of f. For the function $g_i(x) = f(a_1, \ldots, a_{i-1}, x, a_{i+1}, \ldots, a_n)$ is defined on an open interval of $\mathscr{R}$ containing a_i, and has a local maximum (or local minimum) at a_i, so $D_i f(\mathbf{a}) = g_i'(a_i) = 0$ by the familiar result from elementary calculus. Later in this chapter we will discuss multivariable maximum–minimum problems in considerable detail.

Equation (10) can be rewritten as a multivariable version of the equation $df = (df/dx)\,dx$ of Section 1. Let $x^1, \ldots, x^n$ be the coordinate functions of the identity mapping of $\mathscr{R}^n$, that is, $x^i : \mathscr{R}^n \to \mathscr{R}$ is defined by $x^i(p_1, \ldots, p_n) = p_i$, $i = 1, \ldots, n$. Then x^i is a linear function, so

$$dx_{\mathbf{a}}{}^i(\mathbf{h}) = x^i(\mathbf{h}) = h_i$$

for all $\mathbf{a} \in \mathscr{R}^n$, by Example 3. If $f : \mathscr{R}^n \to \mathscr{R}$ is differentiable at $\mathbf{a}$, then Theorem 2.1 and Eq. (10) therefore give

$$\begin{aligned} df_{\mathbf{a}}(\mathbf{h}) &= D_{\mathbf{h}} f(\mathbf{a}) \\ &= \nabla f(\mathbf{a}) \cdot \mathbf{h} \\ &= \sum_{i=1}^{n} D_i f(\mathbf{a}) h_i \\ &= \sum_{i=1}^{n} D_i f(\mathbf{a})\, dx_{\mathbf{a}}{}^i(\mathbf{h}), \end{aligned}$$

so the linear functions $df_{\mathbf{a}}$ and $\sum_{i=1}^{n} D_i f(\mathbf{a})\, dx_{\mathbf{a}}{}^i$ are equal. If we delete the subscript $\mathbf{a}$, and write $\partial f/\partial x^i$ for $D_i f(\mathbf{a})$, we obtain the classical formula

$$df = \frac{\partial f}{\partial x^1}\, dx^1 + \frac{\partial f}{\partial x^2}\, dx^2 + \cdots + \frac{\partial f}{\partial x^n}\, dx^n. \tag{11}$$

The mapping $\mathbf{a} \to df_{\mathbf{a}}$, which associates with each point $\mathbf{a} \in \mathscr{R}^n$ the linear function $df_{\mathbf{a}} : \mathscr{R}^n \to \mathscr{R}$, is called a *differential form*, and Eq. (11) is the historical reason for this terminology. In Chapter V we shall discuss differential forms in detail.

We now apply Theorem 2.2 to finish the computation of the derivative matrix $F'(\mathbf{a})$. First we need the following lemma on "componentwise differentiation."

Lemma 2.3 The mapping $F : \mathscr{R}^n \to \mathscr{R}^m$ is differentiable at $\mathbf{a}$ if and only if each of its component functions $F^1, \ldots, F^m$ is, and

$$dF_{\mathbf{a}} = (dF_{\mathbf{a}}{}^1, \ldots, dF_{\mathbf{a}}{}^m).$$

(Here we have labeled the component functions with superscripts, rather than subscripts as usual, merely to avoid double subscripts.)

This lemma follows immediately from a componentwise reading of the vector equation (4).

Theorem 2.4 If $F : \mathscr{R}^n \to \mathscr{R}^m$ is differentiable at $\mathbf{a}$, then the matrix $F'(\mathbf{a})$ of $dF_{\mathbf{a}}$ is

$$F'(\mathbf{a}) = (D_j F^i(\mathbf{a})).$$

[That is, $D_j F^i(\mathbf{a})$ is the element in the ith row and jth column of $F'(\mathbf{a})$.]

$$F' = \begin{pmatrix} \dfrac{\partial F^1}{\partial x_1} & \cdots & \dfrac{\partial F^1}{\partial x_n} \\ \vdots & & \vdots \\ \dfrac{\partial F^m}{\partial x_1} & \cdots & \dfrac{\partial F^m}{\partial x_n} \end{pmatrix}.$$

PROOF

$$dF_{\mathbf{a}}(\mathbf{v}) = \begin{pmatrix} dF_{\mathbf{a}}{}^1(\mathbf{v}) \\ \vdots \\ dF_{\mathbf{a}}{}^m(\mathbf{v}) \end{pmatrix} \qquad \text{(by Lemma 2.3)}$$

$$= \begin{pmatrix} \sum_{j=1}^n D_j F^1(\mathbf{a}) v_j \\ \vdots \\ \sum_{j=1}^n D_j F^m(\mathbf{a}) v_j \end{pmatrix} \qquad \text{(by Theorem 2.2)}$$

$$= (D_j F^i(\mathbf{a})) \begin{pmatrix} v_1 \\ \vdots \\ v_n \end{pmatrix}$$

by the definition of matrix multiplication. ∎

Finally we formulate a sufficient condition for differentiability. The mapping $F : \mathscr{R}^n \to \mathscr{R}^m$ is said to be *continuously differentiable* at $\mathbf{a}$ if the partial derivatives $D_1 F, \ldots, D_n F$ all exist at each point of some open set containing $\mathbf{a}$, and are continuous at $\mathbf{a}$.

Theorem 2.5 If F is continuously differentiable at $\mathbf{a}$, then F is differentiable at $\mathbf{a}$.

PROOF By Lemma 2.3, it suffices to consider a continuously differentiable real-valued function $f : \mathscr{R}^n \to \mathscr{R}$. Given $\mathbf{h} = (h_1, \ldots, h_n)$, let $\mathbf{h}_0 = \mathbf{0}$, $\mathbf{h}_i = (h_1, \ldots, h_i, 0, \ldots, 0)$, $i = 1, \ldots, n$ (see Fig. 2.9). Then

$$f(\mathbf{a} + \mathbf{h}) - f(\mathbf{a}) = \sum_{i=1}^n [f(\mathbf{a} + \mathbf{h}_i) - f(\mathbf{a} + \mathbf{h}_{i-1})].$$

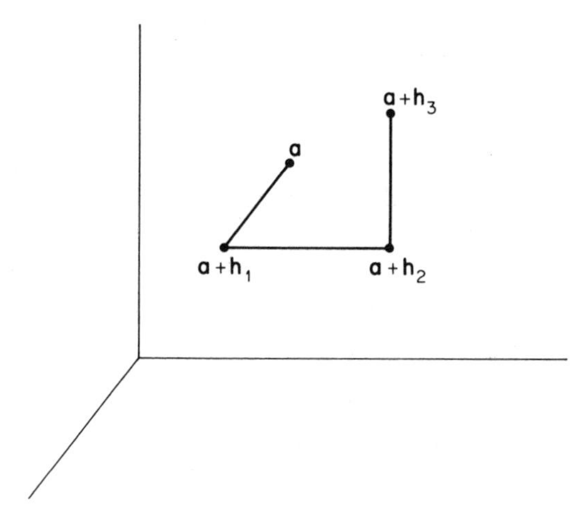

Figure 2.9

The single-variable mean value theorem gives

$$\begin{aligned} f(\mathbf{a} + \mathbf{h}_i) - f(\mathbf{a} + \mathbf{h}_{i-1}) &= f(a_1 + h_1, \ldots, a_{i-1} + h_{i-1}, a_i + h_i, \\ &\qquad a_{i+1}, \ldots, a_n) - f(a_1 + h_1, \ldots, a_{i-1} \\ &\qquad + h_{i-1}, a_i, \ldots, a_n) \\ &= D_i f(a_1 + h_1, \ldots, a_{i-1} + h_{i-1}, \\ &\qquad c_i, a_{i+1}, \ldots, a_n) h_i \end{aligned}$$

for some $c_i \in (a_i, a_i + h_i)$, since $D_i f$ is the derivative of the function

$$g(x) = f(a_1, \ldots, a_{i-1}, x, a_{i+1}, \ldots, a_n).$$

Thus $f(\mathbf{a} + \mathbf{h}_i) - f(\mathbf{a} + \mathbf{h}_{i-1}) = h_i D_i f(\mathbf{b}_i)$ for some point $\mathbf{b}_i$ which approaches $\mathbf{a}$ as $\mathbf{h} \to \mathbf{0}$. Consequently

$$\begin{aligned} \lim_{\mathbf{h} \to \mathbf{0}} \frac{|f(\mathbf{a} + \mathbf{h}) - f(\mathbf{a}) - \sum_{i=1}^n D_i f(\mathbf{a}) h_i|}{|\mathbf{h}|} &= \lim_{\mathbf{h} \to \mathbf{0}} \frac{|\sum_{i=1}^n [D_i f(\mathbf{b}_i) - D_i f(\mathbf{a})] h_i|}{|\mathbf{h}|} \\ &\leqq \lim_{\mathbf{h} \to \mathbf{0}} \sum_{i=1}^n |D_i f(\mathbf{b}_i) - D_i f(\mathbf{a})| \frac{|h_i|}{|\mathbf{h}|} \\ &\leqq \lim_{\mathbf{h} \to \mathbf{0}} \sum_{i=1}^n |D_i f(\mathbf{b}_i) - D_i f(\mathbf{a})| \\ &= 0 \end{aligned}$$

as desired, since each $\mathbf{b}_i \to \mathbf{a}$ as $\mathbf{h} \to \mathbf{0}$, and each $D_i f$ is continuous at $\mathbf{a}$. ∎

Let us now summarize what has thus far been said about differentiability for functions of several variables, and in particular point out that the rather complicated concept of differentiability, as defined by Eq. (4), has now been justified.

For the importance of directional derivatives (rates of change) is obvious enough and, if a mapping is differentiable, then Theorem 2.2 gives a pleasant expression for its directional derivatives in terms of its partial derivatives, which are comparatively easy to compute; Theorem 2.4 similarly describes the derivative matrix. Finally Theorem 2.5 provides an effective test for the differentiability of a function in terms of its partial derivatives, thereby eliminating (in most cases) the necessity of verifying that it satisfies the definition of differentiability. In short, every continuously differentiable function is differentiable, and every differentiable function has directional derivatives; in general, neither of these implications may be reversed (see Example 4 and Exercise 2.5).

We began this section with a general discussion of tangent planes, which served to motivate the definition of differentiability. It is appropriate to conclude with an example in which our results are applied to actually compute a tangent plane.

Example 6 Let $F : \mathscr{R}^2 \to \mathscr{R}^4$ be defined by

$$F(x_1, x_2) = (x_2, x_1, x_1 x_2, x_2{}^2 - x_1{}^2).$$

Then F is obviously continuously differentiable, and therefore differentiable (Theorem 2.5). Let $\mathbf{a} = (1, 2)$, and suppose we want to determine the tangent plane to the image S of F at the point $F(\mathbf{a}) = (2, 1, 2, 3)$. By Theorem 2.4, the matrix of the linear mapping $dF_{\mathbf{a}} : \mathscr{R}^2 \to \mathscr{R}^4$ is the 4×2 matrix

$$F'(\mathbf{a}) = \begin{pmatrix} 0 & 1 \\ 1 & 0 \\ 2 & 1 \\ -2 & 4 \end{pmatrix}$$

The image $\mathscr{L}_{\mathbf{a}}$ of $dF_{\mathbf{a}}$ is that subspace of $\mathscr{R}^4$ which is generated by the column vectors $\mathbf{b}_1 = (0, 1, 2, -2)$ and $\mathbf{b}_2 = (1, 0, 1, 4)$ of $F'(\mathbf{a})$ (see Theorem I.5.2). Since $\mathbf{b}_1$ and $\mathbf{b}_2$ are linearly independent, $\mathscr{L}_{\mathbf{a}}$ is 2-dimensional, and so is its orthogonal complement (Theorem I.3.4). In order to write $\mathscr{L}_{\mathbf{a}}$ in the form $A\mathbf{x} = \mathbf{0}$, we therefore need to find two linearly independent vectors $\mathbf{a}_1$ and $\mathbf{a}_2$ which are orthogonal to both $\mathbf{b}_1$ and $\mathbf{b}_2$; they will then be the row vectors of the matrix A. Two such vectors $\mathbf{a}_1$ and $\mathbf{a}_2$ are easily found by solving the equations

$$\begin{aligned} x_1 + x_3 + 4x_4 &= 0 \qquad (\mathbf{b}_1 \cdot \mathbf{x} = \mathbf{0}), \\ x_2 + 2x_3 - 2x_4 &= 0 \qquad (\mathbf{b}_2 \cdot \mathbf{x} = \mathbf{0}); \end{aligned}$$

for example, $\mathbf{a}_1 = (5, 0, -1, -1)$ and $\mathbf{a}_2 = (0, 10, -4, 1)$.

The desired tangent plane T to S at the point $F(\mathbf{a}) = (2, 1, 2, 3)$ is now the parallel translate of $\mathscr{L}_{\mathbf{a}}$ to $F(\mathbf{a})$. That is, T is the set of all points $\mathbf{x} \in \mathscr{R}^4$ such that $A(\mathbf{x} - F(\mathbf{a})) = \mathbf{0}$,

$$\begin{pmatrix} 5 & 0 & -1 & -1 \\ 0 & 10 & -4 & 1 \end{pmatrix} \begin{pmatrix} x_1 - 2 \\ x_2 - 1 \\ x_3 - 2 \\ x_4 - 3 \end{pmatrix} = \begin{pmatrix} 0 \\ 0 \end{pmatrix}.$$

Upon simplification, we obtain the two equations

$$\begin{aligned} 5x_1 - x_3 - x_4 &= 5, \\ 10x_2 - 4x_3 + x_4 &= 5. \end{aligned}$$

The solution set of each of these equations is a 3-dimensional hyperplane in $\mathscr{R}^4$; the intersection of these two hyperplanes is the desired (2-dimensional) tangent plane T.

Exercises

2.1 If $F: \mathscr{R}^n \to \mathscr{R}^m$ is differentiable at $\mathbf{a}$, show that F is continuous at $\mathbf{a}$. *Hint:* Let

$$R(\mathbf{h}) = \frac{F(\mathbf{a}+\mathbf{h}) - F(\mathbf{a}) - dF_{\mathbf{a}}(\mathbf{h})}{|\mathbf{h}|} \quad \text{if} \quad \mathbf{h} \neq \mathbf{0}.$$

Then

$$F(\mathbf{a}+\mathbf{h}) = F(\mathbf{a}) + dF_{\mathbf{a}}(\mathbf{h}) + |\mathbf{h}|\,R(\mathbf{h}).$$

2.2 If $p: \mathscr{R}^2 \to \mathscr{R}$ is defined by $p(x, y) = xy$, show that p is differentiable everywhere with $dp_{(a,b)}(x, y) = bx + ay$. *Hint:* Let $L(x, y) = bx + ay$, $\mathbf{a} = (a, b)$, $\mathbf{h} = (h, k)$. Then show that $p(\mathbf{a}+\mathbf{h}) - p(\mathbf{a}) - L(\mathbf{h}) = hk$. But $|hk| \leqq h^2 + k^2$ because $|hk| \leqq l^2$, $l = \max(|h|, |k|)$.

2.3 If $f: \mathscr{R}^2 \to \mathscr{R}$ is defined by $f(x, y) = xy^2/(x^2 + y^2)$ unless $x = y = 0$, and $f(0, 0) = 0$, show that $D_{\mathbf{v}}f(0, 0)$ exists for all $\mathbf{v}$, but f is *not* differentiable at $(0, 0)$. *Hint:* Note first that $f(t\mathbf{v}) = tf(\mathbf{v})$ for all $t \in \mathscr{R}$ and $\mathbf{v} \in \mathscr{R}^2$. Then show that $D_{\mathbf{v}}f(0, 0) = f(\mathbf{v})$ for all $\mathbf{v}$. Hence $D_1 f(0, 0) = D_2 f(0, 0) = 0$ but $D_{(1,1)} f(0, 0) = \frac{1}{2}$.

2.4 Do the same as in the previous problem with the function $f: \mathscr{R}^2 \to \mathscr{R}$ defined by $f(x, y) = (x^{1/3} + y^{1/3})^3$.

2.5 Let $f: \mathscr{R}^2 \to \mathscr{R}$ be defined by $f(x, y) = x^3 \sin(1/x) + y^2$ for $x \neq 0$, and $f(0, y) = y^2$.

(a) Show that f is continuous at $(0, 0)$.
(*b*) Find the partial derivatives of f at $(0, 0)$.
(*c*) Show that f is differentiable at $(0, 0)$.
(*d*) Show that $D_1 f$ is *not* continuous at $(0, 0)$.

2.6 Use the approximation $\nabla f_{\mathbf{a}} \approx df_{\mathbf{a}}$ to estimate the value of

(a) $[(3.02)^2 + (1.97)^2 + (5.98)^2]$,
(b) $(e^4)^{1/10} = e^{0.4} = e^{(1.1)^2 - (0.9)^2}$.

2.7 As in Exercise 1.3, a potential function for the vector field $F: \mathscr{R}^n \to \mathscr{R}^n$ is a differentiable function $V: \mathscr{R}^n \to \mathscr{R}$ such that $F = -\nabla V$. Find a potential function for the vector field F defined for all $\mathbf{x} \neq 0$ by the formula

(a) $F(\mathbf{x}) = r^n \mathbf{x}$, where $r = |\mathbf{x}|$. Treat separately the cases $n = 2$ and $n \neq 2$.
(b) $F(\mathbf{x}) = [g'(r)/r]\mathbf{x}$, where g is a differentiable function of one variable.

2.8 Let $f: \mathscr{R}^n \to \mathscr{R}$ be differentiable. If $f(\mathbf{0}) = 0$ and $f(t\mathbf{x}) = tf(\mathbf{x})$ for all $t \in \mathscr{R}$ and $\mathbf{x} \in \mathscr{R}^n$, prove that $f(\mathbf{x}) = \nabla f(\mathbf{0}) \cdot \mathbf{x}$ for all $\mathbf{x} \in \mathscr{R}^n$. In particular f is linear. Consequently any *homogeneous* function $g: \mathscr{R}^n \to \mathscr{R}$ [meaning that $g(t\mathbf{x}) = tg(\mathbf{x})$], which is *not* linear, must fail to be differentiable at the origin, although it has directional derivatives there (why?).

2.9 If $f: \mathscr{R}^n \to \mathscr{R}^m$ and $g: \mathscr{R}^n \to \mathscr{R}^k$ are both differentiable at $\mathbf{a} \in \mathscr{R}^n$, prove directly from the definition that the mapping $h: \mathscr{R}^n \to \mathscr{R}^{m+k}$, defined by $h(\mathbf{x}) = (f(\mathbf{x}), g(\mathbf{x}))$, is differentiable at $\mathbf{a}$.

2.10 Let the mapping $F: \mathscr{R}^2 \to \mathscr{R}^2$ be defined by $F(x_1, x_2) = (\sin(x_1 - x_2), \cos(x_1 + x_2))$. Find the linear equations of the tangent plane in $\mathscr{R}^4$ to the *graph* of F at the point $(\pi/4, \pi/4, 0, 0)$.

2.11 Let $f: \mathscr{R}^2{}_{uv} \to \mathscr{R}^3{}_{xyz}$ be the differentiable mapping defined by

$$x = uv, \qquad y = u^2 - v^2, \qquad z = u + v.$$

Let $\mathbf{p} = (1, 1) \in \mathscr{R}^2$ and $\mathbf{q} = f(\mathbf{p}) = (1, 0, 2) \in \mathscr{R}^3$. Given a *unit* vector $\mathbf{u} = (u, v)$, let $\varphi_\mathbf{u} : \mathscr{R} \to \mathscr{R}^2$ be the straight line through $\mathbf{p}$, and $\psi_\mathbf{u} : \mathscr{R} \to \mathscr{R}^3$ the curve through $\mathbf{q}$, defined by $\varphi_\mathbf{u}(t) = \mathbf{p} + t\mathbf{u}$, $\psi_\mathbf{u}(t) = f(\varphi_\mathbf{u}(t))$, respectively. Then

$$\psi_\mathbf{u}'(0) = D_\mathbf{u} f(\mathbf{p})$$

by the definition of the directional derivative.

(a) For what *unit* vector(s) $\mathbf{u}$ is the speed $|\psi_\mathbf{u}'(0)|$ maximal?
(b) Suppose that $g : \mathscr{R}^3 \to \mathscr{R}$ is a differentiable function such that $\nabla g(\mathbf{q}) = (1, 1, -1)$, and define

$$h_\mathbf{u}(t) = g(f(\mathbf{p} + t\mathbf{u})).$$

Assuming the chain rule result that

$$h_\mathbf{u}'(0) = \nabla g(\mathbf{q}) \cdot D_\mathbf{u} f(\mathbf{p}),$$

find the *unit* vector $\mathbf{u}$ that maximizes $h_\mathbf{u}'(0)$.
(c) Write the equation of the tangent plane to the image surface of f at the point $f(1, 1) = (1, 0, 2)$.

3 THE CHAIN RULE

Consider the composition $H = G \circ F$ of two differentiable mappings $F : \mathscr{R}^n \to \mathscr{R}^m$ and $G : \mathscr{R}^m \to \mathscr{R}^k$. For example, $F(\mathbf{x}) \in \mathscr{R}^m$ might be the price vector of m intermediate products that are manufactured at a factory from n raw materials whose cost vector is $\mathbf{x}$ (that is, the components of $\mathbf{x} \in \mathscr{R}^n$ are the prices of the n raw materials), and $H(\mathbf{x}) = G(F(\mathbf{x}))$ the resulting price vector of k final products that are manufactured at a second factory from the m intermediate products. We might wish to estimate the change $\Delta H_\mathbf{a}(\mathbf{h}) = H(\mathbf{a} + \mathbf{h}) - H(\mathbf{a})$ in the prices of the final products, resulting from a change from $\mathbf{a}$ to $\mathbf{a} + \mathbf{h}$ in the costs of the raw materials. Using the approximations $\Delta F \approx dF$ and $\Delta G \approx dG$, without initially worrying about the accuracy of our estimates, we obtain

$$\begin{aligned}
\Delta H_\mathbf{a}(\mathbf{h}) &= G(F(\mathbf{a} + \mathbf{h})) - G(F(\mathbf{a})) \\
&= G(F(\mathbf{a}) + [F(\mathbf{a} + \mathbf{h}) - F(\mathbf{a})]) - G(F(\mathbf{a})) \\
&= \Delta G_{F(\mathbf{a})}(F(\mathbf{a} + \mathbf{h}) - F(\mathbf{a})) \\
&\approx dG_{F(\mathbf{a})}(\Delta F_\mathbf{a}(\mathbf{h})) \\
&\approx dG_{F(\mathbf{a})}(dF_\mathbf{a}(\mathbf{h})).
\end{aligned}$$

This heuristic "argument" suggests the possibility that the multivariable chain rule takes the form $dH_\mathbf{a} = dG_{F(\mathbf{a})} \circ dF_\mathbf{a}$, analogous to the restatement in Section 1 of the familiar single-variable chain rule.

Theorem 3.1 (The Chain Rule) Let U and V be open subsets of $\mathscr{R}^n$ and $\mathscr{R}^m$ respectively. If the mappings $F : U \to \mathscr{R}^m$ and $G : V \to \mathscr{R}^k$ are differentiable at $\mathbf{a} \in U$ and $F(\mathbf{a}) \in V$ respectively, then their composition $H = G \circ F$ is differentiable at $\mathbf{a}$, and

$$dH_{\mathbf{a}} = dG_{F(\mathbf{a})} \circ dF_{\mathbf{a}} \qquad \begin{pmatrix}\text{composition of}\\ \text{linear mappings}\end{pmatrix}. \tag{1}$$

In terms of derivatives, we therefore have

$$H'(\mathbf{a}) = G'(F(\mathbf{a})) \cdot F'(\mathbf{a}) \qquad \begin{pmatrix}\text{matrix}\\ \text{multiplication}\end{pmatrix}. \tag{2}$$

In brief, the differential of the composition is the composition of the differentials; the derivative of the composition is the product of the derivatives.

PROOF We must show that

$$\lim_{\mathbf{h}\to\mathbf{0}} \frac{H(\mathbf{a}+\mathbf{h}) - H(\mathbf{a}) - dG_{F(\mathbf{a})} \circ dF_{\mathbf{a}}(\mathbf{h})}{|\mathbf{h}|} = \mathbf{0}.$$

If we define

$$\varphi(\mathbf{h}) = \frac{F(\mathbf{a}+\mathbf{h}) - F(\mathbf{a}) - dF_{\mathbf{a}}(\mathbf{h})}{|\mathbf{h}|} \qquad \text{for} \quad \mathbf{h} \neq \mathbf{0} \tag{3}$$

and

$$\psi(\mathbf{k}) = \frac{G(F(\mathbf{a})+\mathbf{k}) - G(F(\mathbf{a})) - dG_{F(\mathbf{a})}(\mathbf{k})}{|\mathbf{k}|} \qquad \text{for} \quad \mathbf{k} \neq \mathbf{0}, \tag{4}$$

then the fact that F and G are differentiable at $\mathbf{a}$ and $F(\mathbf{a})$, respectively, implies that

$$\lim_{\mathbf{h}\to\mathbf{0}} \varphi(\mathbf{h}) = \lim_{\mathbf{k}\to\mathbf{0}} \psi(\mathbf{k}) = \mathbf{0}.$$

Then

$$\begin{aligned} H(\mathbf{a}+\mathbf{h}) - H(\mathbf{a}) &= G(F(\mathbf{a}+\mathbf{h})) - G(F(\mathbf{a})) \\ &= G(F(\mathbf{a}) + (F(\mathbf{a}+\mathbf{h}) - F(\mathbf{a}))) - G(F(\mathbf{a})) \\ &= dG_{F(\mathbf{a})}(F(\mathbf{a}+\mathbf{h}) - F(\mathbf{a})) \\ &\quad + |F(\mathbf{a}+\mathbf{h}) - F(\mathbf{a})|\, \psi(F(\mathbf{a}+\mathbf{h}) - F(\mathbf{a})), \end{aligned}$$

by Eq. (4) with $\mathbf{k} = F(\mathbf{a}+\mathbf{h}) - F(\mathbf{a})$. Using (3) we then obtain

$$\begin{aligned} H(\mathbf{a}+\mathbf{h}) - H(\mathbf{a}) &= dG_{F(\mathbf{a})}(dF_{\mathbf{a}}(\mathbf{h}) + |\mathbf{h}|\,\varphi(\mathbf{h})) \\ &\quad + |F(\mathbf{a}+\mathbf{h}) - F(\mathbf{a})|\, \psi(F(\mathbf{a}+\mathbf{h}) - F(\mathbf{a})) \\ &= dG_{F(\mathbf{a})} \circ dF_{\mathbf{a}}(\mathbf{h}) + |\mathbf{h}|\, dG_{F(\mathbf{a})}(\varphi(\mathbf{h})) \\ &\quad + |\mathbf{h}| \, \left| dF_{\mathbf{a}}\left(\frac{\mathbf{h}}{|\mathbf{h}|}\right) + \varphi(\mathbf{h}) \right| \psi(F(\mathbf{a}+\mathbf{h}) - F(\mathbf{a})). \end{aligned}$$

Therefore

$$\frac{H(\mathbf{a}+\mathbf{h}) - H(\mathbf{a}) - dG_{F(\mathbf{a})} \circ dF_{\mathbf{a}}(\mathbf{h})}{|\mathbf{h}|}$$

$$= dG_{F(\mathbf{a})}(\varphi(\mathbf{h})) + \left| dF_{\mathbf{a}}\left(\frac{\mathbf{h}}{|\mathbf{h}|}\right) + \varphi(\mathbf{h}) \right| \psi(F(\mathbf{a}+\mathbf{h}) - F(\mathbf{a})). \tag{5}$$

But $\lim_{\mathbf{h}\to\mathbf{0}} dG_{F(\mathbf{a})}(\varphi(\mathbf{h})) = \mathbf{0}$ because $\lim_{\mathbf{h}\to\mathbf{0}} \varphi(\mathbf{h}) = \mathbf{0}$ and the linear mapping $dG_{F(\mathbf{a})}$ is continuous. Also $\lim_{\mathbf{h}\to\mathbf{0}} \psi(F(\mathbf{a}+\mathbf{h}) - F(\mathbf{a})) = \mathbf{0}$ because F is continuous at $\mathbf{a}$ and $\lim_{\mathbf{k}\to\mathbf{0}} \psi(\mathbf{k}) = 0$. Finally the number $|dF_{\mathbf{a}}(\mathbf{h}/|\mathbf{h}|) + \varphi(\mathbf{h})|$ remains bounded, because $\mathbf{h}/|\mathbf{h}| \in S^{n-1}$ and the component functions of the linear mapping $dF_{\mathbf{a}}$ are continuous and therefore bounded on the unit sphere $S^{n-1} \subset \mathscr{R}^n$ (Theorem I.8.8).

Consequently the limit of (5) is zero as desired. Of course Eq. (2) follows immediately from (1), since the matrix of the composition of two linear mappings is the product of their matrices. ∎

We list in the following examples some typical chain rule formulas obtained by equating components of the matrix equation (2) for various values of n, m, and k. It is the formulation of the chain rule in terms of differential linear mappings which enables us to give a single proof for all of these formulas, despite their wide variety.

Example 1 If $n = k = 1$, so we have differentiable mappings $\mathscr{R} \xrightarrow{f} \mathscr{R}^m \xrightarrow{g} \mathscr{R}$, then $h = g \circ f: \mathscr{R} \to \mathscr{R}$ is differentiable with

$$h'(t) = g'(f(t))f'(t).$$

Here $g'(f(t))$ is a $1 \times m$ row matrix, and $f'(t)$ is an $m \times 1$ column matrix. In terms of the gradient of g, we have

$$h'(t) = \nabla g(f(t)) \cdot f'(t) \qquad \text{(dot product)}. \tag{6}$$

This is a generalization of the fact that $D_{\mathbf{v}} g(\mathbf{a}) = \nabla g(\mathbf{a}) \cdot \mathbf{v}$ [see Eq. (10) of Section 2]. If we think of $f(t)$ as the position vector of a particle moving in $\mathscr{R}^m$, with g a temperature function on $\mathscr{R}^m$, then (6) gives the rate of change of the temperature of the particle. In particular, we see that this rate of change depends only upon the velocity vector of the particle.

In terms of the component functions $f_1, \ldots, f_m$ of f and the partial derivatives of g, (6) becomes

$$\frac{dh}{dt} = \frac{\partial g}{\partial x_1}\frac{\partial f_1}{\partial t} + \frac{\partial g}{\partial x_2}\frac{\partial f_2}{\partial t} + \cdots + \frac{\partial g}{\partial x_m}\frac{\partial f_m}{\partial t}.$$

If we write $x_i = f_i(t)$ and $u = g(\mathbf{x})$, following the common practice of using the symbol for a typical value of a function to denote the function itself, then the above equation takes the easily remembered form

$$\frac{du}{dt} = \frac{\partial u}{\partial x_1}\frac{dx_1}{dt} + \frac{\partial u}{\partial x_2}\frac{dx_2}{dt} + \cdots + \frac{\partial u}{\partial x_m}\frac{dx_m}{dt}.$$

Example 2 Given differentiable mappings $\mathscr{R}^2 \xrightarrow{F} \mathscr{R}^3 \xrightarrow{G} \mathscr{R}^2$ with composition $H : \mathscr{R}^2 \to \mathscr{R}^2$, the chain rule gives

$$\begin{pmatrix} D_1 H_1(\mathbf{a}) & D_2 H_1(\mathbf{a}) \\ D_1 H_2(\mathbf{a}) & D_2 H_2(\mathbf{a}) \end{pmatrix}$$

$$= \begin{pmatrix} D_1 G_1(F(\mathbf{a})) & D_2 G_1(F(\mathbf{a})) & D_3 G_1(F(\mathbf{a})) \\ D_1 G_2(F(\mathbf{a})) & D_2 G_2(F(\mathbf{a})) & D_3 G_2(F(\mathbf{a})) \end{pmatrix} \begin{pmatrix} D_1 F_1(\mathbf{a}) & D_2 F_1(\mathbf{a}) \\ D_1 F_2(\mathbf{a}) & D_2 F_2(\mathbf{a}) \\ D_1 F_3(\mathbf{a}) & D_2 F_3(\mathbf{a}) \end{pmatrix}.$$

If we write $F(s, t) = (x, y, z)$ and $G(x, y, z) = (u, v)$, this equation can be rewritten

$$\begin{pmatrix} \dfrac{\partial H_1}{\partial s} & \dfrac{\partial H_1}{\partial t} \\ \dfrac{\partial H_2}{\partial s} & \dfrac{\partial H_2}{\partial t} \end{pmatrix} = \begin{pmatrix} \dfrac{\partial G_1}{\partial x} & \dfrac{\partial G_1}{\partial y} & \dfrac{\partial G_1}{\partial z} \\ \dfrac{\partial G_2}{\partial x} & \dfrac{\partial G_2}{\partial y} & \dfrac{\partial G_2}{\partial z} \end{pmatrix} \begin{pmatrix} \dfrac{\partial F_1}{\partial s} & \dfrac{\partial F_1}{\partial t} \\ \dfrac{\partial F_2}{\partial s} & \dfrac{\partial F_2}{\partial t} \\ \dfrac{\partial F_3}{\partial s} & \dfrac{\partial F_3}{\partial t} \end{pmatrix}.$$

For example, we have

$$\frac{\partial H_1}{\partial t} = \frac{\partial G_1}{\partial x}\frac{\partial F_1}{\partial t} + \frac{\partial G_1}{\partial y}\frac{\partial F_2}{\partial t} + \frac{\partial G_1}{\partial z}\frac{\partial F_3}{\partial t}.$$

Writing

$$\frac{\partial u}{\partial s} = D_1 H_1(s, t), \qquad \frac{\partial u}{\partial x} = D_1 G_1(F(s, t)), \qquad \frac{\partial x}{\partial t} = D_2 F_1(s, t), \qquad \text{etc.,}$$

to go all the way with variables representing functions, we obtain formulas such as

$$\frac{\partial u}{\partial t} = \frac{\partial u}{\partial x}\frac{\partial x}{\partial t} + \frac{\partial u}{\partial y}\frac{\partial y}{\partial t} + \frac{\partial u}{\partial z}\frac{\partial z}{\partial t}$$

and

$$\frac{\partial v}{\partial s} = \frac{\partial v}{\partial x}\frac{\partial x}{\partial s} + \frac{\partial v}{\partial y}\frac{\partial y}{\partial s} + \frac{\partial v}{\partial z}\frac{\partial z}{\partial s}.$$

The obvious nature of the formal pattern of chain rule formulas expressed in terms of variables, as above, often compensates for their disadvantage of not containing explicit reference to the points at which the various derivatives are evaluated.

Example 3 Let $T: \mathscr{R}^2 \to \mathscr{R}^2$ be the familiar "polar coordinate mapping" defined by $T(r, \theta) = (r \cos \theta, r \sin \theta)$ (Fig. 2.10). Given a differentiable function $f: \mathscr{R}^2 \to \mathscr{R}$, define $g = f \circ T$, so $g(r, \theta) = f(r \cos \theta, r \sin \theta)$. Then the chain rule gives

$$\begin{pmatrix} \dfrac{\partial g}{\partial r} & \dfrac{\partial g}{\partial \theta} \end{pmatrix} = \begin{pmatrix} \dfrac{\partial f}{\partial x} & \dfrac{\partial f}{\partial y} \end{pmatrix} \begin{pmatrix} \cos \theta & -r \sin \theta \\ \sin \theta & r \cos \theta \end{pmatrix},$$

so

$$\frac{\partial g}{\partial r} = \frac{\partial f}{\partial x} \cos \theta + \frac{\partial f}{\partial y} \sin \theta, \qquad \frac{\partial g}{\partial \theta} = -\frac{\partial f}{\partial x} r \sin \theta + \frac{\partial f}{\partial y} r \cos \theta.$$

Thus we have expressed the partial derivatives of g in terms of those of f, that is, in terms of $\partial f/\partial x = D_1 f(r \cos \theta, r \sin \theta)$ and $\partial f/\partial y = D_2 f(r \cos \theta, r \sin \theta)$.

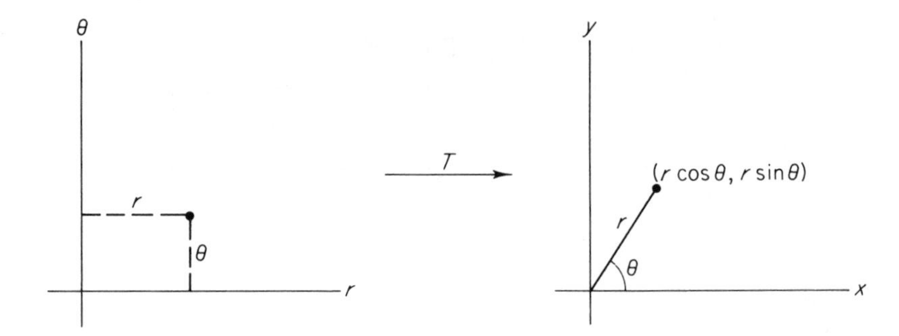

Figure 2.10

The same can be done for the second order partial derivatives. Given a differentiable mapping $F: \mathscr{R}^n \to \mathscr{R}^m$, the partial derivative $D_i F$ is again a mapping from $\mathscr{R}^n$ to $\mathscr{R}^m$. If *it* is differentiable at $\mathbf{a}$, we can consider the *second partial derivative*

$$D_j D_i F(\mathbf{a}) = D_j(D_i F)(\mathbf{a}).$$

The classical notation is

$$D_j D_i F = \frac{\partial^2 F}{\partial x_j \partial x_i}.$$

For example, the function $f: \mathscr{R}^2 \to \mathscr{R}$ has second-order partial derivatives

$$\frac{\partial^2 f}{\partial x^2} = D_1 D_1 f, \qquad \frac{\partial^2 f}{\partial x \, \partial y} = D_1 D_2 f,$$

$$\frac{\partial^2 f}{\partial y \, \partial x} = D_2 D_1 f, \qquad \text{and} \qquad \frac{\partial^2 f}{\partial y^2} = D_2 D_2 f.$$

Continuing Example 3 we have

$$\begin{aligned}
\frac{\partial^2 g}{\partial\theta\,\partial r} &= \frac{\partial}{\partial\theta}\left(\frac{\partial f}{\partial x}\right)\cos\theta - \frac{\partial f}{\partial x}\sin\theta + \frac{\partial}{\partial\theta}\left(\frac{\partial f}{\partial y}\right)\sin\theta + \frac{\partial f}{\partial y}\cos\theta \\
&= \left(-\frac{\partial^2 f}{\partial x^2} r\sin\theta + \frac{\partial^2 f}{\partial y\,\partial x} r\cos\theta\right)\cos\theta - \frac{\partial f}{\partial x}\sin\theta \\
&\quad + \left(-\frac{\partial^2 f}{\partial x\,\partial y} r\sin\theta + \frac{\partial^2 f}{\partial y^2} r\cos\theta\right)\sin\theta + \frac{\partial f}{\partial y}\cos\theta \\
&= r\cos\theta\sin\theta\left(\frac{\partial^2 f}{\partial y^2} - \frac{\partial^2 f}{\partial x^2}\right) + r(\cos^2\theta - \sin^2\theta)\frac{\partial^2 f}{\partial x\,\partial y} \\
&\quad - \frac{\partial f}{\partial x}\sin\theta + \frac{\partial f}{\partial y}\cos\theta.
\end{aligned}$$

In the last step we have used the fact that the "mixed partial derivatives" $\partial^2 f/\partial x\,\partial y$ and $\partial^2 f/\partial y\,\partial x$ are equal, which will be established at the end of this section under the hypothesis that they are continuous.

In Exercise 3.9, the student will continue in this manner to show that *Laplace's equation*

$$\nabla^2 u = \frac{\partial^2 u}{\partial x^2} + \frac{\partial^2 u}{\partial y^2} = 0 \tag{7}$$

transforms to

$$\frac{\partial^2 u}{\partial r^2} + \frac{1}{r}\frac{\partial u}{\partial r} + \frac{1}{r^2}\frac{\partial^2 u}{\partial\theta^2} = 0 \tag{8}$$

in polar coordinates.

As a standard application of this fact, consider a uniform circular disk of radius 1, whose boundary is heated in such a way that its temperature on the boundary is given by the function $g : [0, 2\pi] \to \mathscr{R}$, that is,

$$u(1, \theta) = g(\theta)$$

for each $\theta \in [0, 2\pi]$; see Fig. 2.11. Then certain physical considerations suggest that the temperature function $u(r, \theta)$ on the disk satisfies Laplace's equation (8) in polar coordinates. Now it is easily verified directly (do this) that, for each positive integer n, the functions $r^n \cos n\theta$ and $r^n \sin n\theta$ satisfy Eq. (8). Therefore, if a Fourier series expansion

$$f(\theta) = \tfrac{1}{2}a_0 + \sum_{n=1}^{\infty}(a_n\cos n\theta + b_n\sin n\theta)$$

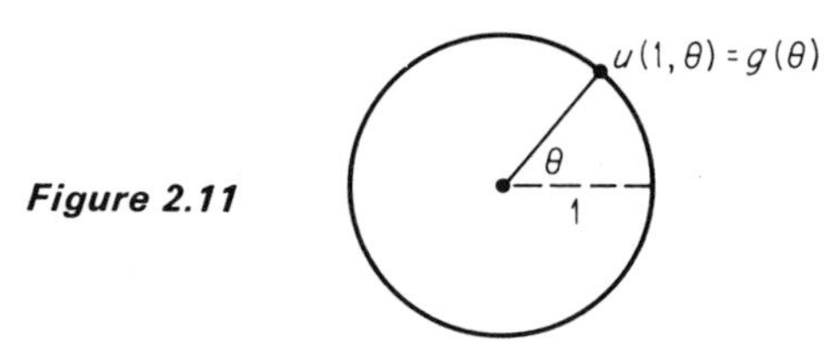

Figure 2.11

for the function g can be found, then the series

$$\tfrac{1}{2}a_0 + \sum_{n=1}^{\infty} (a_n r^n \cos n\theta + b_n r^n \sin n\theta)$$

is a plausible candidate for the temperature function $u(r, \theta)$—it reduces to $g(\theta)$ when $r = 1$, and satisfies Eq. (8), if it converges for all $r \in [0, 1]$ and if its first and second order derivatives can be computed by termwise differentiation.

Example 4 Consider an infinitely long vibrating string whose equilibrium position lies along the x-axis, and denote by $f(x, t)$ the displacement of the point x at time t (Fig. 2.12). Then physical considerations suggest that f satisfies the one-dimensional *wave equation*

$$\frac{\partial^2 f}{\partial x^2} = \frac{1}{a^2} \frac{\partial^2 f}{\partial t^2}, \tag{9}$$

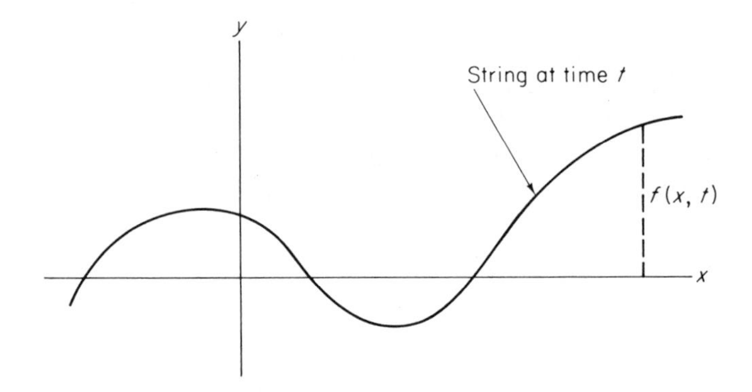

Figure 2.12

where a is a certain constant. In order to solve this partial differential equation, we make the substitution

$$x = Au + Bv, \qquad t = Cu + Dv,$$

where A, B, C, D are constants to be determined. Writing $g(u, v) = f(Au + Bv, Cu + Dv)$, we find that

$$\frac{\partial^2 g}{\partial v\, \partial u} = AB \frac{\partial^2 f}{\partial x^2} + (AD + BC) \frac{\partial^2 f}{\partial x\, \partial t} + CD \frac{\partial^2 f}{\partial t^2}$$

(see Exercise 3.7). If we choose $A = B = \frac{1}{2}$, $C = 1/2a$, $D = -1/2a$, then it follows from this equation and (9) that

$$\frac{\partial^2 g}{\partial v\, \partial u} = 0.$$

This implies that there exist functions $\varphi, \psi : \mathscr{R} \to \mathscr{R}$ such that

$$g(u, v) = \varphi(u) + \psi(v). \qquad \text{(Why?)}$$

In terms of x and t, this means that

$$f(x, t) = \varphi(x + at) + \psi(x - at). \tag{10}$$

Suppose now that we are given the initial position

$$f(x, 0) = F(x)$$

and the initial velocity $D_2 f(x, 0) = G(x)$ of the string. Then from (10) we obtain

$$\varphi(x) + \psi(x) = F(x) \tag{11}$$

and

$$a\varphi'(x) - a\psi'(x) = G(x),$$

so

$$a\varphi(x) - a\psi(x) = \int_0^x G(s)\, ds + K \tag{12}$$

by the fundamental theorem of calculus. We then solve (11) and (12) for $\varphi(x)$ and $\psi(x)$:

$$\varphi(x) = \frac{1}{2} F(x) + \frac{1}{2a} \int_0^x G(s)\, ds + \frac{K}{2a}, \tag{13}$$

$$\psi(x) = \frac{1}{2} F(x) - \frac{1}{2a} \int_0^x G(s)\, ds - \frac{K}{2a}. \tag{14}$$

Upon substituting $x + at$ for x in (13), and $x - at$ for x in (14), and adding, we obtain

$$f(x, t) = \frac{F(x + at) + F(x - at)}{2} + \frac{1}{2a} \int_{x-at}^{x+at} G(s)\, ds.$$

This is "d'Alembert's solution" of the wave equation. If $G(x) \equiv 0$, the picture looks like Fig. 2.13. Thus we have two "waves" moving in opposite directions.

The last two examples illustrate the use of chain rule formulas to "transform" partial differential equations so as to render them more amenable to solution.

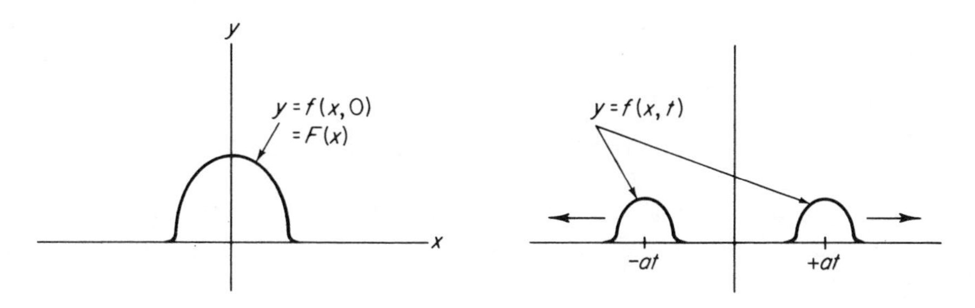

Figure 2.13

We shall now apply the chain rule to generalize some of the basic results of single-variable calculus. First consider the fact that a function defined on an open interval is constant if and only if its derivative is zero there. Since the function $f(x, y)$ defined for $x \neq 0$ by

$$f(x, y) = \begin{cases} +1 & \text{if } x > 0, \\ -1 & \text{if } x < 0, \end{cases}$$

has zero derivative (or gradient) where it is defined, it is clear that some restriction must be placed on the domain of definition of a mapping if we are to generalize this result correctly.

The open set $U \subset \mathscr{R}^n$ is said to be *connected* if and only if given any two points $\mathbf{a}$ and $\mathbf{b}$ of U, there is a differentiable mapping $\varphi : \mathscr{R} \to U$ such that $\varphi(0) = \mathbf{a}$ and $\varphi(1) = \mathbf{b}$ (Fig. 2.14). Of course the mapping $F : U \to \mathscr{R}^m$ is said to be *constant* on U if $F(\mathbf{a}) = F(\mathbf{b})$ for any two points $\mathbf{a}, \mathbf{b} \in U$, so that there exists $\mathbf{c} \in \mathscr{R}^m$ such that $F(\mathbf{x}) = \mathbf{c}$ for all $\mathbf{x} \in U$.

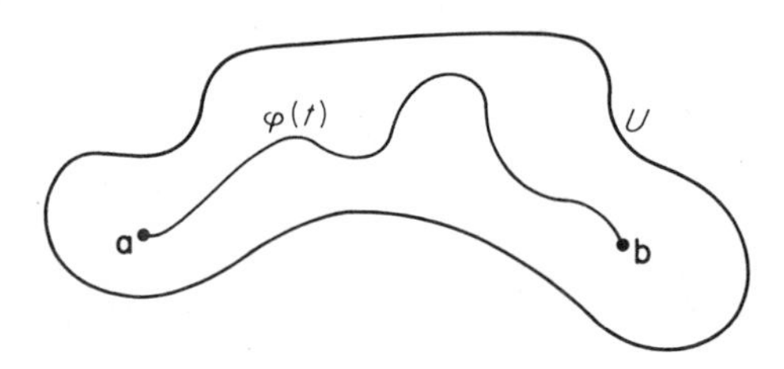

Figure 2.14

Theorem 3.2 Let U be a connected open subset of $\mathscr{R}^n$. Then the differentiable mapping $F : U \to \mathscr{R}^m$ is constant on U if and only if $F'(\mathbf{x}) = 0$ (that is, the zero matrix) for all $\mathbf{x} \in U$.

PROOF Since F is constant if and only if each of its component functions is, and the matrix $F'(\mathbf{x})$ is zero if and only if each of its rows is, we may assume that F is real valued, $F = f : U \to \mathscr{R}$. Since we already know that $f'(\mathbf{x}) = \mathbf{0}$ if f is constant, suppose that $f'(\mathbf{x}) = \nabla f(\mathbf{x}) = \mathbf{0}$ for all $\mathbf{x} \in U$.

Given $\mathbf{a}$ and $\mathbf{b} \in U$, let $\varphi : \mathscr{R} \to U$ be a differentiable mapping with $\varphi(0) = \mathbf{a}$, $\varphi(1) = \mathbf{b}$.

If $g = f \circ \varphi : \mathscr{R} \to \mathscr{R}$, then

$$g'(t) = \nabla f(\varphi(t)) \cdot \varphi'(t) = 0$$

for all $t \in \mathscr{R}$, by Eq. (6) above. Therefore g is constant on $[0, 1]$, so

$$f(\mathbf{a}) = f(\varphi(0)) = g(0) = g(1) = f(\varphi(1)) = f(\mathbf{b}).$$ ∎

Corollary 3.3 Let F and G be two differentiable mappings of the connected set $U \subset \mathscr{R}^n$ into $\mathscr{R}^m$. If $F'(\mathbf{x}) = G'(\mathbf{x})$ for all $\mathbf{x} \in U$, then there exists $\mathbf{c} \in \mathscr{R}^m$ such that

$$F(\mathbf{x}) = G(\mathbf{x}) + \mathbf{c}$$

for all $\mathbf{x} \in U$. That is, F and G differ only by a constant.

PROOF Apply Theorem 3.2 to the mapping $F - G$. ∎

Now consider a differentiable function $f: U \to \mathscr{R}$, where U is a connected open set in $\mathscr{R}^2$. We say that f is *independent of* y if there exists a function $g: \mathscr{R} \to \mathscr{R}$ such that $f(x, y) = g(x)$ if $(x, y) \in U$. At first glance it might seem that f is independent of y if $D_2 f = 0$ on U. To see that this is not so, however, consider the function f defined on

$$U = \{(x, y) \in \mathscr{R}^2 : x > 0 \quad \text{or} \quad y \neq 0\}$$

by $f(x, y) = x^2$ if $x > 0$ or $y > 0$, and $f(x, y) = -x^2$ if $x \leqq 0$ and $y < 0$. Then $D_2 f(x, y) = 0$ on U. But $f(-1, 1) = 1$, while $f(-1, -1) = -1$. We leave it as an exercise for the student to formulate a condition on U under which $D_2 f = 0$ *does* imply that f is independent of y.

Let us recall here the statement of the *mean value theorem* of elementary single-variable calculus. If $f: [a, b] \to \mathscr{R}$ is a differentiable function, then there exists a point $\xi \in (a, b)$ such that

$$f(b) - f(a) = f'(\xi)(b - a).$$

The mean value theorem generalizes to *real-valued* functions on $\mathscr{R}^n$ (however it is in general false for vector-valued functions—see Exercise 1.12). In the following statement of the mean value theorem in $\mathscr{R}^n$, by the *line segment* from $\mathbf{a}$ to $\mathbf{b}$ is meant the set of all points in $\mathscr{R}^n$ of the form $(1 - t)\mathbf{a} + t\mathbf{b}$ for $t \in [0, 1]$.

Theorem 3.4 (Mean Value Theorem) Suppose that U is an open set in $\mathscr{R}^n$, and that $\mathbf{a}$ and $\mathbf{b}$ are two points of U such that U contains the line segment L from $\mathbf{a}$ to $\mathbf{b}$. If f is a differentiable real-valued function on U, then

$$f(\mathbf{b}) - f(\mathbf{a}) = f'(\mathbf{c})(\mathbf{b} - \mathbf{a}) = \nabla f(\mathbf{c}) \cdot (\mathbf{b} - \mathbf{a})$$

for some point $\mathbf{c} \in L$.

PROOF Let φ be the mapping of $[0, 1]$ onto L defined by

$$\varphi(t) = \mathbf{a} + t(\mathbf{b} - \mathbf{a}) = (1 - t)\mathbf{a} + t\mathbf{b}, \qquad t \in [0, 1].$$

Then φ is differentiable with $\varphi'(t) = \mathbf{b} - \mathbf{a}$. Hence the composition $g = f \circ \varphi$ is differentiable by the chain rule. Since $g: [0, 1] \to \mathscr{R}$, the single-variable mean value theorem gives a point $\xi \in [0, 1]$ such that $g(1) - g(0) = g'(\xi)$. If $\mathbf{c} = \varphi(\xi) \in L$, we then have

$$\begin{aligned} f(\mathbf{b}) - f(\mathbf{a}) &= g(1) - g(0) \\ &= g'(\xi) \\ &= \nabla f(\varphi(\xi)) \cdot \varphi'(\xi) \qquad \text{[by Eq. (6)]} \\ &= \nabla f(\mathbf{c}) \cdot (\mathbf{b} - \mathbf{a}). \end{aligned}$$

∎

Note that here we have employed the chain rule to deduce the mean value theorem for functions of several variables from the single-variable mean value theorem.

Next we are going to use the mean value theorem to prove that the second partial derivatives $D_j D_i f$ and $D_i D_j f$ are equal under appropriate conditions. First note that, if we write $\mathbf{b} = \mathbf{a} + \mathbf{h}$ in the mean value theorem, then its conclusion becomes

$$f(\mathbf{a} + \mathbf{h}) - f(\mathbf{a}) = f'(\mathbf{a} + \theta\mathbf{h})\mathbf{h} = \nabla f(\mathbf{a} + \theta\mathbf{h}) \cdot \mathbf{h}$$

for some $\theta \in (0, 1)$.

Recall the notation $\Delta f_{\mathbf{a}}(\mathbf{h}) = f(\mathbf{a} + \mathbf{h}) - f(\mathbf{a})$. The mapping $\Delta f_{\mathbf{a}} : \mathscr{R}^n \to \mathscr{R}$ is sometimes called the "first difference" of f at $\mathbf{a}$. The "second difference" of f at $\mathbf{a}$ is a function of two points $\mathbf{h}, \mathbf{k}$ defined by (see Fig. 2.15)

$$\Delta^2 f_{\mathbf{a}}(\mathbf{h}, \mathbf{k}) = f(\mathbf{a} + \mathbf{h} + \mathbf{k}) - f(\mathbf{a} + \mathbf{h}) - f(\mathbf{a} + \mathbf{k}) + f(\mathbf{a}).$$

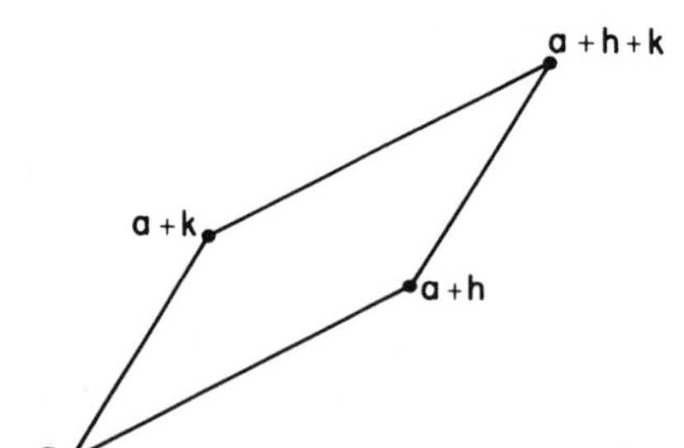

Figure 2.15

The desired equality of second order partial derivatives will follow easily from the following lemma, which expresses $\Delta^2 f_{\mathbf{a}}(\mathbf{h}, \mathbf{k})$ in terms of the second order directional derivative $D_{\mathbf{k}} D_{\mathbf{h}} f(\mathbf{x})$, which is by definition the derivative with respect to $\mathbf{k}$ of the function $D_{\mathbf{h}} f$ at $\mathbf{x}$, that is

$$D_{\mathbf{k}} D_{\mathbf{h}} f(\mathbf{x}) = \lim_{t \to 0} \frac{D_{\mathbf{h}} f(\mathbf{x} + t\mathbf{k}) - D_{\mathbf{h}} f(\mathbf{x})}{t}.$$

Lemma 3.5 Let U be an open set in $\mathscr{R}^n$ which contains the parallelogram determined by the points $\mathbf{a}$, $\mathbf{a} + \mathbf{h}$, $\mathbf{a} + \mathbf{k}$, $\mathbf{a} + \mathbf{h} + \mathbf{k}$. If the real-valued function f and its directional derivative $D_{\mathbf{h}} f$ are both differentiable on U, then there exist numbers $\alpha, \beta \in (0, 1)$ such that

$$\Delta^2 f_{\mathbf{a}}(\mathbf{h}, \mathbf{k}) = D_{\mathbf{k}} D_{\mathbf{h}} f(\mathbf{a} + \alpha\mathbf{h} + \beta\mathbf{k}).$$

PROOF Define $g(\mathbf{x})$ in a neighborhood of the line segment from $\mathbf{a}$ to $\mathbf{a} + \mathbf{h}$ by

$$g(\mathbf{x}) = f(\mathbf{x} + \mathbf{k}) - f(\mathbf{x}).$$

Then g is differentiable, with

$$dg_{\mathbf{x}} = df_{\mathbf{x}+\mathbf{k}} - df_{\mathbf{x}},$$

and

$$\begin{aligned}
\Delta^2 f_{\mathbf{a}}(\mathbf{h}, \mathbf{k}) &= g(\mathbf{a} + \mathbf{h}) - g(\mathbf{a}) \\
&= g'(\mathbf{a} + \alpha\mathbf{h})\mathbf{h} && \text{(for some } \alpha \in (0, 1)\text{, by the MVT)} \\
&= \nabla g(\mathbf{a} + \alpha\mathbf{h}) \cdot \mathbf{h} \\
&= D_{\mathbf{h}} g(\mathbf{a} + \alpha\mathbf{h}) && \text{(by Eq. (10) of Section 2)} \\
&= dg_{\mathbf{a}+\alpha\mathbf{h}}(\mathbf{h}) && \text{(by Theorem 2.1)} \\
&= df_{\mathbf{a}+\alpha\mathbf{h}+\mathbf{k}}(\mathbf{h}) - df_{\mathbf{a}+\alpha\mathbf{h}}(\mathbf{h}) \\
&= D_{\mathbf{h}} f(\mathbf{a} + \alpha\mathbf{h} + \mathbf{k}) - D_{\mathbf{h}}(\mathbf{a} + \alpha\mathbf{h}) \\
&= (D_{\mathbf{h}} f)'(\mathbf{a} + \alpha\mathbf{h} + \beta\mathbf{k})(\mathbf{k}) && \text{(for some } \beta \in (0, 1)\text{,} \\
& && \text{by the MVT)} \\
&= \nabla(D_{\mathbf{h}} f)(\mathbf{a} + \alpha\mathbf{h} + \beta\mathbf{k}) \cdot \mathbf{k} \\
&= D_{\mathbf{k}} D_{\mathbf{h}} f(\mathbf{a} + \alpha\mathbf{h} + \beta\mathbf{k}) && \text{(Section 2, Eq. (10)).}
\end{aligned}$$

∎

Theorem 3.6 Let f be a real-valued function defined on the open set U in $\mathscr{R}^n$. If the first and second partial derivatives of f exist and are continuous on U, then $D_i D_j f = D_j D_i f$ on U.

PROOF Theorem 2.5 implies that both f and its partial derivatives $D_1 f$ and $D_2 f$ are differentiable on U. We can therefore apply Lemma 3.5 with $\mathbf{h} = h\mathbf{e}_i$ and $\mathbf{k} = k\mathbf{e}_j$, provided that h and k are sufficiently small that U contains the rectangle with vertices $\mathbf{a}$, $\mathbf{a} + h\mathbf{e}_i$, $\mathbf{a} + k\mathbf{e}_j$, $\mathbf{a} + h\mathbf{e}_i + k\mathbf{e}_j$. We obtain $\alpha_1, \beta_1 \in (0, 1)$ such that

$$\Delta^2 f_{\mathbf{a}}(h\mathbf{e}_i, k\mathbf{e}_j) = D_{k\mathbf{e}_j} D_{h\mathbf{e}_i} f(\mathbf{a} + \alpha_1 h\mathbf{e}_i + \beta_1 k\mathbf{e}_j).$$

If we apply Lemma 3.5 again with $\mathbf{h}$ and $\mathbf{k}$ interchanged, we obtain α_2, $\beta_2 \in (0, 1)$ such that

$$\Delta^2 f_{\mathbf{a}}(k\mathbf{e}_j, h\mathbf{e}_i) = D_{h\mathbf{e}_i} D_{k\mathbf{e}_j} f(\mathbf{a} + \alpha_2 k\mathbf{e}_j + \beta_2 h\mathbf{e}_j).$$

But it is clear from the definition of $\Delta^2 f_{\mathbf{a}}$ that

$$\Delta^2 f_{\mathbf{a}}(h\mathbf{e}_i, k\mathbf{e}_j) = \Delta^2 f_{\mathbf{a}}(k\mathbf{e}_j, h\mathbf{e}_j),$$

so we conclude that

$$hk D_j D_i f(\mathbf{a} + \alpha_1 h\mathbf{e}_i + \beta_1 k\mathbf{e}_j) = hk D_i D_j f(\mathbf{a} + \alpha_2 k\mathbf{e}_j + \beta_2 h\mathbf{e}_i),$$

using the facts that $D_{h\mathbf{e}_i} f = h D_1 f$ and $D_{k\mathbf{e}_j} f = k D_2 f$.

If we now divide the previous equation by hk, and take the limit as $h \to 0$, $k \to 0$, we obtain

$$D_j D_i f(\mathbf{a}) = D_i D_j f(\mathbf{a})$$

because both are continuous at $\mathbf{a}$. ∎

REMARK In this proof we actually used only the facts that f, $D_1 f$, $D_2 f$ are differentiable on U, and that $D_j D_i f$ and $D_i D_j f$ are continuous at the point $\mathbf{a}$. Exercise 3.16 shows that the continuity of $D_j D_i f$ and $D_i D_j f$ at $\mathbf{a}$ are necessary for their equality there.

Exercises

3.1 Let $f: \mathscr{R}^2 \to \mathscr{R}$ be differentiable at each point of the unit circle $x^2 + y^2 = 1$. Show that, if $\mathbf{u}$ is the unit tangent vector to the circle which points in the counterclockwise direction, then

$$D_{\mathbf{u}} f(x, y) = -y D_1 f(x, y) + x D_2 f(x, y).$$

3.2 If f and g are differentiable real-valued functions on $\mathscr{R}^n$, show that
(a) $\nabla(f + g) = \nabla f + \nabla g$,
(b) $\nabla(fg) = f \nabla g + g \nabla f$,
(c) $\nabla(f^n) = n f^{n-1} \nabla f$.

3.3 Let $F, G : \mathscr{R}^n \to \mathscr{R}^n$ be differentiable mappings, and $h : \mathscr{R}^n \to \mathscr{R}$ a differentiable function. Given $\mathbf{u}, \mathbf{v} \in \mathscr{R}^n$, show that
(a) $D_{\mathbf{u}}(F + G) = D_{\mathbf{u}} F + D_{\mathbf{u}} G$,
(b) $D_{\mathbf{u}+\mathbf{v}} F = D_{\mathbf{u}} F + D_{\mathbf{v}} F$,
(c) $D_{\mathbf{u}}(hF) = (D_{\mathbf{u}} h)F + h(D_{\mathbf{u}} F)$.

3.4 Show that each of the following two functions is a solution of the *heat equation* $\partial u/\partial t = k^2\, \partial^2 u/\partial x^2$ (k a constant).
(a) $e^{-k^2 a^2 t} \sin ax$.
(b) $(1/\sqrt{t}) \exp(-x^2/4k^2 t)$.

3.5 Suppose that $f: \mathscr{R}^2 \to \mathscr{R}$ has continuous second order partial derivatives. Set $x = s + t$, $y = s - t$ to obtain $g : \mathscr{R}^2 \to \mathscr{R}$ defined by $g(s, t) = f(s + t, s - t)$. Show that

$$\frac{\partial^2 f}{\partial x^2} - \frac{\partial^2 f}{\partial y^2} = \frac{\partial^2 g}{\partial t\, \partial s},$$

that is, that

$$D_1 D_1 f(s + t, s - t) - D_2 D_2 f(s + t, s - t) = D_2 D_1 g(x, t).$$

3.6 Show that

$$5\frac{\partial^2 u}{\partial x^2} + 2\frac{\partial^2 u}{\partial x\, \partial y} + 2\frac{\partial^2 u}{\partial y^2} \qquad \text{becomes} \qquad \frac{\partial^2 u}{\partial s^2} + \frac{\partial^2 u}{\partial t^2}$$

if we set $x = 2s + t$, $y = s - t$. First state what this actually means, in terms of functions.

3.7 If $g(u, v) = f(Au + Bv, Cu + Dv)$, where A, B, C, D are constants, show that

$$\frac{\partial^2 g}{\partial v\, \partial u} = AB\frac{\partial^2 f}{\partial x^2} + (AD + BC)\frac{\partial^2 f}{\partial x\, \partial y} + CD\frac{\partial^2 f}{\partial y^2}.$$

3.8 Let $f: \mathscr{R}^2 \to \mathscr{R}$ be a function with continuous second partial derivatives, so that $\partial^2 f/\partial x\, \partial y = \partial^2 f/\partial y\, \partial x$. If $g : \mathscr{R}^2 \to \mathscr{R}$ is defined by $g(r, \theta) = f(r \cos \theta, r \sin \theta)$, show that

$$\left(\frac{\partial g}{\partial r}\right)^2 + \frac{1}{r^2}\left(\frac{\partial g}{\partial \theta}\right)^2 = \left(\frac{\partial f}{\partial x}\right)^2 + \left(\frac{\partial f}{\partial y}\right)^2 = |\nabla f|^2.$$

This gives the length of the gradient vector in polar coordinates.

3.9 If f and g are as in the previous problem, show that

$$\frac{\partial^2 g}{\partial r^2}+\frac{1}{r^2}\frac{\partial^2 g}{\partial \theta^2}+\frac{1}{r}\frac{\partial g}{\partial r}=\frac{\partial^2 f}{\partial x^2}+\frac{\partial^2 f}{\partial y^2}.$$

This gives the 2-dimensional Laplacian in polar coordinates.

3.10 Given a function $f: \mathscr{R}^3 \to \mathscr{R}$ with continuous second partial derivatives, define

$$F(\rho, \theta, \phi)=f(\rho \cos \theta \sin \phi, \rho \sin \theta \sin \phi, \rho \cos \phi),$$

where ρ, θ, ϕ are the usual spherical coordinates. We want to express the 3-dimensional Laplacian

$$\nabla^2 f=\frac{\partial^2 f}{\partial x^2}+\frac{\partial^2 f}{\partial y^2}+\frac{\partial^2 f}{\partial z^2}$$

in spherical coordinates, that is, in terms of partial derivatives of F.

(a) First define $g(r, \theta, z)=f(r \cos \theta, r \sin \theta, z)$ and conclude from Exercise 3.9 that

$$\nabla^2 f=\frac{\partial^2 g}{\partial r^2}+\frac{1}{r^2}\frac{\partial^2 g}{\partial \theta^2}+\frac{1}{r}\frac{\partial g}{\partial r}+\frac{\partial^2 g}{\partial z^2}.$$

(b) Now define $F(\rho, \theta, \phi)=g(\rho \sin \phi, \theta, \rho \cos \phi)$. Noting that, except for a change in notation, this is the same transformation as before, deduce that

$$\nabla^2 f=\frac{\partial^2 F}{\partial \rho^2}+\frac{2}{\rho}\frac{\partial F}{\partial \rho}+\frac{1}{\rho^2}\frac{\partial^2 F}{\partial \phi^2}+\frac{\cos \phi}{\rho^2 \sin \phi}\frac{\partial F}{\partial \phi}+\frac{1}{\rho^2 \sin^2 \phi}\frac{\partial^2 F}{\partial \theta^2}.$$

3.11 (a) If $f(\mathbf{x})=g(r)$, $r=|\mathbf{x}|$, and $n \geqq 3$, show that

$$\nabla^2 f=\frac{\partial^2 f}{\partial {x_1}^2}+\cdots+\frac{\partial^2 f}{\partial {x_n}^2}=\frac{n-1}{r}g'(r)+g''(r)$$

for $\mathbf{x} \neq 0$.

(b) Deduce from (a) that, if $\nabla^2 f=0$, then

$$f(\mathbf{x})=\frac{a}{|\mathbf{x}|^{n-2}}+b, \qquad \mathbf{x} \neq 0,$$

where a and b are constants.

3.12 Verify that the functions $r^n \cos n\theta$ and $r^n \sin n\theta$ satisfy the 2-dimensional Laplace equation in polar coordinates.

3.13 If $f(x, y, z)=(1/r)\, g(t-r/c)$, where c is constant and $r=(x^2+y^2+z^2)^{1/2}$, show that f satisfies the 3-dimensional wave equation

$$\frac{\partial^2 f}{\partial x^2}+\frac{\partial^2 f}{\partial y^2}+\frac{\partial^2 f}{\partial z^2}=\frac{1}{c^2}\frac{\partial^2 f}{\partial t^2}.$$

3.14 The following example illustrates the hazards of denoting functions by real variables. Let $w=f(x, y, z)$ and $z=g(x, y)$. Then

$$\frac{\partial w}{\partial x}=\frac{\partial w}{\partial x}\frac{\partial x}{\partial x}+\frac{\partial w}{\partial y}\frac{\partial y}{\partial x}+\frac{\partial w}{\partial z}\frac{\partial z}{\partial x}=\frac{\partial w}{\partial x}+\frac{\partial w}{\partial z}\frac{\partial z}{\partial x},$$

since $\partial x/\partial x=1$ and $\partial y/\partial x=0$. Hence $\partial w/\partial z\; \partial z/\partial x=0$. But if $w=x+y+z$ and $z=x+y$, then $\partial w/\partial z=\partial z/\partial x=1$, so we have $1=0$. Where is the mistake?

3.15 Use the mean value theorem to show that $5.18<[(4.1)^2+(3.2)^2]^{1/2}<5.21$. *Hint:* Note first that $(5)^2<x^2+y^2<(5.5)^2$ if $4<x<4.1$ and $3<y<3.2$.

3.16 Define $f: \mathscr{R}^2 \to \mathscr{R}$ by $f(x, y) = xy(x^2 - y^2)/(x^2 + y^2)$ unless $x = y = 0$, and $f(0, 0) = 0$.
(a) Show that $D_1 f(0, y) = -y$ and $D_2 f(x, 0) = x$ for all x and y.
(b) Conclude that $D_1 D_2 f(0, 0)$ and $D_2 D_1 f(0, 0)$ exist but are *not* equal.

3.17 The object of this problem is to show that, by an appropriate transformation of variables, the general homogeneous second order partial differential equation

$$a\frac{\partial^2 u}{\partial x^2} + 2b\frac{\partial^2 u}{\partial x\, \partial y} + c\frac{\partial^2 u}{\partial y^2} = 0 \tag{$*$}$$

with constant coefficients can be reduced to either Laplace's equation, the wave equation, or the heat equation.
(a) If $ac - b^2 > 0$, show that the substitution $s = (bx - ay)/(ac - b^2)^{1/2}$, $t = y$ changes $(*)$ to

$$\frac{\partial^2 u}{\partial s^2} + \frac{\partial^2 u}{\partial t^2} = 0.$$

(b) If $ac - b^2 = 0$, show that the substitution $s = bx - ay$, $t = y$ changes $(*)$ to

$$\frac{\partial^2 u}{\partial t^2} = 0.$$

(c) If $ac - b^2 < 0$, show that the substitution

$$s = \left[\frac{-b + (b^2 - ac)^{1/2}}{a}\right]x + ay, \qquad t = \left[\frac{-b - (b^2 - ac)^{1/2}}{a}\right]x + ay$$

changes $(*)$ to $\partial^2 u/\partial s\, \partial t = 0$.

3.18 Let $F: \mathscr{R}^n \to \mathscr{R}^m$ be differentiable at $\mathbf{a} \in \mathscr{R}^n$. Given a differentiable curve $\varphi: \mathscr{R} \to \mathscr{R}^n$ with $\varphi(0) = \mathbf{a}$, $\varphi'(0) = \mathbf{v}$, define $\psi = F \circ \varphi$, and show that $\psi'(0) = dF_{\mathbf{a}}(\mathbf{v})$. Hence, if $\tilde{\varphi}: \mathscr{R} \to \mathscr{R}^n$ is a second curve with $\tilde{\varphi}(0) = a$, $\tilde{\varphi}'(0) = \mathbf{v}$, and $\tilde{\psi} = F \circ \tilde{\varphi}$, then $\tilde{\psi}'(0) = \psi'(0)$, because both are equal to $dF_{\mathbf{a}}(\mathbf{v})$. Consequently F maps curves through $\mathbf{a}$, with the same velocity vector, to curves through $F(\mathbf{a})$ with the same velocity vector.

3.19 Let $\varphi: \mathscr{R} \to \mathscr{R}^n$, $f: \mathscr{R}^n \to \mathscr{R}^m$, and $g: \mathscr{R}^m \to \mathscr{R}$ be differentiable mappings. If $h = g \circ f \circ \varphi$ show that $h'(t) = \nabla g(f(\varphi(t))) \cdot D_{\varphi'(t)} f(\varphi(t))$.

4 LAGRANGE MULTIPLIERS AND THE CLASSIFICATION OF CRITICAL POINTS FOR FUNCTIONS OF TWO VARIABLES

We saw in Section 2 that a necessary condition, that the differentiable function $f: \mathscr{R}^2 \to \mathscr{R}$ have a local extremum at the point $\mathbf{p} \in \mathscr{R}^2$, is that $\mathbf{p}$ be a critical point for f, that is, that $\nabla f(\mathbf{p}) = \mathbf{0}$. In this section we investigate *sufficient* conditions for local maxima and minima of functions of two variables. The general case (functions of n variables) will be treated in Section 8.

It turns out that we must first consider the special problem of maximizing or minimizing a function of the form

$$f(x, y) = ax^2 + 2bxy + cy^2,$$

called a *quadratic form*, at the points of the unit circle $x^2 + y^2 = 1$. This is a special case of the general problem of maximizing or minimizing one function on the "zero set" of another function. By the *zero set* $g(x, y) = 0$ of the function $g : \mathscr{R}^2 \to \mathscr{R}$ is naturally meant the set $\{\mathbf{p} \in \mathscr{R}^2 : g(\mathbf{p}) = 0\}$. The important fact about a zero set is that, under appropriate conditions, it looks, at least locally, like the image of a curve.

Theorem 4.1 Let S be the zero set of the continuously differentiable function $g : \mathscr{R}^2 \to \mathscr{R}$, and suppose $\mathbf{p}$ is a point of S where $\nabla g(\mathbf{p}) \neq \mathbf{0}$. Then there is a rectangle Q centered at $\mathbf{p}$, and a differentiable curve $\varphi : \mathscr{R} \to \mathscr{R}^2$ with $\varphi(0) = \mathbf{p}$ and $\varphi'(0) \neq \mathbf{0}$, such that S and the image of φ agree inside Q. That is, a point of Q lies on the zero set S of g if and only if it lies on the image of the curve φ (Fig. 2.16).

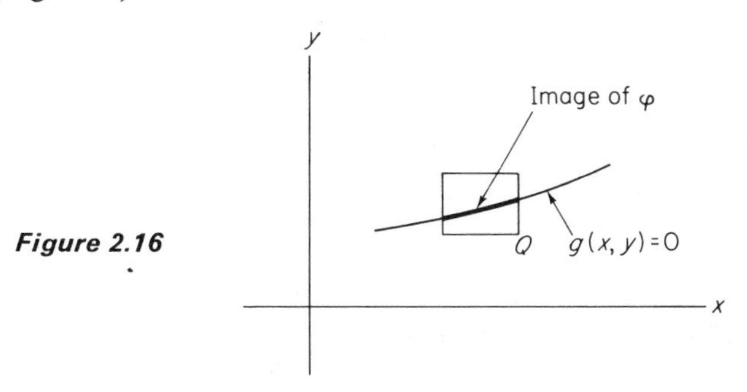

Figure 2.16

Theorem 4.1 is a consequence of the *implicit function theorem* which will be proved in Chapter III. This basic theorem asserts that, if $g : \mathscr{R}^2 \to \mathscr{R}$ is a continuously differentiable function and $\mathbf{p}$ a point where $g(\mathbf{p}) = 0$ and $D_2 g(\mathbf{p}) \neq 0$, then in some neighborhood of $\mathbf{p}$ the equation $g(x, y) = 0$ can be "solved for y as a continuously differentiable function of x." That is, there exists a $\mathscr{C}^1$ function $y = h(x)$ such that, inside some rectangle Q centered at $\mathbf{p}$, the zero set S of g agrees with the *graph* of h. Note that, in this case, the curve $\varphi(t) = (t, h(t))$ satisfies the conclusion of Theorem 4.1.

The roles of x and y in the implicit function theorem can be reversed. If $D_1 g(\mathbf{p}) \neq 0$, the conclusion is that, in some neighborhood of $\mathbf{p}$, the equation $g(x, y) = 0$ can be "solved for x as a function of y." If $x = k(y)$ is this solution, then $\varphi(t) = (k(t), t)$ is the desired curve in Theorem 4.1.

Since the hypothesis $\nabla g(\mathbf{p}) \neq \mathbf{0}$ in Theorem 4.1 implies that either $D_1 g(\mathbf{p}) \neq 0$ or $D_2 g(\mathbf{p}) \neq 0$, we see that Theorem 4.1 does follow from the implicit function theorem.

For example, suppose that $g(x, y) = x^2 + y^2 - 1$, so the zero set S is the unit circle. Then, near $(1, 0)$, S agrees with the graph of $x = (1 - y^2)^{1/2}$, while near $(0, -1)$ it agrees with the graph of $y = -(1 - x^2)^{1/2}$ (see Fig. 2.17).

The condition $\nabla g(\mathbf{p}) \neq \mathbf{0}$ is necessary for the conclusion of Theorem 4.1. For example, if $g(x, y) = x^2 + y^2$, or if $g(x, y) = x^2 - y^2$, then $\mathbf{0}$ is a critical

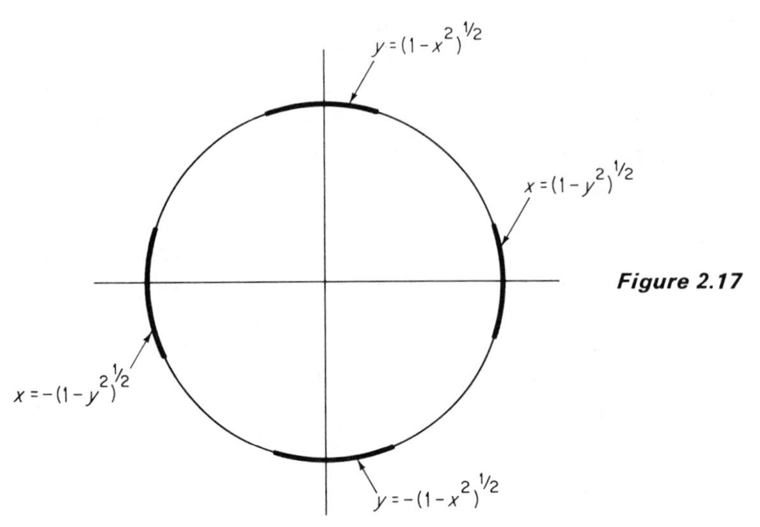

Figure 2.17

point in the zero set of **g**, and S does *not* look like the image of a curve near **0** (see Fig. 2.18).

We are now ready to study the extreme values attained by the function f on the zero set S of the function g. We say that f attains its maximum value (respectively, minimum value) on S at the point $\mathbf{p} \in S$ if $f(\mathbf{p}) \geqq f(\mathbf{q})$ (respectively, $f(\mathbf{p}) \leqq f(\mathbf{q})$) for all $\mathbf{q} \in S$.

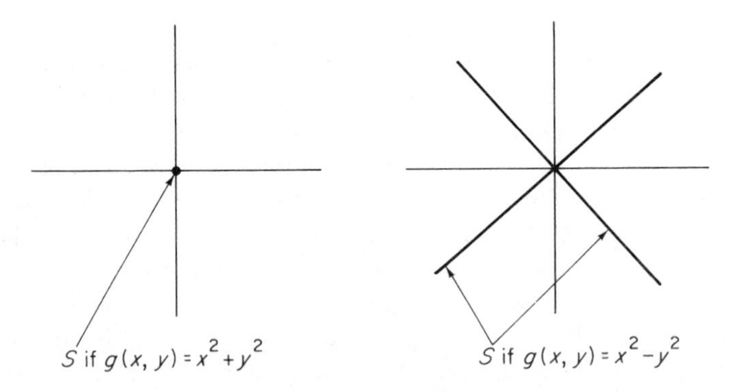

Figure 2.18

Theorem 4.2 Let f and g be continuously differentiable functions on $\mathscr{R}^2$. Suppose that f attains its maximum or minimum value on the zero set S of g at the point $\mathbf{p}$ where $\nabla g(\mathbf{p}) \neq \mathbf{0}$. Then

$$\nabla f(\mathbf{p}) = \lambda \, \nabla g(\mathbf{p}) \tag{1}$$

for some number λ.

The number λ is called a "Lagrange multiplier."

PROOF Let $\varphi : \mathscr{R} \to \mathscr{R}^2$ be the differentiable curve given by Theorem 4.1. Then $g(\varphi(t)) = 0$ for t sufficiently small, so the chain rule gives

$$0 = \nabla g(\mathbf{p}) \cdot \varphi'(\mathbf{0}).$$

Since f attains a maximum or minimum on S at $\mathbf{p}$, the function $h(t) = f(\varphi(t))$ attains a maximum or minimum at 0. Hence $h'(0) = 0$, so the chain rule gives

$$0 = h'(0) = \nabla f(\mathbf{p}) \cdot \varphi'(0).$$

Thus the vectors $\nabla f(\mathbf{p})$ and $\nabla g(\mathbf{p}) \neq \mathbf{0}$ are both orthogonal to the nonzero vector $\varphi'(0)$, and are therefore collinear. This implies that (1) holds for some $\lambda \in \mathscr{R}$. ▮

Theorem 4.2 provides a recipe for locating the points $\mathbf{p} = (x, y)$ at which f attains its maximum and minimum values (if any) on the zero set of g (provided they are attained at points where the gradient vector $\nabla g \neq \mathbf{0}$). The vector equation (1) gives two scalar equations in the unknowns x, y, λ, while $g(x, y) = 0$ is a third equation. In principle these three equations can be solved for x, y, λ. Each solution (x, y, λ) gives a candidate (x, y) for an extreme point. We can finally compare the values of f at these candidate points to ascertain where its maximum and minimum values on S are attained.

Example 1 Suppose we want to find the point(s) of the rectangular hyperbola $xy = 1$ which are closest to the origin. Take $f(x, y) = x^2 + y^2$ (the square of the distance from $\mathbf{0}$) and $g(x, y) = xy - 1$. Then $\nabla f = (2x, 2y)$ and $\nabla g = (y, x)$, so our three equations are

$$2x = \lambda y, \qquad 2y = \lambda x, \qquad xy = 1.$$

From the third equation, we see that $x \neq 0$ and $y \neq 0$, so we obtain

$$\lambda = \frac{2x}{y} = \frac{2y}{x}$$

from the first two equations. Hence $x^2 = y^2$. Since $xy = 1$, we obtain the two solutions $(1, 1)$ and $(-1, -1)$.

Example 2 Suppose we want to find the maximum and minimum values of $f(x, y) = xy$ on the circle $g(x, y) = x^2 + y^2 - 1 = 0$. Then $\nabla f = (y, x)$ and $\nabla g = (2x, 2y)$, so our three equations are

$$y = 2\lambda x, \qquad x = 2\lambda y, \qquad x^2 + y^2 = 1.$$

From the first two we see that if either x or y is zero, then so is the other. But the third equation implies that not both are zero, so it follows that neither is. We therefore obtain

$$\lambda = \frac{y}{2x} = \frac{x}{2y}$$

from the first two equations. Hence $x^2 = y^2 = \frac{1}{2}$ (since $x^2 + y^2 = 1$). This gives four solutions: $(1/\sqrt{2}, 1/\sqrt{2})$, $(-1/\sqrt{2}, 1/\sqrt{2})$, $(-1/\sqrt{2}, -1/\sqrt{2})$, $(1/\sqrt{2}, -1/\sqrt{2})$. Evaluating f at these four points, we find that f attains its maximum value of $\frac{1}{2}$ at the first and third of these points, and its minimum value of $-\frac{1}{2}$ at the other two (Fig. 2.19).

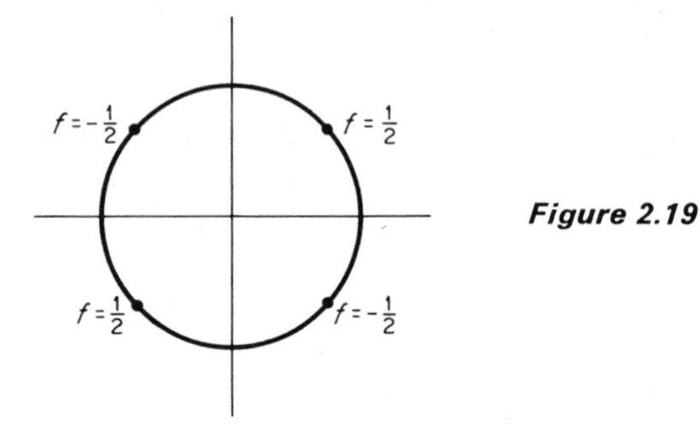

Figure 2.19

Example 3 The general quadratic form $f(x, y) = ax^2 + 2bxy + cy^2$ attains both a maximum value and a minimum value on the unit circle $g(x, y) = x^2 + y^2 - 1 = 0$, because f is continuous and the circle is closed and bounded (see Section I.8). Applying Theorem 4.2, we obtain the three equations

$$ax + by = \lambda x, \qquad bx + cy = \lambda y, \qquad x^2 + y^2 = 1. \tag{2}$$

The first two of these can be written in the form

$$(a - \lambda)x + by = 0, \qquad bx + (c - \lambda)y = 0.$$

Thus the two vectors $(a - \lambda, b)$ and $(b, c - \lambda)$ are both orthogonal to the vector (x, y), which is nonzero because $x^2 + y^2 = 1$. Hence they are collinear, and it follows easily that

$$(a - \lambda)(c - \lambda) - b^2 = 0.$$

This is a quadratic equation whose two roots are

$$\lambda_1, \lambda_2 = \tfrac{1}{2}\{a + c \pm [(a - c)^2 + 4b^2]^{1/2}\}.$$

It is easily verified that

$$\lambda_1 + \lambda_2 = a + c \qquad \text{and} \qquad \lambda_1\lambda_2 = ac - b^2. \tag{3}$$

If $ac - b^2 < 0$, then λ_1 and λ_2 have different signs. If $ac - b^2 > 0$, then they have the same sign, the sign of a and c, which have the same sign because $ac \geqq ac - b^2 > 0$.

Now, instead of proceeding to solve for x and y, let us just consider a solution (x_i, y_i, λ_i) of (2). Then

$$\begin{aligned} f(x_i, y_i) &= ax_i^2 + 2bx_iy_i + cy_i^2 \\ &= (ax_i + by_i)x_i + (bx_i + cy_i)y_i \\ &= \lambda_i x_i^2 + \lambda_i y_i^2 \\ &= \lambda_i. \end{aligned}$$

Thus λ_1 and λ_2 are the maximum and minimum values of $f(x, y)$ on $x^2 + y^2 = 1$, so we do not need to solve for x and y explicitly after all.

To summarize Example 3, we see that the maximum and minimum values of $(x, y) = ax^2 + 2bxy + cy^2$ on the unit circle

(i) are both positive if $a > 0$ and $ac - b^2 > 0$,
(ii) are both negative if $a < 0$ and $ac - b^2 > 0$,
(iii) have different signs if $ac - b^2 < 0$.

A quadratic form is called *positive-definite* if $f(x, y) > 0$ unless $x = y = 0$, *negative-definite* if $f(x, y) < 0$ unless $x = y = 0$, and *nondefinite* if it has both positive and negative values. If (x, y) is a point not necessarily on the unit circle, then

$$f(x, y) = (x^2 + y^2)f\left(\frac{x}{(x^2 + y^2)^{1/2}}, \frac{y}{(x^2 + y^2)^{1/2}}\right).$$

Consequently we see that the character of a quadratic form is determined by the signs of the maximum and minimum values which it attains on the unit circle. Combining this remark with (i), (ii), (iii) above, we have proved the following theorem.

Theorem 4.3 The quadratic form $f(x, y) = ax^2 + 2bxy + cy^2$ is

(i) positive definite if $a > 0$ and $ac - b^2 > 0$,
(ii) negative-definite if $a < 0$ and $ac - b^2 > 0$,
(iii) nondefinite if $ac - b^2 < 0$.

Now we want to use Theorem 4.3 to derive a "second-derivative test" for functions of two variables. In addition to Theorem 4.3, we will need a certain type of Taylor expansion for twice continuously differentiable functions of two variables. The function $f: \mathscr{R}^2 \to \mathscr{R}$ is said to be *twice continuously differentiable* at $\mathbf{p}$ if it has continuous partial derivatives in a neighborhood of $\mathbf{p}$, which are themselves continuously differentiable at $\mathbf{p}$. The partial derivatives of $D_1 f$ are $D_1(D_1 f) = D_1{}^2 f$ and $D_2 D_1 f$, and the partial derivatives of $D_2 f$ are $D_1 D_2 f$ and $D_2 D_2 f = D_2{}^2 f$.

Now suppose that f is twice continuously differentiable on some disk centered at the point $\mathbf{p} = (a, b)$, and let $\mathbf{q} = (a + h, b + k)$ be a point of this disk. Define $\varphi : [0, 1] \to \mathscr{R}$ by $\varphi(t) = f(a + th, b + tk)$, that is, $\varphi = f \circ \mathbf{c}$ where $\mathbf{c}(t) = (a + th, b + tk)$ (see Fig. 2.20).

Taylor's formula for single-variable functions will be reviewed in Section 6. Applied to φ on the interval $[0, 1]$, it gives

$$\varphi(1) = \varphi(0) + \varphi'(0) + \tfrac{1}{2}\varphi''(\tau) \tag{4}$$

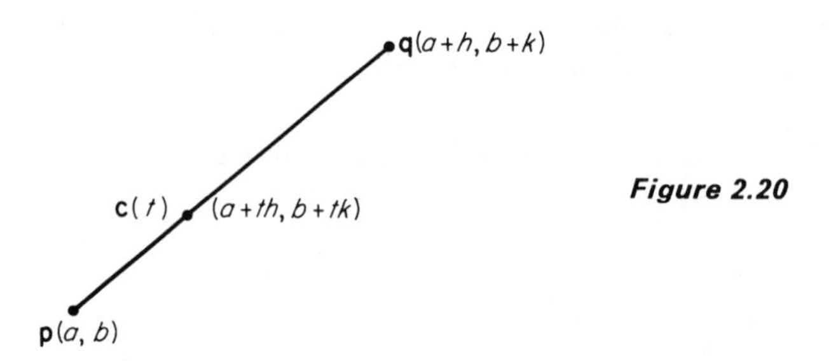

Figure 2.20

for some $\tau \in (0, 1)$. An application of the chain rule first gives

$$\varphi'(t) = \nabla f(\mathbf{c}(t)) \cdot \mathbf{c}'(t) = D_1 f(\mathbf{c}(t))h + D_2 f(\mathbf{c}(t))k,$$

and then

$$\varphi''(t) = [D_1 D_1 f(\mathbf{c}(t))h + D_2 D_1 f(\mathbf{c}(t))k]h + [D_1 D_2 f(\mathbf{c}(t))h + D_2 D_2 f(\mathbf{c}(t))k]k,$$
$$\varphi''(t) = D_1{}^2 f(\mathbf{c}(t))h^2 + 2D_1 D_2 f(\mathbf{c}(t))hk + D_2{}^2 f(\mathbf{c}(t))k^2.$$

Since $\mathbf{c}(0) = (a, b)$ and $\mathbf{c}(\tau) = (a + \tau h, b + \tau k)$, (4) now becomes

$$f(a + h, b + k) = f(a, b) + [D_1 f(a, b)h + D_2 f(a, b)k] + \tfrac{1}{2} q_\tau(h, k), \tag{5}$$

where

$$q_\tau(h, k) = D_1{}^2 f(a + \tau h, b + \tau k)h^2 + 2D_1 D_2 f(a + \tau h, b + \tau k)hk$$
$$+ D_2{}^2(a + \tau h, b + \tau k)k^2.$$

This is the desired Taylor expansion of f. We could proceed in this manner to derive the general kth degree Taylor formula for a function of two variables, but will instead defer this until Section 7 since (5) is all that is needed here.

We are finally ready to state the "second-derivative test" for functions of two variables. Its proof will involve an application of Theorem 4.3 to the quadratic form

$$q(h, k) = D_1{}^2 f(a, b)h^2 + 2D_1 D_2 f(a, b)hk + D_2{}^2 f(a, b)k^2.$$

Let us write

$$\Delta = D_1{}^2 f(a, b) D_2{}^2 f(a, b) - (D_1 D_2 f(a, b))^2 \tag{6}$$

and call Δ the *determinant* of the quadratic form q.

Theorem 4.4 Let $f : \mathscr{R}^2 \to \mathscr{R}$ be twice continuously differentiable in a neighborhood of the critical point $\mathbf{p} = (a, b)$. Then f has

(i) a local minimum at $\mathbf{p}$ if $\Delta > 0$ and $D_1{}^2 f(\mathbf{p}) > 0$,
(ii) a local maximum at $\mathbf{p}$ if $\Delta > 0$ and $D_1{}^2 f(\mathbf{p}) < 0$,
(iii) neither a local minimum nor a local maximum at $\mathbf{p}$ if $\Delta < 0$ (so in this case $\mathbf{p}$ is a "saddle point" for f).

If $\Delta = 0$, then the theorem does not apply.

PROOF Since the functions $D_1{}^2f(x, y)$ and

$$\Delta(x, y) = D_1{}^2f(x, y)D_2{}^2f(x, y) - (D_1 D_2 f(x, y))^2$$

are continuous and nonzero at $\mathbf{p}$, we can choose a circular disk centered at $\mathbf{p}$ and so small that each has the same sign at every point of this disk. If $(a + h, b + k)$ is a point of this disk, then (5) gives

$$f(a + h, b + k) = f(a, b) + \tfrac{1}{2}q_\tau(h, k) \tag{7}$$

because $D_1 f(a, b) = D_2 f(a, b) = 0$.

In case (i), both $D_1{}^2f(a + \tau h, b + \tau k)$ and the determinant $\Delta(a + \tau h, b + \tau k)$ of q_τ are positive, so Theorem 4.3(i) implies that the quadratic form q_τ is positive-definite. We therefore see from (7) that $f(a + h, b + k) > f(a, b)$. This being true for all sufficiently small h and k, we conclude that f has a local minimum at $\mathbf{p}$.

The proof in case (ii) is the same, except that we apply Theorem 4.3(ii) to show that q_τ is negative-definite, so $f(a + h, b + k) < f(a, b)$ for all h and k sufficiently small.

In case (iii), $\Delta(a + \tau h, b + \tau k) < 0$, so q_τ is nondefinite by Theorem 4.3(iii). Therefore $q_\tau(h, k)$ assumes both positive and negative values for arbitrarily small values of h and k, so it is clear from (7) that f has neither a local minimum nor a local maximum at $\mathbf{p}$. ∎

Example 4 Let $f(x, y) = xy + 2x - y$. Then

$$D_1 f(x, y) = y + 2 \qquad \text{and} \qquad D_2 f(x, y) = x - 1,$$

so $(1, -2)$ is the only critical point. Since $D_1{}^2f = D_2{}^2f = 0$ and $D_1 D_2 f = 1$, $\Delta < 0$, so f has neither a local minimum nor a local maximum at $(1, -2)$.

The character of a given critical point can often be ascertained without application of Theorem 4.4. Consider, for example, a function f which is defined on a set D in the plane that consists of all points on and inside some simple closed curve C (that is, C is a closed curve with no self-intersections). We proved in Section I.8 that, if f is *continuous* on such a set D, then it attains both a maximum value and a minimum value at points of D (why?).

Now suppose in addition that f is zero at each point of C, and positive at each point inside C. Its maximum value must then be attained at an interior point which must be a critical point. If it happens that f has only a single critical point $\mathbf{p}$ inside C, then f *must* attain its (local and absolute) maximum value at at $\mathbf{p}$, so we need not apply Theorem 4.4.

Example 5 Suppose that we want to prove that the cube of edge 10 is the rectangular solid of value 1000 which has the least total surface area A. If x, y, z are the dimensions of the box, then $xyz = 1000$ and $A = 2xy + 2xz + 2yz$. Hence the function

$$f(x, y) = 2xy + \frac{2000}{x} + \frac{2000}{y},$$

defined on the open first quadrant $x > 0$, $y > 0$, gives the total surface area of the rectangular solid with volume 1000 whose base has dimensions x and y.

It is clear that we need not consider either very small or very large values of x and y. For instance $f(x, y) \geqq 2000$ if either $x \leqq 1$ or $y \leqq 1$ or $xy \geqq 1000$, while the cube of edge 10 has total surface area 600. So we consider f on the set D pictured in Fig. 2.21. Since $f(x, y) \geqq 2000$ at each point of the boundary C of D,

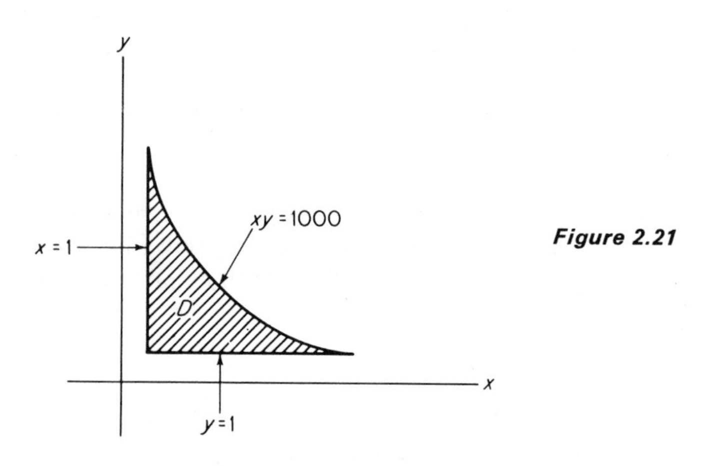

Figure 2.21

and since f attains values less than 2000 inside C [$f(10, 10) = 600$], it follows that f must attain its minimum value at a critical point interior to D. Now

$$D_1 f = 2y - \frac{2000}{x^2} \qquad \text{and} \qquad D_2 f = 2x - \frac{2000}{y^2}.$$

We find easily that the only critical point is (10, 10), so $f(10, 10) = 600$ (the total surface area of the cube of edge 10) *must* be the minimum value of f.

In general, if f is a differentiable function on a region D bounded by a simple closed curve C, then f may attain its maximum and minimum values on D either at interior points of D or at points of the boundary curve C. The procedure for maximizing or minimizing f on D is therefore to locate both the critical points of f that are interior to C, and the possible maximum–minimum points on C (by the Lagrange multiplier method), and finally to compare the values of f at all of the candidate points so obtained.

Example 6 Suppose we want to find the maximum and minimum values of $f(x, y) = xy$ on the unit disk $D = \{(x, y) : x^2 + y^2 \leq 1\}$. In Example 2 we have seen that the maximum and minimum values of $f(x, y)$ on the boundary $x^2 + y^2 = 1$ of D are $f(1/\sqrt{2}, 1/\sqrt{2}) = f(-1/\sqrt{2}, -1/\sqrt{2}) = \frac{1}{2}$ and $f(1/\sqrt{2}, -1/\sqrt{2}) = f(-1/\sqrt{2}, 1/\sqrt{2}) = -\frac{1}{2}$, respectively. The only interior critical point is the origin where $f(0, 0) = 0$. Thus $\frac{1}{2}$ and $-\frac{1}{2}$ are the extreme values of f on D.

Exercises

4.1 Find the shortest distance from the point (1, 0) to a point of the parabola $y^2 = 4x$.

4.2 Find the points of the ellipse $x^2/9 + y^2/4 = 1$ which are closest to and farthest from the point (1, 0).

4.3 Find the maximal area of a rectangle (with vertical and horizontal sides) inscribed in the ellipse $x^2/a^2 + y^2/b^2 = 1$.

4.4 The equation $73x^2 + 72xy + 52y^2 = 100$ defines an ellipse which is centered at the origin, but has been rotated about it. Find the semiaxes of this ellipse by maximizing and minimizing $f(x, y) = x^2 + y^2$ on it.

4.5 (a) Show that $xy \leq \frac{1}{4}$ if (x, y) is a point of the line segment $x + y = 1$, $x \geq 0$, $y \geq 0$.
(b) If a and b are positive numbers, show that $(ab)^{1/2} \leq \frac{1}{2}(a + b)$. *Hint:* Apply (a) with $x = a/(a + b)$, $y = b/(a + b)$.

4.6 (a) Show that $\log xy \leq \log \frac{1}{2}$ if (x, y) is a point of the unit circle $x^2 + y^2 = 1$ in the open first quadrant $x > 0$, $y > 0$. Hence $xy \leq \frac{1}{2}$ for such a point.
(b) Apply (a) with $x = a^{1/2}/(a + b)^{1/2}$, $y = b^{1/2}/(a + b)^{1/2}$ to show again that $(ab)^{1/2} \leq \frac{1}{2}(a + b)$ if $a > 0$, $b > 0$.

4.7 (a) Show that $|ax + by| \leq (a^2 + b^2)^{1/2}$ if $x^2 + y^2 = 1$ by finding the maximum and minimum values of $f(x, y) = ax + by$ on the unit circle.
(b) Prove the Cauchy–Schwarz inequality

$$|(a, b) \cdot (c, d)| \leq |(a, b)| \cdot |(c, d)|$$

by applying (a) with $x = c/(c^2 + d^2)^{1/2}$, $y = d/(c^2 + d^2)^{1/2}$.

4.8 Let C be the curve defined by $g(x, y) = 0$, and suppose $\nabla g \neq 0$ at each point of C. Let $\mathbf{p}$ be a fixed point not on C, and let $\mathbf{q}$ be a point of C that is closer to $\mathbf{p}$ than any other point of C. Prove that the line through $\mathbf{p}$ and $\mathbf{q}$ is orthogonal to the curve C at $\mathbf{q}$.

4.9 Prove that, among all closed rectangular boxes with total surface area 600, the cube with edge 10 is the one with the largest volume.

4.10 Show that the rectangular solid of largest volume that can be inscribed in the unit sphere is a cube.

4.11 Find the maximum volume of a rectangular box without top, which has the combined area of its sides and bottom equal to 100.

4.12 Find the maximum of the product of the sines of the three angles of a triangle, and show that the desired triangle is equilateral.

4.13 If a triangle has sides of lengths x, y, z, so its perimeter is $2s = x + y + z$, then its area A is given by $A^2 = s(s - x)(s - y)(s - z)$. Show that, among all triangles with a given perimeter, the one with the largest area is equilateral.

4.14 Find the maximum and minimum values of the function $f(x, y) = x^2 - y^2$ on the elliptical disk $x^2/16 + y^2/9 \leq 1$.

4.15 Find and classify the critical points of the function $f(x, y) = (x^2 + y^2)e^{x^2 - y^2}$.

4.16 If x, y, z are any three positive numbers whose sum is a fixed number s, show that $xyz \leq (s/3)^3$ by maximizing $f(x, y) = xy(s - x - y)$ on the appropriate set. Conclude from this that $(xyz)^{1/3} \leqq \frac{1}{3}(x + y + z)$, another case of the "arithmetic–geometric means inequality."

4.17 A wire of length 100 is cut into three pieces of lengths x, y, and $100 - x - y$. The first piece is bent into the shape of an equilateral triangle, the second is bent into the shape of a rectangle whose base is twice its height, and the third is made into a square. Find the minimum of the sum of the areas if $x > 0$, $y > 0$, $100 - x - y > 0$ (use Theorem 4.4 to check that you have a minimum). Find the maximum of the sum of the areas if we allow $x = 0$ or $y = 0$ or $100 - x - y = 0$, or any two of these.

The remaining three exercises deal with the quadratic form

$$f(x, y) = ax^2 + 2bxy + cy^2$$

of Example 3 and Theorem 4.3.

4.18 Let (x_1, y_1, λ_1) and (x_2, y_2, λ_2) be two solutions of the equations

$$ax + by = \lambda x, \qquad bx + cy = \lambda y, \qquad x^2 + y^2 = 1, \tag{2}$$

which were obtained in Example 3. If $\lambda_1 \neq \lambda_2$, show that the vectors $\mathbf{v}_1 = (x_1, y_1)$ and $\mathbf{v}_2 = (x_2, y_2)$ are orthogonal. *Hint:* Substitute (x_1, y_1, λ_1) into the first two equations of (2), multiply the two equations by x_2 and y_2 respectively, and then add. Now substitute (x_2, y_2, λ_2) into the two equations, multiply them by x_1 and y_1 and then add. Finally subtract the results to obtain $(\lambda_1 - \lambda_2)\mathbf{v}_1 \cdot \mathbf{v}_2 = 0$.

4.19 Define the linear mapping $L : \mathscr{R}^2 \to \mathscr{R}^2$ by

$$L(x, y) = (ax + by, bx + cy) \in \mathscr{R}^2,$$

and note that $f(\mathbf{x}) = \mathbf{x} \cdot L(\mathbf{x})$ for all $\mathbf{x} \in \mathscr{R}^2$. If $\mathbf{v}_1$ and $\mathbf{v}_2$ are as in the previous problem, show that

$$L(\mathbf{v}_1) = \lambda_1 \mathbf{v}_1 \qquad \text{and} \qquad L(\mathbf{v}_2) = \lambda_2 \mathbf{v}_2.$$

A vector, whose image under the linear mapping $L : \mathscr{R}^2 \to \mathscr{R}^2$ is a scalar multiple of itself, is called an *eigenvector* of L.

4.20 Let $\mathbf{v}_1$ and $\mathbf{v}_2$ be the eigenvectors of the previous problem. Given $\mathbf{x} \in \mathscr{R}^2$, let (u_1, u_2) be the coordinates of $\mathbf{x}$ with respect to axes through $\mathbf{v}_1$ and $\mathbf{v}_2$ (see Fig. 2.22). That is, u_1 and u_2 are the (unique) numbers such that

$$\mathbf{x} = u_1\mathbf{v}_1 + u_2\mathbf{v}_2.$$

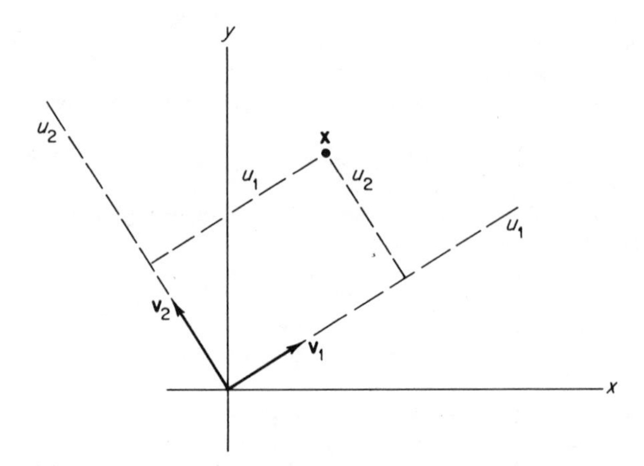

Figure 2.22

Substitute this equation into $f(\mathbf{x}) = \mathbf{x} \cdot L(\mathbf{x})$, and then apply the fact that $\mathbf{v}_1$ and $\mathbf{v}_2$ are eigenvectors of L to deduce that

$$f(\mathbf{x}) = \lambda_1 u_1^2 + \lambda_2 u_2^2.$$

Thus, in the new coordinate system, f is a sum or difference of squares.

4.21 Deduce from the previous problem that the graph of equation $ax^2 + 2bxy + cy^2 = 1$ is
(a) an ellipse if $ac - b^2 > 0$,
(b) a hyperbola if $ac - b^2 < 0$.

5 MAXIMA AND MINIMA, MANIFOLDS, AND LAGRANGE MULTIPLIERS

In this section we generalize the Lagrange multiplier method to $\mathscr{R}^n$. Let D be a compact (that is, by Theorem I.8.6, closed and bounded) subset of $\mathscr{R}^n$. If the function $f: D \to \mathscr{R}$ is continuous then, by Theorem I.8.8, there exists a point $\mathbf{p} \in D$ at which f attains its (absolute) maximum value on D (and similarly f attains an absolute minimum value at some point of D). The point $\mathbf{p}$ may be either a boundary point or an interior point of D. Recall that $\mathbf{p}$ is a *boundary point* of D if and only if every open ball centered at $\mathbf{p}$ contains both points of D and points of $\mathscr{R}^n - D$; an *interior point* of D is a point of D that is not a boundary point. Thus $\mathbf{p}$ is an interior point of D if and only if D contains some open ball centered at $\mathbf{p}$. The set of all boundary (interior) points of D is called its *boundary* (*interior*).

For example, the open ball $B_r(\mathbf{p})$ is the interior of the closed ball $\bar{B}_r(\mathbf{p})$; the sphere $S_r(\mathbf{p}) = \bar{B}_r(\mathbf{p}) - B_r(\mathbf{p})$ is its boundary.

We say that the function $f: D \to \mathscr{R}$ has a *local maximum* (respectively, *local minimum*) on D at the point $\mathbf{p} \in D$ if and only if there exists an open ball B centered at $\mathbf{p}$ such that $f(\mathbf{x}) \leqq f(\mathbf{p})$ [respectively, $f(\mathbf{x}) \geqq f(\mathbf{p})$] for all points $\mathbf{x} \in B \cap D$. Thus f has a local maximum on D at the point $\mathbf{p}$ if its value at $\mathbf{p}$ is at least as large as at any "nearby" point of D.

In applied maximum–minimum problems the set D is frequently the set of points on or within some closed and bounded $(n-1)$-dimensional surface S in $\mathscr{R}^n$; S is then the boundary of D. We will see in Corollary 5.2 that, if the differentiable function $f: D \to \mathscr{R}$ has a local maximum or minimum at the interior point $\mathbf{p} \in D$, then $\mathbf{p}$ must be a *critical point* of f, that is, a point at which all of the 1st partial derivatives of f vanish, so $\nabla f(\mathbf{p}) = \mathbf{0}$. The critical points of f can (in principle) be found by "setting the partial derivatives of f all equal to zero and solving for the coordinates $x_1, \ldots, x_n$." The location of critical points in higher dimensions does not differ essentially from their location in 2-dimensional problems of the sort discussed at the end of the previous section.

If, however, $\mathbf{p}$ is a *boundary* point of D at which f has a local maximum or minimum on D, then the situation is quite different—the location of such points is a Lagrange multiplier type of problem; this section is devoted to such problems. Our methods will be based on the following result (see Fig. 2.23).

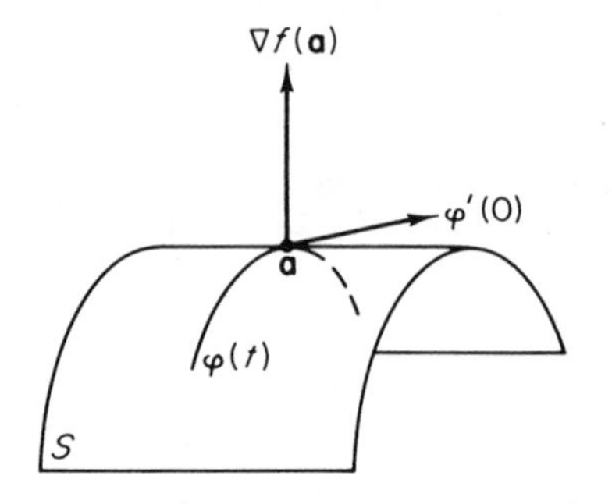

Figure 2.23

Theorem 5.1 Let S be a set in $\mathscr{R}^n$, and $\varphi : \mathscr{R} \to S$ a differentiable curve with $\varphi(0) = \mathbf{a}$. If f is a differentiable real-valued function defined on some open set containing S, and f has a local maximum (or local minimum) on S at $\mathbf{a}$, then the gradient vector $\nabla f(\mathbf{a})$ is orthogonal to the velocity vector $\varphi'(0)$.

PROOF The composite function $h = f \circ \varphi : \mathscr{R} \to \mathscr{R}$ is differentiable at $0 \in \mathscr{R}$, and attains a local maximum (or local minimum) there. Therefore $h'(0) = 0$, so the chain rule gives

$$\begin{aligned}\nabla f(\mathbf{a}) \cdot \varphi'(0) &= \nabla f(\varphi(0)) \cdot \varphi'(0) \\ &= h'(0) \\ &= 0.\end{aligned}$$

∎

It is an immediate corollary that *interior* local maximum–minimum points are critical points.

Corollary 5.2 If U is an open subset of $\mathscr{R}^n$, and $\mathbf{a} \in U$ is a point at which the differentiable function $f: U \to \mathscr{R}$ has a local maximum or local minimum, then $\mathbf{a}$ is a critical point of f. That is, $\nabla f(\mathbf{a}) = \mathbf{0}$.

PROOF Given $\mathbf{v} \in \mathscr{R}^n$, define $\varphi : \mathscr{R} \to \mathscr{R}^n$ by $\varphi(t) = \mathbf{a} + t\mathbf{v}$, so $\varphi'(t) \equiv \mathbf{v}$. Then Theorem 5.1 gives

$$\nabla f(\mathbf{a}) \cdot \mathbf{v} = \nabla f(\mathbf{a}) \cdot \varphi'(0) = 0.$$

Since $\nabla f(\mathbf{a})$ is thus orthogonal to every vector $\mathbf{v} \in \mathscr{R}^n$, it follows that $\nabla f(\mathbf{a}) = \mathbf{0}$. ∎

Example 1 Suppose we want to find the maximum and minimum values of the function $f(x, y, z) = x + y + z$ on the unit sphere

$$S^2 = \{(x, y, z) \in \mathscr{R}^3 : x^2 + y^2 + z^2 = 1\}$$

in $\mathscr{R}^3$. Theorem 5.1 tells us that, if $\mathbf{a} = (x, y, z)$ is a point at which f attains its maximum or minimum on S^2, then $\nabla f(\mathbf{a})$ is orthogonal to every curve on S^2 passing through $\mathbf{a}$, and is therefore orthogonal to the sphere at $\mathbf{a}$ (that is, to its tangent plane at $\mathbf{a}$). But it is clear that $\mathbf{a}$ itself is orthogonal to S^2 at $\mathbf{a}$. Hence $\nabla f(\mathbf{a})$ and $\mathbf{a} \neq 0$ are collinear vectors (Fig. 2.24), so $\nabla f(\mathbf{a}) = \lambda \mathbf{a}$ for some number λ. But $\nabla f = (1, 1, 1)$, so

$$(1, 1, 1) = \lambda(x, y, z),$$

so $x = y = z = 1/\lambda$. Since $x^2 + y^2 + z^2 = 1$, the two possibilities are $\mathbf{a} = (1/\sqrt{3}, 1/\sqrt{3}, 1/\sqrt{3})$ and $\mathbf{a} = (-1/\sqrt{3}, -1/\sqrt{3}, -1/\sqrt{3})$. Since Theorem I.8.8 implies that f *does* attain maximum and minimum values on S^2, we conclude that these are $f(1/\sqrt{3}, 1/\sqrt{3}, 1/\sqrt{3}) = \sqrt{3}$ and $f(-1/\sqrt{3}, -1/\sqrt{3}, -1/\sqrt{3}) = -\sqrt{3}$, respectively.

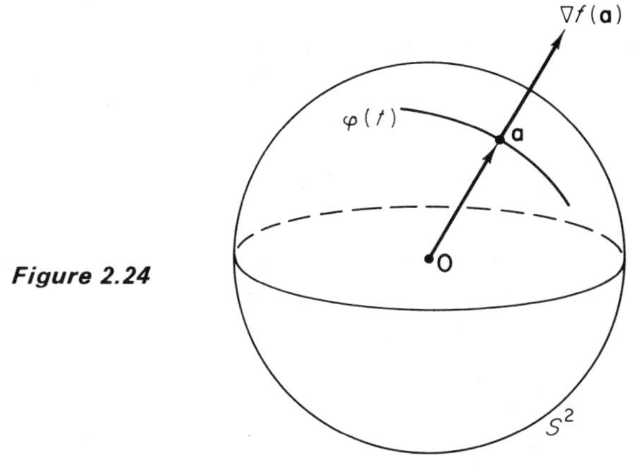

Figure 2.24

The reason we were able to solve this problem so easily is that, at every point, the unit sphere S^2 in $\mathscr{R}^3$ has a 2-dimensional tangent plane which can be described readily in terms of its normal vector. We want to generalize (and systematize) the method of Example 1, so as to be able to find the maximum and minimum values of a differentiable real-valued function on an $(n-1)$-dimensional surface in $\mathscr{R}^n$.

We now need to make precise our idea of what an $(n-1)$-dimensional surface is. To start with, we want an $(n-1)$-dimensional surface (called an $(n-1)$-manifold in the definition below) to be a set in $\mathscr{R}^n$ which has at each point an $(n-1)$-dimensional tangent plane. The set M in $\mathscr{R}^n$ is said to have a *k-dimensional tangent plane* at the point $\mathbf{a} \in M$ if the union of all tangent lines at $\mathbf{a}$, to differentiable curves on M passing through $\mathbf{a}$, is a k-dimensional plane (that is, is a translate of a k-dimensional subspace of $\mathscr{R}^n$). See Fig. 2.25.

A manifold will be defined as a union of sets called "patches." Let $\pi_i : \mathscr{R}^n \to \mathscr{R}^{n-1}$ denote the projection mapping which simply deletes the ith coordinate, that is,

$$\pi_i(x_1, \ldots, x_n) = (x_1, \ldots, \hat{x}_i, \ldots, x_n),$$

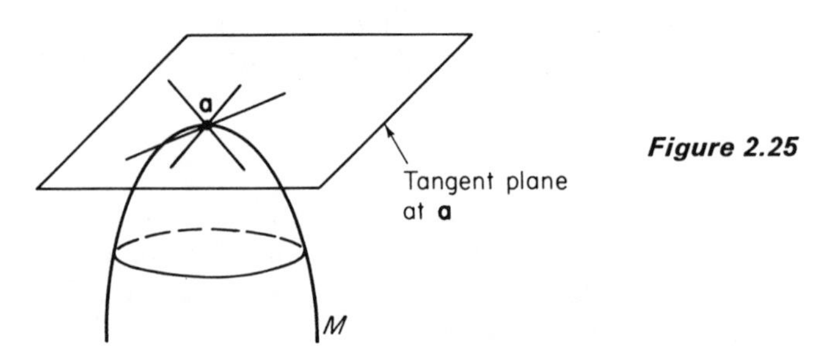

Figure 2.25

where the symbol $\hat{x}_i$ means that the coordinate x_i has been deleted from the n-tuple, leaving an $(n-1)$-tuple, or point of $\mathscr{R}^{n-1}$. The set P in $\mathscr{R}^n$ is called an $(n-1)$-*dimensional patch* if and only if for some positive integer $i \leqq n$, there exists a differentiable function $h : U \to \mathscr{R}$, on an open subset $U \subset \mathscr{R}^{n-1}$, such that

$$P = \{\text{all } \mathbf{x} \in \mathscr{R}^n : \pi_i(\mathbf{x}) \in U \qquad \text{and} \qquad x_i = h(\pi_i(\mathbf{x}))\}.$$

In other words, P is the graph in $\mathscr{R}^n$ of the differentiable function h, regarding h as defined on an open subset of the $(n-1)$-dimensional coordinate plane $\pi_i(\mathscr{R}^n)$ that is spanned by the unit basis vectors $\mathbf{e}_1, \ldots, \mathbf{e}_{i-1}, \mathbf{e}_{i+1}, \ldots, \mathbf{e}_n$ (see Fig. 2.26). To put it still another way, the ith coordinate of a point of P is a differentiable function of its remaining $n-1$ coordinates. We will see in the proof of Theorem 5.3 that every $(n-1)$-dimensional patch in $\mathscr{R}^n$ has an $(n-1)$-dimensional tangent plane at each of its points.

The set M in $\mathscr{R}^n$ is called an $(n-1)$-*dimensional manifold*, or simply an

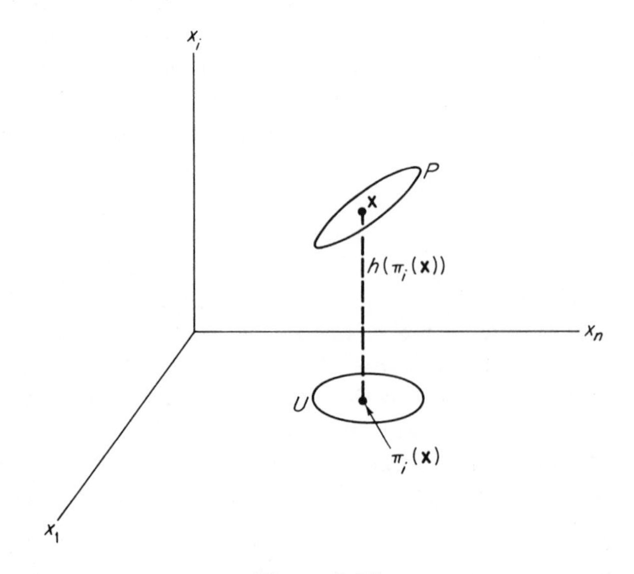

Figure 2.26

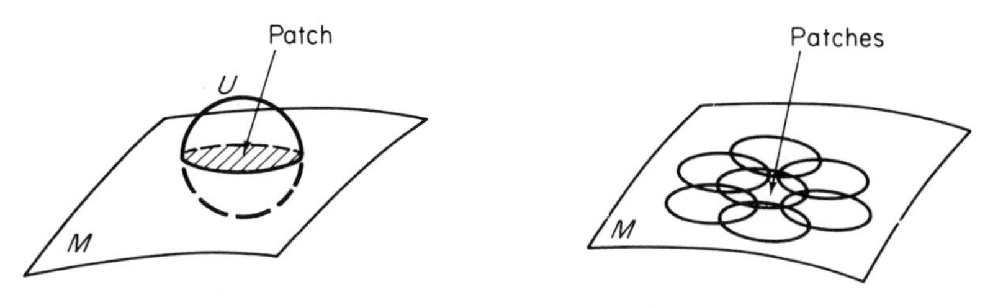

Figure 2.27

$(n-1)$-*manifold*, if and only if each point $\mathbf{a} \in M$ lies in an open subset U of $\mathscr{R}^n$ such that $U \cap M$ is an $(n-1)$-dimensional patch (see Fig. 2.27). Roughly speaking, a manifold is simply a union of patches, although this is not quite right, because in an arbitrary union of patches two of them might intersect "wrong."

Example 2 The unit circle $x^2 + y^2 = 1$ is a 1-manifold in $\mathscr{R}^2$, since it is the union of the 1-dimensional patches corresponding to the *open* semicircles $x > 0$, $x < 0$, $y > 0$, $y < 0$ (see Fig. 2.28). Similarly the unit sphere $x^2 + y^2 + z^2 = 1$ is a 2-manifold in $\mathscr{R}^3$, since it is covered by six 2-dimensional patches—the upper and lower, front and back, right and left open hemispheres determined by $z > 0$, $z < 0$, $x > 0$, $x < 0$, $y > 0$, $y < 0$, respectively. The student should be able to generalize this approach so as to see that the unit sphere S^{n-1} in $\mathscr{R}^n$ is an $(n-1)$-manifold.

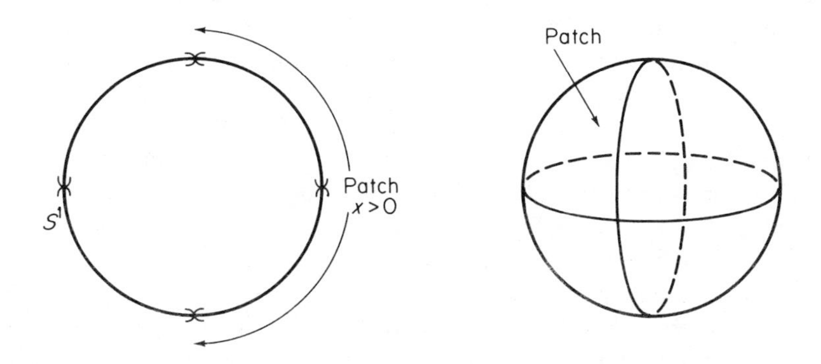

Figure 2.28

Example 3 The "torus" T in $\mathscr{R}^3$, obtained by rotating about the z-axis the circle $(y-1)^2 + z^2 = 4$ in the yz-plane, is a 2-manifold (Fig. 2.29). The upper and lower open halves of T, determined by the conditions $z > 0$ and $z < 0$ respectively, are 2-dimensional patches; each is clearly the graph of a differentiable function defined on the open "annulus" $1 < x^2 + y^2 < 9$ in the xy-plane. These two patches cover all of T except for the points on the circles $x^2 + y^2 = 1$ and $x^2 + y^2 = 9$ in the xy-plane. Additional patches in T, covering these two circles,

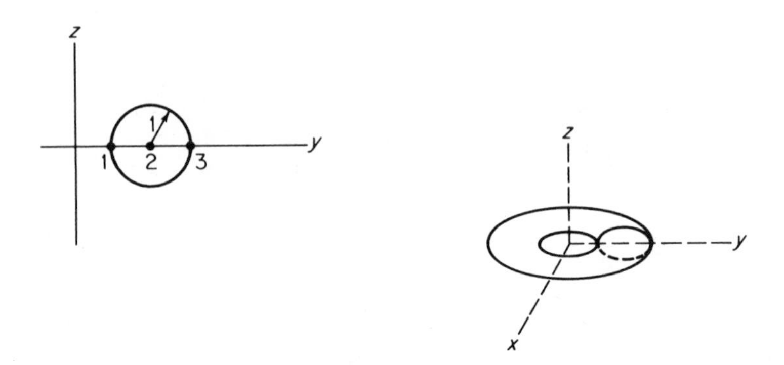

Figure 2.29

must be defined in order to complete the proof that T is a 2-manifold (see Exercise 5.1).

The following theorem gives the particular property of $(n-1)$-manifolds in $\mathscr{R}^n$ which is important for maximum–minimum problems.

Theorem 5.3 If M is an $(n-1)$-dimensional manifold in $\mathscr{R}^n$, then, at each of its points, M has an $(n-1)$-dimensional tangent plane.

PROOF Given $\mathbf{a} \in M$, we want to show that the union of all tangent lines at $\mathbf{a}$, to differentiable curves through $\mathbf{a}$ on M, is an $(n-1)$-dimensional plane or, equivalently, that the set of all velocity vectors of such curves is an $(n-1)$-dimensional subspace of $\mathscr{R}^n$.

The fact that M is an $(n-1)$-manifold means that, near $\mathbf{a}$, M coincides with the graph of some differentiable function $h : \mathscr{R}^n \to \mathscr{R}$. That is, for some $i \leqq n$, $x_i = h(x_1, \ldots, \hat{x}_i, \ldots, x_n)$ for all points $(x_1, \ldots, x_n)$ of M sufficiently close to $\mathbf{a}$. Let us consider the case $i = n$ (from which the other cases differ only by a permutation of the coordinates).

Let $\varphi : \mathscr{R} \to M$ be a differentiable curve with $\varphi(0) = \mathbf{a}$, and define $\psi : \mathscr{R} \to \mathscr{R}^{n-1}$ by $\psi = \pi \circ \varphi$, where $\pi : \mathscr{R}^n \to \mathscr{R}^{n-1}$ is the usual projection. If $\psi(0) = \mathbf{b} \in \mathscr{R}^{n-1}$, then the image of φ near $\mathbf{a}$ lies directly "above" the image of ψ near $\mathbf{b}$. That is, $\varphi(t) = (\psi(t), h(\psi(t))$ for t sufficiently close to 0. Applying the chain rule, we therefore obtain

$$\begin{aligned} \varphi'(0) &= (\psi'(0), \nabla h(\mathbf{b}) \cdot \psi'(0)) \\ &= \sum_{i=1}^{n-1} \psi_i'(0)(\mathbf{e}_i, D_i h(\mathbf{b})), \end{aligned} \tag{1}$$

where $\mathbf{e}_1, \ldots, \mathbf{e}_{n-1}$ are the unit basis vectors *in* $\mathscr{R}^{n-1}$. Consequently $\varphi'(0)$ lies in the $(n-1)$-dimensional subspace of $\mathscr{R}^n$ spanned by the $n-1$ (clearly linearly independent) vectors

$$(\mathbf{e}_1, D_1 h(\mathbf{b})), \ldots, (\mathbf{e}_{n-1}, D_{n-1} h(\mathbf{b})).$$

Conversely, given a vector $\mathbf{v} = \sum_{i=1}^{n-1} v_i(\mathbf{e}_i, D_i h(\mathbf{b}))$ of this $(n-1)$-dimensional space, consider the differentiable curve $\varphi : \mathscr{R} \to M$ defined by

$$\varphi(t) = (\mathbf{b} + t\mathbf{w}, h(\mathbf{b} + t\mathbf{w})),$$

where $\mathbf{w} = (v_1, \ldots, v_{n-1}) \in \mathscr{R}^{n-1}$. It is then clear from (1) that $\varphi'(0) = \mathbf{v}$. Thus every point of our $(n-1)$-dimensional subspace is the velocity vector of some curve through $\mathbf{a}$ on M. ∎

In order to apply Theorem 5.3, we need to be able to recognize an $(n-1)$-manifold (as such) when we see one. We give in Theorem 5.4 below a useful sufficient condition that a set $M \subset \mathscr{R}^n$ be an $(n-1)$-manifold. For its proof we need the following basic theorem, which will be established in Chapter III. It asserts that if $g : \mathscr{R}^n \to \mathscr{R}$ is a continuously differentiable function, and $g(\mathbf{a}) = 0$ with some partial derivative $D_i g(\mathbf{a}) \neq 0$, then near $\mathbf{a}$ the equation

$$g(x_1, \ldots, x_n) = 0$$

can be "solved for x_i as a function of the remaining variables." This implies that, near $\mathbf{a}$, the set $S = g^{-1}(0)$ looks like an $(n-1)$-dimensional patch, hence like an $(n-1)$-manifold (see Fig. 2.30). We state this theorem with $i = n$.

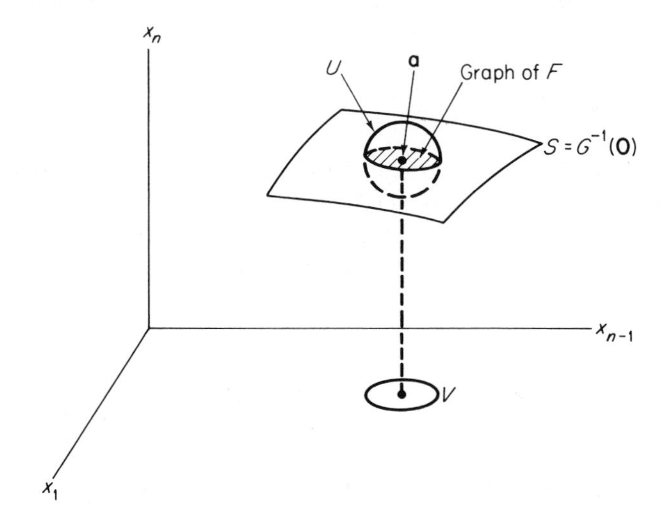

Figure 2.30

Implicit Function Theorem Let $G : \mathscr{R}^n \to \mathscr{R}$ be continuously differentiable, and suppose that $G(\mathbf{a}) = 0$ while $D_n G(\mathbf{a}) \neq 0$. Then there exists a neighborhood U of $\mathbf{a}$ and a differentiable function F defined on a neighborhood V of $(a_1, \ldots, a_{n-1}) \in \mathscr{R}^{n-1}$, such that

$$U \cap G^{-1}(0) = \{\mathbf{x} \in \mathscr{R}^n : (x_1, \ldots, x_{n-1}) \in V \text{ and } x_n = F(x_1, \ldots, x_{n-1})\}.$$

In particular,

$$G(x_1, \ldots, x_{n-1}, F(x_1, \ldots, x_{n-1})) = 0$$

for all $(x_1, \ldots, x_{n-1}) \in V$.

Theorem 5.4 Suppose that $g : \mathscr{R}^n \to \mathscr{R}$ is continuously differentiable. If M is the set of all those points $\mathbf{x} \in S = g^{-1}(0)$ at which $\nabla g(\mathbf{x}) \neq \mathbf{0}$, then M is an $(n-1)$-manifold. Given $\mathbf{a} \in M$, the gradient vector $\nabla g(\mathbf{a})$ is orthogonal to the tangent plane to M at $\mathbf{a}$.

PROOF Let $\mathbf{a}$ be a point of M, so $g(\mathbf{a}) = 0$ and $\nabla g(\mathbf{a}) \neq \mathbf{0}$. Then $D_i g(\mathbf{a}) = 0$ for some $i \leqq n$. Define $G : \mathscr{R}^n \to \mathscr{R}$ by

$$G(x_1, \ldots, x_n) = g(x_1, \ldots, x_{i-1}, x_n, x_i, \ldots, x_{n-1}).$$

Then $G(\mathbf{b}) = 0$ and $D_n G(\mathbf{b}) \neq 0$, where $\mathbf{b} = (a_1, \ldots, a_{i-1}, a_{i+1}, \ldots, a_n, a_i)$. Let $U \subset \mathscr{R}^n$ and $V \subset \mathscr{R}^{n-1}$ be the open sets, and $F : V \to \mathscr{R}$ the implicitly defined function, supplied by the implicit function theorem, so that

$$U \cap G^{-1}(0) = \{\mathbf{x} \in \mathscr{R}^n : (x_1, \ldots, x_{n-1}) \in V \text{ and } x_n = F(x_1, \ldots, x_{n-1})\}.$$

Now let W be the set of all points $(x_1, \ldots, x_n) \in \mathscr{R}^n$ such that $(x_1, \ldots, x_{i-1}, x_{i+1}, \ldots, x_n, x_i) \in U$. Then $W \cap M$ is clearly an $(n-1)$-dimensional patch; in particular,

$$W \cap M = \{\mathbf{x} \in W : x_i = F(x_1, \ldots, x_{i-1}, x_{i+1}, \ldots, x_n)\}.$$

To prove that $\nabla g(\mathbf{a})$ is orthogonal to the tangent plane to M at $\mathbf{a}$, we need to show that, if $\varphi : \mathscr{R} \to M$ is a differentiable curve with $\varphi(0) = \mathbf{a}$, then the vectors $\nabla g(\mathbf{a})$ and $\varphi'(0)$ are orthogonal. But the composite function $g \circ \varphi : \mathscr{R} \to \mathscr{R}$ is identically zero, so the chain rule gives

$$\nabla g(\mathbf{a}) \circ \varphi'(0) = (g \circ \varphi)'(0) = 0. \qquad \blacksquare$$

For example, if $g(\mathbf{x}) = x_1^2 + \cdots + x_n^2 - 1$, then M is the unit sphere S^{n-1} in $\mathscr{R}^n$, so Theorem 5.4 provides a quick proof that S^{n-1} is an $(n-1)$-manifold.

We are finally ready to "put it all together."

Theorem 5.5 Suppose $g : \mathscr{R}^n \to \mathscr{R}$ is continuously differentiable, and let M be the set of all those points $\mathbf{x} \in \mathscr{R}^n$ at which both $g(\mathbf{x}) = 0$ and $\nabla g(\mathbf{x}) \neq \mathbf{0}$. If the differentiable function $f : \mathscr{R}^n \to \mathscr{R}$ attains a local maximum or minimum on M at the point $\mathbf{a} \in M$, then

$$\nabla f(\mathbf{a}) = \lambda \, \nabla g(\mathbf{a}) \tag{2}$$

for some number λ (called the "Lagrange multiplier").

PROOF By Theorem 5.4, M is an $(n-1)$-manifold, so M has an $(n-1)$-dimensional tangent plane by Theorem 5.3. The vectors $\nabla f(\mathbf{a})$ and $\nabla g(\mathbf{a})$ are both orthogonal to this tangent plane, by Theorems 5.1 and 5.4, respectively. Since the orthogonal complement to an $(n-1)$-dimensional subspace of $\mathscr{R}^n$ is 1-dimensional, by Theorem I.3.4, it follows that $\nabla f(\mathbf{a})$ and $\nabla g(\mathbf{a})$ are collinear. Since $\nabla g(\mathbf{a}) \neq \mathbf{0}$, this implies that $\nabla f(\mathbf{a})$ is a multiple of $\nabla g(\mathbf{a})$. ∎

According to this theorem, in order to maximize or minimize the function $f : \mathscr{R}^n \to \mathscr{R}$ subject to the "constraint equation"

$$g(x_1, \ldots, x_n) = 0,$$

it suffices to solve the $n+1$ scalar equations

$$g(\mathbf{x}) = 0, \qquad \text{and} \qquad \nabla f(\mathbf{x}) = \lambda \nabla g(\mathbf{x})$$

for the $n+1$ "unknowns" $x_1, \ldots, x_n, \lambda$. If these equations have several solutions, we can determine which (if any) gives a maximum and which gives a minimum by computing the value of f at each. This in brief is the "Lagrange multiplier method."

Example 4 Let us reconsider Example 5 of Section 4. We want to find the rectangular box of volume 1000 which has the least total surface area A. If

$$f(x, y, z) = 2xy + 2xz + 2yz \qquad \text{and} \qquad g(x, y, z) = xyz - 1000,$$

our problem is to minimize the function f on the 2-manifold in $\mathscr{R}^3$ given by $g(x, y, z) = 0$. Since $\nabla f = (2y + 2z, 2x + 2z, 2x + 2y)$ and $\nabla g = (yz, xz, xy)$, we want to solve the equations

$$\begin{aligned} 2y + 2z &= \lambda yz, \\ 2x + 2z &= \lambda xz, \\ 2x + 2y &= \lambda xy, \\ xyz &= 1000. \end{aligned}$$

Upon multiplying the first three equations by x, y, and z respectively, and then substituting $xyz = 1000$ on the right hand sides, we obtain

$$xy + xz = xy + yz = xz + yz = 500\lambda,$$

from which it follows easily that $x = y = z$. Since $xyz = 1000$, our solution gives a cube of edge 10.

Now we want to generalize the Lagrange multiplier method so as to be able to maximize or minimize a function $f : \mathscr{R}^n \to \mathscr{R}$ subject to *several* constraint equations

$$g_1(\mathbf{x}) = 0, \ldots, g_m(\mathbf{x}) = 0, \tag{3}$$

where $m < n$. For example, suppose that we wish to maximize the function $f(x, y, z) = x^2 + y^2 + z^2$ subject to the conditions $x^2 + y^2 = 1$ and $x + y + z = 0$. The intersection of the cylinder $x^2 + y^2 = 1$ and the plane $x + y + z = 0$ is an ellipse in $\mathscr{R}^3$, and we are simply asking for the maximum distance (squared) from the origin to a point of this ellipse.

If $G : \mathscr{R}^n \to \mathscr{R}^m$ is the mapping whose component functions are the functions $g_1, \ldots, g_m$, then equations (3) may be rewritten as $G(\mathbf{x}) = \mathbf{0}$. Experience suggests that the set $S = G^{-1}(\mathbf{0})$ may (in some sense) be an $(n - m)$-dimensional surface in $\mathscr{R}^n$. To make this precise, we need to define k-manifolds in $\mathscr{R}^n$, for all $k < n$.

Our definition of $(n - 1)$-dimensional patches can be rephrased to say that $P \subset \mathscr{R}^n$ is an $(n - 1)$-dimensional patch if and only if there exists a permutation $x_{i_1}, \ldots, x_{i_n}$ of the coordinates $x_1, \ldots, x_n$ in $\mathscr{R}^n$, and a differentiable function $h : U \to \mathscr{R}$ on an open set $U \subset \mathscr{R}^{n-1}$, such that

$$P = \{\mathbf{x} \in \mathscr{R}^n : (x_{i_1}, \ldots, x_{i_{n-1}}) \in U \text{ and } x_{i_n} = h(x_{i_1}, \ldots, x_{i_{n-1}})\}.$$

Similarly we say that the set $P \subset \mathscr{R}^n$ is a *k-dimensional patch* if and only if there exists a permutation $x_{i_1}, \ldots, x_{i_n}$ of $x_1, \ldots, x_n$, and a differentiable mapping $h : U \to \mathscr{R}^{n-k}$ defined on an open set $U \subset \mathscr{R}^k$, such that,

$$P = \{\mathbf{x} \in \mathscr{R}^n : (x_{i_1}, \ldots, x_{i_k}) \in U \text{ and } (x_{i_{k+1}}, \ldots, x_{i_n}) = h(x_{i_1}, \ldots, x_{i_k})\}.$$

Thus P is simply the graph of h, regarded as a function of $x_{i_1}, \ldots, x_{i_k}$, rather than $x_1, \ldots, x_k$ as usual; the coordinates $x_{i_{k+1}}, \ldots, x_{i_n}$ of a point of P are differentiable functions of its remaining k coordinates (see Fig. 2.31).

The set $M \subset \mathscr{R}^n$ is called a *k-dimensional manifold in $\mathscr{R}^n$*, or *k-manifold*, if every point of M lies in an open subset V of $\mathscr{R}^n$ such that $V \cap M$ is a k-dimensional patch. Thus a k-manifold is a set which is made up of k-dimensional

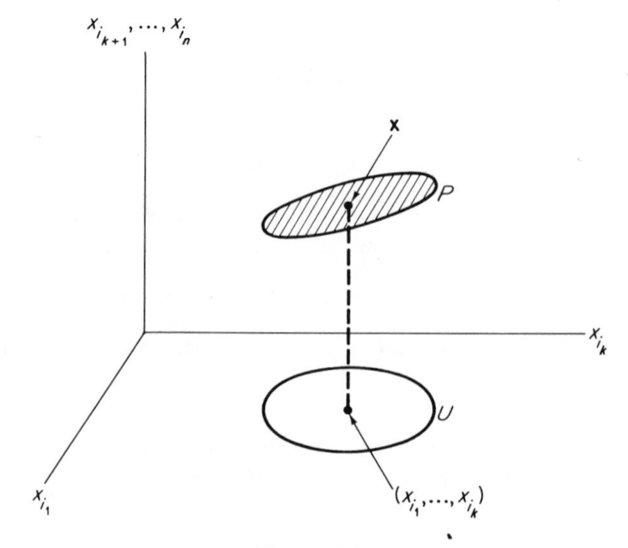

Figure 2.31

patches, in the same way that an $(n-1)$-manifold is made up of $(n-1)$-dimensional patches. For example, it is easily verified that the circle $x^2 + y^2 = 1$ in the xy-plane is a 1-manifold in $\mathscr{R}^3$. This is a special case of the fact that, if M is a k-manifold in $\mathscr{R}^n$, and $\mathscr{R}^n$ is regarded as a subspace of $\mathscr{R}^p$ $(p > n)$, then M is a k-manifold in $\mathscr{R}^p$ (Exercise 5.2).

In regard to both its statement and its proof (see Exercise 5.16), the following result is the expected generalization of Theorem 5.3.

Theorem 5.6 If M is a k-dimensional manifold in $\mathscr{R}^n$ then, at each of its points, M has a k-dimensional tangent plane.

In order to generalize Theorem 5.4, the following generalization of the implicit function theorem is required; its proof will be given in Chapter III.

Implicit Mapping Theorem Let $G : \mathscr{R}^n \to \mathscr{R}^m$ $(m < n)$ be a continuously differentiable mapping. Suppose that $G(\mathbf{a}) = \mathbf{0}$ and that the rank of the derivative matrix $G'(\mathbf{a})$ is m. Then there exists a permutation $x_{i_1}, \ldots, x_{i_n}$ of the coordinates in $\mathscr{R}^n$, an open subset U of $\mathscr{R}^n$ containing $\mathbf{a}$, an open subset V of $\mathscr{R}^{n-m}$ containing $\mathbf{b} = (a_{i_1}, \ldots, a_{i_{n-m}})$, and a differentiable mapping $h : V \to \mathscr{R}^m$ such that each point $\mathbf{x} \in U$ lies on $S = G^{-1}(\mathbf{0})$ if and only if $(x_{i_1}, \ldots, x_{i_{n-m}}) \in V$ and

$$(x_{i_{n-m+1}}, \ldots, x_{i_n}) = h(x_{i_1}, \ldots, x_{i_{n-m}}).$$

Recall that the $m \times n$ matrix $G'(\mathbf{a})$ has rank m if and only if its m row vectors $\nabla G_1(\mathbf{a}), \ldots, \nabla G_m(\mathbf{a})$ (the gradient vectors of the component functions of G) are linearly independent (Section I.5). If $m = 1$, so $G = g : \mathscr{R}^n \to \mathscr{R}$, this is just the condition that $\nabla g(\mathbf{a}) \neq \mathbf{0}$, so some partial derivative $D_i g(\mathbf{a}) \neq 0$. Thus the implicit mapping theorem is indeed a generalization of the implicit function theorem.

The conclusion of the implicit mapping theorem asserts that, near $\mathbf{a}$, the m equations

$$G_1(\mathbf{x}) = 0, \ldots, G_m(\mathbf{x}) = 0$$

can be solved (uniquely) for the m variables $x_{i_{n-m+1}}, \ldots, x_{i_n}$ as differentiable functions of the variables $x_{i_1}, \ldots, x_{i_{n-m}}$. Thus the set $S = G^{-1}(\mathbf{0})$ looks, near $\mathbf{a}$, like an $(n-m)$-dimensional manifold. Using the implicit mapping theorem in place of the implicit function theorem, the proof of Theorem 5.4 translates into a proof of the following generalization.

Theorem 5.7 Suppose that the mapping $G : \mathscr{R}^n \to \mathscr{R}^m$ is continuously differentiable. If M is the set of all those points $\mathbf{x} \in S = \mathbf{G}^{-1}(\mathbf{0})$ for which the rank of $G'(\mathbf{x})$ is m, then M is an $(n-m)$-manifold. Given $\mathbf{a} \in M$, the gradient vectors $\nabla G_1(\mathbf{a}), \ldots, \nabla G_m(\mathbf{a})$, of the component functions of G, are all orthogonal to the tangent plane to M at $\mathbf{a}$ (see Fig. 2.32).

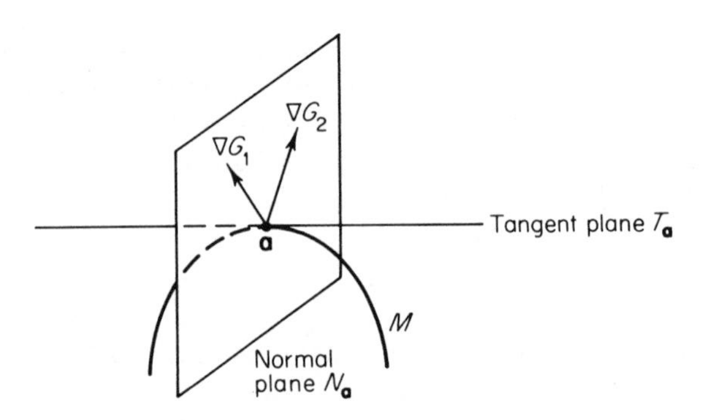

Figure 2.32

In brief, this theorem asserts that the solution set of m equations in $n > m$ variables is, in general, an $(n - m)$-dimensional manifold in $\mathscr{R}^n$. Here the phrase "in general" means that, if our equations are

$$\begin{gathered} G_1(x_1, \ldots, x_n) = 0, \\ \vdots \\ G_m(x_1, \ldots, x_n) = 0, \end{gathered}$$

we must know that the functions $G_1, \ldots, G_m$ are continuously differentiable, and also that the gradient vectors $\nabla G_1, \ldots, \nabla G_m$ are linearly independent at each point of $M = G^{-1}(\mathbf{0})$, and finally that M is nonempty to start with.

Example 5 If $G : \mathscr{R}^3 \to \mathscr{R}^2$ is defined by

$$G_1(x, y, z) = x^2 + y^2 + z^2 - 1, \qquad G_2(x, y, z) = x + y + z - 1,$$

then $G^{-1}(\mathbf{0})$ is the intersection M of the unit sphere $x^2 + y^2 + z^2 = 1$ and the plane $x + y + z = 1$. Of course it is obvious that M is a circle. However, to conclude from Theorem 5.7 that M is a 1-manifold, we must first verify that $\nabla G_1 = (2x, 2y, 2z)$ and $\nabla G_2 = (1, 1, 1)$ are linearly independent (that is, not collinear) at each point of M. But the only points of the unit sphere, where ∇G_1 is collinear with $(1, 1, 1)$, are $(1/\sqrt{3}, 1/\sqrt{3}, 1/\sqrt{3})$ and $(-1/\sqrt{3}, -1/\sqrt{3}, -1/\sqrt{3})$, neither of which lies on the plane $x + y + z = 1$.

Example 6 If $G : \mathscr{R}^4 \to \mathscr{R}^2$ is defined by

$$G_1(\mathbf{x}) = {x_1}^2 + {x_2}^2 - 1 \qquad \text{and} \qquad G_2(\mathbf{x}) = {x_3}^2 + {x_4}^2 - 1,$$

the gradient vectors

$$\nabla G_1(\mathbf{x}) = (2x_1, 2x_2, 0, 0) \qquad \text{and} \qquad \nabla G_2(\mathbf{x}) = (0, 0, 2x_3, 2x_4)$$

are linearly independent at each point of $M = G^{-1}(\mathbf{0})$ (Why?), so M is a 2-manifold in $\mathscr{R}^4$ (it is a torus).

Example 7 If $g(x, y, z) = x^2 + y^2 - z^2$, then $S = g^{-1}(0)$ is a double cone which fails to be a 2-manifold only at the origin. Note that $(0, 0, 0)$ is the only point of S where $\nabla g = (2x, 2y, -2z)$ is zero.

We are finally ready for the general version of the Lagrange multiplier method.

Theorem 5.8 Suppose $G : \mathscr{R}^n \to \mathscr{R}^m$ $(m < n)$ is continuously differentiable, and denote by M the set of all those points $\mathbf{x} \in \mathscr{R}^n$ such that $G(\mathbf{x}) = \mathbf{0}$, and also the gradient vectors $\nabla G_1(\mathbf{x}), \ldots, \nabla G_m(\mathbf{x})$ are linearly independent. If the differentiable function $f : \mathscr{R}^n \to \mathscr{R}$ attains a local maximum or minimum on M at the point $\mathbf{a} \in M$, then there exist real numbers $\lambda_1, \ldots, \lambda_m$ (called Lagrange multipliers) such that

$$\nabla f(\mathbf{a}) = \lambda_1 \nabla G_1(\mathbf{a}) + \cdots + \lambda_m \nabla G_m(\mathbf{a}). \tag{4}$$

PROOF By Theorem 5.7, M is an $(n - m)$-manifold, and therefore has an $(n - m)$-dimensional tangent plane $T_\mathbf{a}$ at $\mathbf{a}$, by Theorem 5.6. If $N_\mathbf{a}$ is the orthogonal complement to the translate of $T_\mathbf{a}$ to the origin, then Theorem I.3.4 implies that $\dim N_\mathbf{a} = m$. The linearly independent vectors $\nabla G_1(\mathbf{a}), \ldots, \nabla G_m(\mathbf{a})$ lie in $N_\mathbf{a}$ (Theorem 5.7), and therefore constitute a basis for $N_\mathbf{a}$. Since, by Theorem 5.1, $\nabla f(\mathbf{a})$ also lies in $N_\mathbf{a}$, it follows that $\nabla f(\mathbf{a})$ is a linear combination of the vectors $\nabla G_1(\mathbf{a}), \ldots, \nabla G_m(\mathbf{a})$. ∎

In short, in order to locate all points $(x_1, \ldots, x_n) \in M$ at which f can attain a maximum or minimum value, it suffices to solve the $n + m$ scalar equations

$$\begin{gathered} G_1(\mathbf{x}) = 0, \\ \vdots \\ G_m(\mathbf{x}) = 0, \\ \nabla f(\mathbf{x}) = \lambda_1 \nabla G_1(\mathbf{x}) + \cdots + \lambda_m \nabla G_m(\mathbf{x}) \end{gathered}$$

for the $n + m$ "unknowns" $x_1, \ldots, x_n, \lambda_1, \ldots, \lambda_m$.

Example 8 Suppose we want to maximize the function $f(x, y, z) = x$ on the circle of intersection of the plane $z = 1$ and the sphere $x^2 + y^2 + z^2 = 4$ (Fig. 2.33). We define $g : \mathscr{R}^3 \to \mathscr{R}^2$ by $g_1(x, y, z) = z - 1$ and $g_2(x, y, z) = x^2 + y^2 + z^2 - 1$. Then $g^{-1}(0)$ is the given circle of intersection.

Since $\nabla f = (1, 0, 0)$, $\nabla g_1 = (0, 0, 1)$, $\nabla g_2 = (2x, 2y, 2z)$, we want to solve the equations

$$\begin{gathered} z = 1, \qquad x^2 + y^2 + z^2 = 4, \\ 1 = 2\lambda_2 x, \qquad 0 = 2\lambda_2 y, \qquad 0 = \lambda_1 + 2\lambda_2 z. \end{gathered}$$

We obtain the two solutions $(\pm\sqrt{3}, 0, 1)$ for (x, y, z), so the maximum is $\sqrt{3}$ and the minimum is $-\sqrt{3}$.

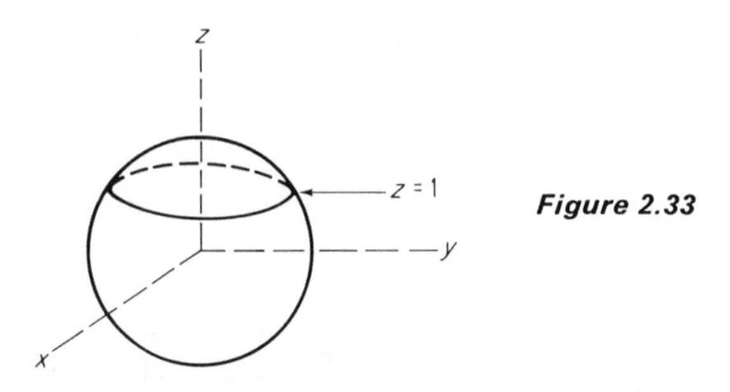

Figure 2.33

Example 9 Suppose we want to find the minimum distance between the circle $x^2 + y^2 = 1$ and the line $x + y = 4$ (Fig. 2.34). Given a point (x, y) on the circle and a point (u, v) in the line, the square of the distance between them is

$$f(x, y, u, v) = (x - u)^2 + (y - v)^2.$$

So we want to minimize f subject to the "constraints" $x^2 + y^2 = 1$ and $u + v = 4$. That is, we want to minimize the function $f : \mathscr{R}^4 \to \mathscr{R}$ on the 2-manifold M in $\mathscr{R}^4$ defined by the equations

$$g_1(x, y, u, v) = x^2 + y^2 - 1 = 0$$

and

$$g_2(x, y, u, v) = u + v - 4 = 0.$$

Note that the gradient vectors $\nabla g_1 = (2x, 2y, 0, 0)$ and $\nabla g_2 = (0, 0, 1, 1)$ are never collinear, so Theorem 5.7 implies that $M = g^{-1}(\mathbf{0})$ *is* a 2-manifold. Since

$$\nabla f = (2(x - u), 2(y - v), -2(x - u), -2(y - v)),$$

Theorem 5.8 directs us to solve the equations

$$\begin{aligned} x^2 + y^2 &= 1, & u + v &= 4, \\ 2(x - u) &= 2\lambda_1 x, & -2(x - u) &= \lambda_2, \\ 2(y - v) &= 2\lambda_1 y, & -2(y - v) &= \lambda_2. \end{aligned}$$

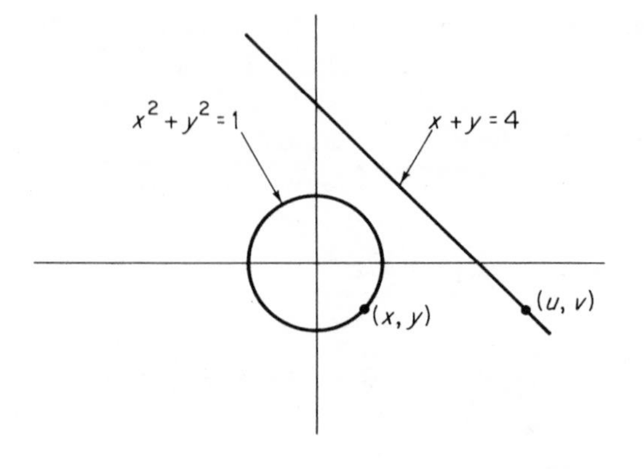

Figure 2.34

From $-2(x-u)=\lambda_2=-2(y-v)$, we see that

$$x-u=y-v.$$

If λ_1 were 0, we would have $(x, y)=(u, v)$ from $2(x-u)=2\lambda_1 x$ and $2(y-v)=2\lambda_1 y$. But the circle and the line have no point in common, so we conclude that $\lambda_1 \neq 0$. Therefore

$$x=\frac{x-u}{\lambda_1}=\frac{y-v}{\lambda_1}=y,$$

so finally $u=v$. Substituting $x=y$ and $u=v$ into $x^2+y^2=1$ and $u+v=4$, we obtain $x=y=\pm 1/\sqrt{2}$, $u=v=2$. Consequently, the closest points on the circle and line are $(1/\sqrt{2}, 1/\sqrt{2})$ and $(2, 2)$.

Example 10 Let us generalize the preceding example. Suppose M and N are two manifolds in $\mathscr{R}^n$, defined by $g(\mathbf{x})=0$ and $h(\mathbf{x})=0$, where

$$g:\mathscr{R}^n \to \mathscr{R}^m \qquad \text{and} \qquad h:\mathscr{R}^n \to \mathscr{R}^k$$

are mappings satisfying the hypotheses of Theorem 5.7. Let $\mathbf{p} \in M$ and $\mathbf{q} \in N$ be two points which are closer together than any other pair of points of M and N.

If $\mathbf{x}=(x_1, \ldots, x_n)$ and $\mathbf{y}=(y_1, \ldots, y_n)$ are any two points of M and N respectively, the square of the distance between them is

$$f(\mathbf{x}, \mathbf{y})=\sum_{i=1}^{n}(x_i-y_i)^2.$$

So to find the points $\mathbf{p}$ and $\mathbf{q}$, we need to minimize the function $f:\mathscr{R}^{2n} \to \mathscr{R}$ on the manifold in $\mathscr{R}^{2n}=\mathscr{R}^n \times \mathscr{R}^n$ defined by the equation $G(\mathbf{x}, \mathbf{y})=0$, where

$$G(\mathbf{x}, \mathbf{y})=(g(\mathbf{x}), h(\mathbf{y})) \in \mathscr{R}^{m+k}=\mathscr{R}^m \times \mathscr{R}^k.$$

That is, $G:\mathscr{R}^{2n} \to \mathscr{R}^{m+k}$ is defined by

$$G(\mathbf{x}, \mathbf{y})=(g_1(\mathbf{x}), \ldots, g_m(\mathbf{x}), h_1(\mathbf{y}), \ldots, h_k(\mathbf{y})) \qquad \text{for} \quad (\mathbf{x}, \mathbf{y}) \in \mathscr{R}^{2n}.$$

Theorem 5.8 implies that $\nabla f=\lambda_1 \nabla G_1+\cdots+\lambda_{m+k}\nabla G_{m+k}$ at $(\mathbf{p}, \mathbf{q})$. Since

$$\begin{aligned}
\nabla f(\mathbf{x}, \mathbf{y}) &= (\mathbf{x}-\mathbf{y}, \mathbf{y}-\mathbf{x}) \in \mathscr{R}^{2n},\\
\nabla G_1(\mathbf{x}, \mathbf{y}) &= (\nabla g_1(\mathbf{x}), \mathbf{0}) \in \mathscr{R}^{2n},\\
&\vdots\\
\nabla G_m(\mathbf{x}, \mathbf{y}) &= (\nabla g_m(\mathbf{x}), \mathbf{0}) \in \mathscr{R}^{2n},\\
\nabla G_{m+1}(\mathbf{x}, \mathbf{y}) &= (\mathbf{0}, \nabla h_1(\mathbf{y})) \in \mathscr{R}^{2n},\\
&\vdots\\
\nabla G_{m+k}(\mathbf{x}, \mathbf{y}) &= (\mathbf{0}, \nabla h_k(\mathbf{y})) \in \mathscr{R}^{2n},
\end{aligned}$$

we conclude that the solution satisfies

$$\mathbf{x}-\mathbf{y}=\sum_{i=1}^{m}\lambda_i \nabla g_i(\mathbf{x})=-\sum_{j=1}^{k}\lambda_{m+j}\nabla h_j(\mathbf{y}).$$

Since $(\mathbf{p}, \mathbf{q})$ is assumed to be the solution, we conclude that *the line joining* $\mathbf{p}$ *and* $\mathbf{q}$ *is both orthogonal to* M *at* $\mathbf{p}$ *and orthogonal to* N *at* $\mathbf{q}$.

Let us apply this fact to find the points on the unit sphere $x^2 + y^2 + z^2 = 1$ and the plane $u + v + w = 3$ which are closest. The vector (x, y, z) is orthogonal to the sphere at (x, y, z), and $(1, 1, 1)$ is orthogonal to the plane at (u, v, w). So the vector $(x - u, y - v, z - w)$ from (x, y, z) to $(u, v \ w)$ must be a multiple of both (x, y, z) and $(1, 1, 1)$:

$$x - u = k = lx, \qquad y - v = k = ly, \qquad z - w = k = lz.$$

Hence $x = y = z$ and $u = v = w$. Consequently the points $(1/\sqrt{3}. \ 1/\sqrt{3}, 1/\sqrt{3})$ and $(1, 1, 1)$ are the closest points on the sphere and plane, respectively.

Exercises

5.1 Complete the proof that the torus in Example 3 is a 2-manifold.

5.2 If M is a k-manifold in $\mathscr{R}^n$, and $\mathscr{R}^n \subset \mathscr{R}^p$, show that M is a k-manifold in $\mathscr{R}^p$.

5.3 If M is a k-manifold in $\mathscr{R}^m$ and N is an l-manifold in $\mathscr{R}^n$, show that $M \times N$ is a $(k + l)$-manifold in $\mathscr{R}^{m+n} = \mathscr{R}^m \times \mathscr{R}^n$.

5.4 Find the points of the ellipse $x^2/9 + y^2/4 = 1$ which are closest to and farthest from the point $(1, 1)$.

5.5 Find the maximal volume of a closed rectangular box whose total surface area is 54.

5.6 Find the dimensions of a box of maximal volume which can be inscribed in the ellipsoid

$$x^2/a^2 + y^2/b^2 + z^2/c^2 = 1.$$

(*Answer:* maximum volume $= 8abc/3\sqrt{3}$.)

5.7 Let the manifold S in $\mathscr{R}^n$ be defined by $g(\mathbf{x}) = 0$. If $\mathbf{p}$ is a point not on S, and $\mathbf{q}$ is the point of S which is closest to $\mathbf{p}$, show that the line from $\mathbf{p}$ to $\mathbf{q}$ is perpendicular to S at $\mathbf{q}$. *Hint:* Minimize $f(\mathbf{x}) = |\mathbf{x} - \mathbf{p}|^2$ on S.

5.8 Show that the maximum value of the function $f(\mathbf{x}) = x_1{}^2 x_2{}^2 \cdots x_n{}^2$ on the sphere $S^{n-1} = \{\mathbf{x} \in \mathscr{R}^n : |\mathbf{x}| = 1\}$ is $(1/n)^n$. That is $(x_1{}^2 \cdots x_n{}^2)^{1/n} \leq 1/n$ if $\mathbf{x} \in S^{n-1}$.
Given n positive numbers $a_1, \ldots, a_n$, define

$$x_i = \frac{a_i{}^{1/2}}{(a_1 + \cdots + a_n)^{1/2}}, \qquad i = 1, \ldots, n.$$

Then $x_1{}^2 + \cdots + x_n{}^2 = 1$, so

$$\left(\frac{a_1 \cdots a_n}{(a_1 + \cdots + a_n)^n}\right)^{1/n} \leq \frac{1}{n} \qquad \text{or} \qquad (a_1 \cdots a_n)^{1/n} \leq \frac{1}{n}(a_1 + \cdots + a_n).$$

Thus the *geometric mean* of n positive numbers is no greater than their *arithmetic mean.*

5.9 Find the minimum value of $f(\mathbf{x}) = n^{-1}(x_1 + \cdots + x_n)$ on the surface $g(\mathbf{x}) = x_1 x_2 \cdots x_n - 1 = 0$. Deduce again the geometric–arithmetic means inequality.

5.10 The planes $x + 2y + z = 4$ and $3x + y + 2z = 3$ intersect in a straight line L. Find the point of L which is closest to the origin.

5.11 Find the highest and lowest points on the ellipse of intersection of the cylinder $x^2 + y^2 = 1$ and the plane $x + y + z = 1$.

5.12 Find the points of the line $x + y = 10$ and the ellipse $x^2 + 2y^2 = 1$ which are closest.

5.13 Find the points of the circle $x^2 + y^2 = 1$ and the parabola $y^2 = 2(4 - x)$ which are closest.

5.14 Find the points of the ellipsoid $x^2 + 2y^2 + 3z^2 = 1$ which are closest to and farthest from the plane $x + y + z = 10$.
5.15 Generalize the proof of Theorem 5.3 so as to prove Theorem 5.6.
5.16 Verify the last assertion of Theorem 5.7.

6 TAYLOR'S FORMULA FOR SINGLE-VARIABLE FUNCTIONS

In order to generalize the results of Section 4, and in particular to apply the Lagrange multiplier method to classify critical points for functions of n variables, we will need Taylor's formula for functions on $\mathscr{R}^n$. As preparation for the treatment in Section 7 of the multivariable Taylor's formula, this section is devoted to the single-variable Taylor's formula.

Taylor's formula provides polynomial approximations to general functions. We will give examples to illustrate both the practical utility and the theoretical applications of such approximations.

If $f: \mathscr{R} \to \mathscr{R}$ is differentiable at a, and $R(h)$ is defined by

$$f(a+h) = f(a) + f'(a)h + R(h), \tag{1}$$

then it follows immediately from the definition of $f'(a)$ that

$$\lim_{h \to 0} \frac{R(h)}{h} = 0. \tag{2}$$

With $x = a + h$, (1) and (2) become

$$f(x) = f(a) + f'(a)(x-a) + R(x-a), \tag{1'}$$

where

$$\lim_{x \to a} \frac{R(x-a)}{x-a} = 0. \tag{2'}$$

The linear function $P(x-a) = f(a) + f'(a)(x-a)$ is simply that first degree polynomial in $(x-a)$ whose value and first derivative at a agree with those of f at a. The kth degree polynomial in $(x-a)$, such that the values of it and of its first k derivatives at a agree with those of f and its first k derivatives f', f'', $f^{(3)}, \ldots, f^{(k)}$ at a, is

$$P_k(x-a) = f(a) + f'(a)(x-a) + \frac{f''(a)}{2!}(x-a)^2 + \cdots + \frac{f^{(k)}(a)}{k!}(x-a)^k. \tag{3}$$

This fact may be easily checked by repeated differentiation of $P_k(x-a)$. The polynomial $P_k(x-a)$ is called the kth *degree Taylor polynomial of f at a*.

The *remainder* $f(x) - P_k(x-a)$ is denoted by $R_k(x-a)$, so

$$f(x) = P_k(x-a) + R_k(x-a). \tag{4}$$

With $x - a = h$, this becomes

$$f(a + h) = P_k(h) + R_k(h), \tag{4'}$$

where

$$P_k(h) = f(a) + f'(a)h + \frac{f''(a)}{2!}h^2 + \cdots + \frac{f^{(k)}(a)}{k!}h^k. \tag{3'}$$

In order to make effective use of Taylor polynomials, we need an explicit formula for $R_k(x - a)$ which will provide information as to how closely $P_k(x - a)$ approximates $f(x)$ near a. For example, whenever we can show that

$$\lim_{k \to \infty} R_k(x - a) = 0,$$

this will mean that f is arbitrarily closely approximated by its Taylor polynomials; they can then be used to calculate $f(x)$ as closely as desired. Equation (4), or (4′), together with such an explicit expression for the remainder R_k, is referred to as *Taylor's formula*. The formula for R_k given in Theorem 6.1 below is known as the *Lagrange form* of the remainder.

Theorem 6.1 Suppose that the $(k + 1)$th derivative $f^{(k+1)}$ of $f : \mathscr{R} \to \mathscr{R}$ exists at each point of the closed interval I with endpoints a and x. Then there exists a point ζ between a and x such that

$$R_k(x - a) = \frac{f^{(k+1)}(\zeta)}{(k + 1)}(x - a)^{k+1}. \tag{5}$$

Hence

$$f(x) = f(a) + f'(a)(x - a) + \cdots + \frac{f^{(k)}(a)}{k!}(x - a)^k + \frac{f^{(k+1)}(\zeta)}{(k + 1)!}(x - a)^{k+1}$$

or

$$f(a + h) = f(a) + f'(a)h + \cdots + \frac{f^{(k)}(a)}{k!}h^k + \frac{f^{(k+1)}(\zeta)}{(k + 1)!}h^{k+1}$$

with $h = x - a$.

REMARK This is a generalization of the mean value theorem; in particular, $P_0(x - a) = f(a)$, so the case $k = 0$ of the theorem is simply the mean value theorem

$$f(a + h) = f(a) + f'(\zeta)h$$

for the function f on the interval I. Moreover the proof which we shall give for Taylor's formula is a direct generalization of the proof of the mean value theorem. So for motivation we review the proof of the mean value theorem (slightly rephrased).

First we define $R_0(t)$ for $t \in [0, h]$ (for convenience we assume $h > 0$) by

$$R_0(t) = f(a + t) - f(a) = f(a + t) - P_0(t),$$

and note that

$$R_0(0) = 0 \tag{6}$$

while

$$R_0'(t) = f'(a + t). \tag{7}$$

Then we define $\varphi : [0, h] \to \mathscr{R}$ by

$$\varphi(t) = R_0(t) - Kt, \tag{8}$$

where the constant K is chosen so that Rolle's theorem [the familiar fact that, if f is a differentiable function on $[a, b]$ with $f(a) = f(b) = 0$, then there exists a point $\xi \in (a, b)$ where $f'(\xi) = 0$] will apply to φ on $[0, h]$, that is,

$$K = \frac{R_0(h)}{h}, \tag{9}$$

so it follows that $\varphi(h) = 0$. Hence Rolle's theorem gives a $\bar{t} \in (0, h)$ such that

$$\begin{aligned} 0 = \varphi'(\bar{t}) &= R_0'(\bar{t}) - K && \text{by (8)} \\ &= f'(a + \bar{t}) - K && \text{by (7)}. \end{aligned}$$

Hence $K = f'(\zeta)$ where $\zeta = a + \bar{t}$, so from (9) we obtain $R_0(h) = f'(\zeta)h$ as desired.

PROOF OF THEOREM 6.1 We generalize the above proof, labeling the formulas with the same numbers (primed) to facilitate comparison.

First we define $R_k(t)$ for $t \in [0, h]$ by

$$R_k(t) = f(a + t) - P_k(t),$$

and note that

$$R_k(0) = R_k'(0) = R_k''(0) = \cdots = R_k^{(k)}(0) = 0 \tag{6'}$$

while

$$R_k^{(k+1)}(t) = f^{(k+1)}(a + t). \tag{7'}$$

The reason for (6′) is that the first k derivatives of $P_k(x - a)$ at a, and hence the first k derivatives of $P_k(t)$ at 0, agree with those of f at a, while (7′) follows from the fact that $P_k^{(k+1)}(t) \equiv 0$ because $P_k(t)$ is a polynomial of degree k.

Now we define $\varphi : [0, h] \to \mathscr{R}$ by

$$\varphi(t) = R_k(t) - Kt^{k+1}, \tag{8'}$$

where the constant K is chosen so that Rolle's theorem will apply to φ on $[0, h]$, that is,

$$K = \frac{R_k(h)}{h^{k+1}}, \tag{9'}$$

so it follows that $\varphi(h) = 0$. Hence Rolle's theorem gives a point $\bar{t}_1 \in (0, h)$ such that $\varphi'(\bar{t}_1) = 0$.

It follows from (6′) and (7′) that

$$\left.\begin{aligned} &\varphi(0) = \varphi'(0) = \varphi''(0) = \cdots = \varphi^{(k)}(0) = 0, \\ &\text{while} \\ &\varphi^{(k+1)}(t) = f^{(k+1)}(a + t) - K(k + 1)!. \end{aligned}\right\} \tag{10}$$

Therefore we can apply Rolle's theorem to φ' on the interval $[0, \bar{t}_1]$ to obtain a point $\bar{t}_2 \in (0, \bar{t}_1)$ such that $\varphi''(\bar{t}_2) = 0$.

$0 \quad \bar{t}_{k+1} \quad \cdots \quad \bar{t}_3 \quad \bar{t}_2 \quad \bar{t}_1 \quad h$

By (10), φ'' satisfies the hypotheses of Rolle's theorem on $[0, \bar{t}_2]$, so we can continue in this way. After $k + 1$ applications of Rolle's theorem, we finally obtain a point $\bar{t}_{k+1} \in (0, h)$ such that $\varphi^{(k+1)}(t_{k+1}) = 0$. From the second equation in (10) we then obtain

$$K = \frac{f^{(k+1)}(\zeta)}{(k + 1)!}$$

with $\zeta = a + \bar{t}_{k+1}$. Finally (9′) gives

$$R_k(h) = \frac{f^{(k+1)}(\zeta)}{(k + 1)!} h^{k+1}$$

as desired. ∎

Corollary 6.2 If, in addition to the hypotheses of Theorem 6.1, $|f^{(k+1)}(\zeta)| \leqq M$ for every $\zeta \in I$, then

$$|R_k(x - a)| \leqq \frac{M}{(k + 1)!} |x - a|^{k+1}. \tag{11}$$

It follows that

$$\lim_{h \to 0} \frac{R_k(h)}{h^k} = \lim_{x \to a} \frac{R_k(x - a)}{(x - a)^k} = 0. \tag{12}$$

In particular, (12) holds if $f^{(k+1)}$ is continuous at a, because it will then necessarily be bounded (by some M) on some open interval containing a.

Example 1 As a standard first example, we take $f(x) = e^x$, $a = 0$. Then $f^{(k)}(x) = e^x$, so $f^{(k)}(0) = 1$ for all k. Then

$$P_k(x) = 1 + x + \frac{x^2}{2!} + \cdots + \frac{x^k}{k!}$$

and

$$R_k(x) = \frac{e^{\zeta} x^{k+1}}{(k + 1)!}$$

for some ζ between 0 and x. Therefore

$$\left.\begin{aligned} 0 < R_k(x) &< \frac{e^x x^{k+1}}{(k+1)!} \qquad \text{if} \quad 0 < \zeta < x, \\ \text{while} \qquad & \\ 0 < |R_k(x)| &< \frac{|x|^{k+1}}{(k+1)!} \qquad \text{if} \quad x < \zeta < 0. \end{aligned}\right\} \tag{13}$$

In either case the elementary fact that

$$\lim_{k\to\infty} \frac{x^k}{k!} = 0 \qquad \text{for all} \quad x \in \mathscr{R}$$

implies that $\lim_{k\to\infty} R_k(x) = 0$ for all x, so

$$\begin{aligned} e^x &= P_k(x) + R_k(x) \qquad\qquad \text{(for all} \quad k) \\ &= \lim_{k\to\infty} [P_k(x) + R_k(x)] \\ &= \lim_{k\to\infty} P_k(x) + \lim_{k\to\infty} R_k(x) \\ &= \lim_{k\to\infty} \sum_{n=0}^{k} \frac{x^n}{n!}, \\ e^x &= \sum_{n=0}^{\infty} \frac{x^n}{n!} \qquad\qquad \text{for all} \quad x. \end{aligned}$$

[To verify the elementary fact used above, choose a fixed integer m such that $|x|/m < \frac{1}{2}$. If $k > m$, then

$$\begin{aligned} \frac{|x|^k}{k!} &= \frac{|x|^m}{m!} \cdot \frac{|x|}{m+1} \cdot \frac{|x|}{m+2} \cdot \cdots \cdot \frac{|x|}{k} \\ &< \frac{1}{2^{k-m}} \frac{|x|^m}{m!} \to 0 \end{aligned}$$

as $k \to \infty$.]

In order to calculate the value of e^x with preassigned accuracy by simply calculating $P_k(x)$, we must be able to estimate the error $R_k(x)$. For this we need the preliminary estimate $e < 4$. Since $\log e = 1$ and $\log x$ is a strictly increasing function, to verify that $e < 4$ it suffices to show that $\log 4 > 1$. But

$$\begin{aligned} \log 4 &= \int_1^4 \frac{dt}{t} = \int_1^2 \frac{dt}{t} + \int_2^3 \frac{dt}{t} + \int_3^4 \frac{dt}{t} \\ &\geqq \int_1^2 \frac{1}{2}\,dt + \int_2^3 \frac{1}{3}\,dt + \int_3^4 \frac{1}{4}\,dt \\ &= \frac{1}{2} + \frac{1}{3} + \frac{1}{4} > 1. \end{aligned}$$

From (13) we now see that $R_k(x) < 4/(k+1)!$ if $x \in [0, 1]$; this can be used to compute e to any desired accuracy (see Exercise 6.1).

Example 2 To calculate $\sqrt{2}$, we take $f(x) = \sqrt{x}$, $a = 1.96$, $h = 0.04$, and consider the first degree Taylor formula

$$f(a+h) = f(a) + f'(a)h + R_1(h),$$

where $R_1(h) = f''(\zeta)h^2/2$ for some $\zeta \in (a, a+h)$. Since $f'(x) = \frac{1}{2}x^{-1/2}$, $f''(x) = -\frac{1}{4}x^{-3/2}$, we see that $R_1(h) < 0$ and

$$\begin{aligned}\sqrt{2} &= f(1.96 + 0.04) = (1.96)^{1/2} + \frac{(0.04)}{2(1.96)^{1/2}} + R_1(0.04) \\ &= 1.4 + \frac{0.04}{2.8} + R_1(0.04) = 1.4143 + R_1(0.04).\end{aligned}$$

Since $|R_1(0.04)| = \frac{1}{2}|f''(\zeta)|(0.04)^2$ with $\zeta > 1.96$,

$$|R_1(0.04)| < \frac{1}{2} \times \frac{1}{4} \times \frac{1}{(1.4)^3}(0.04)^2 < 0.0001,$$

so we conclude that

$$1.4142 < \sqrt{2} < 1.4143$$

(actually $\sqrt{2} = 1.41421$ to five places).

The next two examples give a good indication of the wide range of application of Taylor's formula.

Example 3 We show that the number e is irrational. To the contrary, suppose that $e = p/q$ where p and q are positive integers. Since $e = 2.718$ to three decimal places (see Exercise 6.1), it is clear that e is not an integral multiple of 1, $\frac{1}{2}$, or $\frac{1}{3}$, so $q > 3$. By Example 1, we can write

$$\frac{p}{q} = e = 1 + 1 + \frac{1}{2!} + \cdots + \frac{1}{q!} + R_q,$$

where

$$0 < R_q = \frac{e^{\zeta}}{(q+1)!} < \frac{e}{(q+1)!} < \frac{3}{(q+1)!}$$

since $0 < \zeta < 1$ and $e < 3$. Upon multiplication of both sides of the above equation by $q!$, we obtain

$$(q-1)!\,p = 2(q!) + q(q-1)\cdot\cdots\cdot(4)(3) + \cdots + 1 + q!\,R_q.$$

But this is a contradiction, because the left-hand side $(q-1)!\,p$ is an integer, but the right-hand side is *not*, because

$$0 < q!\,R_q < q!\frac{3}{(q+1)!} = \frac{3}{q+1} < 1$$

since $q > 3$.

Example 4 We use Taylor's formula to prove that, *if* $f'' - f = 0$ *on* R *and* $f(0) = f'(0) = 0$, *then* $f = 0$ on $\mathscr{R}$.

Since $f'' = f$, we see by repeated differentiation that $f^{(k)}$ exists for all k; in particular,

$$f^{(k)} = \begin{cases} f & \text{if } k \text{ is even,} \\ f' & \text{if } k \text{ is odd.} \end{cases}$$

Since $f(0) = f'(0) = 0$, it follows that $f^{(k)}(0) = 0$ for all k. Consequently Theorem 4.1 gives, for each k, a point $\zeta_k \in (0, x)$ such that

$$f(x) = R_k(x) = \frac{f^{(k+1)}(\zeta_k)x^{k+1}}{(k+1)!}.$$

Since there are really only two different derivatives involved, and each is continuous because it is differentiable, there exists a constant M such that

$$|f^{(k+1)}(t)| \leqq M \qquad \text{for} \quad t \in [0, x] \text{ and all } k.$$

Hence $|f(x)| \leqq M|x|^{k+1}/(k+1)!$ *for all* k. But $\lim_{k \to \infty} x^{k+1}/(k+1)! = 0$, so we conclude that $f(x) = 0$.

Now we apply Taylor's formula to give sufficient conditions for local maxima and minima of real-valued single-variable functions.

Theorem 6.3 Suppose that $f^{(k+1)}$ exists in a neighborhood of a and is continuous at a. Suppose also that

$$f'(a) = f''(a) = \cdots = f^{(k-1)}(a) = 0,$$

but $f^{(k)}(a) \neq 0$. Then

(a) f has a local minimum at a if k is even, and $f^{(k)}(a) > 0$;
(b) f has a local maximum at a if k is even and $f^{(k)}(a) < 0$;
(c) f has neither a maximum nor a minimum at a if k is odd.

This is a generalization of the familiar "second derivative test" which asserts that, if $f'(a) = 0$, then f has a local minimum at a if $f''(a) > 0$, and a local maximum at a if $f''(a) < 0$. The three cases can be remembered by thinking of the three graphs in Fig. 2.35.

If $f^{(k)}(a) = 0$ *for all* k, then Theorem 6.3 provides no information as to the behavior of f in a neighborhood of a. For instance, if

$$f(x) = \begin{cases} e^{-1/x^2} & \text{if } x \neq 0, \\ 0 & \text{if } x = 0, \end{cases}$$

then it turns out that $f^{(k)}(0) = 0$ for all k, so Theorem 6.3 does not apply. However it is obvious that f has a local minimum at 0, since $f(x) > 0$ for $x \neq 0$ (Fig. 2.36).

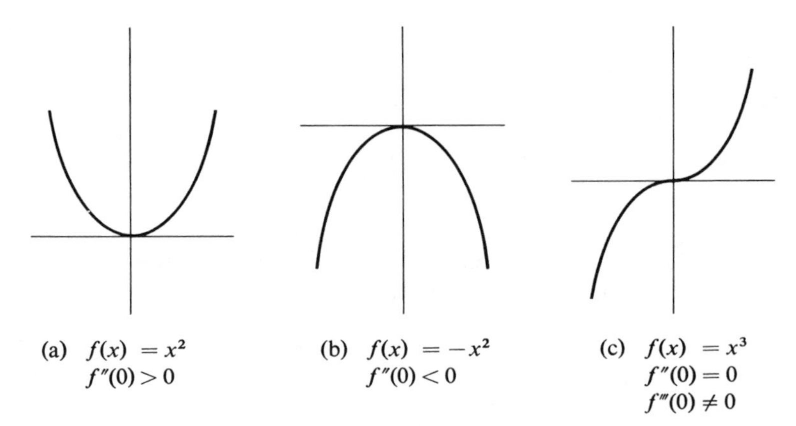

Figure 2.35

As motivation for the proof of Theorem 6.3, let us consider first the "second-derivative test." If $f'(a) = 0$, then Taylor's formula with $k = 2$ is

$$f(x) = f(a) + \tfrac{1}{2}f''(a)(x-a)^2 + R_2(x-a),$$

where $\lim_{x\to a} R_2(x-a)/(x-a)^2 = 0$ by Corollary 6.2 (assuming that $f^{(3)}$ is continuous at a). By transposing $f(a)$ and dividing by $(x-a)^2$, we obtain

$$\frac{f(x) - f(a)}{(x-a)^2} = \frac{1}{2}f''(a) + \frac{R_2(x-a)}{(x-a)^2},$$

so it follows that

$$\lim_{x\to a} \frac{f(x) - f(a)}{(x-a)^2} = \frac{1}{2}f''(a).$$

If $f''(a) > 0$, this implies that $f(x) - f(a) > 0$ if x is sufficiently close to a, since $(x-a)^2 > 0$ for all $x \neq a$. Thus $f(a)$ is a local minimum. Similarly $f(a)$ is a local maximum if $f''(a) < 0$.

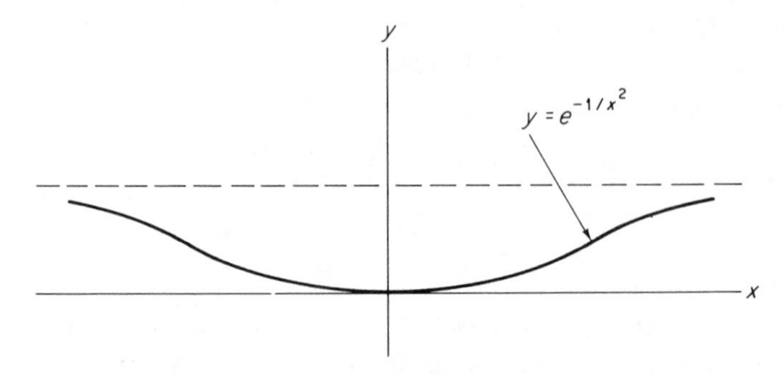

Figure 2.36

In similar fashion we can show that, if $f'(a) = f''(a) = 0$ while $f^{(3)}(a) \neq 0$, then f has neither a maximum nor a minimum at a (this fact might be called the "third-derivative test"). To see this, we look at Taylor's formula with $k = 3$,

$$f(x) = f(a) + \tfrac{1}{6} f^{(3)}(a)(x-a)^3 + R_3(x-a),$$

where $\lim_{x \to a} R_3(x-a)/(x-a)^3 = 0$. Transposing $f(a)$ and then dividing by $(x-a)^3$, we obtain

$$\frac{f(x) - f(a)}{(x-a)^3} = \frac{1}{6} f^{(3)}(a) + \frac{R_3(x-a)}{(x-a)^3},$$

so it follows that

$$\lim_{x \to a} \frac{f(x) - f(a)}{(x-a)^3} = \frac{1}{6} f^{(3)}(a).$$

If, for instance, $f^{(3)}(a) > 0$, we see that $[f(x) - f(a)]/(x-a)^3 > 0$ if x is sufficiently close to a. Since $(x-a)^3 > 0$ if $x > a$ and $(x-a)^3 < 0$ if $x < a$, it follows that, for x sufficiently close to a, $f(x) - f(a) > 0$ if $x > a$, and $f(x) - f(a) < 0$ if $x < a$. These inequalities are reversed if $f^{(3)}(a) < 0$. Consequently $f(a)$ is neither a local maximum nor a local minimum.

The proof of Theorem 6.3 simply consists of replacing 2 and 3 in the above discussion by k, the order of the first nonzero derivative of f at the critical point a. If k is even the argument is the same as when $k = 2$, while if k is odd it is the same as when $k = 3$.

PROOF OF THEOREM 6.3 Because of the hypotheses, Taylor's formula takes the form

$$f(x) = f(a) + \frac{f^{(k)}(a)}{k!}(x-a)^k + R_k(x-a),$$

where $\lim_{x \to a} R_k(x-a)/(x-a)^k = 0$ by Corollary 6.2. If we transpose $f(a)$, divide by $(x-a)^k$, and then take limits as $x \to a$, we therefore obtain

$$\begin{aligned} \lim_{x \to a} \frac{f(x) - f(a)}{(x-a)^k} &= \lim_{x \to a} \left(\frac{f^{(k)}(a)}{k!} + \frac{R_k(x-a)}{(x-a)^k} \right) \\ &= \frac{f^{(k)}(a)}{k!}. \end{aligned} \tag{14}$$

In case (a), $\lim_{x \to a} [f(x) - f(a)]/(x-a)^k > 0$ by (14), so it follows that there exists a $\delta > 0$ such that

$$0 < |x - a| < \delta \Rightarrow \frac{f(x) - f(a)}{(x-a)^k} > 0.$$

Since k is *even* in this case, $(x-a)^k > 0$ whether $x > a$ or $x < a$, so

$$0 < |x-a| < \delta \Rightarrow f(x) - f(a) > 0, \qquad \text{or} \qquad f(x) > f(a).$$

Therefore $f(a)$ is a local minimum.

The proof in case (b) is the same except for reversal of the inequalities.

In case (c), supposing $f^{(k)}(a) > 0$, there exists (just as above) a $\delta > 0$ such that

$$0 < |x-a| < \delta \Rightarrow \frac{f(x)-f(a)}{(x-a)^k} > 0,$$

But now, since k is *odd*, the sign of $(x-a)^k$ depends upon whether $x < a$ or $x > a$. The same is then true of $f(x) - f(a)$, so $f(x) < f(a)$ if $x > a$, and $f(x) > f(a)$ if $x > a$; the situation is reversed if $f^{(k)}(a) < 0$. In either event it is clear that $f(a)$ is neither a local maximum nor a local minimum. ∎

Let us look at the case $k = 2$ of Theorem 6.3 in a bit more detail. We have

$$f(x) = f(a) + \tfrac{1}{2}f''(a)(x-a)^2 + R_2(x-a), \tag{15}$$

where $\lim_{x \to a} R_2(x-a)/(x-a)^2 = 0$. Therefore, given $\varepsilon < \frac{1}{2}|f''(a)|$, there exists a $\delta > 0$ such that

$$0 < |x-a| < \delta \Rightarrow \frac{|R_2(x-a)|}{(x-a)^2} < \varepsilon,$$

which implies that

$$-\varepsilon(x-a)^2 < R_2(x-a) < +\varepsilon(x-a)^2. \tag{16}$$

Substituting (16) into (15), we obtain

$$f(a) + [\tfrac{1}{2}f''(a) - \varepsilon](x-a)^2 < f(x) < f(a) + [\tfrac{1}{2}f''(a) + \varepsilon](x-a)^2 \tag{17}$$

If $f''(a) > 0$, then $\frac{1}{2}f''(a) - \varepsilon$ and $\frac{1}{2}f''(a) + \varepsilon$ are both positive because $\varepsilon < \frac{1}{2}|f''(a)|$. It follows that the graphs of the equations

$$y = f(a) + [\tfrac{1}{2}f''(a) - \varepsilon](x-a)^2 \qquad \text{and} \qquad y = f(a) + [\tfrac{1}{2}f''(a) + \varepsilon](x-a)^2$$

are then parabolas opening upwards with vertex $(a, f(a))$ (see Fig. 2.37). The fact that inequality (17) holds if $0 < |x-a| < \delta$ means that the part of the graph of $y = f(x)$, over the interval $(a-\delta, a+\delta)$, lies between these two parabolas. This makes it clear that f has a local minimum at a if $f''(a) > 0$.

The situation is similar if $f''(a) < 0$, except that the parabolas open downward, so f has a local maximum at a.

In the case $k = 3$, these parabolas are replaced by the cubic curves

$$y = f(a) + [\tfrac{1}{6}f'''(a) \pm \varepsilon](x-a)^3,$$

which look like Fig. 2.35c above, so f has neither a maximum nor a minimum at a.

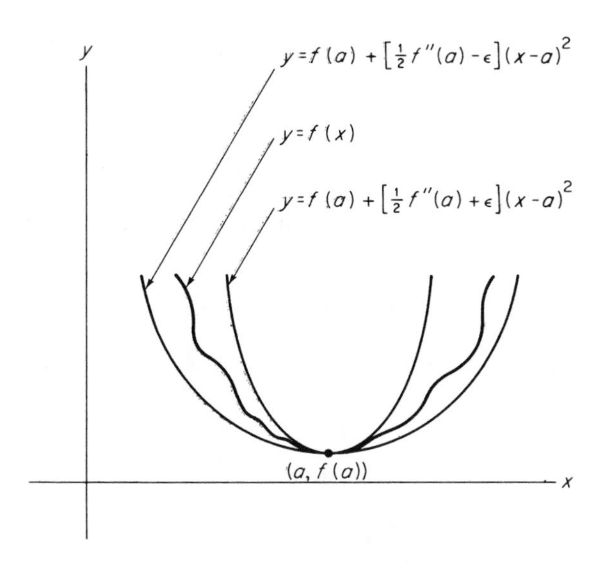

Figure 2.37

Exercises

6.1 Show that $e = 2.718$, accurate to three decimal places. *Hint:* Refer to the error estimate at the end of Example 1; choose k such that $4/(k+1)! < 10^{-4}$.

6.2 Prove that, if we compute e^x by the approximation

$$e^x \approx 1 + x + \frac{x^2}{2} + \frac{x^3}{6} + \frac{x^4}{24},$$

then the error will not exceed 0.001 if $x \in [0, \frac{1}{2}]$. Then compute $\sqrt[3]{e}$ accurate to two decimal places.

6.3 If $Q(x) = \sum_{n=0}^{k} b_n(x-a)^n$, show that $Q^{(n)}(a) = n!\,b_n$ for $n \leqq k$. Conclude that

$$P_k(x-a) = \sum_{n=0}^{k} \frac{f^{(n)}(a)}{n!}(x-a)^n$$

is the only kth degree polynomial in $(x-a)$ such that the values of it and its first k derivatives at a agree with those of f at a.

6.4 (a) Show that the values of the sine function for angles between 40° and 50° can be computed by means of the approximation

$$\sin\left(\frac{\pi}{4} + h\right) \approx \frac{\sqrt{2}}{2}\left(1 + h - \frac{h^2}{2} - \frac{h^3}{6}\right)$$

with 4-place accuracy. *Hint:* With $f(x) = \sin x$, $a = \pi/4$, $k = 3$, show that the error is less than 10^{-5}, since $5° = \pi/36 < 1/10$ rad.
(b) Compute sin 50°, accurate to four decimal places.

6.5 Show that

$$\sin x = \sum_{m=0}^{\infty} \frac{(-1)^m x^{2m+1}}{(2m+1)!} \qquad \text{and} \qquad \cos x = \sum_{m=0}^{\infty} \frac{(-1)^m x^{2m}}{(2m)!}$$

for all x.

6.6 Show that the kth degree Taylor polynomial of $f(x) = \log x$ at $a = 1$ is

$$P_k(x-1) = (x-1) - \frac{(x-1)^2}{2} + \cdots + \frac{(-1)^{k-1}(x-1)^k}{k}$$

and that $\lim_{k\to\infty} R_k(x-1) = 0$ *if* $x \in (1, 2)$. Then compute $\log \frac{3}{2}$ with error $< 10^{-3}$ *Hint:* Show by induction that $f^{(k)}(x) = (-1)^{k-1}(k-1)!/x^k$.

6.7 If $f''(x) = f(x)$ for all x, show that there exist constants a and b so that

$$f(x) = a\,e^x + b\,e^{-x} \qquad \text{for all} \quad x.$$

Hint: Let $g(x) = f(x) - a\,e^x - b\,e^{-x}$, show how to choose a and b so that $g(0) = g'(0) = 0$. Then apply Example 4.

6.8 If α is a fixed real number and n is a positive integer, show that the nth degree Taylor polynomial at $a = 0$ for

$$f(x) = (1+x)^\alpha$$

is $P_n(x) = \sum_{j=0}^{j=n} \binom{\alpha}{j} x^j$, where the "binomial coefficient" $\binom{\alpha}{j}$ is defined by

$$\binom{\alpha}{j} = \frac{\alpha(\alpha-1)\cdots(\alpha-j+1)}{j!}$$

(remember that $0! = 1$). If $\alpha = n$, then

$$\binom{n}{j} = \frac{n(n-1)\cdots(n-j+1)}{j!} = \frac{n!}{j!(n-j)!},$$

so it follows that

$$f(x) = (1+x)^n = \sum_{j=0}^{n} \frac{n!}{j!(n-j)!} x^j,$$

since $R_n(x) \equiv 0$, because $f^{(n+1)}(x) \equiv 0$.
If α is *not* an integer, then $\binom{\alpha}{j} \neq 0$ for all j, so the series $\sum_0^\infty \binom{\alpha}{j} x^j$ is infinite. The *binomial theorem* asserts that this infinite series converges to $f(x) = (1+x)^\alpha$ *if* $|x| < 1$, and can be proved by showing that $\lim_{n\to\infty} R_n(x) = 0$ for $|x| < 1$.

6.9 Locate the critical points of

$$f(x) = x^3(x-1)^4$$

and apply Theorem 6.3 to determine the character of each. *Hint:* Do not expand before differentiating.

6.10 Let $f(x) = x \tan^{-1} x - \sin^2 x$. Assuming the fact that the sixth degree Taylor polynomials at $a = 0$ of $\tan^{-1} x$ and $\sin^2 x$ are

$$x - \frac{x^3}{3} + \frac{x^5}{5} \qquad \text{and} \qquad x^2 - \frac{x^4}{3} + \frac{2}{45}x^6,$$

respectively, prove that

$$f(x) = \tfrac{7}{45}x^6 + R(x),$$

where $\lim_{x\to 0} R(x) = 0$. Deduce by the *proof* of Theorem 6.3 that f has a local minimum at 0.
Contemplate the tedium of computing the first six derivatives of f. If one could endure it, he would find that

$$f'(0) = f''(0) = f^{(3)}(0) = f^{(4)}(0) = f^{(5)}(0) = 0$$

but $f^{(6)}(0) = 112 > 0$, so the *statement* of Theorem 6.3(a) would then give the above result.

6.11 (a) This problem gives a form of "l'Hospital's rule." Suppose that f and g have $k+1$ continuous derivatives in a neighborhood of a, and that both f and g and their first $k-1$ derivatives vanish at a. If $g^{(k)}(a) \neq 0$, prove that

$$\lim_{x \to a} \frac{f(x)}{g(x)} = \frac{f^{(k)}(a)}{g^{(k)}(a)}.$$

Hint: Substitute the kth degree Taylor expansions of $f(x)$ and $g(x)$, then divide numerator and denominator by $(x-a)^k$ before taking the limit as $x \to a$.

(b) Apply (a) with $k = 2$ to evaluate

$$\lim_{x \to 0} \frac{(\sin x)^2}{e^{x^2} - 1}.$$

6.12 In order to determine the character of $f(x) = (e^{-x} - 1)(\tan^{-1}(x) - x)$ at the critical point 0, substitute the fourth degree Taylor expansions of e^{-x} and $\tan^{-1} x$ to show that

$$f(x) = \tfrac{1}{3}x^4 + R_4(x),$$

where $\lim_{x \to 0} R_4(x)/x^4 = 0$. What is your conclusion?

7 TAYLOR'S FORMULA IN SEVERAL VARIABLES

Before generalizing Taylor's formula to higher dimensions, we need to discuss higher-order partial derivatives. Let f be a real-valued function defined on an open subset U of $\mathscr{R}^n$. Then we say that f is of class $\mathscr{C}^k$ on U if all iterated partial derivatives of f, of order at most k, exist and are continuous on U. More precisely, this means that, given any sequence $i_1, i_2, \ldots, i_q$, where $q \leqq k$ and each i_j is one of the integers 1 through n, the iterated partial derivative

$$D_{i_1} D_{i_2} \cdots D_{i_q} f$$

exists and is continuous on U.

If this is the case, then it makes no difference in which order the partial derivatives are taken. That is, if $i_1', i_2', \ldots, i_q'$ is a permutation of the sequence $i_1, \ldots, i_q$ (meaning simply that each of the integers 1 through n occurs the same number of times in the two sequences), then

$$D_{i_1} D_{i_2} \cdots D_{i_q} f = D_{i_1'} D_{i_2'} \cdots D_{i_q'} f.$$

This fact follows by induction on q from Theorem 3.6, which is the case $q = 2$ of this result (see Exercise 7.1).

In particular, if for each $r = 1, 2, \ldots, n$, j_r is the number of times that r appears in the sequence $i_1, \ldots, i_q$, then

$$D_{i_1} D_{i_2} \cdots D_{i_q} f = D_1^{j_1} D_2^{j_2} \cdots D_n^{j_n} f.$$

If f is of class $\mathscr{C}^k$ on U, and $j_1 + \cdots + j_n < k$, then $D_1^{j_1} \cdots D_n^{j_n} f$ is differentiable on U by Theorem 2.5. Therefore, given $\mathbf{h} = (h_1, \ldots, h_n)$, its directional derivative with respect to $\mathbf{h}$ exists, with

$$\begin{aligned} D_{\mathbf{h}}\, D_1^{j_1} \cdots D_n^{j_n} f &= \sum_{r=1}^{n} h_r\, D_r\, D_1^{j_1} \cdots D_n^{j_n} f \\ &= \sum_{r=1}^{n} h_r D_1^{j_1} \cdots D_r^{j_r+1} \cdots D_n^{j_n} f \end{aligned}$$

by Theorem 2.2 and the above remarks. It follows that the iterated directional derivative $D_{\mathbf{h}}{}^k f = D_{\mathbf{h}} \cdots D_{\mathbf{h}} f$ exists, with

$$\begin{aligned} D_{\mathbf{h}}{}^k f &= (h_1 D_1 + \cdots + h_n D_n)^k f \\ &= \sum_{j_1 + \cdots + j_n = k} \begin{pmatrix} & k & \\ j_1 & \cdots & j_n \end{pmatrix} h_1^{j_1} \cdots h_n^{j_n} D_1^{j_1} \cdots D_n^{j_n} f, \end{aligned} \tag{1}$$

the summation being taken over all n-tuples $j_1, \ldots, j_n$ of nonnegative integers whose sum is k. Here the symbol

$$\begin{pmatrix} & k & \\ j_1 & \cdots & j_n \end{pmatrix}$$

denotes the "multinomial coefficient" appearing in the general multinomial formula

$$(x_1 + \cdots + x_n)^k = \sum_{j_1 + \cdots + j_n = k} \begin{pmatrix} & k & \\ j_1 & \cdots & j_n \end{pmatrix} x_1^{j_1} \cdots x_n^{j_n}.$$

Actually

$$\begin{pmatrix} & k & \\ j_1 & \cdots & j_n \end{pmatrix} = \frac{k!}{j_1! \cdots j_n!};$$

see Exercise 7.2.

For example, if $n = 2$ we have ordinary binomial coefficients and (1) gives

$$\begin{aligned} D_{\mathbf{h}}{}^k &= (h_1 D_1 + h_2 D_2)^k f \\ &= \sum_{j_1 + j_2 = k} \frac{k!}{j_1!\, j_2!} h_1^{j_1} h_2^{j_2} D_1^{j_1} D_2^{j_2} f \\ &= \sum_{j=1}^{k} \frac{k!}{j!\,(k-j)!} h_1{}^j h_2^{k-j} D_1{}^j D_2^{k-j} f. \end{aligned}$$

For instance, we obtain

$$\begin{aligned} D_{\mathbf{h}}{}^2 f = (h_1 D_1 + h_2 D_2)^2 f &= \left(h_1 \frac{\partial}{\partial x} + h_2 \frac{\partial}{\partial y}\right)^2 f \\ &= h_1{}^2 \frac{\partial^2 f}{\partial x^2} + 2h_1 h_2 \frac{\partial^2 f}{\partial x\, \partial y} + h_2{}^2 \frac{\partial^2 f}{\partial y^2} \end{aligned}$$

and

$$D_{\mathbf{h}}{}^3 f = \left(h_1 \frac{\partial}{\partial x} + h_2 \frac{\partial}{\partial y}\right)^3 f$$

$$= h_1{}^3 \frac{\partial^3 f}{\partial x^3} + 3h_1{}^2 h_2 \frac{\partial^2 f}{\partial x^2\, \partial y} + 3h_1 h_2{}^2 \frac{\partial^2 f}{\partial x\, \partial y^2} + h_2{}^3 \frac{\partial^3 f}{\partial y^3}$$

with $k = 2$ and $k = 3$, respectively.

These iterated directional derivatives are what we need to state the multidimensional Taylor's formula. To see why this should be, consider the 1-dimensional Taylor's formula in the form

$$f(a + h) = f(a) + f'(a)h + \frac{f''(a)}{2!} h^2 + \cdots + \frac{f^{(k)}(a)}{k!} h^k + \frac{f^{(k+1)}(\xi)}{(k+1)!} h^{k+1} \tag{2}$$

given in Theorem 6.1. Since $f^{(r)}(a)h^r = h^r D_1{}^r f(a) = (hD_1)^r f(a) = D_h f(a)$ by Theorem 2.2 with $n = 1$, so that $D_h = hD_1$, (2) can be rewritten

$$f(a + h) = \sum_{r=0}^{k} \frac{D_h{}^r f(a)}{r!} + \frac{D_h^{k+1} f(\xi)}{(k+1)!}, \tag{3}$$

where $D_h{}^0 f(a) = f(a)$. But now (3) is a meaningful equation if f is a real-valued function on $\mathscr{R}^n$. If f is of class $\mathscr{C}^k$ in a neighborhood of $\mathbf{a}$, then we see from Eq. (1) above that

$$P_k(\mathbf{h}) = \sum_{r=0}^{k} \frac{D_{\mathbf{h}}{}^r(\mathbf{a})}{r!}$$

$$= f(\mathbf{a}) + D_1 f(\mathbf{a})h_1 + \cdots + D_n f(\mathbf{a})h_n$$

$$+ \cdots + \sum_{j_1 + \cdots + j_n = k} \binom{k}{j_1 \cdots j_n} D_1^{j_1} \cdots D_n^{j_n} f(\mathbf{a}) h_1^{j_1} \cdots h_n^{j_n}$$

is a polynomial of degree at most k in the components $h_1, \ldots, h_n$ of $\mathbf{h}$. Therefore $P_k(\mathbf{h})$ is called the *kth degree Taylor polynomial of f at* $\mathbf{a}$. The *kth degree remainder of f at* $\mathbf{a}$ is defined by

$$R_k(\mathbf{h}) = f(\mathbf{a} + \mathbf{h}) - P_k(\mathbf{h}).$$

The notation $P_k(\mathbf{h})$ and $R_k(\mathbf{h})$ is incomplete, since it contains explicit reference to neither the function f nor the point $\mathbf{a}$, but this will cause no confusion in what follows.

Theorem 7.1 (Taylor's Formula) If f is a real-valued function of class $\mathscr{C}^{k+1}$ on an open set containing the line segment L from $\mathbf{a}$ to $\mathbf{a} + \mathbf{h}$, then there exists a point $\boldsymbol{\xi} \in L$ such that

$$R_k(\mathbf{h}) = \frac{D_{\mathbf{h}}^{k+1}(\boldsymbol{\xi})}{(k+1)!},$$

so

$$f(\mathbf{a} + \mathbf{h}) = \sum_{r=0}^{k} \frac{D_{\mathbf{h}}^{r} f(\mathbf{a})}{r!} + \frac{D_{\mathbf{h}}^{k+1} f(\xi)}{(k+1)!}.$$

PROOF If $\varphi : \mathscr{R} \to \mathscr{R}^n$ is defined by

$$\varphi(t) = \mathbf{a} + t\mathbf{h},$$

and $g(t) = f(\varphi(t)) = f(\mathbf{a} + t\mathbf{h})$, then $g(0) = f(\mathbf{a})$, $g(1) = f(\mathbf{a} + \mathbf{h})$, and Taylor's formula in one variable gives

$$g(1) = g(0) + \sum_{r=1}^{k} \frac{g^{(r)}(0)}{r!} + \frac{g^{(k+1)}(c)}{(k+1)!}$$

for some $c \in (0, 1)$. To complete the proof, it therefore suffices to show that

$$g^{(r)}(t) = D_{\mathbf{h}}^{r} f(\mathbf{a} + t\mathbf{h}) \tag{4}$$

for $r \leqq k + 1$, because then we have

$$g^{(r)}(0) = D_{\mathbf{h}}^{r} f(\mathbf{a})$$

and $g^{(k+1)}(c) = D_{\mathbf{h}}^{k+1} f(\xi)$ with $\xi = \varphi(c) \in L$ (see Fig. 2.38).

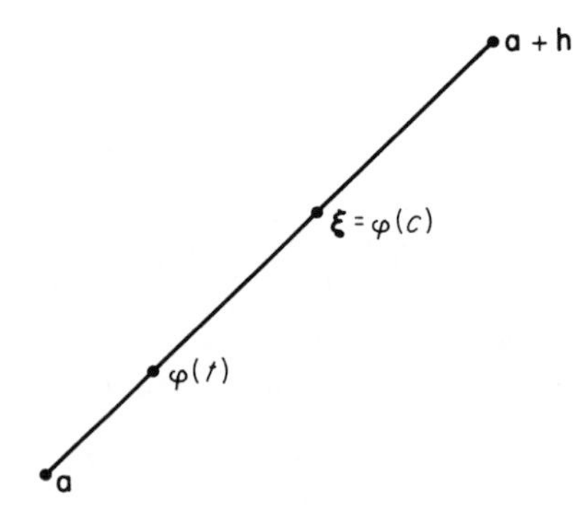

Figure 2.38

Actually, in order to apply the single-variable Taylor's formula, we need to know that g is of class $\mathscr{C}^{k+1}$ on $[0, 1]$, but this will follow from (4) because f is of class $\mathscr{C}^{k+1}$.

Note first that

$$\begin{aligned} g'(t) &= \nabla f(\varphi(t)) \cdot \varphi'(t) \qquad \text{(chain rule)} \\ &= \nabla f(\mathbf{a} + t\mathbf{h}) \cdot \mathbf{h} \\ &= D_{\mathbf{h}} f(\mathbf{a} + t\mathbf{h}). \end{aligned}$$

Then assume inductively that (4) holds for $r \leqq k$, and let $f_1(\mathbf{x}) = D_{\mathbf{h}}^r f(\mathbf{x})$. Then $g^{(r)} = f_1 \circ \varphi$, so

$$\begin{aligned} g^{(r+1)}(t) &= Dg^{(r)}(t) \\ &= \nabla f_1(\varphi(t)) \cdot \varphi'(t) \\ &= \nabla f_1(\mathbf{a} + t\mathbf{h}) \cdot \mathbf{h} \\ &= D_{\mathbf{h}} f_1(\mathbf{a} + t\mathbf{h}) \\ &= D_{\mathbf{h}} D_{\mathbf{h}}^r f(\mathbf{a} + t\mathbf{h}) \\ &= D_{\mathbf{h}}^{r+1} f(\mathbf{a} + t\mathbf{h}). \end{aligned}$$

Thus (4) holds by induction for $r \leqq k + 1$ as desired. ∎

If we write $\mathbf{x} = \mathbf{a} + \mathbf{h}$, then we obtain

$$f(\mathbf{x}) = P_k(\mathbf{x} - \mathbf{a}) + R_k(\mathbf{x} - \mathbf{a}),$$

where $P_k(\mathbf{x} - \mathbf{a})$ is a kth degree polynomial in the components $x_1 - a_1, x_2 - a_2, \ldots, x_n - a_n$ of $\mathbf{h} = \mathbf{x} - \mathbf{a}$, and

$$R_k(\mathbf{x} - \mathbf{a}) = \frac{D_{\mathbf{x}-\mathbf{a}}^{k+1} f(\mathbf{a} + \tau\mathbf{h})}{(k+1)!}$$

for some $\tau \in (0, 1)$.

Example 1 Let $f(\mathbf{x}) = e^{x_1 + \cdots + x_n}$. Then each $D_1^{j_1} \cdots D_n^{j_n} f(\mathbf{0}) = 1$, so

$$\begin{aligned} D_{\mathbf{x}}^r f(\mathbf{0}) &= \sum_{j_1 + \cdots + j_n = r} \binom{r}{j_1 \ \cdots \ j_n} x_1^{j_1} \cdots x_n^{j_n} D_1^{j_1} \cdots D_n^{j_n} f(\mathbf{0}) \\ &= \sum_{j_1 + \cdots + j_n = r} \binom{r}{j_1 \ \cdots \ j_n} x_1^{j_1} \cdots x_n^{j_n} \\ &= (x_1 + \cdots + x_n)^r. \end{aligned}$$

Hence the kth degree Taylor polynomial of $e^{x_1 + \cdots + x_n}$ is

$$P_k(\mathbf{x}) = \sum_{r=0}^{k} \frac{(x_1 + \cdots + x_n)^r}{r!}$$

as expected.

Example 2 Suppose we want to expand $f(x, y) = xy$ in powers of $x - 1$ and $y - 1$. Of course the result will be

$$xy = 1 + (x - 1) + (y - 1) + (x - 1)(y - 1),$$

but let us obtain this result by calculating the second degree Taylor polynomial $P_2(\mathbf{h})$ of $f(x, y)$ at $\mathbf{a} = (1, 1)$ with $\mathbf{h} = (h_1, h_2) = (x - 1, y - 1)$. We will then have $f(x, y) = P_2(x-1, y-1)$, since $R_3(\mathbf{h}) = 0$ because all third order partial derivatives of f vanish. Now

$$f(1, 1) = 1,$$

$$D_{\mathbf{h}} f(x, y) = \left(h_1 \frac{\partial}{\partial x} + h_2 \frac{\partial}{\partial y}\right) f(x, y)$$

$$= h_1 y + h_2 x.$$

So $D_{\mathbf{h}} f(1, 1) = h_1 + h_2 = (x-1) + (y-1)$, and

$$D_{\mathbf{h}}{}^2 f(x, y) = \left(h_1 \frac{\partial}{\partial x} + h_2 \frac{\partial}{\partial y}\right)^2 f(x, y)$$

$$= \left(h_1 \frac{\partial}{\partial x} + h_2 \frac{\partial}{\partial y}\right)(h_1 y + h_2 x)$$

$$= 2h_1 h_2$$

$$= 2(x-1)(y-1).$$

Hence

$$P_2(x-1, y-1) = f(1, 1) + D_{\mathbf{h}}{}^1 f(1, 1) + \frac{1}{2!} D_{\mathbf{h}}{}^2 f(1, 1)$$

$$= 1 + (x-1) + (y-1) + (x-1)(y-1)$$

as predicted.

Next we generalize the error estimate of Corollary 6.2.

Corollary 7.2 If f is of class $\mathscr{C}^{k+1}$ in a neighborhood U of $\mathbf{a}$, and $R_k(\mathbf{h})$ is the kth degree remainder of f at $\mathbf{a}$, then

$$\lim_{\mathbf{h} \to \mathbf{0}} \frac{R_k(\mathbf{h})}{|\mathbf{h}|^k} = 0.$$

PROOF If $\mathbf{h}$ is sufficiently small that the line segment from $\mathbf{a}$ to $\mathbf{a} + \mathbf{h}$ lies in U, then Theorem 7.1 gives

$$R_k(\mathbf{h}) = \frac{D_{\mathbf{h}}^{k+1} f(\mathbf{a} + \tau\mathbf{h})}{(k+1)!} \qquad [\text{for some } \tau \in (0, 1)]$$

$$= \frac{1}{(k+1)!} \sum \binom{k+1}{j_1 \cdots j_n} h_1^{j_1} \cdots h_n^{j_n} D_1^{j_1} \cdots D_n^{j_n} f(\mathbf{a} + \tau\mathbf{h}).$$

But $\lim_{\mathbf{h} \to \mathbf{0}} D_1^{j_1} \cdots D_n^{j_n} f(\mathbf{a} + \tau\mathbf{h}) = D_1^{j_1} \cdots D_n^{j_n} f(\mathbf{a})$ because f is of class $\mathscr{C}^{k+1}$. It therefore suffices to see that

$$\lim_{\mathbf{h} \to \mathbf{0}} \frac{h_1^{j_1} \cdots h_n^{j_n}}{|\mathbf{h}|^k} = 0$$

if $j_1 + \cdots + j_n = k + 1$. But this is so because each $|h_i|/|\mathbf{h}| \leqq 1$, and there is one more factor in the numerator than in the denominator. ∎

Rewriting the conclusion of Corollary 7.2 in terms of the kth degree Taylor polynomial, we have

$$\lim_{\mathbf{x}\to\mathbf{a}} \frac{f(\mathbf{x}) - P_k(\mathbf{x}-\mathbf{a})}{|\mathbf{x}-\mathbf{a}|^k} = 0. \tag{5}$$

We shall see that this result characterizes $P_k(\mathbf{x}-\mathbf{a})$; that is, $P_k(\mathbf{x}-\mathbf{a})$ is the *only* kth degree polynomial in $x_1 - a_1, \ldots, x_n - a_n$ satisfying condition (5).

Lemma 7.3 If $Q(\mathbf{x})$ and $Q^*(\mathbf{x})$ are two kth degree polynomials in $x_1, \ldots, x_n$ such that

$$\lim_{\mathbf{x}\to\mathbf{0}} \frac{Q(\mathbf{x}) - Q^*(\mathbf{x})}{|\mathbf{x}|^k} = 0,$$

then $Q = Q^*$.

PROOF Supposing to the contrary that $Q \neq Q^*$, let

$$Q(\mathbf{x}) - Q^*(\mathbf{x}) = F(\mathbf{x}) + G(\mathbf{x}),$$

where $F(\mathbf{x})$ is the polynomial consisting of all those terms of the lowest degree l which actually appear in $Q(\mathbf{x}) - Q^*(\mathbf{x})$, and $G(\mathbf{x})$ consists of all those terms of degree higher than l.

Choose a fixed point $\mathbf{b} \neq \mathbf{0}$ such that $F(\mathbf{b}) \neq \mathbf{0}$. Since $|t\mathbf{b}|^l \geqq |t\mathbf{b}|^k$ for t sufficiently small, we have

$$\begin{aligned} 0 &= \lim_{t\to 0} \frac{Q(t\mathbf{b}) - Q^*(t\mathbf{b})}{|t\mathbf{b}|^l} \\ &= \lim_{t\to 0} \frac{F(t\mathbf{b}) + G(t\mathbf{b})}{t^l|\mathbf{b}|^l} \\ &= \frac{F(\mathbf{b})}{|\mathbf{b}|^l} \neq 0, \end{aligned}$$

since each term of F is of degree l, while each term of G is of degree $> l$. This contradiction proves that $Q = Q^*$. ∎

We can now establish the converse to Corollary 7.2.

Theorem 7.4 If $f : \mathscr{R}^n \to \mathscr{R}$ is of class $\mathscr{C}^{k+1}$ in a neighborhood of $\mathbf{a}$, and Q is a polynomial of degree k such that

$$\lim_{\mathbf{x}\to\mathbf{a}} \frac{f(\mathbf{x}) - Q(\mathbf{x}-\mathbf{a})}{|\mathbf{x}-\mathbf{a}|^k} = 0, \tag{6}$$

then Q is the kth degree Taylor polynomial of f at $\mathbf{a}$.

PROOF Since

$$\left|\frac{Q(\mathbf{h}) - P_k(\mathbf{h})}{|\mathbf{h}|^k}\right| \leqq \left|\frac{f(\mathbf{a}+\mathbf{h}) - Q(\mathbf{h})}{|\mathbf{h}|^k}\right| + \left|\frac{f(\mathbf{a}+\mathbf{h}) - P_k(\mathbf{h})}{|\mathbf{h}|^k}\right|$$

by the triangle inequality, (5) and (6) imply that

$$\lim_{\mathbf{h}\to\mathbf{0}} \frac{Q(\mathbf{h}) - P_k(\mathbf{h})}{|\mathbf{h}|^k} = 0.$$

Since Q and P_k are both kth degree polynomials, Lemma 7.3 gives $Q = P_k$ as desired. ∎

The above theorem can often be used to discover Taylor polynomials of a function when the explicit calculation of its derivatives is inconvenient. In order to verify that a given kth degree polynomial Q is the kth degree Taylor polynomial of the class $\mathscr{C}^{k+1}$ function f at $\mathbf{a}$, we need only verify that Q satisfies condition (6).

Example 3 We calculate the third degree Taylor polynomial of $e^x \sin x$ by multiplying together those of e^x and $\sin x$. We know that

$$e^x = 1 + x + \frac{x^2}{2} + \frac{x^3}{6} + R(x) \qquad \text{and} \qquad \sin x = x - \frac{x^3}{6} + \bar{R}(x),$$

where $\lim_{x\to 0} R(x)/x^3 = \lim_{x\to 0} \bar{R}(x)/x^3 = 0$. Hence

$$e^x \sin x = x + x^2 + \frac{x^3}{3} + R^*(x),$$

where

$$R^*(x) = -\frac{x^5}{12} - \frac{x^6}{36} + \left(x - \frac{x^3}{6}\right)R(x) + \left(1 + x + \frac{x^2}{2} + \frac{x^3}{6}\right)\bar{R}(x) + R(x)\bar{R}(x).$$

Since it is clear that

$$\lim_{x\to 0} \frac{R^*(x)}{x^3} = 0,$$

it follows from Theorem 7.4 that $x + x^2 + \frac{1}{3}x^3$ is the third degree Taylor polynomial of $e^x \sin x$ at 0.

Example 4 Recall that

$$e^t = 1 + t + \frac{t^2}{2} + R(t), \qquad \lim_{t\to 0} \frac{R(t)}{t^2} = 0.$$

If we substitute $t = x$ and $t = y$, and multiply the results, we obtain

$$e^{x+y} = 1 + (x + y) + \frac{1}{2!}(x + y)^2 + R^*(x, y),$$

where

$$R^*(x,y) = (\tfrac{1}{2})xy(x + y + \tfrac{1}{2}xy) + (1 + x + \tfrac{1}{2}x^2)R(y) + (1 + y + \tfrac{1}{2}y^2)R(x) + R(x)R(y).$$

Since it is easily verified that

$$\lim_{(x,y)\to(0,0)} \frac{R^*(x, y)}{x^2 + y^2} = 0,$$

it follows from Theorem 7.4 that $1 + (x + y) + \frac{1}{2}(x + y)^2$ is the second degree Taylor polynomial of e^{x+y} at $(0, 0)$.

We shall next give a first application of Taylor polynomials to multivariable maximum–minimum problems. Given a critical point $\mathbf{a}$ for $f: \mathscr{R}^n \to \mathscr{R}$, we would like to have a criterion for determining whether f has a local maximum or a local minimum, or neither, at $\mathbf{a}$. If f is of class $\mathscr{C}^3$ in a neighborhood of $\mathbf{a}$, we can write its second degree Taylor expansion in the form

$$f(\mathbf{a} + \mathbf{h}) - f(\mathbf{a}) = q(\mathbf{h}) + R_2(\mathbf{h}),$$

where

$$q(\mathbf{h}) = \tfrac{1}{2}D_{\mathbf{h}}{}^2 f(\mathbf{a}) = \tfrac{1}{2}(h_1 D_1 + \cdots + h_n D_n)^2 f(\mathbf{a})$$

and

$$\lim_{\mathbf{h}\to\mathbf{0}} \frac{R_2(\mathbf{h})}{|\mathbf{h}|^2} = 0.$$

If not all of the second partial derivatives of f vanish at $\mathbf{a}$, then $q(\mathbf{h})$ is a (non-trivial) homogeneous second degree polynomial in $h_1, \ldots, h_n$ of the form

$$q(\mathbf{h}) = \sum_{1 \leqq i \leqq j \leqq n} a_{ij} h_i h_j,$$

and is called the *quadratic form* of f at the critical point $\mathbf{a}$. Note that

$$q(\mathbf{h}) = \begin{cases} |\mathbf{h}|^2 q\left(\dfrac{\mathbf{h}}{|\mathbf{h}|}\right) & \text{if} \quad \mathbf{h} \neq \mathbf{0}, \\ 0 & \text{if} \quad \mathbf{h} = \mathbf{0}. \end{cases}$$

Since $\mathbf{h}/|\mathbf{h}|$ is a point of the unit sphere S^{n-1} in $\mathscr{R}^n$, it follows that the quadratic form q is completely determined by its values on S^{n-1}. As in the 2-dimensional case of Section 4, a quadratic form is called *positive-definite* (respectively *negative-definite*) if and only if it is positive (respectively negative) at every point of S^{n-1} (and hence everywhere except $\mathbf{0}$), and is called *nondefinite* if it assumes both positive and negative values on S^{n-1} (and hence in every neighborhood of $\mathbf{0}$).

For example, $x^2 + y^2$ is positive-definite and $-x^2 - y^2$ is negative-definite, while $x^2 - y^2$ and xy are nondefinite. Note that y^2, regarded as a quadratic form in x and y whose coefficients of both x^2 and xy are zero, is neither positive-definite nor negative-definite nor nondefinite (it is not negative anywhere, but is zero on the x-axis).

In the case $n = 1$, when f is a function of a single variable with critical point a, the quadratic form of f at a is simply

$$q(h) = \tfrac{1}{2}f''(a)h^2.$$

Note that q is positive-definite if $f''(a) > 0$, and is negative-definite if $f''(a) < 0$ (the nondefinite case cannot occur if $n = 1$). Therefore the "second derivative test" for a single-variable function states that f has a local minimum at a if its quadratic form $q(h)$ at a is positive-definite, and a local maximum at a if $q(h)$ is negative-definite. If $f''(a) = 0$, then $q(h) = 0$ is neither positive-definite nor negative-definite (nor nondefinite), and we cannot determine (without considering higher derivatives) the character of the critical point a. The following theorem is the multivariable generalization of the "second derivative test."

Theorem 7.5 Let f be of class $\mathscr{C}^3$ in a neighborhood of the critical point $\mathbf{a}$. Then f has

(a) a local minimum at $\mathbf{a}$ if its quadratic form $q(\mathbf{h})$ is positive-definite,
(b) a local maximum at $\mathbf{a}$ if $q(\mathbf{h})$ is negative-definite,
(c) neither if $q(\mathbf{h})$ is nondefinite.

PROOF Since $f(\mathbf{a} + \mathbf{h}) - f(\mathbf{a}) = q(\mathbf{h}) + R_2(\mathbf{h})$, it suffices for (a) to find a $\delta > 0$ such that

$$0 < |\mathbf{h}| < \delta \Rightarrow \frac{q(\mathbf{h}) + R_2(\mathbf{h})}{|\mathbf{h}|^2} > 0.$$

Note that

$$\frac{q(\mathbf{h}) + R_2(\mathbf{h})}{|\mathbf{h}|^2} = q\left(\frac{\mathbf{h}}{|\mathbf{h}|}\right) + \frac{R_2(\mathbf{h})}{|\mathbf{h}|^2}.$$

Since $\mathbf{h}/|\mathbf{h}| \in S^{n-1}$, and S^{n-1} is a closed and bounded set, $q(\mathbf{h}/|\mathbf{h}|)$ attains a minimum value m, and $m > 0$ because q is positive-definite. Then

$$\frac{q(\mathbf{h}) + R_2(\mathbf{h})}{|\mathbf{h}|^2} \geqq m + \frac{R_2(\mathbf{h})}{|\mathbf{h}|^2}.$$

But $\lim_{h \to 0} R_2(\mathbf{h})/|\mathbf{h}|^2 = 0$, so there exists $\delta > 0$ such that

$$0 < |\mathbf{h}| < \delta \Rightarrow \left|\frac{R_2(\mathbf{h})}{|\mathbf{h}|^2}\right| < \frac{m}{2}.$$

Then

$$0 < |\mathbf{h}| < \delta \Rightarrow \frac{q(\mathbf{h}) + R_2(\mathbf{h})}{|\mathbf{h}|^2} > m - \frac{m}{2} > 0$$

as desired.

The proof for case (b) is similar.

If $q(\mathbf{h})$ is nondefinite, choose two fixed points $\mathbf{h}_1$ and $\mathbf{h}_2$ in $\mathscr{R}^n$ such that $q(\mathbf{h}_1) > 0$ and $q(\mathbf{h}_2) < 0$. Then

$$\begin{aligned} f(\mathbf{a} + t\mathbf{h}_i) - f(\mathbf{a}) &= q(t\mathbf{h}_i) + R_2(t\mathbf{h}_i) \\ &= t^2 q(\mathbf{h}_i) + t^2 |\mathbf{h}_i|^2 \frac{R_2(t\mathbf{h}_i)}{|t\mathbf{h}_i|^2} \\ &= t^2 \left[q(\mathbf{h}_i) + |\mathbf{h}_i|^2 \frac{R_2(t\mathbf{h}_i)}{|t\mathbf{h}_i|^2} \right]. \end{aligned}$$

Since $\lim_{t \to 0} R_2(t\mathbf{h}_i)/|t\mathbf{h}_i|^2 = 0$, it follows that, for t sufficiently small, $f(\mathbf{a} + t\mathbf{h}_i) - f(\mathbf{a})$ is positive if $i = 1$, and negative if $i = 2$. Hence f has neither a local minimum nor a local maximum at $\mathbf{a}$. ∎

For example, suppose that $\mathbf{a}$ is a critical point of $f(x, y)$. Then f has a local minimum at $\mathbf{a}$ if $q(x, y) = x^2 + y^2$, a local maximum at $\mathbf{a}$ if $q(x, y) = -x^2 - y^2$, and neither if $q(x, y) = x^2 - y^2$. If $q(x, y) = y^2$ or $q(x, y) = 0$, then Theorem 7.5 does not apply. In the cases where the theorem does apply, it says simply that the character of f at $\mathbf{a}$ is the same as that of q at $\mathbf{0}$ (see Fig. 2.39).

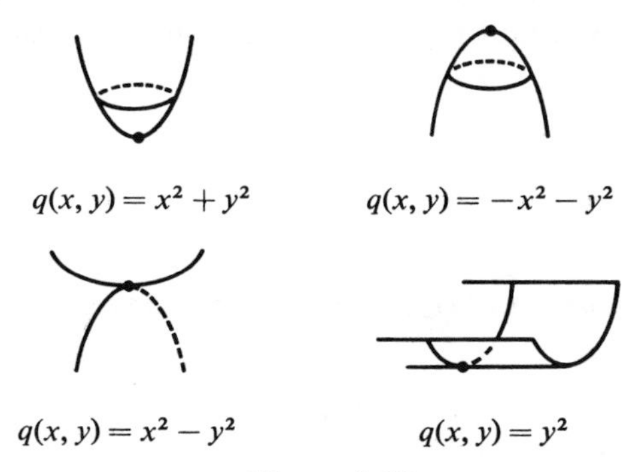

Figure 2.39

Example 5 Let $f(x, y) = x^2 + y^2 + \cos x$. Then $(0, 0)$ is a critical point of f. Since

$$\cos x = 1 - \frac{x^2}{2} + R_2(x), \qquad \lim_{x \to 0} \frac{R_2(x)}{x^2} = 0,$$

Theorem 7.4 implies that the second degree Taylor polynomial of f at $(0, 0)$ is

$$P_2(x, y) = (x^2 + y^2) + (1 - \tfrac{1}{2}x^2) = 1 + \tfrac{1}{2}x^2 + y^2,$$

so the quadratic form of f at $(0, 0)$ is

$$q(x, y) = \tfrac{1}{2}x^2 + y^2.$$

Since $q(x, y)$ is obviously positive-definite, it follows from Theorem 7.5 that f has a local minimum at $(0, 0)$.

Example 6 The point $\mathbf{a} = (2, \pi/2)$ is a critical point for the function $f(x, y) = x^2 \sin y - 4x$. Since

$$D_1 D_1 f\left(2, \frac{\pi}{2}\right) = 2, \qquad D_1 D_2 f\left(2, \frac{\pi}{2}\right) = 0, \qquad D_2 D_2 f\left(2, \frac{\pi}{2}\right) = -4,$$

the second degree Taylor expansion of f at $(2, \pi/2)$ is

$$\begin{aligned} f(\mathbf{a} + \mathbf{h}) &= f(\mathbf{a}) + \tfrac{1}{2}[D_2{}^2 f(\mathbf{a})h_1{}^2 + 2D_1 D_2 f(\mathbf{a})h_1 h_2 + D_2{}^2 f(\mathbf{a})] + R_2(\mathbf{h}) \\ &= -4 + h_1{}^2 - 2h_2{}^2 + R(\mathbf{h}). \end{aligned}$$

Consequently the quadratic form of f at $(2, \pi/2)$ is defined by

$$q(x, y) = x^2 - 2y^2$$

(writing q in terms of x and y through habit). Since q is clearly nondefinite, it follows from Theorem 7.5 that f has neither a local maximum nor a local minimum at $(2, \pi/2)$.

Consequently, in order to determine the character of the critical point $\mathbf{a}$ of f, we need to maximize and minimize its quadratic form on the unit sphere S^{n-1}. This is a Lagrange multiplier problem with which we shall deal in the next section.

Exercises

7.1 If f is a function of class $\mathscr{C}^q$, and $i_1', \ldots, i_q'$ is a permutation of $i_1, \ldots, i_q$, prove by induction on q that

$$D_{i_1} \cdots D_{i_q} f = D_{i_1'} \cdots D_{i_q'} f.$$

7.2 Let $f(\mathbf{x}) = (x_1 + \cdots + x_n)^k$.

(a) Show that $D_1^{j_1} \cdots D_n^{j_n} f(\mathbf{x}) = k!$ if $j_1 + \cdots + j_n = k$.

(b) Show that

$$D_1^{j_1} \cdots D_n^{j_n} x_1^{i_1} \cdots x_n^{i_n} = \begin{cases} j_1! \cdots j_n! & \text{if } \ i_1 = j_1, \ldots, i_n = j_n, \\ 0 & \text{otherwise.} \end{cases}$$

(c) Conclude that

$$\binom{k}{j_1 \cdots j_n} = \frac{k!}{j_1! \cdots j_n!}.$$

7.3 Find the third degree Taylor polynomial of $f(x, y) = (x + y)^3$ at (0, 0) and (1, 1).

7.4 Find the third degree Taylor polynomial of $f(x, y, z) = xy^2z^3$ at $(1, 0, -1)$.

7.5 Assuming the facts that the sixth degree Taylor polynomials of $\tan^{-1} x$ and $\sin^2 x$ are

$$x - \frac{x^3}{3} + \frac{x^5}{5} \quad \text{and} \quad x^2 - \frac{x^4}{3} + \frac{2}{45}x^6,$$

respectively, use Theorem 7.4 to show that the sixth degree Taylor polynomial of

$$f(x) = x \tan^{-1} x - \sin^2 x$$

at 0 is $\frac{7}{45}x^6$. Why does it now follow from Theorem 6.3 that f has a local minimum at 0?

7.6 Let $f(x, y) = e^{xy} \sin(x + y)$. Multiply the Taylor expansions

$$e^{xy} = 1 + xy + \frac{x^2y^2}{2!} + R_2(x, y)$$

and

$$\sin(x + y) = (x + y) - \frac{1}{3!}(x + y)^3 + R_3(x, y)$$

together, and apply Theorem 7.4 to show that

$$x + y + \frac{1}{3!}(-x^3 + 3x^2y + 3xy^2 - y^3)$$

is the third degree Taylor polynomial of f at 0. Conclude that

$$D_1{}^3f(0) = D_2{}^3f(0) = -1,$$
$$D_1D_2{}^2f(0) = D_1{}^2D_2 f(0) = 1.$$

7.7 Apply Theorem 7.4 to prove that the Taylor polynomial of degree $4n + 1$ for $f(x) = \sin(x^2)$ at 0 is

$$x^2 - \frac{x^6}{3!} + \cdots + (-1)^n \frac{x^{4n}}{(2n)!}.$$

Hint: $\sin x = P_{2n+1}(x) + R_{2n+1}(x)$, where

$$P_{2n+1}(x) = x - \frac{x^3}{3!} + \cdots + (-1)^n \frac{x^{2n+1}}{(2n+1)!} \quad \text{and} \quad \lim_{x \to 0} \frac{R_{2n+1}(x)}{x^{2n+1}} = 0.$$

Hence $\lim_{x \to 0} R_{2n+1}(x^2)/x^{4n+1} = 0$.

7.8 Apply Theorem 7.4 to show that the sixth degree Taylor polynomial of $f(x) = \sin^2 x$ at 0 is $x^2 - \frac{1}{3}x^4 + \frac{2}{45}x^6$.

7.9 Apply Theorem 7.4 to prove that the kth degree Taylor polynomial of $f(x)g(x)$ can be obtained by multiplying together those of $f(x)$ and $g(x)$, and then deleting from the product all terms of degree greater than k.

7.10 Find and classify the critical points of $f(x, y) = (x^2 + y^2)e^{x^2 - y^2}$.

7.11 Classify the critical point $(-1, \pi/2, 0)$ of $f(x, y, z) = x \sin z - z \sin y$.

7.12 Use the Taylor expansions given in Exercises 7.5 and 7.6 to classify the critical point (0, 0, 0) of the function $f(x, y, z) = x^2 + y^2 + e^{xy} - y \tan^{-1} x + \sin^2 z$.

8 THE CLASSIFICATION OF CRITICAL POINTS

Let f be a real-valued function of class $\mathscr{C}^3$ on an open set $U \subset \mathscr{R}^n$, and let $\mathbf{a} \in U$ be a critical point of f. In order to apply Theorem 7.5 to determine the nature of the critical point $\mathbf{a}$, we need to decide the character of the quadratic form $q : \mathscr{R}^n \to \mathscr{R}$ defined by

$$\begin{aligned} q(\mathbf{x}) &= \tfrac{1}{2} D_{\mathbf{x}}{}^2 f(\mathbf{a}) = \tfrac{1}{2}(x_1 D_1 + \cdots + x_n D_n)^2 f(\mathbf{a}) \\ &= \tfrac{1}{2} \sum_{i,\, j=1}^{n} D_i D_j f(\mathbf{a}) x_i x_j \\ &= \mathbf{x}^t A \mathbf{x}, \end{aligned}$$

where the $n \times n$ matrix $A = (a_{ij})$ is defined by $a_{ij} = \frac{1}{2} D_i D_j f(\mathbf{a})$, and

$$\mathbf{x} = \begin{pmatrix} x_1 \\ \vdots \\ x_n \end{pmatrix}, \qquad \mathbf{x}^t = (x_1 \cdots x_n).$$

Note that the matrix A is *symmetric*, that is $a_{ij} = a_{ji}$, since $D_i D_j = D_j D_i$.

We have seen that, in order to determine whether q is positive-definite, negative-definite, or nondefinite, it suffices to determine the maximum and minimum values of q on the unit sphere S^{n-1}. The following result rephrases this problem in terms of the linear mapping $L : \mathscr{R}^n \to \mathscr{R}^n$ defined by

$$L(\mathbf{x}) = A\mathbf{x}.$$

We shall refer to the quadratic form q, the symmetric matrix A, and the linear mapping L as being "associated" with one another.

Theorem 8.1 If the quadratic form q attains its maximum or minimum value on S^{n-1} at the point $\mathbf{v} \in S^{n-1}$, then there exists a real number λ such that

$$L(\mathbf{v}) = \lambda \mathbf{v},$$

where L is the linear mapping associated with q.

Such a *nonzero* vector $\mathbf{v}$, which upon application of the linear mapping $L : \mathscr{R}^n \to \mathscr{R}^n$ is simply multiplied by some real number λ, $L(\mathbf{v}) = \lambda \mathbf{v}$, is called an *eigenvector* of L, with associated *eigenvalue* λ.

PROOF We apply the Lagrange multiplier method. According to Theorem 5.5, there exists a number $\lambda \in \mathscr{R}$ such that

$$\nabla q(\mathbf{v}) = \lambda \nabla g(\mathbf{v}), \tag{1}$$

where $g(\mathbf{x}) = \sum_1^n x_i^2 - 1 = 0$ defines the sphere S^{n-1}.

But $D_k g(\mathbf{x}) = 2x_k$, and

$$\begin{aligned} D_k q(\mathbf{x}) &= (0 \cdots 1 \cdots 0)A\mathbf{x} + \mathbf{x}^t A \begin{pmatrix} 0 \\ \vdots \\ 1 \\ \vdots \\ 0 \end{pmatrix} \\ &= \sum_{j=1}^n a_{kj} x_j + \sum_{j=1}^n a_{jk} x_j \\ &= 2 \sum_{j=1}^n a_{kj} x_j \qquad \text{(because } a_{kj} = a_{jk}) \\ &= 2A_k \mathbf{x}, \end{aligned}$$

where the row vector A_k is the kth row of A, that is, $A_k = (a_{k1} \cdots a_{kn})$. Consequently Eq. (1) implies that

$$A_k \mathbf{v} = \lambda v_k$$

for each $i = 1, \ldots, n$. But $A_k \mathbf{v}$ is the ith row of $A\mathbf{v}$, while λv_k is the kth element of the column vector $\lambda \mathbf{v}$, so we have

$$L(\mathbf{v}) = A\mathbf{v} = \lambda \mathbf{v}$$

as desired. ∎

Thus the problem of maximizing and minimizing q on S^{n-1} involves the eigenvectors and eigenvalues of the associated linear mapping. Actually, it will turn out that if we can determine the eigenvalues without locating the eigenvectors first, then we need not be concerned with the latter.

Lemma 8.2 If $\mathbf{v} \in S^{n-1}$ is an eigenvector of the linear mapping $L : \mathscr{R}^n \to \mathscr{R}^n$ associated with the quadratic form q, then

$$q(\mathbf{v}) = \lambda,$$

where λ is the eigenvalue associated with $\mathbf{v}$.

PROOF

$$\begin{aligned} q(\mathbf{v}) &= \mathbf{v}^t A \mathbf{v} = \mathbf{v}^t(\lambda \mathbf{v}) \\ &= \lambda \mathbf{v}^t \mathbf{v} = \lambda \sum_1^n v_i^2 \\ &= \lambda \end{aligned}$$

since $\mathbf{v} \in S^{n-1}$. ∎

The fact that we need only determine the eigenvalues of L (and not the eigenvectors) is a significant advantage, because the following theorem asserts that we therefore need only solve the so-called *characteristic equation*

$$|A - \lambda I| = 0.$$

Here $|A - \lambda I|$ denotes the determinant of the matrix $A - \lambda I$ (where I is the $n \times n$ identity matrix).

Theorem 8.3 The real number λ is an eigenvalue of the linear mapping $L : \mathscr{R}^n \to \mathscr{R}^n$, defined by $L(\mathbf{x}) = A\mathbf{x}$, if and only if λ satisfies the equation

$$|A - \lambda I| = 0.$$

PROOF Suppose first that λ is an eigenvalue of L, that is,

$$L(\mathbf{v}) = \lambda\mathbf{v} = \lambda I\mathbf{v} = A\mathbf{v}$$

for some $\mathbf{v} \neq 0$. Then $(A - \lambda I)\mathbf{v} = 0$.

Enumerating the rows of the matrix $A - \lambda I$, we obtain

$$(A_1 - \lambda\mathbf{e}_1) \cdot \mathbf{v} = 0,$$
$$\vdots$$
$$(A_n - \lambda\mathbf{e}_n) \cdot \mathbf{v} = 0,$$

where $\mathbf{e}_1, \ldots, \mathbf{e}_n$ are the standard unit vectors in $\mathscr{R}^n$. Hence the n vectors $A_1 - \lambda\mathbf{e}_1, \ldots, A_n - \lambda\mathbf{e}_n$ all lie in the $(n-1)$-dimensional hyperplane through $\mathbf{0}$ perpendicular to $\mathbf{v}$, and are therefore linearly dependent, so it follows from Theorem I.6.1 that

$$|A - \lambda I| = 0$$

as desired.

Conversely, if λ is a solution of $|A - \lambda I| = 0$, then the row vectors $A_1 - \lambda e_1, \ldots, A_n - \lambda\mathbf{e}_n$ of $A - \lambda I$ are linearly dependent, and therefore all lie in some $(n-1)$-dimensional hyperplane through the origin. If $\mathbf{v}$ is a unit vector perpendicular to this hyperplane, then

$$(A_i - \lambda\mathbf{e}_i) \cdot \mathbf{v} = 0, \qquad i = 1, \ldots, n.$$

Hence $(A - \lambda I)\mathbf{v} = 0$, so

$$L(\mathbf{v}) = A\mathbf{v} = \lambda I\mathbf{v} = \lambda\mathbf{v}$$

as desired. ∎

Corollary 8.4 The maximum (minimum) value attained by the quadratic form $q(\mathbf{x}) = \mathbf{x}^t A\mathbf{x}$ on S^{n-1} is the largest (smallest) real root of the equation

$$|A - \lambda I| = 0.$$

PROOF If $q(\mathbf{v})$ is the maximum (minimum) value of q on S^{n-1}, then $\mathbf{v}$ is an eigenvector of $L(\mathbf{x}) = A\mathbf{x}$ by Theorem 8.1, and $q(\mathbf{v}) = \lambda$, the associated eigenvalue, by Lemma 8.2. Theorem 8.3 therefore implies that $q(\mathbf{v})$ is a root of $|A - \lambda I| = 0$. On the other hand, it follows from 8.2 and 8.3 that every root of $|A - \lambda I| = 0$ is a value of q on S^{n-1}. Consequently $q(\mathbf{v})$ must be the largest (smallest) root of $|A - \lambda I| = 0$. ∎

Example 1 Suppose $\mathbf{a}$ is a critical point of the function $f: \mathscr{R}^3 \to \mathscr{R}$ and that the quadratic form of f at $\mathbf{a}$ is

$$q(x, y, z) = x^2 + y^2 + z^2 + 4yz, \text{ or } q(x_1, x_2, x_3) = {x_1}^2 + {x_2}^2 + {x_3}^2 + 4x_2 x_3,$$

so the matrix of q is

$$A = \begin{pmatrix} 1 & 0 & 0 \\ 0 & 1 & 2 \\ 0 & 2 & 1 \end{pmatrix}.$$

The characteristic equation of A is then

$$\begin{vmatrix} 1-\lambda & 0 & 0 \\ 0 & 1-\lambda & 2 \\ 0 & 2 & 1-\lambda \end{vmatrix} = (1-\lambda)[(1-\lambda)^2 - 4] = 0,$$

with roots $\lambda = -1, 1, 3$. It follows from Corollary 8.4 that the maximum and minimum values of q in S^2 are $+3$ and -1 respectively. Since q attains both positive and negative values, it is nondefinite. Hence Theorem 7.5 implies that f has *neither* a maximum nor a minimum at $\mathbf{a}$.

The above example indicates how Corollary 8.4 reduces the study of the quadratic form $q(x) = \mathbf{x}^t A\mathbf{x}$ to the problem of finding the largest and smallest real roots of the characteristic equation $|A - \lambda I| = 0$. However this problem itself can be difficult. It is therefore desirable to carry our analysis of quadratic forms somewhat further.

Recall that the matrix A of a quadratic form is symmetric, that is, $a_{ij} = a_{ji}$. We say that the linear mapping $L: \mathscr{R}^n \to \mathscr{R}^n$ is *symmetric* if and only if $L(\mathbf{x}) \cdot \mathbf{y} = \mathbf{x} \cdot L(\mathbf{y})$ for all $\mathbf{x}$ and $\mathbf{y}$ in $\mathscr{R}^n$. The motivation for this terminology is the following result.

Lemma 8.5 If A is the matrix of the linear mapping $L: \mathscr{R}^n \to \mathscr{R}^n$, then L is symmetric if and only if A is symmetric.

PROOF If $\mathbf{x} = \sum_1^n x_i \mathbf{e}_i$, $\mathbf{y} = \sum_1^n y_i \mathbf{e}_i$, then

$$L(\mathbf{x}) \cdot \mathbf{y} = \sum_{i,j=1}^n x_i y_j L(\mathbf{e}_i) \cdot \mathbf{e}_j \qquad \text{and} \qquad \mathbf{x} \cdot L(\mathbf{y}) = \sum_{i,j=1}^n x_i y_j \mathbf{e}_i \cdot L(\mathbf{e}_j).$$

Hence L is symmetric if and only if

$$L(\mathbf{e}_i) \cdot \mathbf{e}_j = \mathbf{e}_i \cdot L(\mathbf{e}_j)$$

for all $i, j = 1, \ldots, n$. But

$$L(\mathbf{e}_i) \cdot \mathbf{e}_j = \mathbf{e}_j \cdot L(\mathbf{e}_i) = \mathbf{e}_j{}^t A \mathbf{e}_i = a_{ji}$$

and $\mathbf{e}_i \cdot L(\mathbf{e}_j) = \mathbf{e}_i{}^t A \mathbf{e}_j = a_{ij}$. ∎

The following theorem gives the precise character of a quadratic form q in terms of the eigenvectors and eigenvalues of its associated linear mapping L.

Theorem 8.6 If q is a quadratic form on $\mathscr{R}^n$, then there exists an orthogonal set of unit eigenvectors $\mathbf{v}_1, \ldots, \mathbf{v}_n$ of the associated linear mapping L. If $y_1, \ldots, y_n$ are the coordinates of $\mathbf{x} \in \mathscr{R}^n$ with respect to the basis $\mathbf{v}_1, \ldots, \mathbf{v}_n$, that is,

$$\mathbf{x} = y_1 \mathbf{v}_1 + \cdots + y_n \mathbf{v}_n,$$

then

$$q(\mathbf{x}) = \lambda_1 {y_1}^2 + \cdots + \lambda_n {y_n}^2, \tag{2}$$

where $\lambda_1, \ldots, \lambda_n$ are the eigenvalues of $\mathbf{v}_1, \ldots, \mathbf{v}_n$.

PROOF This is trivially true if $n = 1$, for then q is of the form $q(x) = \lambda x^2$, with $L(x) = \lambda x$. We therefore assume by induction that it is true in dimension $n - 1$.

If q attains its maximum value on the unit sphere S^{n-1} at $\mathbf{v}_n \in S^{n-1}$, then Theorem 8.1 says that $\mathbf{v}_n$ is an eigenvector of L. See Fig. 2.40.

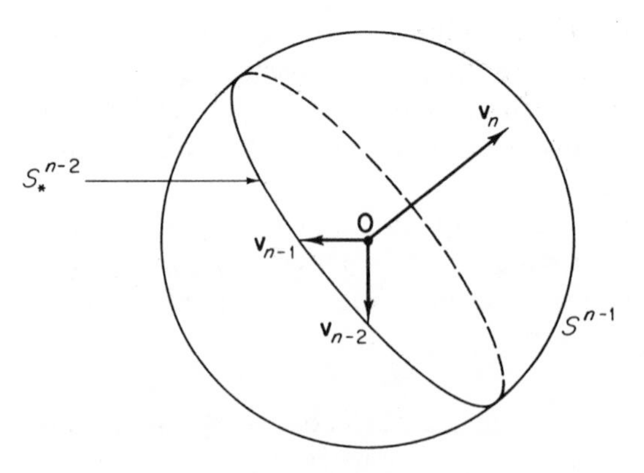

$q(\mathbf{v}_n) =$ maximum of q on S^{n-1}
$q(\mathbf{v}_{n-1}) =$ maximum of q on $S_*^{n-2} = S^{n-1} \cap R_*^{n-1}$
$q(\mathbf{v}_{n-2}) =$ maximum of q on $S_*^{n-3} = \{\mathbf{x} \in S_*^{n-2} : \mathbf{x} \cdot \mathbf{v}_{n-1} = 0\}$

Figure 2.40

Let $\mathscr{R}_*^{n-1}$ be the subspace of $\mathscr{R}^n$ perpendicular to $\mathbf{v}_n$. Then L maps $\mathscr{R}_*^{n-1}$ into itself, because, given $\mathbf{x} \in \mathscr{R}_*^{n-1}$, we have

$$\begin{aligned} L(\mathbf{x}) \cdot \mathbf{v}_n &= \mathbf{x} \cdot L(\mathbf{v}_n) \\ &= \lambda_n \mathbf{x} \cdot \mathbf{v}_n \\ &= 0, \end{aligned}$$

using the symmetry of L (Lemma 8.5).

Let $L^* : \mathscr{R}_*^{n-1} \to \mathscr{R}_*^{n-1}$ and $q^* : \mathscr{R}_*^{n-1} \to \mathscr{R}$ be the "restrictions" of L and q to $\mathscr{R}_*^{n-1}$, that is,

$$L^*(\mathbf{x}) = L(\mathbf{x}) \qquad \text{and} \qquad q^*(\mathbf{x}) = q(\mathbf{x}), \qquad \mathbf{x} \in \mathscr{R}_*^{n-1}.$$

If $(u_1, \ldots, u_{n-1})$ are the coordinates of $\mathbf{x}$ with respect to an orthonormal basis for $\mathscr{R}^{n-1}$, then by Lemma 8.5,

$$L^*(\mathbf{u}) = B\mathbf{u},$$

where B is a symmetric $(n-1) \times (n-1)$ matrix. Then

$$\begin{aligned} q^*(\mathbf{u}) &= q(\mathbf{x}) \\ &= \mathbf{x} \cdot L(\mathbf{x}) \\ &= \mathbf{u} \cdot L^*(\mathbf{u}) \\ &= \mathbf{u}^{\mathrm{t}} B\mathbf{u}, \end{aligned}$$

so q^* is a quadratic form in $u_1, \ldots, u_{n-1}$.

The inductive hypothesis therefore gives an orthogonal set of unit eigenvectors $\mathbf{v}_1, \ldots, \mathbf{v}_{n-1}$ for L^*. Then $\mathbf{v}_1, \ldots, \mathbf{v}_n$ is the desired orthogonal set of unit eigenvectors for L.

To prove the last statement of the theorem, let $\mathbf{x} = \sum_1^n y_i \mathbf{v}_i$. Then

$$\begin{aligned} q(\mathbf{x}) &= \mathbf{x}^{\mathrm{t}} A\mathbf{x} \\ &= \mathbf{x} \cdot L(\mathbf{x}) \\ &= \left(\sum_1^n y_i \mathbf{v}_i\right) \cdot L\left(\sum_1^n y_j \mathbf{v}_j\right) \\ &= \left(\sum_1^n y_i \mathbf{v}_i\right) \cdot \left(\sum_{j=1}^n \lambda_j y_j \mathbf{v}_j\right) \\ &= \sum_{i,\, j=1}^n \lambda_j y_i y_j (\mathbf{v}_i \cdot \mathbf{v}_j) \\ &= \sum_{i=1}^n \lambda_i y_i^2, \end{aligned}$$

since $\mathbf{v}_i \cdot \mathbf{v}_j = 0$ unless $i = j$, and $\mathbf{v}_i \cdot \mathbf{v}_i = 1$. ∎

Equation (2) may be written in the form

$$q\left(\sum_{1}^{n} y_i \mathbf{v}_i\right) = (y_1 \cdots y_n)\begin{pmatrix} \lambda_1 & & & 0 \\ & \lambda_2 & & \\ & & \ddots & \\ 0 & & & \lambda_n \end{pmatrix}\begin{pmatrix} y_1 \\ \vdots \\ y_n \end{pmatrix}.$$

The process of finding an orthogonal set of unit eigenvectors, with respect to which the matrix of q is diagonal, is sometimes called *diagonalization* of the quadratic form q.

Note that, independently of Corollary 8.4, it follows immediately from Eq. (2) that

(a) q is positive-definite if all the eigenvalues are positive,
(b) q is negative-definite if all the eigenvalues are negative, and
(c) q is nondefinite if some are positive and some are negative.

If one or more eigenvalues are zero, while the others are either all positive or all negative, then q is neither positive-definite nor negative-definite nor nondefinite. According to Lemma 8.7 below, this can occur only if the determinant of the matrix of q vanishes.

Example 2 If $q(x, y, z) = xy + yz$, the matrix of q is

$$A = \begin{pmatrix} 0 & \frac{1}{2} & 0 \\ \frac{1}{2} & 0 & \frac{1}{2} \\ 0 & \frac{1}{2} & 0 \end{pmatrix}.$$

The characteristic equation is then

$$|A - \lambda I| = -\lambda^3 + \tfrac{1}{2}\lambda = 0$$

with solutions $\lambda_1 = -1/\sqrt{2}$, $\lambda_2 = 0$, $\lambda_3 = +1/\sqrt{2}$. By substituting each of these eigenvalues in turn into the matrix equation

$$(A - \lambda I)\mathbf{v} = 0$$

(from the proof of Theorem 8.3), and solving for a unit vector $\mathbf{v}$, we obtain the eigenvectors

$$\mathbf{v}_1 = \left(\frac{1}{2}, -\frac{1}{\sqrt{2}}, \frac{1}{2}\right), \qquad \mathbf{v}_2 = \left(\frac{1}{\sqrt{2}}, 0, -\frac{1}{\sqrt{2}}\right), \qquad \mathbf{v}_3 = \left(\frac{1}{2}, \frac{1}{\sqrt{2}}, \frac{1}{2}\right).$$

If u, v, w are coordinates with respect to coordinate axes determined by $\mathbf{v}_1, \mathbf{v}_2, \mathbf{v}_3$, that is, $\mathbf{x} = u\mathbf{v}_1 + v\mathbf{v}_2 + w\mathbf{v}_3$, then Eq. (2) gives

$$q(\mathbf{x}) = -\frac{1}{\sqrt{2}}u^2 + \frac{1}{\sqrt{2}}w^2.$$

Lemma 8.7 If $\lambda_1, \ldots, \lambda_n$ are eigenvalues of the symmetric $n \times n$ matrix A, corresponding to orthogonal unit eigenvectors $\mathbf{v}_1, \ldots, \mathbf{v}_n$, then

$$|A| = \lambda_1 \lambda_2 \cdots \lambda_n. \tag{3}$$

PROOF Let

$$\mathbf{v}_j = \begin{pmatrix} v_{1j} \\ \vdots \\ v_{nj} \end{pmatrix}, \qquad V = (v_{ij}).$$

Then, since $A\mathbf{v}_j = \lambda_i \mathbf{v}_j$, we have

$$AV = (\lambda_j v_{ij}).$$

It follows that

$$|A| \cdot |V| = \lambda_1 \lambda_2 \cdots \lambda_n |V|.$$

Since $|V| \neq 0$ because the vectors $\mathbf{v}_1, \ldots, \mathbf{v}_n$ are orthogonal and therefore linearly independent, this gives Eq. (3). ∎

We are finally ready to give a complete analysis of a quadratic form $q(\mathbf{x}) = \mathbf{x}^t A \mathbf{x}$ on $\mathscr{R}^n$ for which $|A| \neq 0$. Writing $A = (a_{ij})$ as usual, we denote by Δ_k the determinant of the upper left-hand $k \times k$ submatrix of A, that is,

$$\Delta_k = \begin{vmatrix} a_{11} & \cdots & a_{1k} \\ \vdots & & \\ a_{k1} & \cdots & a_{kk} \end{vmatrix}.$$

Thus

$$\Delta_1 = a_{11}, \quad \Delta_2 = \begin{vmatrix} a_{11} & a_{12} \\ a_{21} & a_{22} \end{vmatrix}, \ldots, \Delta_n = |A|.$$

Theorem 8.8 Let $q(\mathbf{x}) = \mathbf{x}^t A \mathbf{x}$ be a quadratic form on $\mathscr{R}^n$ with $|A| \neq 0$. Then q is

(a) positive-definite if and only if $\Delta_k > 0$ for each $k = 1, \ldots, n$,
(b) negative-definite if and only if $(-1)^k \Delta_k > 0$ for each $k = 1, \ldots, n$,
(c) nondefinite if neither of the previous two conditions is satisfied.

PROOF OF (a) This is obvious if $n = 1$; assume by induction its truth in dimension $n - 1$.

Since $0 < \Delta_n = |A| = \lambda_1 \lambda_2 \cdots \lambda_n$ by Lemma 8.7, all the eigenvalues $\lambda_1, \ldots, \lambda_n$ of A are nonzero. Either they are all positive, in which case we are finished, or an *even* number of them are negative. In the latter case, let λ_i and λ_j be two *negative* eigenvalues, corresponding to eigenvectors $\mathbf{v}_i$ and $\mathbf{v}_j$.

Let $\mathscr{P}$ be the 2-plane in $\mathscr{R}^n$ spanned by $\mathbf{v}_i$ and $\mathbf{v}_j$. Then, given $\mathbf{x} = y_i \mathbf{v}_i + y_j \mathbf{v}_j \in \mathscr{P}$, we have

$$\begin{aligned} q(\mathbf{x}) &= \mathbf{x} \cdot A\mathbf{x} \\ &= (y_i \mathbf{v}_i + y_j \mathbf{v}_j) \cdot (\lambda_i y_i \mathbf{v}_i + \lambda_j y_j \mathbf{v}_j) \\ &= \lambda_i y_i^2 + \lambda_j y_j^2 \\ &< 0 \end{aligned}$$

so q is *negative* on $\mathscr{P}$ minus the origin.

Now let $q_* : \mathscr{R}^{n-1} \to \mathscr{R}$ be the restriction of q to $\mathscr{R}^{n-1}$, defined by

$$\begin{aligned} q_*(x_1, \ldots, x_{n-1}) &= q(x_1, \ldots, x_{n-1}, 0) \\ &= (x_1 \cdots x_{n-1}) A_* \begin{pmatrix} x_1 \\ \vdots \\ x_{n-1} \end{pmatrix}, \end{aligned}$$

where

$$A_* = \begin{pmatrix} a_{11} & \cdots & a_{1,n-1} \\ \vdots & & \\ a_{n-1,1} & \cdots & a_{n-1,n-1} \end{pmatrix}.$$

Since $\Delta_1 > 0, \ldots, \Delta_{n-1} > 0$, the inductive assumption implies that q_* is positive-definite on $\mathscr{R}^{n-1}$. But this is a contradiction, since the $(n-1)$-plane $\mathscr{R}^{n-1}$ and the 2-plane $\mathscr{P}$, on which q is negative, must have at least a line in common. We therefore conclude that q has *no* negative eigenvalues, and hence is positive-definite.

Conversely, if q is positive-definite, we want to prove that each Δ_k is positive. If m is the minimum value of q on S^{n-1}, then $m > 0$.

Let $q_k : \mathscr{R}^k \to \mathscr{R}$ be the restriction of q to $\mathscr{R}^k$, that is,

$$\begin{aligned} q_k(x_1, \ldots, x_k) &= q(x_1, \ldots, x_k, 0, \ldots, 0) \\ &= (x_1 \cdots x_k) A_k \begin{pmatrix} x_1 \\ \vdots \\ x_k \end{pmatrix}, \end{aligned}$$

where A_k is the upper left-hand $k \times k$ submatrix of A, so $\Delta_k = |A_k|$, and define $L_k : \mathscr{R}^k \to \mathscr{R}^k$ by $L_k(\mathbf{x}) = A_k \mathbf{x}$. Then L_k is symmetric by Lemma 8.5.

Let $\mu_1, \ldots, \mu_k$ be the eigenvalues of L_k given by Theorem 8.6. Since by Lemma 8.2 each μ_i is the value of q_k at some point of S^{k-1}, the unit sphere in $\mathscr{R}^k$, and $q_k(\mathbf{x}) = q(\mathbf{x}, \mathbf{0})$, it follows that each $\mu_i \geqq m > 0$. Finally Lemma 8.7 implies that

$$\Delta_k = |A_k| = \mu_1 \mu_2 \cdots \mu_k \geqq m^k > 0.$$

PROOF OF (b) Define the quadratic form $q^* : \mathscr{R}^n \to \mathscr{R}$ by $q^*(\mathbf{x}) = -q(\mathbf{x})$. Then clearly q is negative-definite if and only if q^* is positive-definite. By part (a), q^*

is positive-definite if and only if each Δ_k^* is positive, where $\Delta_k^* = |-A_k|$, A_k being the upper left-hand $k \times k$ submatrix of A. But, since each of the k rows of $-A_k$ is -1 times the corresponding row of A_k, it follows that

$$\Delta_k^* = |-A_k| = (-1)^k |A_k| = (-1)^k \Delta_k.$$

PROOF OF (c) Since $\Delta_n = \lambda_1 \lambda_2 \cdots \lambda_n \neq 0$, all of the eigenvalues $\lambda_1, \ldots, \lambda_n$ are nonzero. If all of them were positive, then part (a) would imply that $\Delta_k > 0$ for all k, while if all of them were negative, then part (b) would imply that $(-1)^k \Delta_k > 0$ for all k. Since neither is the case, we conclude that q has both positive eigenvalues and negative eigenvalues, and is therefore nondefinite. ∎

Recalling that, if $\mathbf{a}$ is a critical point of the class $\mathscr{C}^3$ function $f: \mathscr{R}^n \to \mathscr{R}$, the quadratic form of f at $\mathbf{a}$ is $q(\mathbf{x}) = \frac{1}{2}\mathbf{x}^t(D_i D_j f(\mathbf{a}))\mathbf{x}$, Theorems 7.5 and 8.8 enable us to determine the behavior of f in a neighborhood of $\mathbf{a}$, provided that the determinant

$$|D_i D_j f(\mathbf{a})|,$$

often called the *Hessian determinant* of f at $\mathbf{a}$, is *nonzero*. But if $|D_i D_j f(\mathbf{a})| = 0$, then these theorems do not apply, and *ad hoc* methods must be applied.

Note that the classical second derivative test for a function of two variables (Theorem 4.4) is an immediate corollary to Theorems 7.5 and 8.8.

It will clarify the picture at this point to generalize the discussion at the end of Section 6. Suppose for the sake of simplicity that the origin is a critical point of the class $\mathscr{C}^3$ function $f: \mathscr{R}^n \to \mathscr{R}$, and that $f(\mathbf{0}) = 0$. Then

$$f(\mathbf{x}) = q(\mathbf{x}) + R_2(\mathbf{x}) \tag{4}$$

with

$$q(\mathbf{x}) = \frac{1}{2}\mathbf{x}^t(D_i D_j f(\mathbf{0}))\mathbf{x} \qquad \text{and} \qquad \lim_{|\mathbf{x}| \to 0} \frac{R_2(\mathbf{x})}{|\mathbf{x}|^2} = 0.$$

Let $\mathbf{v}_1, \ldots, \mathbf{v}_n$ be an orthogonal set of unit eigenvectors (provided by Theorem 8.6), with associated eigenvalues $\lambda_1, \ldots, \lambda_n$. Assume that $|D_i D_j f(0)| \neq 0$, so Lemma 8.7 implies that each λ_i is nonzero. Let $\lambda_1, \ldots, \lambda_k$ be the positive eigenvalues, and $\lambda_{k+1}, \ldots, \lambda_n$ the negative ones (writing $k = 0$ if all the eigenvalues are negative). Since

$$q(\mathbf{x}) = \lambda_1 {y_1}^2 + \cdots + \lambda_n {y_n}^2 \tag{5}$$

if $\mathbf{x} = y_1\mathbf{v}_1 + \cdots + y_n\mathbf{v}_n$, it is clear that q has a minimum at $\mathbf{0}$ if $k = n$ and a maximum if $k = 0$.

If $1 \leqq k < n$, so that q has both positive and negative eigenvalues, denote by $\mathscr{R}_*^k$ the subspace of $\mathscr{R}^n$ spanned by $\mathbf{v}_1, \ldots, \mathbf{v}_k$, and by $\mathscr{R}_*^{n-k}$ the subspace of $\mathscr{R}^n$ spanned by $\mathbf{v}_{k+1}, \ldots, \mathbf{v}_n$. Then the graph of $z = q(\mathbf{x})$ in $\mathscr{R}^{n+1} = \{(\mathbf{x}, z) | \mathbf{x} \in \mathscr{R}^n$

and $z \in \mathscr{R}\}$ looks like Fig. 2.41; q has a minimum on $\mathscr{R}_*^k$ and a maximum on $\mathscr{R}_*^{n-k}$.

Given $\varepsilon < \min_i |\lambda_i|$, choose $\delta > 0$ such that

$$0 < |x| < \delta \Rightarrow \frac{|R_2(\mathbf{x})|}{|\mathbf{x}|^2} < \varepsilon,$$

or

$$-\varepsilon \sum_{i=1}^{n} y_i^2 < R_2(\mathbf{x}) < \varepsilon \sum_{i=1}^{n} y_i^2$$

since $|\mathbf{x}|^2 = \sum_{i=1}^n y_i^2$. This inequality and Eq. (4) yield

$$q(\mathbf{x}) - \varepsilon \sum y_i^2 < f(\mathbf{x}) < q(\mathbf{x}) + \varepsilon \sum y_i^2.$$

Using (5), we obtain

$$\sum_{i=1}^{k} (\lambda_i - \varepsilon) y_i^2 - \sum_{i=k+1}^{n} (\mu_i + \varepsilon) y_i^2 < f(\mathbf{x}) < \sum_{i=1}^{k} (\lambda_i + \varepsilon) y_i^2 - \sum_{i=k+1}^{n} (\mu_i - \varepsilon) y_i^2, \tag{6}$$

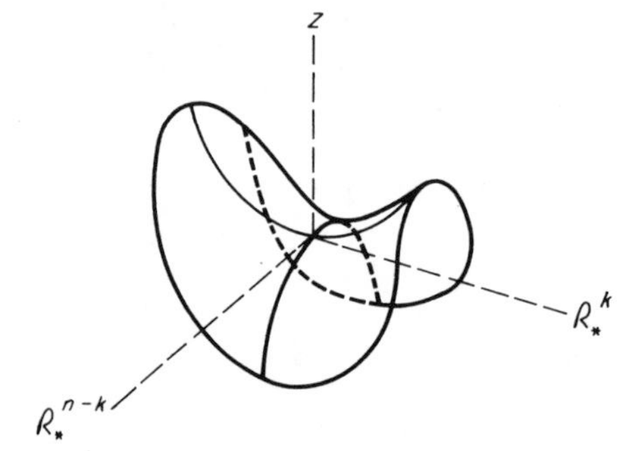

Figure 2.41

where $\mu_i = -\lambda_i > 0$, $i = k+1, \ldots, n$. Since $\lambda_i - \varepsilon > 0$ and $\mu_i - \varepsilon > 0$ because $\varepsilon < \min |\lambda_i|$, inequalities (6) show that, for $|\mathbf{x}| < \delta$, the graph of $z = f(\mathbf{x})$ lies between two very close quadratic surfaces (hyperboloids) of the same type as $z = q(\mathbf{x})$. In particular, if $1 \leqq k < n$, then f has a local minimum on the subspace $\mathscr{R}_*^k$ and a local maximum on the subspace $\mathscr{R}_*^{n-k}$. This makes it clear that the origin is a "saddle point" for f if q has both positive and negative eigenvalues. Thus the general case is just like the 2-dimensional case except that, when there is a saddle point, the subspace on which the critical point is a maximum point may have any dimension from 1 to $n - 1$.

It should be emphasized that, if the Hessian determinant $|D_i D_j f(\mathbf{a})|$ is *zero*, then the above results yield no information about the critical point $\mathbf{a}$. This will be the situation if the quadratic form q of f at $\mathbf{a}$ is either positive-semidefinite but not positive-definite, or negative-semidefinite but not negative-definite. The quadratic form q is called *positive-semidefinite* if $q(x) \geqq 0$ for all $\mathbf{x}$, and *negative-semidefinite* if $q(\mathbf{x}) \leqq 0$ for all $\mathbf{x}$. Notice that the terminology "q is nondefinite," which we have been using, actually means that q is neither positive-semidefinite nor negative-semidefinite (so we might more descriptively have said "nonsemidefinite").

Example 3 Let $f(x, y) = x^2 - 2xy + y^2 + x^4 + y^4$, and $g(x, y) = x^2 - 2xy + y^2 - x^4 - y^4$. Then the quadratic form for both at the critical point $(0, 0)$ is

$$q(x, y) = x^2 - 2xy + y^2 = (x - y)^2,$$

which is positive-semidefinite but not positive-definite. Since

$$f(x, y) = (x - y)^2 + x^4 + y^4,$$

f has a relative minimum at $(0, 0)$. However, since

$$g(t, t) = -2t^4, \qquad g(t, -t) = 2t^2(2 - t^2),$$

it is clear that g has a saddle point at $(0, 0)$.

Thus, if the Hessian determinant vanishes, then anything can happen.

We conclude this section with discussion of a "second derivative test" for constrained (Lagrange multiplier) maximum–minimum problems. We first recall the setting for the general Lagrange multiplier problem: We have a differentiable function $f: \mathscr{R}^n \to \mathscr{R}$ and a continuously differentiable mapping $G: \mathscr{R}^n \to \mathscr{R}^m$ $(m < n)$, and denote by M the set of all those points $\mathbf{x} \in G^{-1}(\mathbf{0})$ at which the gradient vectors $\nabla G_1(\mathbf{x}), \ldots, \nabla G_m(\mathbf{x})$, of the component functions of G, are linearly independent. By Theorem 5.7, M is an $(n - m)$-dimensional manifold in $\mathscr{R}^n$, and by Theorem 5.6 has an $(n - m)$-dimensional tangent space $T_{\mathbf{x}}$ at each point $\mathbf{x} \in M$. Recall that, by definition, $T_{\mathbf{x}}$ is the subspace of $\mathscr{R}^n$ consisting of all velocity vectors (at $\mathbf{x}$) to differentiable curves on M which pass through $\mathbf{x}$. *This notation will be maintained throughout the present discussion.*

We are interested in locating and classifying those points of M at which the function f attains its local maxima and minima on M. According to Theorem 5.1, in order for f to have a local maximum or minimum on M at $\mathbf{a} \in M$, it is necessary that $\mathbf{a}$ be a critical point for f on M in the following sense. The point $\mathbf{a} \in M$ is called a *critical point for f on M* if and only if the gradient vector $\nabla f(\mathbf{a})$ is orthogonal to the tangent space $T_{\mathbf{a}}$ of M at $\mathbf{a}$. Since the linearly independent gradient vectors $\nabla G_1(\mathbf{a}), \ldots, \nabla G_m(\mathbf{a})$ generate the orthogonal complement to $T_{\mathbf{a}}$ (see Fig. 2.42), it follows as in Theorem 5.8 that there exist unique numbers $\lambda_1, \ldots, \lambda_m$ such that

$$\nabla f(\mathbf{a}) = \sum_{\lambda=1}^{m} \lambda_i \nabla G_i(\mathbf{a}). \tag{7}$$

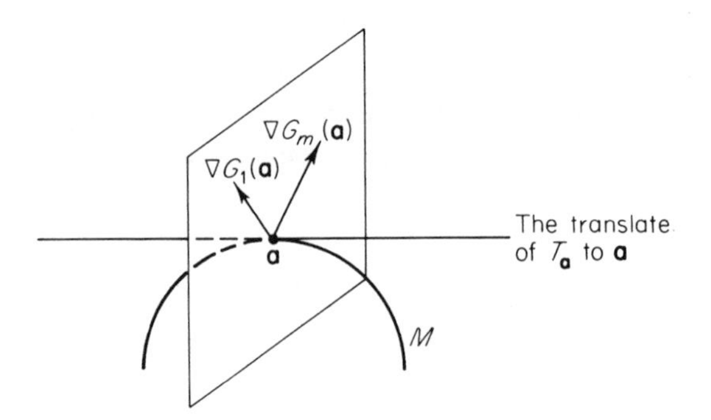

Figure 2.42

We will obtain *sufficient* conditions for f to have a local extremum at $\mathbf{a}$ by considering the *auxiliary function* $H : \mathscr{R}^n \to \mathscr{R}$ *for* f *at* $\mathbf{a}$ defined by

$$H(\mathbf{x}) = f(\mathbf{x}) - \sum_{i=1}^{m} \lambda_i G_i(\mathbf{x}). \tag{8}$$

Notice that (7) simply asserts that $\nabla H(\mathbf{a}) = 0$, so $\mathbf{a}$ is an (ordinary) critical point for H. We are, in particular, interested in the quadratic form

$$q(\mathbf{h}) = \frac{1}{2} \sum_{i,j=1}^{n} D_i D_j H(\mathbf{a}) h_i h_j$$

of H at $\mathbf{a}$.

Theorem 8.9 Let $\mathbf{a}$ be a critical point for f on M, and denote by $q : \mathscr{R}^n \to \mathscr{R}$ the quadratic form at $\mathbf{a}$ of the auxiliary function $H = f - \sum_{i=1}^{m} \lambda_i G_i$, as above. If f and G are both of class $\mathscr{C}^3$ in a neighborhood of $\mathbf{a}$, then f has

(a) a local minimum on M at $\mathbf{a}$ if q is positive-definite on the tangent space $T_{\mathbf{a}}$ to M at $\mathbf{a}$,
(b) a local maximum on M at $\mathbf{a}$ if q is negative-definite on $T_{\mathbf{a}}$,
(c) neither if q is nondefinite on $T_{\mathbf{a}}$.

The statement, that "q is positive-definite on $T_{\mathbf{a}}$," naturally means that $q(\mathbf{h}) > 0$ for all nonzero vectors $\mathbf{h} \in T_{\mathbf{a}}$; similarly for the other two cases.

PROOF The proof of each part of this theorem is an extension of the proof of the corresponding part of Theorem 7.5. We give the details only for part (a).

We start with the Taylor expansion

$$H(\mathbf{a} + \mathbf{h}) - H(\mathbf{a}) = q(\mathbf{h}) + R_2(\mathbf{h})$$

for the auxiliary function H. Since $H(\mathbf{x}) = f(\mathbf{x})$ if $\mathbf{x} \in M$, it clearly suffices to find a $\delta > 0$ such that

$$0 < |\mathbf{h}| < \delta \Rightarrow \frac{q(\mathbf{h}) + R_2(\mathbf{h})}{|\mathbf{h}|^2} > 0 \tag{9}$$

if $\mathbf{a} + \mathbf{h} \in M$.

Let m be the minimum value attained by q on the unit sphere $S^* = S^{n-1} \cap T_{\mathbf{a}}$ in the tangent plane $T_{\mathbf{a}}$. Then $m > 0$ because q is positive-definite on $T_{\mathbf{a}}$. Noting that

$$\frac{q(\mathbf{h}) + R_2(\mathbf{h})}{|\mathbf{h}|^2} = q\left(\frac{\mathbf{h}}{|\mathbf{h}|}\right) + \frac{R_2(\mathbf{h})}{|\mathbf{h}|^2},$$

we choose $\delta > 0$ so small that

$$0 < |\mathbf{h}| < \delta \Rightarrow \frac{|R_2(\mathbf{h})|}{|\mathbf{h}|^2} < \frac{m}{2},$$

and also so small that $0 < |\mathbf{h}| < \delta$ and $\mathbf{a} + \mathbf{h} \in M$ together imply that $\mathbf{h}/|\mathbf{h}|$ is sufficiently near to S^* that

$$q\left(\frac{\mathbf{h}}{|\mathbf{h}|}\right) > \frac{m}{2}.$$

Then (9) follows as desired. The latter condition on δ can be satisfied because S^* is (by definition) the set of all tangent vectors at $\mathbf{a}$ to unit-speed curves on M which pass through $\mathbf{a}$. Therefore, if $\mathbf{a} + \mathbf{h} \in M$ and $\mathbf{h}$ is sufficiently small, it follows that $\mathbf{h}/|\mathbf{h}|$ is close to a vector in S^*. ∎

The following two examples provide illustrative applications of Theorem 8.9.

Example 4 In Example 4 of Section 5 we wanted to find the box with volume 1000 having least total surface area, and this involved minimizing the function

$$f(x, y, z) = 2xy + 2xz + 2yz$$

on the 2-manifold $M \subset \mathscr{R}^3$ defined by the constraint equation

$$g(x, y, z) = xyz - 1000 = 0.$$

We found the single critical point $\mathbf{a} = (10, 10, 10)$ for f on M, with $\lambda = \frac{2}{5}$. The auxiliary function is then defined by

$$\begin{aligned} h(x, y, z) &= f(x, y, z) - \lambda g(x, y, z) \\ &= 2xy + 2xz + 2yz - \tfrac{2}{5}xyz + 400. \end{aligned}$$

We find by routine computation that the matrix of second partial derivatives of h at $\mathbf{a}$ is

$$\begin{pmatrix} 0 & -2 & -2 \\ -2 & 0 & -2 \\ -2 & -2 & 0 \end{pmatrix},$$

so the quadratic form of h at $\mathbf{a}$ is

$$q(x, y, z) = -2xy - 2xz - 2yz. \tag{10}$$

It is clear that q is nondefinite on $\mathscr{R}^3$. However it is the behavior of q on the tangent plane $T_{\mathbf{a}}$ that interests us.

Since the gradient vector $\nabla g(\mathbf{a}) = (100, 100, 100)$ is orthogonal to M at $\mathbf{a} = (10, 10, 10)$, the tangent plane $T_{\mathbf{a}}$ is generated by the vectors

$$\mathbf{v}_1 = (1, -1, 0) \qquad \text{and} \qquad \mathbf{v}_2 = (1, 0, -1).$$

Given $\mathbf{v} \in T_{\mathbf{a}}$, we may therefore write

$$\mathbf{v} = s\mathbf{v}_1 + t\mathbf{v}_2 = (s + t, -s, -t),$$

and then find by substitution into (10) that

$$q(\mathbf{v}) = 2s^2 + 2st + 2t^2.$$

Since the quadratic form $s^2 + st + t^2$ is positive-definite (by the $ac - b^2$ test), it follows that q is *positive-definite* on $T_{\mathbf{a}}$. Therefore Theorem 8.9 assures us that f does indeed have a local *minimum* on M at the critical point $\mathbf{a} = (10, 10, 10)$.

Example 5 In Example 9 of Section 5 we sought the minimum distance between the circle $x^2 + y^2 = 1$ and the straight line $x + y = 4$. This involved minimizing the function

$$f(x, y, u, v) = (x - u)^2 + (y - v)^2$$

on the 2-manifold $M \subset \mathscr{R}^4$ defined by the constraint equations

$$G_1(x, y, u, v) = x^2 + y^2 - 1 = 0,$$
$$G_2(x, y, u, v) = u + v - 4 = 0.$$

We found the geometrically obvious critical points (see Fig. 2.43), $\mathbf{a} = (1/\sqrt{2}, 1/\sqrt{2}, 2, 2)$ with $\lambda_1 = 1 - 2\sqrt{2}$ and $\lambda_2 = 4 - \sqrt{2}$, and $\mathbf{b} = (-1/\sqrt{2}, -1/\sqrt{2}, 2, 2)$ with $\lambda_1 = 1 + 2\sqrt{2}$ and $\lambda_2 = 4 + \sqrt{2}$. It is obvious that f has a minimum at $\mathbf{a}$, and neither a maximum nor a minimum at $\mathbf{b}$, but we want to verify this using Theorem 8.9.

The auxiliary function H is defined by

$$H(x, y, u, v) = (x - u)^2 + (y - v)^2 - \lambda_1(x^2 + y^2 - 1) - \lambda_2(u + v - 4).$$

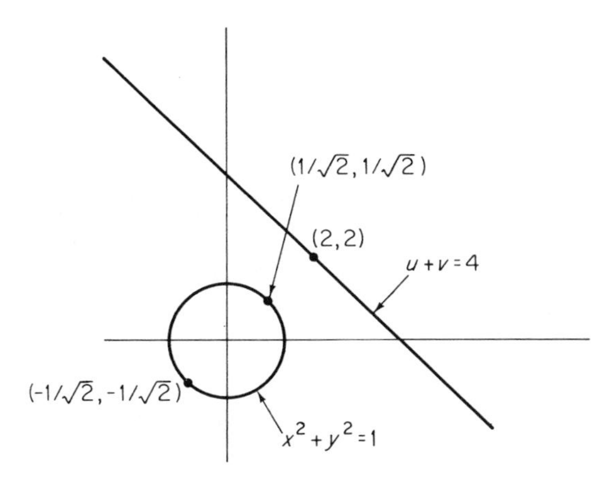

Figure 2.43

By routine computation, we find that the matrix of second partial derivatives of H is

$$\begin{pmatrix} 2-2\lambda_1 & 0 & -2 & 0 \\ 0 & 2-2\lambda_1 & 0 & -2 \\ -2 & 0 & 2 & 0 \\ 0 & -2 & 0 & 2 \end{pmatrix}.$$

Substituting $\lambda_1 = 1 - 2\sqrt{2}$ and $\lambda_2 = 4 - \sqrt{2}$, we find that the quadratic form of H at the critical point $\mathbf{a} = (1/\sqrt{2}, 1/\sqrt{2}, 2, 2)$ is

$$q(x, y, u, v) = 4\sqrt{2}\,x^2 + 4\sqrt{2}\,y^2 + 2u^2 + 2v^2 - 4xu - 4yv.$$

Computing the subdeterminants $\Delta_1, \Delta_2, \Delta_3, \Delta_4$ of Theorem 8.8, we find that all are positive, so this quadratic form q is positive-definite on $\mathscr{R}^4$, and hence on the tangent plane $T_\mathbf{a}$ (this could also be verified by the method of the previous example). Hence Theorem 8.9 implies that f does indeed attain a local minimum at $\mathbf{a}$.

Substituting $\lambda_1 = 1 + 2\sqrt{2}$ and $\lambda_2 = 4 + 2\sqrt{2}$, we find that the quadratic form of H at the other critical point $\mathbf{b} = (-1/\sqrt{2}, -1/\sqrt{2}, 2, 2)$ is

$$q(x, y, u, v) = -4\sqrt{2}\,x^2 - 4\sqrt{2}\,y^2 + 2u^2 + 2v^2 - 4xu - 4yv.$$

Now the vectors $\mathbf{v}_1 = (1, -1, 0, 0)$ and $\mathbf{v}_2 = (0, 0, 1, -1)$ obviously lie in the tangent plane $T_\mathbf{b}$. Since

$$q(\mathbf{v}_1) = -8\sqrt{2} < 0 \qquad \text{while} \qquad q(\mathbf{v}_2) = 8 > 0,$$

we see that q is nondefinite on $T_\mathbf{b}$, so Theorem 8.9 implies that f does not have a local extremum at $\mathbf{b}$.

Exercises

8.1 This exercise gives an alternative proof of Theorem 8.8 in the 3-dimensional case. If

$$q(x, y, z) = (x\ y\ z)\ A \begin{pmatrix} x \\ y \\ z \end{pmatrix},$$

where $A = (a_{ij})$ is a symmetric 3×3 matrix with $\Delta_1 \neq 0$, $\Delta_2 \neq 0$, show that

$$q(x, y, z) = \Delta_1 \left(x + \frac{a_{12}}{a_{11}} y + \frac{a_{13}}{a_{11}} z \right)^2 + \frac{\Delta_2}{\Delta_1} \left(y + \frac{\begin{vmatrix} a_{11} & a_{12} \\ a_{31} & a_{32} \end{vmatrix}}{\Delta_3} z \right)^2 + \frac{\Delta_3}{\Delta_2} z^2.$$

Conclude that q is positive-definite if $\Delta_1, \Delta_2, \Delta_3$ are all positive, while q is negative-definite if $\Delta_1 < 0$, $\Delta_2 > 0$, $\Delta_3 < 0$.

8.2 Show that $q(x, y, z) = 2x^2 + 5y^2 + 2z^2 + 2xz$ is positive-definite by
(a) applying the previous exercise,
(b) solving the characteristic equation.

8.3 Use the method of Example 2 to diagonalize the quadratic form of Exercise 8.2, and find its eigenvectors. Then sketch the graph of the equation $2x^2 + 5y^2 + 2z^2 + 2xz = 1$.

8.4 This problem deals with the function $f : \mathscr{R}^3 \to \mathscr{R}$ defined by

$$f(x, y, z) = x^2 + 4y^2 + z^2 + 2xz + (x^2 + y^2 + z^2) \cos xyz.$$

Note that $\mathbf{0} = (0, 0, 0)$ is a critical point of f.
(a) Show that $q(x, y, z) = 2x^2 + 5y^2 + 2z^2 + 2xz$ is the quadratic form of f at $\mathbf{0}$ by substituting the expansion $\cos t = 1 - \frac{1}{2}t^2 + R(t)$, collecting all second degree terms, and *verifying* the appropriate condition on the remainder $\bar{R}(x, y, z)$. State the uniqueness theorem which you apply.
(b) Write down the symmetric 3×3 matrix A such that

$$q(x, y, z) = (x\ y\ z)\ A \begin{pmatrix} x \\ y \\ z \end{pmatrix}.$$

By calculating determinants of appropriate submatrices of A, determine the behavior of f at $\mathbf{0}$.
(c) Find the eigenvalues $\lambda_1, \lambda_2, \lambda_3$ of q. Let $\mathbf{v}_1, \mathbf{v}_2, \mathbf{v}_3$ be the eigenvectors corresponding to $\lambda_1, \lambda_2, \lambda_3$ (do not solve for them). Express q in the uvw-coordinate system determined by $\mathbf{v}_1, \mathbf{v}_2, \mathbf{v}_3$. That is, write $q(u\mathbf{v}_1 + v\mathbf{v}_2 + w\mathbf{v}_3)$ in terms of u, v, w. Then give a geometric description (or sketch) of the surface

$$2x^2 + 5y^2 + 2z^2 + 2xz = 1.$$

8.5 Suppose you want to design an ice-cube tray of minimal cost, which will make one dozen ice "cubes" (not necessarily actually cubes) from 12 in.^{3} of water. Assume that the tray is divided into 12 square compartments in a 2×6 pattern as shown in Fig. 2.44, and that

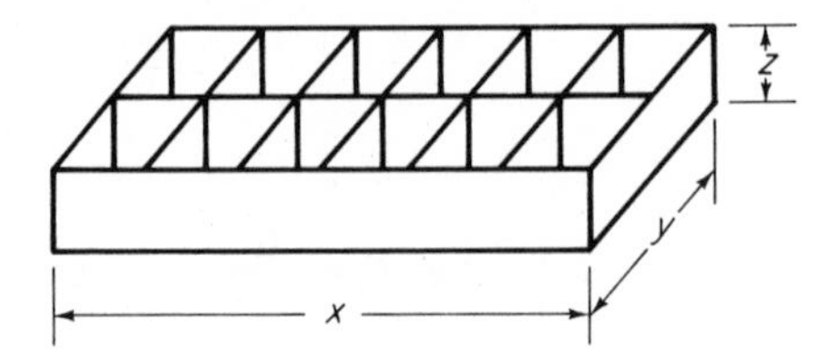

Figure 2.44

the material used costs 1 cent per square inch. Use the Lagrange multiplier method to minimize the cost function $f(x, y, z) = xy + 3xz + 7yz$ subject to the constraints $x = 3y$ and $xyz = 12$. Apply Theorem 8.9 to verify that you do have a local minimum.

8.6 Apply Theorem 8.8 to show that the quadratic form

$$q(x_1, x_2, x_2, x_4) = x_1{}^2 + x_2{}^2 - x_3{}^2 - x_4{}^2 + 4x_1x_2 - 2x_1x_4 + 2x_3x_4$$

is nondefinite. Solve the characteristic equation for its eigenvalues, and then find an orthonormal set of eigenvectors.

8.7 Let $f(\mathbf{x}) = \mathbf{x}^t A\mathbf{x}$ be a quadratic form on $\mathscr{R}^n$ (with A symmetric), and let $\mathbf{v}_1, \ldots, \mathbf{v}_n$ be the orthonormal eigenvectors given by Theorem 8.6. If $P = (\mathbf{v}_1, \ldots, \mathbf{v}_n)$ is the $n \times n$ matrix having these eigenvectors as its column vectors, show that

$$P^{-1}AP = P^tAP = \begin{pmatrix} \lambda_1 & & & 0 \\ & \lambda_2 & & \\ & & \ddots & \\ 0 & & & \lambda_n \end{pmatrix}.$$

Hint: Let $\mathbf{x} = P\mathbf{y} = y_1\mathbf{v}_1 + \cdots + y_n\mathbf{v}_n$. Then

$$f(\mathbf{x}) = \mathbf{y}^t(P^tAP)\mathbf{y}$$

by the definition of f, while

$$f(\mathbf{x}) = \sum_{i=1}^{n} \lambda_i y_i{}^2$$

by Theorem 8.6. Writing $B = P^tAP$, we have

$$\sum_{i,\, j=1}^{n} b_{ij} y_i y_j = \sum_{i=1}^{n} \lambda_i y_i{}^2$$

for all $\mathbf{y} = (y_1, \ldots, y_n)$. Using the obvious symmetry of B prove that $b_{ii} = \lambda_i$, while $b_{ij} = 0$ if $i \neq j$. Finally note that $P^{-1} = P^t$ by Exercise 6.11 of Chapter I.

Successive Approximations and Implicit Functions

Many important problems in mathematics involve the solution of equations or systems of equations. In this chapter we study the central existence theorems of multivariable calculus that pertain to such problems. The inverse mapping theorem (Theorem 3.3) deals with the problem of solving a system of n equations in n unknowns, while the implicit mapping theorem (Theorem 3.4) deals with a system of n equations in $m + n$ variables $x_1, \ldots, x_m, y_1, \ldots, y_n$, the problem being to solve for $y_1, \ldots, y_n$ as functions of $x_1, \ldots, x_m$.

Our method of treatment of these problems is based on a fixed point theorem known as the contraction mapping theorem. This method not only suffices to establish (under appropriate conditions) the existence of a solution of a system of equations, but also yields an explicitly defined sequence of successive approximations converging to that solution.

In Section 1 we discuss the special cases $n = 1$, $m = 1$, that is, the solution of a single equation $f(x) = 0$ or $G(x, y) = 0$ in one unknown. In Section 2 we give a multivariable generalization of the mean value theorem that is needed for the proofs in Section 3 of the inverse and implicit mapping theorems. In Section 4 these basic theorems are applied to complete the discussion of manifolds that was begun in Chapter II (in connection with Lagrange multipliers).

1 NEWTON'S METHOD AND CONTRACTION MAPPINGS

There is a simple technique of elementary calculus known as Newton's method, for approximating the solution of an equation $f(x) = 0$, where $f: \mathscr{R} \to \mathscr{R}$ is a $\mathscr{C}^1$ function. Let $[a, b]$ be an interval on which $f'(x)$ is nonzero and $f(x)$

changes sign, so the equation $f(x) = 0$ has a single root $x_* \in [a, b]$. Given an arbitrary point $x_0 \in [a, b]$, linear approximation gives

$$f(x_0) - f(x_*) \approx f'(x_0)(x_0 - x_*),$$

from which we obtain

$$x_* \approx x_1 = x_0 - \frac{f(x_0)}{f'(x_0)}.$$

So $x_1 = x_0 - [f(x_0)/f'(x_0)]$ is our first approximation to the root x_* (see Fig. 3.1). Similarly, we obtain the $(n + 1)$th approximation from the nth approximation x_n,

$$x_{n+1} = x_n - \frac{f(x_n)}{f'(x_n)}. \tag{1}$$

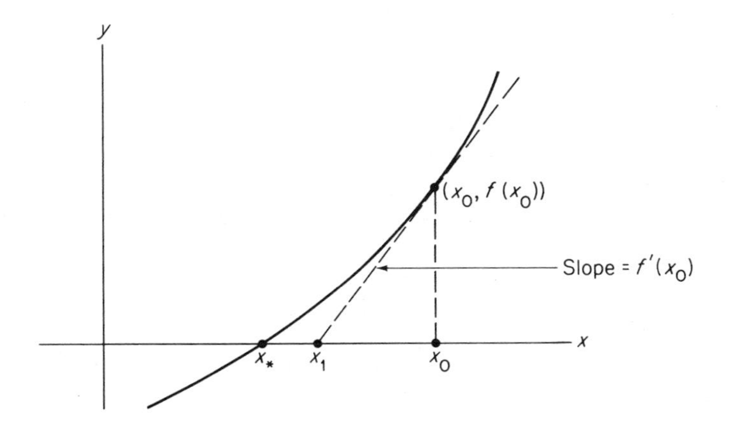

Figure 3.1

Under appropriate hypotheses it can be proved that the sequence $\{x_n\}_0^\infty$ defined inductively by (1) converges to the root x_*. We shall not include such a proof here, because our main interest lines in a certain modification of Newton's method, rather than in Newton's method itself.

In order to avoid the repetitious computation of the derivatives $f'(x_0)$, $f'(x_1), \ldots, f'(x_n), \ldots$ required in (1), it is tempting to simplify the method by considering instead of (1) the alternative sequence defined inductively by

$$x_{n+1} = x_n - \frac{f(x_n)}{f'(x_0)}.$$

But this sequence may fail to converge, as illustrated by Fig. 3.2. However there is a similar simplification which "works"; we consider the sequence $\{x_n\}_0^\infty$ defined inductively by

$$x_{n+1} = x_n - \frac{f(x_n)}{M}, \tag{2}$$

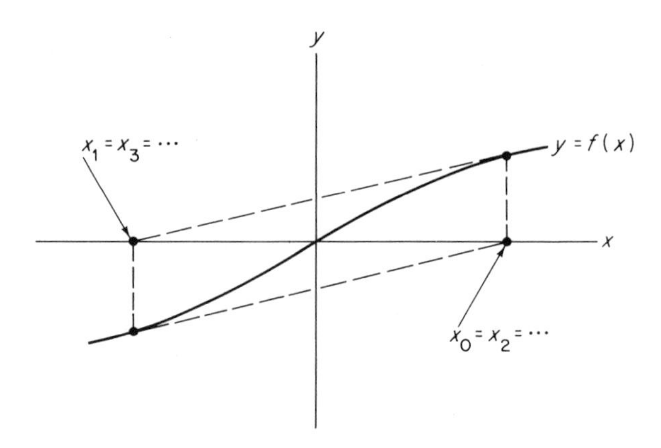

Figure 3.2

where $M = \max |f'(x)|$ if $f'(x) > 0$ on $[a, b]$, and $M = -\max |f'(x)|$ if $f'(x) < 0$ on $[a, b]$. The proof that $\lim x_n = x_*$ if $\{x_n\}_0^\infty$ is defined by (2) makes use of a widely applicable technique which we summarize in the contraction mapping theorem below.

The mapping $\varphi : [a, b] \to [a, b]$ is called a *contraction mapping with contraction constant* $k < 1$ if

$$|\varphi(x) - \varphi(y)| \leqq k|x - y| \tag{3}$$

for all $x, y \in [a, b]$. The contraction mapping theorem asserts that each contraction mapping has a unique *fixed point*, that is, a point $x_* \in [a, b]$ such that $\varphi(x_*) = x_*$, and at the same time provides a sequence of successive approximations to x_*.

Theorem 1.1 Let φ: $[a, b] \to [a, b]$ be a contraction mapping with contraction constant $k < 1$. Then φ has a unique fixed point x_*. Moreover, given $x_0 \in [a, b]$, the sequence $\{x_n\}_0^\infty$ defined inductively by

$$x_{n+1} = \varphi(x_n)$$

converges to x_*. In particular,

$$|x_n - x_*| \leqq \frac{k^n|x_0 - x_1|}{1 - k} \tag{4}$$

for each n.

PROOF Application of condition (3) gives

$$|x_{n+1} - x_n| = |\varphi(x_n) - \varphi(x_{n-1})| \leqq k|x_n - x_{n-1}|,$$

so it follows easily by induction that $|x_{n+1} - x_n| \leqq k^n|x_1 - x_0|$. From this it follows in turn that, if $0 < n < m$, then

$$\begin{aligned}|x_n - x_m| &\leqq |x_n - x_{n+1}| + \cdots + |x_{m-1} - x_m| \\ &\leqq (k^n + \cdots + k^{m-1})|x_1 - x_0| \\ &\leqq k^n|x_1 - x_0|(1 + k + k^2 + \cdots),\end{aligned}$$

$$|x_n - x_m| \leqq \frac{k^n|x_0 - x_1|}{1 - k}, \tag{5}$$

using in the last step the formula for the sum of a geometric series.

Thus the sequence $\{x_n\}_0^\infty$ is a Cauchy sequence, and therefore converges to a point $x_* \in [a, b]$. Note now that (4) follows directly from (5), letting $m \to \infty$.

Condition (3) immediately implies that φ is continuous on $[a, b]$, so

$$\varphi(x_*) = \lim_{n \to \infty} \varphi(x_n) = \lim_{n \to \infty} x_{n+1} = x_*$$

as desired.

Finally, if x_{**} were another fixed point of φ, we would have

$$|x_* - x_{**}| = |\varphi(x_*) - \varphi(x_{**})| \leqq k|x_* - x_{**}|.$$

Since $k < 1$, it follows that $x_* = x_{**}$, so x_* is the unique fixed point of φ. ∎

We are now ready to consider the simplification of Newton's method described by Eq. (2). See Fig. 3.3 for a picture of the sequence defined by this equation.

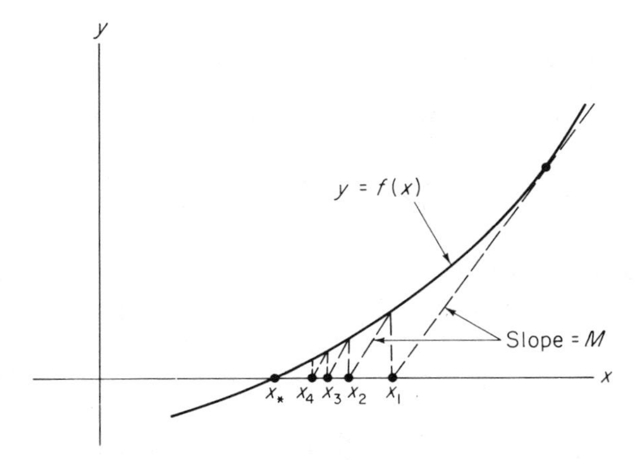

Figure 3.3

Theorem 1.2 Let $f\colon [a, b] \to \mathscr{R}$ be a differentiable function with $f(a) < 0 < f(b)$ and $0 < m < f'(x) \leqq M$ for $x \in [a, b]$. Given $x_0 \in [a, b]$, the sequence $\{x_n\}_0^\infty$ defined inductively by

$$x_{n+1} = x_n - \frac{f(x_n)}{M} \tag{2}$$

converges to the unique root $x_* \in [a, b]$ of the equation $f(x) = 0$. In particular,

$$|x_n - x_*| \leqq \frac{|f(x_0)|}{m}\left(1 - \frac{m}{M}\right)^n \tag{6}$$

for each n.

PROOF Define $\varphi : [a, b] \to \mathscr{R}$ by $\varphi(x) = x - [f(x)/M]$. We want to show that φ is a contraction mapping of $[a, b]$. Since $\varphi'(x) = 1 - [f'(x)/M]$, we see that

$$0 \leqq \varphi'(x) \leqq 1 - \frac{m}{M} = k < 1,$$

so φ is a nondecreasing function. Therefore $a < a - [f(a)/M] = \varphi(a) \leqq \varphi(x) \leqq \varphi(b) = b - [f(b)/M] < b$ for all $x \in [a, b]$, because $f(a) < 0 < f(b)$. Thus $\varphi([a, b]) \subset [a, b]$, and φ is a contraction mapping of $[a, b]$.

Therefore Theorem 1.1 implies that φ has a unique fixed point

$$x_* = \varphi(x_*) = x_* - \frac{f(x_*)}{M},$$

which is clearly the unique root of $f(x) = 0$, and the sequence $\{x_n\}_0^\infty$ defined by

$$x_{n+1} = \varphi(x_n) = x_n - \frac{f(x_n)}{M}$$

converges to x_*. Moreover Eq. (4) gives

$$|x_n - x_*| \leqq \frac{k^n|x_0 - x_1|}{1 - k} = \frac{|f(x_0)|}{m}\left(1 - \frac{m}{M}\right)^n$$

upon substitution of $x_1 = x_0 - [f(x_0)/M]$ and $k = 1 - (m/M)$. ∎

Roughly speaking, Theorem 1.2 says that, if the point x_0 is sufficiently near to a root x_* of $f(x) = 0$ where $f'(x_*) > 0$, then the sequence defined by (2) converges to x_*, with M being an upper bound for $|f'(x)|$ near x_*.

Now let us turn the question around. Given a point x_* where $f(x_*) = 0$, and a number y close to 0, can we find a point x near x_* such that $f(x) = y$? More generally, supposing that $f(a) = b$, and that y is close to b, can we find x close to a such that $f(x) = y$? If so, then

$$y - b = f(x) - f(a) \approx f'(a)(x - a).$$

If $f'(a) \neq 0$, we can solve for

$$x \approx a - \frac{f(a) - y}{f'(a)}.$$

Writing $a = x_0$, the point

$$x_1 = x_0 - \frac{f(x_0) - y}{f'(x_0)},$$

is then a first approximation to a point x such that $f(x) = y$. This suggests the conjecture that the sequence $\{x_n\}_0^\infty$ defined by

$$x_0 = a, \qquad x_{n+1} = x_n - \frac{f(x_n) - y}{f'(x_n)}$$

converges to a point x such that $f(x) = y$. In fact we will now prove that the slightly simpler sequence, with $f'(x_n)$ replaced by $f'(a)$, converges to the desired point x if y is sufficiently close to b.

Theorem 1.3 Let $f : \mathscr{R} \to \mathscr{R}$ be a $\mathscr{C}^1$ function such that $f(a) = b$ and $f'(a) \neq 0$. Then there exist neighborhoods $U = [a - \delta, a + \delta]$ of a and $V = [b - \varepsilon, b + \varepsilon]$ of b such that, given $y_* \in V$, the sequence $\{x_n\}_0^\infty$ defined inductively by

$$x_0 = a, \qquad x_{n+1} = x_n - \frac{f(x_n) - y_*}{f'(a)} \tag{7}$$

converges to a (unique) point $x_* \in U$ such that $f(x_*) = y_*$.

PROOF Choose $\delta > 0$ so small that

$$|f'(a) - f'(x)| \leqq \tfrac{1}{2}|f'(a)| \qquad \text{if} \quad x \in U = [a - \delta, a + \delta].$$

Then let $\varepsilon = \frac{1}{2}\delta|f'(a)|$. It suffices to show that

$$\varphi(x) = x - \frac{f(x) - y_*}{f'(a)}$$

is a contraction mapping of U if $y_* \in V = [b - \varepsilon, b + \varepsilon]$, since $\varphi(x_*) = x_*$ clearly implies that $f(x_*) = y_*$.

First note that

$$|\varphi'(x)| = \left|1 - \frac{f'(x)}{f'(a)}\right| = \frac{|f'(a) - f'(x)|}{|f'(a)|} \leqq \frac{1}{2}$$

if $x \in U$, so φ has contraction constant $\frac{1}{2}$. It remains to show that, if $y_* \in V$, then φ maps the interval U into itself. But

$$\begin{aligned}|\varphi(x) - a| &\leqq |\varphi(x) - \varphi(a)| + |\varphi(a) - a| \\ &\leqq \frac{1}{2}|x - a| + \frac{|f(a) - y_*|}{|f'(a)|} \\ &\leqq \frac{\delta}{2} + \frac{\varepsilon}{|f'(a)|} \\ &= \delta\end{aligned}$$

if $x \in U = [a - \delta, a + \delta]$ and $y_* \in V = [b - \varepsilon, b + \varepsilon]$. Thus $\varphi(x) \in U$ as desired. ∎

This theorem provides a method for computing local inverse functions by successive approximations. Given $y \in V$, define $g(y)$ to be that point $x \in U$ given by the theorem, such that $f(x) = y$. Then f and g are local inverse functions. If we define a sequence $\{g_n\}_0^\infty$ of functions on V by

$$g_0(y) \equiv a, \qquad g_{n+1}(y) = g_n(y) - \frac{f(g_n(y)) - y}{f'(a)}, \tag{8}$$

then we see [by setting $x_n = g_n(y)$] that this sequence of functions converges to g.

Example 1 Let $f(x) = x^2 - 1$, $a = 1$, $b = 0$. Then (8) gives

$$g_0(y) \equiv 1,$$

$$g_1(y) = 1 - \frac{[(1)^2 - 1] - y}{2} = 1 + \frac{y}{2},$$

$$g_2(y) = \left(1 + \frac{y}{2}\right) - \frac{[(1 + y/2)^2 - 1] - y}{2} = 1 + \frac{y}{2} - \frac{y^2}{8},$$

$$g_3(y) = \left(1 + \frac{y}{2} - \frac{y^2}{8} + \frac{y^4}{16}\right) - \frac{y^4}{128}.$$

Thus it appears that we are generating partial sums of the binomial expansion

$$(1 + y)^{1/2} = 1 + \frac{y}{2} - \frac{y^2}{8} + \frac{y^3}{16} + \cdots$$

of the inverse function $g(y) = (1 + y)^{1/2}$.

Next we want to discuss the problem of solving an equation of the form $G(x, y) = 0$ for y as a function of x. We will say that $y = f(x)$ *solves the equation* $G(x, y) = 0$ in a neighborhood of the point (a, b) if $G(a, b) = 0$ and the graph of f agrees with the zero set of G near (a, b). By the latter we mean that there exists a neighborhood W of (a, b), such that a point $(x, y) \in W$ lies on the zero set of G

if and only if $y = f(x)$. That is, if $(x, y) \in W$, then $G(x, y) = 0$ if and only if $y = f(x)$.

Theorem 1.4 below is the implicit function theorem for the simplest case $m = n = 1$. Consideration of equations such as $x^2 - y^2 = 0$ near $(0, 0)$, and $x^2 + y^2 - 1 = 0$ near $(1, 0)$, shows that the hypothesis $D_2 G(a, b) \neq 0$ is necessary.

Theorem 1.4 Let $G : \mathscr{R}^2 \to \mathscr{R}$ be a $\mathscr{C}^1$ function, and $(a, b) \in \mathscr{R}^2$ a point such that $G(a, b) = 0$ and $D_2 G(a, b) \neq 0$. Then there exists a continuous function $f : J \to \mathscr{R}$, defined on a closed interval centered at a, such that $y = f(x)$ solves the equation $G(x, y) = 0$ in a neighborhood of the point (a, b).

In particular, if the functions $\{f_n\}_0^\infty$ are defined inductively by

$$f_0(x) \equiv b, \qquad f_{n+1}(x) = f_n(x) - \frac{G(x, f_n(x))}{D_2 G(a, b)}, \tag{9}$$

then the sequence $\{f_n\}_0^\infty$ converges uniformly to f on J.

MOTIVATION Suppose, for example, that $D_2 G(a, b) > 0$. By the continuity of G and $D_2 G$, we can choose a rectangle (see Fig. 3.4) $Q = [c_1, c_2] \times [d_1, d_2]$ centered at (a, b), such that

$$\begin{aligned} G(x, d_1) &< 0 \qquad \text{for all} \quad x \in [c_1, c_2], \\ G(x, d_2) &> 0 \qquad \text{for all} \quad x \in [c_1, c_2], \\ D_2 G(x, y) &> 0 \qquad \text{for all} \quad (x, y) \in Q. \end{aligned}$$

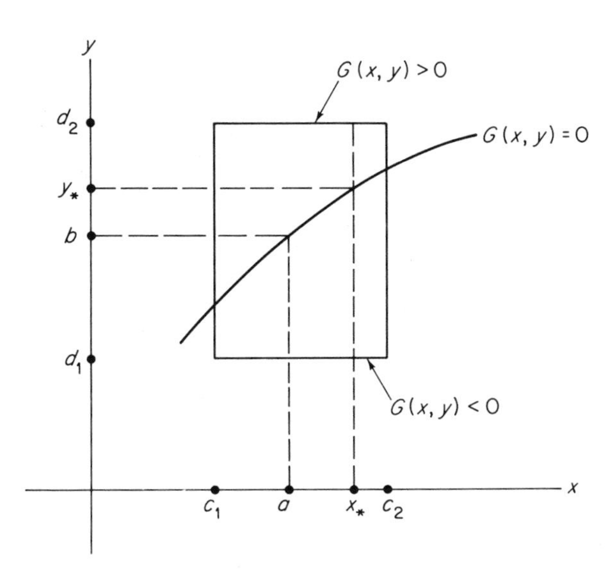

Figure 3.4

Given $x_* \in [c_1, c_2]$, define $G_{x_*}: [d_1, d_2] \to \mathscr{R}$ by

$$G_{x_*}(y) = G(x_*, y).$$

Since $G_{x_*}(d_1) < 0 < G_{x_*}(d_2)$ and $G'_{x_*}(y) > 0$ on $[d_1, d_2]$ the intermediate value theorem (see the Appendix) yields a unique point $y_* \in [d_1, d_2]$ such that

$$G(x_*, y_*) = G_{x_*}(y_*) = 0.$$

Defining $f: [c_1, c_2] \to \mathscr{R}$ by $f(x_*) = y_*$ for each $x_* \in [c_1, c_2]$, we have a function such that $y = f(x)$ solves the equation $G(x, y) = 0$ in the neighborhood Q of (a, b).

We want to compute $y_* = f(x_*)$ by successive approximations. By linear approximation we have

$$G(x_*, y_*) - G(x_*, b) \approx D_2 G(x_*, b)(y_* - b).$$

Recalling that $G(x_*, y_*) = 0$, and setting $y_0 = b$, this gives

$$y_* \approx y_0 - \frac{G(x_*, y_0)}{D_2 G(x_*, y_0)},$$

so

$$y_1 = y_0 - \frac{G(x_*, y_0)}{D_2 G(x_*, y_0)}$$

is our first approximation to y_*. Similarly we obtain

$$y_{n+1} = y_n - \frac{G(x_*, y_n)}{D_2 G(x_*, y_n)}$$

as the $(n + 1)$th approximation.

What we shall prove is that, if the rectangle Q is sufficiently small, then the slightly simpler sequence $\{y_n\}_0^\infty$ defined inductively by

$$y_0 = b, \qquad y_{n+1} = y_n - \frac{G(x_*, y_n)}{D_2 G(a, b)}$$

converges to a point y_* such that $G(x_*, y_*) = 0$.

PROOF First choose $\varepsilon > 0$ such that

$$|D_2 G(x, y) - D_2 G(a, b)| \leqq \tfrac{1}{2}|D_2 G(a, b)|$$

if $|x - a| \leqq \varepsilon$ and $|y - b| \leqq \varepsilon$. Then choose $\delta > 0$ less than ε such that

$$|G(x, b)| \leqq \tfrac{1}{2}\varepsilon |D_2 G(a, b)|$$

if $|x - a| \leqq \delta$. We will work inside the rectangle $W = [a - \delta, a + \delta] \times [b - \varepsilon, b + \varepsilon]$, assuming in addition that δ and ε are sufficiently small that $W \subset Q$.

Let x_* be a fixed point of the interval $[a-\delta, a+\delta]$. In order to prove that the sequence $\{y_n\}_0^\infty$ defined above converges to a point $y_* \in [b-\varepsilon, b+\varepsilon]$ with $G(x_*, y_*) = 0$, it suffices to show that the function $\varphi : [b-\varepsilon, b+\varepsilon] \to \mathscr{R}$, defined by

$$\varphi(y) = y - \frac{G(x_*, y)}{D_2 G(a, b)},$$

is a contraction mapping of the interval $[b-\varepsilon, b+\varepsilon]$. First note that

$$\begin{aligned}|\varphi'(y)| &= \left|1 - \frac{D_2 G(x_*, y)}{D_2 G(a, b)}\right| \\ &= \frac{|D_2 G(a, b) - D_2 G(x_*, y)|}{|D_2 G(a, b)|} \leqq \frac{1}{2}\end{aligned}$$

since $(x_*, y) \in W$. Thus the contraction constant is $\frac{1}{2}$.

It remains to show that $|\varphi(y) - b| \leqq \varepsilon$ if $|y - b| \leqq \varepsilon$. But

$$\begin{aligned}|\varphi(y) - b| &\leqq |\varphi(y) - \varphi(b)| + |\varphi(b) - b| \\ &\leqq \frac{1}{2}|y - b| + \frac{|G(x_*, b)|}{|D_2 G(a, b)|} \\ &\leqq \frac{\varepsilon}{2} + \frac{\varepsilon}{2} = \varepsilon,\end{aligned}$$

since $|x_* - a| \leqq \delta$ implies $|G(x_*, b)| \leqq \frac{1}{2}\varepsilon |D_2 G(a, b)|$.

If we write

$$f_0(x) = b, \qquad f_{n+1}(x) = f_n(x) - \frac{G(x, f_n(x))}{D_2 G(a, b)},$$

the above argument proves that the sequence $\{f_n(x)\}_0^\infty$ converges to the unique number $f(x) \in [b-\varepsilon, b+\varepsilon]$ such that $G(x, f(x)) = 0$. It remains to prove that this convergence is uniform.

To see this, we apply the error estimate of the contraction mapping theorem. Since $|f_0(x) - f_1(x)| \leqq \varepsilon$ because $f_0(x) = b$ and $f_1(x) \in [b-\varepsilon, b+\varepsilon]$, and $k = \frac{1}{2}$, we find that

$$|f_n(x) - f(x)| \leqq \frac{\varepsilon}{2^{n-1}},$$

so the convergence is indeed uniform. The functions $\{f_n\}_0^\infty$ are clearly continuous, so this implies that our solution $f: [a-\delta, a+\delta] \to \mathscr{R}$ is also continuous (by Theorem VI.1.2). ∎

REMARK We will see later (in Section 3) that the fact that G is $\mathscr{C}^1$ implies that f is $\mathscr{C}^1$.

Example 2 Let $G(x, y) = x^2 + y^2 - 1$, so $G(x, y) = 0$ is the unit circle. Let us solve for $y = f(x)$ in a neighborhood of $(0, 1)$ where $D_2 G(0, 1) = 2$. The successive approximations given by (9) are

$$f_0(x) \equiv 1,$$

$$f_1(x) = 1 - \frac{x^2 + (1)^2 - 1}{2} = 1 - \frac{x^2}{2},$$

$$f_2(x) = \left(1 - \frac{x^2}{2}\right) - \frac{x^2 + (1 - \frac{1}{2}x^2)^2 - 1}{2}$$

$$= 1 - \frac{x^2}{2} - \frac{x^4}{8},$$

$$f_3(x) = \left(1 - \frac{x^2}{2} - \frac{x^4}{8}\right) - \frac{1}{2}\left[x^2 + \left(1 - \frac{x^2}{2} - \frac{x^4}{8}\right)^2 - 1\right]$$

$$= \left(1 - \frac{x^2}{2} - \frac{x^4}{8} - \frac{x^6}{16}\right) - \frac{x^8}{128},$$

$$f_4(x) = 1 - \frac{x^2}{2} - \frac{x^4}{8} - \frac{x^6}{16} - \frac{5x^8}{128} + \text{higher order terms.}$$

Thus we are generating partial sums of the binomial series

$$f(x) = (1 - x^2)^{1/2} = 1 - \frac{x^2}{2} - \frac{x^4}{8} - \frac{x^6}{16} - \frac{5x^8}{128} - \cdots.$$

The preceding discussion is easily generalized to treat the problem of solving an equation of the form

$$G(\mathbf{x}, y) = G(x_1, \ldots, x_m, y) = 0,$$

where $G : \mathscr{R}^{m+1} \to \mathscr{R}$, for y as a function $\mathbf{x} = (x_1, \ldots, x_m)$. Given $f : \mathscr{R}^m \to \mathscr{R}$, we say that $y = f(\mathbf{x})$ *solves the equation* $G(\mathbf{x}, y) = 0$ in a neighborhood of $(\mathbf{a}, b)$ if $G(\mathbf{a}, b) = 0$ and the graph of f agrees with the zero set of G near $(\mathbf{a}, b) \in \mathscr{R}^{m+1}$. Theorem 1.4 is simply the case $m = 1$ of the following result.

Theorem 1.5 Let $G : \mathscr{R}^{m+1} \to \mathscr{R}$ be a $\mathscr{C}^1$ function, and $(\mathbf{a}, b) \in \mathscr{R}^m \times \mathscr{R} = \mathscr{R}^{m+1}$ a point such that $G(\mathbf{a}, b) = 0$ and $D_{m+1}G(\mathbf{a}, b) \neq 0$. Then there exists a continuous function $f : J \to \mathscr{R}$, defined on a closed cube $J \subset \mathscr{R}^m$ centered at $\mathbf{a} \in \mathscr{R}^m$, such that $y = f(\mathbf{x})$ solves the equation $G(\mathbf{x}, y) = 0$ in a neighborhood of the point $(\mathbf{a}, b)$.

In particular, if the functions $\{f_n\}_0^\infty$ are defined inductively by

$$f_0(\mathbf{x}) \equiv b, \qquad f_{n+1}(\mathbf{x}) = f_n(\mathbf{x}) - \frac{G(\mathbf{x}, f_n(\mathbf{x}))}{D_{m+1}G(\mathbf{a}, b)},$$

then the sequence $\{f_n\}_0^\infty$ converges uniformly to f on J.

To generalize the proof of Theorem 1.4, simply replace x, a, x_* by $\mathbf{x}$, $\mathbf{a}$, $\mathbf{x}_*$ respectively, and the interval $[a - \delta, a + \delta] \subset \mathscr{R}$ by the m-dimensional interval $\{\mathbf{x} \in \mathscr{R}^m : |\mathbf{x} - \mathbf{a}| \leqq \delta\}$.

Recall that Theorem 1.5, with the strengthened conclusion that the implicitly defined function f is continuously differentiable (which will be established in Section 3), is all that was needed in Section II.5 to prove that the zero set of a $\mathscr{C}^1$ function $G : \mathscr{R}^n \to \mathscr{R}$ is an $(n-1)$-dimensional manifold near each point where the gradient vector ∇G is nonzero. That is, the set

$$\{\mathbf{x} \in \mathscr{R}^n : G(\mathbf{x}) = 0 \text{ and } \nabla G(\mathbf{x}) \neq \mathbf{0}\}$$

is an $(n-1)$-manifold (Theorem II.5.4).

Exercises

1.1 Show that the equation $x^3 + xy + y^3 = 1$ can be solved for $y = f(x)$ in a neighborhood of $(1, 0)$.

1.2 Show that the set of all points $(x, y) \in \mathscr{R}^2$ such that $(x + y)^5 - xy = 1$ is a 1-manifold.

1.3 Show that the equation $y^3 + x^2 y - 2x^3 - x + y = 0$ defines y as a function of x (for all x). Apply Theorem 1.4 to conclude that y is a continuous function of x.

1.4 Let $f : \mathscr{R} \to \mathscr{R}$ and $G : \mathscr{R}^2 \to \mathscr{R}$ be $\mathscr{C}^2$ functions such that $G(x, f(x)) = 0$ for all $x \in \mathscr{R}$. Apply the chain rule to show that

$$f'(x) = -\frac{D_1 G(x, f(x))}{D_2 G(x, f(x))}, \qquad \text{or} \qquad \frac{dy}{dx} = -\frac{\partial G/\partial x}{\partial G/\partial y},$$

and

$$\frac{d^2 y}{dx^2} = -\frac{1}{\partial G/\partial y}\left[\frac{\partial^2 G}{\partial x^2} + 2\frac{dy}{dx}\frac{\partial^2 G}{\partial x\, \partial y} + \left(\frac{dy}{dx}\right)^2 \frac{\partial^2 G}{\partial y^2}\right].$$

1.5 Suppose that $G : \mathscr{R}^{m+1} \to \mathscr{R}$ is a $\mathscr{C}^1$ function satisfying the hypotheses of Theorem 1.5. Assuming that the implicitly defined function $y = f(x_1, \ldots, x_m)$ is also $\mathscr{C}^1$, show that

$$\frac{\partial f}{\partial x_i} = -\frac{D_i G}{D_{m+1} G}, \qquad i = 1, \ldots, m.$$

1.6 Application of Newton's method [Eq. (1)] to the function $f(x) = x^2 - a$, where $a > 0$, gives the formula $x_{n+1} = \frac{1}{2}(x_n + a/x_n)$. If $x_0 > \sqrt{a}$, show that the sequence $\{x_n\}_0^\infty$ converges to $\sqrt{a}$, by proving that $\varphi(x) = \frac{1}{2}(x + a/x)$ is a contraction mapping of $[\sqrt{a}, x_0]$. Then calculate $\sqrt{2}$ accurate to three decimal places.

1.7 Prove that the equation $2 - x - \sin x = 0$ has a unique real root, and that it lies in the interval $[\pi/6, \pi/2]$. Show that $\varphi(x) = 2 - \sin x$ is a contraction mapping of this interval, and then apply Theorem 1.1 to find the root, accurate to three decimal places.

1.8 Show that the equation $x + y - z + \cos xyz = 0$ can be solved for $z = f(x, y)$ in a neighborhood of the origin.

1.9 Show that the equation $z^3 + ze^{x+y} + 2 = 0$ has a unique solution $z = f(x, y)$ defined for all $(x, y) \in \mathscr{R}^2$. Conclude from Theorem 1.5 that f is continuous everywhere.

1.10 If a planet moves along the ellipse $x = a \cos\theta$, $y = b \sin\theta$, with the sun at the focus $((a^2 - b^2)^{1/2}, 0)$, and if t is the time measured from the instant the planet passes through $(a, 0)$, then it follows from Kepler's laws that θ and t satisfy Kepler's equation

$$kt = \theta - \varepsilon \sin\theta,$$

where k is a positive constant and $\varepsilon = c/a$, so $\varepsilon \in (0, 1)$.
(a) Show that Kepler's equation can be solved for $\theta = f(t)$.
(b) Show that $d\theta/dt = k/(1 - \varepsilon \cos \theta)$.
(c) Conclude that $d\theta/dt$ is maximal at the "perihelion" $(a, 0)$, and is minimal at the "aphelion" $(-a, 0)$.

2 THE MULTIVARIABLE MEAN VALUE THEOREM

The mean value theorem for *real-valued* functions states that, if the open set $U \subset \mathscr{R}^n$ contains the line segment L joining the points $\mathbf{a}$ and $\mathbf{b}$, and $f: U \to \mathscr{R}$ is differentiable, then

$$f(\mathbf{b}) - f(\mathbf{a}) = \nabla f(\mathbf{c}) \cdot (\mathbf{b} - \mathbf{a}) = df_{\mathbf{c}}(\mathbf{b} - \mathbf{a}) \tag{1}$$

for some point $\mathbf{c} \in L$ (Theorem II.3.4). We have seen (Exercise II.1.12) that this important result does not generalize to vector-valued functions. However, in many applications of the mean value theorem, all that is actually needed is the numerical estimate

$$|f(\mathbf{b}) - f(\mathbf{a})| \leqq |\mathbf{b} - \mathbf{a}| \max_{\mathbf{x} \in L} |\nabla f(\mathbf{x})|, \tag{2}$$

which follows immediately from (1) and the Cauchy–Schwarz inequality (if f is $\mathscr{C}^1$ so the maximum on the right exists). Fortunately inequality (2) does generalize to the case of $\mathscr{C}^1$ mappings from $\mathscr{R}^n$ to $\mathscr{R}^m$, and we will see that this result, the multivariable mean value theorem, plays a key role in the generalization to higher dimensions of the results of Section 1.

Recall from Section I.3 that a *norm* on the vector space V is a real-valued function $x \to |x|$ such that $|x| > 0$ if $x \neq 0$, $|ax| = |a| \cdot |x|$, and $|x + y| \leqq |x| + |y|$ for all $x, y \in V$ and $a \in \mathscr{R}$. Given a norm on V, by the *ball* of radius r with respect to this norm, centered at $\mathbf{a} \in V$, is meant the set $\{x \in V : |x| \leqq r\}$.

Thus far we have used mainly the Euclidean norm

$$|\mathbf{x}|_2 = (x_1{}^2 + \cdots + x_n{}^2)^{1/2}$$

on $\mathscr{R}^n$. In this section we will find it more convenient to use the "sup norm"

$$|\mathbf{x}|_0 = \max\{|x_1|, \ldots, |x_n|\}$$

which was introduced in Example 3 of Section I.3. The "unit ball" with respect to the sup norm is the cube $C_1{}^n = \{\mathbf{x} \in \mathscr{R}^n : |\mathbf{x}|_0 \leqq 1\}$, which is symmetric with respect to the coordinate planes in $\mathscr{R}^n$, and has the point $(1, 1, \ldots, 1)$ as one of its vertices. The cube $C_r{}^n = \{\mathbf{x} \in \mathscr{R}^n : |x|_0 \leqq r\}$ will be referred to as the "cube of radius r" centered at $\mathbf{0}$. We will delete the dimensional superscript when it is not needed for clarity.

We will see in Section VI.1 that any two norms on $\mathscr{R}^n$ are equivalent, in the sense that every ball with respect to one of the norms contains a ball with respect to the other, centered at the same point. Of course this is "obvious" for the Euclidean norm and the sup norm (Fig. 3.5). Consequently it makes no difference which norm we use in the definitions of limits, continuity, etc. (Why?)

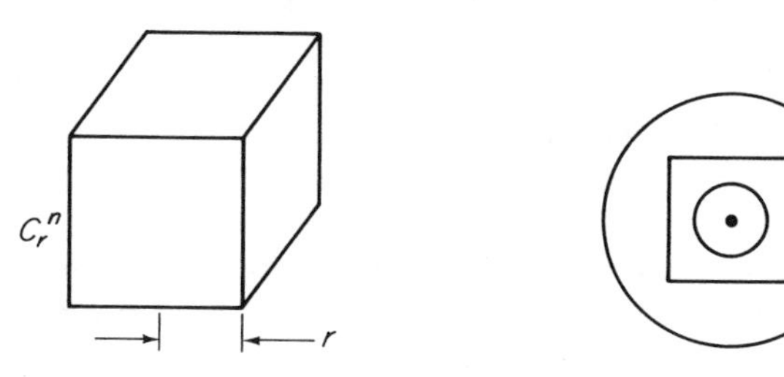

Figure 3.5

We will also need the concept of the norm of a linear mapping $L : \mathscr{R}^n \to \mathscr{R}^m$. The *norm* $\|L\|$ of L is defined by

$$\|L\| = \max_{x \in \partial C_1} |L(\mathbf{x})|_0 .$$

We will show presently that $\| \ \|$ is indeed a norm on the vector space $\mathscr{L}_{mn}$ of all linear mappings from $\mathscr{R}^n$ to $\mathscr{R}^m$—the only property of a norm that is not obvious is the triangle inequality.

We have seen (in Section I.7) that every linear mapping is continuous. This fact, together with the fact that the function $\mathbf{x} \to |\mathbf{x}_0|$ is clearly continuous on $\mathscr{R}^m$, implies that the composite $\mathbf{x} \to |L(\mathbf{x})|_0$ is continuous on the compact set $\partial C_1{}^n$, so the maximum value $\|L\|$ exists. Note that, if $\mathbf{x} \in \mathscr{R}^n$, then $\mathbf{x}/|\mathbf{x}|_0 \in \partial C_1{}^n$, so

$$\left| L\left(\frac{\mathbf{x}}{|\mathbf{x}|_0}\right)\right|_0 \leqq \|L\| \Rightarrow |L(\mathbf{x})|_0 \leqq \|L\| \cdot |\mathbf{x}|_0 .$$

This is half of the following result, which provides an important interpretation of $\|L\|$.

Proposition 2.1 If $L : \mathscr{R}^n \to \mathscr{R}^m$ is a linear mapping, then $\|L\|$ is the least number M such that $|L(\mathbf{x})|_0 \leqq M|\mathbf{x}|_0$ for all $\mathbf{x} \in \mathscr{R}^n$.

PROOF It remains only to be shown that, if $|L(\mathbf{x})|_0 \leqq M|\mathbf{x}|_0$ for all $\mathbf{x} \in \mathscr{R}^n$, then $M \geqq \|L\|$. But this follows immediately from the fact that the inequality $|L(\mathbf{x})|_0 \leqq M|\mathbf{x}|_0$ reduces to $|L(\mathbf{x})|_0 \leqq M$ if $\mathbf{x} \in \partial C_1$, while $\|L\| = \max |L(\mathbf{x})|_0$ for $\mathbf{x} \in \partial C_1$. ∎

In our proof of the mean value theorem we will need the elementary fact that the norm of a component function of the linear mapping L is no greater than the norm $\|L\|$ of L itself.

Lemma 2.2 If $L = (L_1, \ldots, L_m) : \mathscr{R}^n \to \mathscr{R}^m$ is linear, then $\|L_i\| \leqq \|L\|$ for each $i = 1, \ldots, m$.

PROOF Let $\mathbf{x}_0$ be the point of ∂C_1 at which $|L_i(\mathbf{x})|$ is maximal. Then

$$\begin{aligned}\|L_i\| &= |L_i(\mathbf{x}_0)| \leqq \max\{|L_1(\mathbf{x}_0)|, \ldots, |L_m(\mathbf{x}_0)|\} \\ &= |L(\mathbf{x}_0)|_0 \leqq \max_{\mathbf{x} \in \partial C_1} |L(\mathbf{x})|_0 = \|L\|. \end{aligned}$$ ∎

Next we give a formula for actually computing the norm of a given linear mapping. For this we need a particular concept of the "norm" of a matrix. If $A = (a_{ij})$ is an $m \times n$ matrix, we define its *norm* $\|A\|$ by

$$\|A\| = \max_{1 \leqq i \leqq m} \left\{ \sum_{j=1}^{n} |a_{ij}| \right\}. \tag{3}$$

Note that, in terms of the "1-norm" defined on $\mathscr{R}^n$ by

$$|\mathbf{x}|_1 = |x_1| + |x_2| + \cdots + |x_n|,$$

$\|A\|$ is simply the maximum of the 1-norms of the row vectors $A_1, \ldots, A_m$ of A,

$$\|A\| = \max\{|A_1|_1, |A_2|_1, \ldots, |A_m|_1\}.$$

To see that this is actually a norm on the vector space $\mathscr{M}_{mn}$ of all $m \times n$ matrices, let us identify $\mathscr{M}_{mn}$ with $\mathscr{R}^{mn}$ in the natural way:

$$(x_{ij}) \sim (x_{11}, \ldots, x_{1n}, x_{21}, \ldots, x_{2n}, \ldots, x_{m1}, \ldots, x_{mn}).$$

In other words, if $\mathbf{x}_1, \ldots, \mathbf{x}_m$ are the row vectors of the $m \times n$ matrix $X = (x_{ij})$, we identify X with the point

$$(\mathbf{x}_1, \ldots, \mathbf{x}_m) \in \underbrace{\mathscr{R}^n \times \cdots \times \mathscr{R}^n}_{(m \text{ factors})} = \mathscr{R}^{mn}.$$

With this notation, what we want to show is that

$$\|\mathbf{x}\| = \max\{|\mathbf{x}_1|_1, |\mathbf{x}_2|_1, \ldots, |\mathbf{x}_m|_1\}$$

defines a norm on $\mathscr{R}^{mn}$. But this follows easily from the fact that $|\ |_1$ is a norm on $\mathscr{R}^n$ (Exercise 2.2). In particular, $\|\ \|$ satisfies the triangle inequality. A ball with respect to the 1-norm is pictured in Fig. 3.6 (for the case $n = 2$); a ball with respect to the above norm $\|\ \|$ on $\mathscr{R}^{mn}$ is the Cartesian product of m such balls, one in each $\mathscr{R}^n$ factor.

We will now show that the norm of a linear mapping is equal to the norm of its matrix. For example, if $L : \mathscr{R}^3 \to \mathscr{R}^3$ is defined by $L(x, y, z) = (x - 3z, 2x - y - 2z, x + y)$, then $\|L\| = \max\{4, 5, 2\} = 5$.

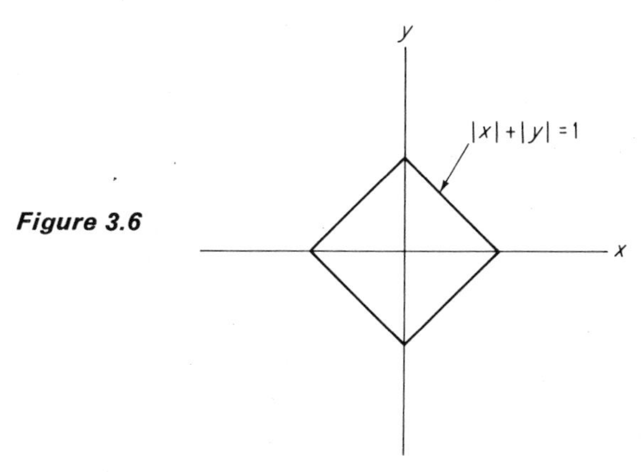

Figure 3.6

Theorem 2.3 Let $A = (a_{ij})$ be the matrix of the linear mapping L: $\mathscr{R}^n \to \mathscr{R}^m$, that is, $L(\mathbf{x}) = A\mathbf{x}$ for all $\mathbf{x} \in \mathscr{R}^n$. Then

$$\|L\| = \|A\|.$$

PROOF Given $\mathbf{x} = (x_1, \ldots, x_n) \in \mathscr{R}^n$, the coordinates $(y_1, \ldots, y_m)$ of $\mathbf{y} = L(\mathbf{x})$ are defined by

$$y_i = \sum_{j=1}^n a_{ij} x_j, \qquad i = 1, \ldots, m.$$

Let $|y_k|$ be the largest of the absolute values of these coordinates of $\mathbf{y}$. Then

$$\begin{aligned} |L(\mathbf{x})|_0 &= \max_{1 \leqq i \leqq m} \left\{ \left| \sum_{j=1}^n a_{ij} x_j \right| \right\} \\ &= \left| \sum_{j=1}^n a_{kj} x_j \right| \\ &\leqq \sum_{j=1}^n |a_{kj} x_j| \\ &\leqq |\mathbf{x}|_0 \sum_{j=1}^n |a_{kj}| \\ &\leqq |\mathbf{x}|_0 \max_{1 \leqq i \leqq m} \left\{ \sum_{j=1}^n |\mathbf{a}_{ij}| \right\} \\ &= \|A\| \cdot |\mathbf{x}|_0 . \end{aligned}$$

Thus $|L(\mathbf{x})|_0 \leqq \|A\| \cdot |\mathbf{x}|_0$ for all $\mathbf{x} \in \mathscr{R}^n$, so it follows from Proposition 2.1 that $\|L\| \leqq \|A\|$.

To prove that $\|L\| \geqq \|A\|$, it suffices to exhibit a point $\mathbf{x} \in \partial C_1$ for which $|L(\mathbf{x})|_0 \geqq \|A\|$. Suppose that the kth row vector $A_k = (a_{k1} \ldots a_{kn})$ is the one whose 1-norm is greatest, so

$$\|A\| = \sum_{j=1}^n |a_{kj}|.$$

For each $j = 1, \ldots, n$, define $\varepsilon_j = \pm 1$ by $a_{kj} = \varepsilon_j |a_{kj}|$. If $\mathbf{x} = (\varepsilon_1, \varepsilon_2, \ldots, \varepsilon_n)$, then $|\mathbf{x}|_0 = 1$, and

$$\begin{aligned}
|L(\mathbf{x})|_0 &= \max_{1 \leqq i \leqq m} \left\{ \left| \sum_{j=1}^{n} a_{ij} \varepsilon_j \right| \right\} \\
&\geqq \left| \sum_{j=1}^{n} a_{kj} \varepsilon_j \right| \\
&= \left| \sum_{j=1}^{n} |a_{kj}| \right| \\
&= \sum_{j=1}^{n} |a_{kj}| \\
&= \|A\|,
\end{aligned}$$

so $|L(\mathbf{x})|_0 \geqq \|A\|$ as desired. ∎

Let $\Phi : \mathscr{L}_{mn} \to \mathscr{M}_{mn}$ be the natural isomorphism from the vector space of all linear mappings $\mathscr{R}^n \to \mathscr{R}^m$ to the vector space of all $m \times n$ matrices, $\Phi(L)$ being the matrix of $L \in \mathscr{L}_{mn}$. Then Theorem 2.3 says simply that the isomorphism Φ is "norm-preserving." Since we have seen that $\| \;\|$ on $\mathscr{M}_{mn}$ satisfies the triangle inequality, it follows easily that the same is true of $\| \;\|$ on $\mathscr{L}_{mn}$. Thus $\| \;\|$ is indeed a *norm* on $\mathscr{L}_{mn}$.

Henceforth we will identify both the linear mapping space $\mathscr{L}_{mn}$ and the matrix space $\mathscr{M}_{mn}$ with Euclidean space $\mathscr{R}^{mn}$, by identifying each linear mapping with its matrix, and each $m \times n$ matrix with a point of $\mathscr{R}^{mn}$ (as above). In other words, we can regard either symbol $\mathscr{L}_{mn}$ as $\mathscr{M}_{mn}$ as denoting $\mathscr{R}^{mn}$ with the norm

$$\|(\mathbf{x}_1, \ldots, \mathbf{x}_m)\| = \max\{|\mathbf{x}_1|_1, \ldots, |\mathbf{x}_m|_1\},$$

where $(\mathbf{x}_1, \ldots, \mathbf{x}_m) \in \mathscr{R}^n \times \cdots \times \mathscr{R}^n = \mathscr{R}^{mn}$.

If $f : \mathscr{R}^n \to \mathscr{R}^m$ is a differentiable mapping, then $df_{\mathbf{x}} \in \mathscr{L}_{mn}$, and $f'(\mathbf{x}) \in \mathscr{M}_{mn}$, so we may regard f' as a mapping form $\mathscr{R}^n$ to $\mathscr{M}_{mn}$,

$$f' : \mathscr{R}^n \to \mathscr{M}_{mn},$$

and similarly df as a mapping from $\mathscr{R}^n$ to $\mathscr{L}_{mn}$. Recall that $f : \mathscr{R}^n \to \mathscr{R}^m$ is $\mathscr{C}^1$ at $\mathbf{a} \in \mathscr{R}^n$ if and only if the first partial derivatives of the component functions of f all exist in a neighborhood of $\mathbf{a}$ and are continuous at $\mathbf{a}$. The following result is an immediate consequence of this definition.

Proposition 2.4 The differentiable mapping $f : \mathscr{R}^n \to \mathscr{R}^m$ is $\mathscr{C}^1$ at $\mathbf{a} \in \mathscr{R}^n$ if and only if $f' : \mathscr{R}^n \to \mathscr{M}_{mn}$ is continuous at $\mathbf{a}$.

We are finally ready for the mean value theorem.

Theorem 2.5 Let $f : U \to \mathscr{R}^m$ be a $\mathscr{C}^1$ mapping, where $U \subset \mathscr{R}^n$ is a neighborhood of the line segment L with endpoints $\mathbf{a}$ and $\mathbf{b}$. Then

$$|f(\mathbf{b}) - f(\mathbf{a})|_0 \leqq |\mathbf{b} - \mathbf{a}|_0 \max_{\mathbf{x} \in L} \|f'(\mathbf{x})\|. \tag{4}$$

PROOF Let $\mathbf{h} = \mathbf{b} - \mathbf{a}$, and define the $\mathscr{C}^1$ curve $\gamma : [0, 1] \to \mathscr{R}^m$ by

$$\gamma(t) = f(\mathbf{a} + t\mathbf{h}).$$

If $f^1, \ldots, f^m$ are the component functions of f, then $\gamma_i(t) = f^i(\mathbf{a} + t\mathbf{h})$ is the ith component function of γ, and

$$\gamma_i'(t) = df^i_{\mathbf{a}+t\mathbf{h}}(\mathbf{h})$$

by the chain rule.

If the maximal (in absolute value) coordinate of $f(\mathbf{b}) - f(\mathbf{a})$ is the kth one, then

$$\begin{aligned}
|f(\mathbf{b}) - f(\mathbf{a})|_0 &= |f^k(\mathbf{b}) - f^k(\mathbf{a})| \\
&= |\gamma_k(1) - \gamma_k(0)| \\
&= \left| \int_0^1 \gamma_k'(t)\, dt \right| \qquad \text{(fundamental theorem of calculus)} \\
&\leqq \int_0^1 |\gamma_k'(t)|\, dt \\
&= \int_0^1 |df^k_{\mathbf{a}+t\mathbf{h}}(\mathbf{h})|\, dt \\
&\leqq \max_{t \in [0,1]} |df^k_{\mathbf{a}+t\mathbf{h}}(\mathbf{h})| \\
&\leqq |\mathbf{h}|_0 \cdot \max_{t \in [0,1]} \|df^k_{\mathbf{a}+t\mathbf{h}}\| \qquad \text{(Proposition 2.1)} \\
&= |\mathbf{h}|_0 \cdot \|df^k_{\mathbf{a}+\tau\mathbf{h}}\| \qquad \text{(maximum for } t = \tau) \\
&\leqq |\mathbf{h}|_0 \cdot \|df_{\mathbf{a}+\tau\mathbf{h}}\| \qquad \text{(Lemma 2.2)} \\
&\leqq |\mathbf{h}|_0 \cdot \max_{t \in [0,1]} \|df_{\mathbf{a}+t\mathbf{h}}\| \\
&= |\mathbf{b} - \mathbf{a}|_0 \cdot \max_{\mathbf{x} \in L} \|f'(\mathbf{x})\|
\end{aligned}$$

as desired. ∎

If U is a convex open set (that is, each line segment joining two points of U lies in U), and $f : U \to \mathscr{R}^m$ is a $\mathscr{C}^1$ mapping such that $\|f'(\mathbf{x})\| \leqq \varepsilon$ for each $x \in U$, then the mean value theorem says that

$$|f(\mathbf{a} + \mathbf{h}) - f(\mathbf{a})|_0 \leqq \varepsilon |\mathbf{h}|_0$$

if $\mathbf{a}$, $\mathbf{a} + \mathbf{h} \in U$. Speaking very roughly, this says that $f(\mathbf{a} + \mathbf{h})$ is approximately equal to the constant $f(\mathbf{a})$ when $|\mathbf{h}|_0$ is very small. The following important corollary to the mean value theorem says (with $\lambda = df_{\mathbf{a}}$) that the actual difference $\Delta f_{\mathbf{a}}(\mathbf{h}) = f(\mathbf{a} + \mathbf{h}) - f(\mathbf{a})$ is approximately equal to the linear difference $df_{\mathbf{a}}(\mathbf{h})$ when $\mathbf{h}$ is very small.

Corollary 2.6 Let $f: U \to \mathscr{R}^m$ be a $\mathscr{C}^1$ mapping, where $U \subset \mathscr{R}^n$ is a neighborhood of the line L with endpoints $\mathbf{a}$ and $\mathbf{a} + \mathbf{h}$. If $\lambda : \mathscr{R}^n \to \mathscr{R}^m$ is a linear mapping, then

$$|f(\mathbf{a} + \mathbf{h}) - f(\mathbf{a}) - \lambda(\mathbf{h})|_0 \leqq |\mathbf{h}|_0 \max_{\mathbf{x} \in L} \|df_{\mathbf{x}} - \lambda\|. \tag{5}$$

PROOF Apply the mean value theorem to the $\mathscr{C}^1$ mapping $g : U \to \mathscr{R}^m$ defined by $g(\mathbf{x}) = f(\mathbf{x}) - \lambda(\mathbf{x})$, noting that $df_{\mathbf{x}} = df_{\mathbf{x}} - \lambda$ because $d\lambda_{\mathbf{x}} = \lambda$ (by Example 3 of Section II.2), and that

$$g(\mathbf{a} + \mathbf{h}) - g(\mathbf{a}) = f(\mathbf{a} + \mathbf{h}) - f(\mathbf{a}) - \lambda(\mathbf{h})$$

because λ is linear. ∎

As a typical application of Corollary 2.6, we can prove in the case $m = n$ that, if U contains the cube C_r of radius r centered at $\mathbf{0}$, and $df_{\mathbf{x}}$ is close (in norm) to the identity mapping $I : \mathscr{R}^n \to \mathscr{R}^n$ for all $\mathbf{x} \in C_r$, then the image under f of the cube C_r is contained in a slightly larger cube (Fig. 3.7). This seems natural enough—if df is sufficiently close to the identity, then f should be also, so no point should be moved very far.

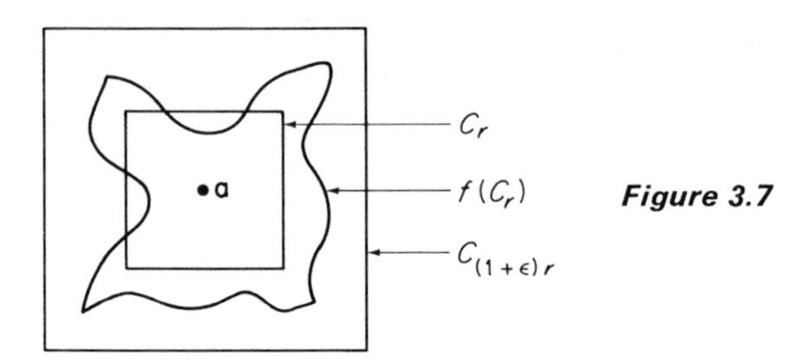

Figure 3.7

Corollary 2.7 Let U be an open set in $\mathscr{R}^n$ containing the cube C_r, and $f: U \to \mathscr{R}^n$ a $\mathscr{C}^1$ mapping such that $f(\mathbf{0}) = \mathbf{0}$ and $df_0 = I$. If

$$\|df_{\mathbf{x}} - I\| < \varepsilon$$

for all $\mathbf{x} \in C_r$, then $f(C_r) \subset C_{(1+\varepsilon)r}$.

PROOF Applying Corollary 2.6 with $\mathbf{a} = \mathbf{0}$, $\lambda = df_0 = I$, and $\mathbf{h} = \mathbf{x} \in C_r$, we obtain

$$|f(\mathbf{x}) - \mathbf{x}|_0 \leqq \varepsilon |\mathbf{x}|_0 .$$

But $||f(\mathbf{x})|_0 - |\mathbf{x}|_0| \leqq |f(\mathbf{x}) - \mathbf{x}|_0$ by the triangle inequality, so it follows that

$$|f(\mathbf{x})|_0 \leqq (1 + \varepsilon)|\mathbf{x}|_0 \leqq (1 + \varepsilon)r$$

as desired. ∎

The following corollary is a somewhat deeper application of the mean value theorem. At the same time it illustrates a general phenomenon which is basic to the linear approximation approach to calculus—the fact that simple properties of the differential of a function often reflect deep properties of the function itself. The point here is that the question as to whether a linear mapping is one-to-one, is a rather simple matter, while for an arbitrary given $\mathscr{C}^1$ mapping this may be a quite complicated question.

Corollary 2.8 Let $f : \mathscr{R}^n \to \mathscr{R}^m$ be $\mathscr{C}^1$ at $\mathbf{a}$. If $df_{\mathbf{a}} : \mathscr{R}^n \to \mathscr{R}^m$ is one-to-one, then f itself is one-to-one on some neighborhood of $\mathbf{a}$.

PROOF Let m be the minimum value of $|df_{\mathbf{a}}(x)|_0$ for $\mathbf{x} \in \partial C_1{}^n$; then $m > 0$ because $df_{\mathbf{a}}$ is one-to-one [otherwise there would be a point $\mathbf{x} \neq \mathbf{0}$ with $df_{\mathbf{a}}(\mathbf{x}) = \mathbf{0}$]. Choose a positive number $\varepsilon < m$.

Since f is $\mathscr{C}^1$ at $\mathbf{a}$, there exists $\delta > 0$ such that

$$|\mathbf{x} - \mathbf{a}|_0 < \delta \Rightarrow \| df_{\mathbf{x}} - df_{\mathbf{a}} \| < \varepsilon.$$

If $\mathbf{x}$ and $\mathbf{y}$ are any two distinct points of the neighborhood

$$U = \{\mathbf{x} \in \mathscr{R}^n : |\mathbf{x} - \mathbf{a}|_0 < \delta\},$$

then an application of Corollary 2.6, with $\lambda = df_{\mathbf{a}}$ and L the line segment from $\mathbf{x}$ to $\mathbf{y}$, yields

$$|f(\mathbf{x}) - f(\mathbf{y}) - df_{\mathbf{a}}(\mathbf{x} - \mathbf{y})|_0 \leqq |\mathbf{x} - \mathbf{y}|_0 \max_{\mathbf{z} \in L} \|df_{\mathbf{z}} - df_{\mathbf{a}}\| < \varepsilon |\mathbf{x} - \mathbf{y}|_0 .$$

The triangle inequality then gives

$$\big| |df_{\mathbf{a}}(\mathbf{x} - \mathbf{y})|_0 - |f(\mathbf{x}) - f(\mathbf{y})|_0 \big| < \varepsilon |\mathbf{x} - \mathbf{y}|_0 ,$$

so

$$\begin{aligned} |f(\mathbf{x}) - f(\mathbf{y})|_0 &> |df_{\mathbf{a}}(\mathbf{x} - \mathbf{y})|_0 - \varepsilon |\mathbf{x} - \mathbf{y}|_0 \\ &\geqq (m - \varepsilon)|\mathbf{x} - \mathbf{y}|_0 > 0. \end{aligned}$$

Thus $f(\mathbf{x}) \neq f(\mathbf{y})$ if $\mathbf{x} \neq \mathbf{y}$. ∎

Corollary 2.8 has the interesting consequence that, if $f : \mathscr{R}^n \to \mathscr{R}^n$ is $\mathscr{C}^1$ with $df_{\mathbf{a}}$ one-to-one (so f is 1–1 in a neighborhood of $\mathbf{a}$), and if f is "slightly perturbed" by means of the addition of a "small" term $g : \mathscr{R}^n \to \mathscr{R}^n$, then the new mapping $h = f + g$ is still one-to-one in a neighborhood of $\mathbf{a}$. See Exercise 2.9 for a precise statement of this result.

In this section we have dealt with the mean value theorem and its corollaries in terms of the sup norms on $\mathscr{R}^n$ and $\mathscr{R}^m$, and the resulting norm

$$\|L\| = \max_{\mathbf{x} \in \partial C^n} |L(\mathbf{x})|_0 \text{ on } \mathscr{L}_{mn} .$$

This will suffice for our purposes. However arbitrary norms $|\ |_m$ on $\mathscr{R}^m$ and $|\ |_n$ on $\mathscr{R}^n$ can be used in the mean value theorem, provided we use the norm

$$\|L\|_{mn} = \max_{\mathbf{x} \in \partial D^n} |L(\mathbf{x})|_m$$

on $\mathscr{L}_{mn}$, where D^n is the unit ball in $\mathscr{R}^n$ with respect to the norm $|\ |_n$. The conclusion of the mean value theorem is then the expected inequality

$$|f(\mathbf{b}) - f(\mathbf{a})|_m \leqq |\mathbf{b} - \mathbf{a}|_n \cdot \max_{\mathbf{x} \in L} \|f'(\mathbf{x})\|_{mn}.$$

In Exercises 2.5 and 2.6 we outline an alternative proof of the mean value theorem which establishes it in this generality.

Exercises

2.1 Let $|\ |_m$ and $|\ |_n$ be norms on $\mathscr{R}^m$ and $\mathscr{R}^n$ respectively. Prove that

$$|(\mathbf{x}, \mathbf{y})|_0 = \max\{|\mathbf{x}|_m, |\mathbf{y}|_n\} \qquad \text{for} \quad (\mathbf{x}, \mathbf{y}) \in \mathscr{R}^{m+n}$$

defines a norm on $\mathscr{R}^{m+n}$. Similarly prove that $|(\mathbf{x}, \mathbf{y})|_1 = |\mathbf{x}|_m + |\mathbf{y}|_n$ defines a norm on $\mathscr{R}^{m+n}$.

2.2 Show that $\|\ \|$, as defined by Eq. (3), is a norm on the space $\mathscr{M}_{mn}$ of $m \times n$ matrices.

2.3 Given $\mathbf{a} \in \mathscr{R}^n$, denote by $L_\mathbf{a}$ the linear function

$$L_\mathbf{a}(\mathbf{x}) = \mathbf{a} \cdot \mathbf{x} = \sum_{i=1}^n a_i x_i.$$

Consider the norms of $L_\mathbf{a}$ with respect to the sup norm $|\ |_0$ and the 1-norm $|\ |_1$ on $\mathscr{R}^n$, defined as in the last paragraph of this section. Show that $\|L_\mathbf{a}\|_1 = |\mathbf{a}|_0$ while $\|L_\mathbf{a}\|_0 = |\mathbf{a}|_1$.

2.4 Let $L : \mathscr{R}^n \to \mathscr{R}^m$ be a linear mapping with matrix (a_{ij}). If we use the 1-norm on $\mathscr{R}^n$ and the sup norm on $\mathscr{R}^m$, show that the corresponding norm on $\mathscr{L}_{mn}$ is

$$\|L\| = \max_{\substack{1 \leqq i \leqq m \\ 1 \leqq j \leqq n}} |a_{ij}|,$$

that is, the sup norm on $\mathscr{R}^{mn}$.

2.5 Let $\gamma : [a, b] \to \mathscr{R}^m$ be a $\mathscr{C}^1$ mapping with $|\gamma'(t)| \leqq M$ for all $t \in [a, b]$, $|\ |$ being an arbitrary norm on $\mathscr{R}^m$. Prove that

$$|\gamma(b) - \gamma(a)| \leqq M(b - a).$$

Outline: Given $\varepsilon > 0$, denote by S_ε the set of points $x \in [a, b]$ such that

$$|\gamma(t) - \gamma(a)| \leqq (M + \varepsilon)(t - a) + \varepsilon$$

for all $t \leqq x$. Let $c = \text{lub } S_\varepsilon$. If $c < b$, then there exists $\delta > 0$ such that

$$|h| \leqq \delta \Rightarrow \left| \frac{\gamma(c + h) - \gamma(c)}{h} \right| < M + \varepsilon.$$

Conclude from this that $c + \delta \in S_\varepsilon$, a contradiction. Therefore $c = b$, so

$$|\gamma(b) - \gamma(a)| \leqq (M + \varepsilon)(b - a) + \varepsilon$$

for all $\varepsilon > 0$.

2.6 Apply the previous exercise to establish the mean value theorem with respect to arbitrary norms on $\mathscr{R}^n$ and $\mathscr{R}^m$. In particular, given $f: U \to \mathscr{R}^m$ where U is a neighborhood in $\mathscr{R}^n$ of the line segment L from $\mathbf{a}$ to $\mathbf{a}+\mathbf{h}$, apply Exercise 2.5 with $\gamma(t) = f(\mathbf{a}+t\mathbf{h})$.

2.7 (a) Show that the linear mapping $T: \mathscr{R}^n \to \mathscr{R}^m$ is one-to-one if and only if $a = \max_{\mathbf{x} \in \partial C^n} |T(\mathbf{x})|_0$ is positive.
(b) Conclude that the linear mapping $T: \mathscr{R}^n \to \mathscr{R}^m$ is one-to-one if and only if there exists $a > 0$ such that $|T(\mathbf{x})|_0 \geqq a|\mathbf{x}|_0$ for all $\mathbf{x} \in \mathscr{R}^n$.

2.8 Let $T: \mathscr{R}^n \to \mathscr{R}^m$ be a one-to-one linear mapping with $|T(\mathbf{x})|_0 \geqq a|\mathbf{x}|_0$ for all $\mathbf{x} \in \mathscr{R}$, where $a > 0$. If $\|S - T\| \leqq \varepsilon < a$, show that $|S(\mathbf{x})|_0 \geqq (a-\varepsilon)|\mathbf{x}|_0$ for all $\mathbf{x} \in \mathscr{R}^n$, so S is also one-to-one. Thus the set of all one-to-one linear mappings $\mathscr{R}^n \to \mathscr{R}^m$ forms an open subset of $\mathscr{L}_{mn} \approx \mathscr{R}^{mn}$.

2.9 Apply Corollary 2.8 and the preceding exercise to prove the following. Let $f: \mathscr{R}^n \to \mathscr{R}^n$ be a $\mathscr{C}^1$ mapping such that $df_{\mathbf{a}}: \mathscr{R}^n \to \mathscr{R}^n$ is one-to-one, so that f is one-to-one in a neighborhood of $\mathbf{a}$. Then there exists $\varepsilon > 0$ such that if $g: \mathscr{R}^n \to \mathscr{R}^n$ is a $\mathscr{C}^1$ mapping with $g(\mathbf{a}) = \mathbf{0}$ and $\|dg_{\mathbf{a}}\| < \varepsilon$, then the mapping $h: \mathscr{R}^n \to \mathscr{R}^n$, defined by $h(\mathbf{x}) = f(\mathbf{x}) + g(\mathbf{x})$, is also one-to-one in a neighborhood of $\mathbf{a}$.

3 THE INVERSE AND IMPLICIT MAPPING THEOREMS

The simplest cases of the inverse and implicit mapping theorems were discussed in Section 1. Theorem 1.3 dealt with the problem of solving an equation of the form $f(x) = y$ for x as a function of y, while Theorem 1.4 dealt with the problem of solving an equation of the form $G(x, y) = 0$ for y as a function of x. In each case we defined a sequence of successive approximations which, under appropriate conditions, converged to a solution.

In this section we establish the analogous higher-dimensional results. Both the statements of the theorems and their proofs will be direct generalizations of those in Section 1. In particular we will employ the method of successive approximations by means of the contraction mapping theorem.

The definition of a contraction mapping in $\mathscr{R}^n$ is the same as on the line. Given a subset C of $\mathscr{R}^n$, the mapping $\varphi: C \to C$ is called a *contraction mapping* with *contraction constant* k if

$$|\varphi(\mathbf{x}) - \varphi(\mathbf{y})|_0 \leqq k|\mathbf{x} - \mathbf{y}|_0$$

for all $\mathbf{x}, \mathbf{y} \in C$. The contraction mapping theorem asserts that, if the set C is closed and bounded, and $k < 1$, then φ has a unique fixed point $\mathbf{x}_* \in C$ such that $\varphi(\mathbf{x}_*) = \mathbf{x}_*$.

Theorem 3.1 Let $\varphi: C \to C$ be a contraction mapping with contraction constant $k < 1$, and with C being a closed and bounded subset of $\mathscr{R}^n$. Then φ has a unique fixed point $\mathbf{x}_*$. Moreover, given $\mathbf{x}_0 \in C$, the sequence $\{\mathbf{x}_m\}_0^\infty$ defined inductively by

$$\mathbf{x}_{m+1} = \varphi(\mathbf{x}_m)$$

converges to x_*. In particular,

$$|\mathbf{x}_m - \mathbf{x}_*|_0 \leqq k^m \frac{|\mathbf{x}_0 - \mathbf{x}_1|_0}{1-k}.$$

The proof given in Section 1 for the case $n = 1$ generalizes immediately, with no essential change in the details. We leave it to the reader to check that the only property of the closed interval $[a, b] \subset \mathscr{R}$, that was used in the proof of Theorem 1.1, is the fact that every Cauchy sequence of points of $[a, b]$ converges to a point of $[a, b]$. But every closed and bounded set $C \subset \mathscr{R}^n$ has this property (see the Appendix).

The inverse mapping theorem asserts that the $\mathscr{C}^1$ mapping $f: \mathscr{R}^n \to \mathscr{R}^n$ is locally invertible in a neighborhood of the point $\mathbf{a} \in \mathscr{R}^n$ if its differential $df_{\mathbf{a}} : \mathscr{R}^n \to \mathscr{R}^n$ at $\mathbf{a}$ is invertible. This means that, if the linear mapping $df_{\mathbf{a}}$: $\mathscr{R}^n \to \mathscr{R}^n$ is one-to-one (and hence onto), then there exists a neighborhood U of $\mathbf{a}$ which f maps one-to-one onto some neighborhood V of $\mathbf{b} = f(\mathbf{a})$, with the inverse mapping $g : V \to U$ also being $\mathscr{C}^1$. Equivalently, if the linear equations

$$\begin{aligned} df_{\mathbf{a}}^1(\mathbf{x}) &= y_1, \\ &\vdots \\ df_{\mathbf{a}}^n(\mathbf{x}) &= y_n \end{aligned}$$

have a unique solution $\mathbf{x} \in \mathscr{R}^n$ for each $\mathbf{y} \in \mathscr{R}^n$, then there exist neighborhoods U of $\mathbf{a}$ and V of $\mathbf{b} = f(\mathbf{a})$, such that the equations

$$\begin{aligned} f^1(x_1, \ldots, x_n) &= y_1, \\ &\vdots \\ f^n(x_1, \ldots, x_n) &= y_n \end{aligned}$$

have a unique solution $(x_1, \ldots, x_n) \in U$ for each $(y_1, \ldots, y_n) \in V$. Here we are writing $f^1, \ldots, f^n$ for the component functions of $f: \mathscr{R}^n \to \mathscr{R}^n$.

It is easy to see that the invertibility of $df_{\mathbf{a}}$ is a necessary condition for the local invertibility of f near $\mathbf{a}$. For if U, V, and g are as above, then the compositions $g \circ f$ and $f \circ g$ are equal to the identity mapping on U and V, respectively. Consequently the chain rule implies that

$$dg_{\mathbf{b}} \circ df_{\mathbf{a}} = df_{\mathbf{a}} \circ dg_{\mathbf{b}} = \text{identity mapping of } \mathscr{R}^n.$$

This obviously means that $df_{\mathbf{a}}$ is invertible with $df_{\mathbf{a}}^{-1} = dg_{\mathbf{b}}$. Equivalently, the derivative matrix $f'(\mathbf{a})$ must be invertible with $f'(\mathbf{a})^{-1} = g'(\mathbf{b})$. But the matrix $f'(\mathbf{a})$ is invertible if and only if its determinant is nonzero. So a necessary condition for the local invertibility of f near $f'(\mathbf{a})$ is that $|f'(\mathbf{a})| \neq 0$.

The following example shows that *local* invertibility is the most that can be hoped for. That is, f may be $\mathscr{C}^1$ on the open set G with $|f'(\mathbf{a})| \neq 0$ for each $\mathbf{a} \in G$, without f being one-to-one on G.

Example 1 Consider the $\mathscr{C}^1$ mapping $f: \mathscr{R}^2 \to \mathscr{R}^2$ defined by

$$f(x, y) = (x^2 - y^2, 2xy).$$

Since $\cos^2 \theta - \sin^2 \theta = \cos 2\theta$ and $2 \sin \theta \cos \theta = \sin 2\theta$, we see that in polar coordinates f is described by

$$f(r \cos \theta, r \sin \theta) = (r^2 \cos 2\theta, r^2 \sin 2\theta).$$

From this it follows that f maps the circle of radius r twice around the circle of radius r^2. In particular, f maps both of the points $(r \cos \theta, r \sin \theta)$ and $(r \cos (\theta + \pi), r \sin(\theta + \pi))$ to the same point $(r^2 \cos 2\theta, r^2 \sin 2\theta)$. Thus f maps the open set $\mathscr{R}^2 - \mathbf{0}$ "two-to-one" onto itself. However, $|f'(x, y)| = 4(x^2 + y^2)$, so $f'(x, y)$ is invertible at each point of $\mathscr{R}^2 - \mathbf{0}$.

We now begin with the proof of the inverse mapping theorem. It will be convenient for us to start with the special case in which the point $\mathbf{a}$ is the origin, with $f(\mathbf{0}) = \mathbf{0}$ and $f'(\mathbf{0}) = I$ (the $n \times n$ identity matrix). This is the following substantial lemma; it contains some additional information that will be useful in Chapter IV (in connection with changes of variables in multiple integrals). Given $\varepsilon > 0$, note that, since f is $\mathscr{C}^1$ with $df_\mathbf{0} = I$, there exists $r > 0$ such that $\|df_\mathbf{x} - I\| < \varepsilon$ for all points $\mathbf{x} \in C_r$ (the cube of radius r centered at C_r).

Lemma 3.2 Let $f: \mathscr{R}^n \to \mathscr{R}^n$ be a $\mathscr{C}^1$ mapping such that $f(\mathbf{0}) = \mathbf{0}$ and $df_\mathbf{0} = I$. Suppose also that

$$\|df_\mathbf{x} - I\| \leqq \varepsilon < 1$$

for all $\mathbf{x} \in C_r$. Then

$$C_{(1-\varepsilon)r} \subset f(C_r) \subset C_{(1+\varepsilon)r}.$$

Moreover, if $V = \operatorname{int} C_{(1-\varepsilon)r}$ and $U = \operatorname{int} C_r \cap f^{-1}(V)$, then $f: U \to V$ is a one-to-one onto mapping, and the inverse mapping $g: V \to U$ is differentiable at $\mathbf{0}$.

Finally, the local inverse mapping $g: V \to U$ is the limit of the sequence of successive approximations $\{g_m\}_0^\infty$ defined inductively on V by

$$g_0(\mathbf{y}) = \mathbf{0}, \qquad g_{m+1}(\mathbf{y}) = g_m(\mathbf{y}) - f(g_m(\mathbf{y})) + \mathbf{y}$$

for $\mathbf{y} \in V$.

PROOF We have already shown that $f(C_r) \subset C_{(1+\varepsilon)r}$ (Corollary 2.7), and it follows from the proof of Corollary 2.8 that f is one-to-one on C_r—the cube C_r satisfies the conditions specified in the proof of Corollary 2.7. Alternatively, we can apply Corollary 2.6 with $\lambda = df_\mathbf{0} = I$ to see that

$$|f(\mathbf{x}) - f(\mathbf{y}) - (\mathbf{x} - \mathbf{y})|_0 \leqq \varepsilon |\mathbf{x} - \mathbf{y}|_0 \tag{1}$$

if $\mathbf{x}, \mathbf{y} \in C_r$. From this inequality it follows that

$$(1 - \varepsilon)|\mathbf{x} - \mathbf{y}|_0 \leqq |f(\mathbf{x}) - f(\mathbf{y})|_0 \leqq (1 + \varepsilon)|\mathbf{x} - \mathbf{y}|_0. \tag{2}$$

The left-hand inequality shows that f is one-to-one on C_r, while the right-hand one (with $\mathbf{y} = \mathbf{0}$) shows that $f(C_r) \subset C_{(1+\varepsilon)r}$.

So it remains to show that $f(C_r)$ contains the smaller cube $C_{(1-\varepsilon)r}$. We will apply the contraction mapping theorem to prove this. Given $\mathbf{y} \in C_{(1-\varepsilon)r}$, define $\varphi : \mathscr{R}^n \to \mathscr{R}^n$ by

$$\varphi(\mathbf{x}) = \mathbf{x} - f(\mathbf{x}) + \mathbf{y}.$$

We want to show that φ is a contraction mapping of C_r; its unique fixed point will then be the desired point $\mathbf{x} \in C_r$ such that $f(\mathbf{x}) = \mathbf{y}$.

To see that φ maps C_r into itself, we apply Corollary 2.6:

$$\begin{aligned}
|\varphi(\mathbf{x})|_0 &\leqq |f(\mathbf{x}) - \mathbf{x}|_0 + |\mathbf{y}|_0 \\
&= |f(\mathbf{x}) - f(\mathbf{0}) - df_{\mathbf{0}}(\mathbf{x} - \mathbf{0})|_0 + |\mathbf{y}|_0 \\
&\leqq |\mathbf{y}|_0 + |\mathbf{x}|_0 \max_{\mathbf{x} \in C_r} \|df_{\mathbf{x}} - df_{\mathbf{0}}\| \\
&\leqq (1 - \varepsilon)r + r\varepsilon \\
&= r,
\end{aligned}$$

so if $\mathbf{x} \in C_r$, then $\varphi(\mathbf{x}) \in C_r$ also. Note here that, if $\mathbf{y} \in \operatorname{int} C_{(1-\varepsilon)r}$, then $|\varphi(\mathbf{x})|_0 < r$, so $\varphi(\mathbf{x}) \in \operatorname{int} C_r$.

To see that $\varphi : C_r \to C_r$ is a contraction mapping, we need only note that

$$|\varphi(\mathbf{x}) - \varphi(\mathbf{y})|_0 = |f(\mathbf{x}) - f(\mathbf{y}) - (\mathbf{x} - \mathbf{y})|_0 \leqq \varepsilon|\mathbf{x} - \mathbf{y}|_0$$

by (1).

Thus $\varphi : C_r \to C_r$ is indeed a contraction mapping, with contraction constant $\varepsilon < 1$, and therefore has a unique fixed point $\mathbf{x}$ such that $f(\mathbf{x}) = \mathbf{y}$. We have noted that φ maps C_r into $\operatorname{int} C_r$ if $\mathbf{y} \in V = \operatorname{int} C_{(1-\varepsilon)r}$. Hence in this case the fixed point $\mathbf{x}$ lies in $\operatorname{int} C_r$. Therefore, if $U = f^{-1}(V) \cap \operatorname{int} C_r$, then U and V are open neighborhoods of $\mathbf{0}$ such that f maps U one-to-one onto V.

The fact that the fixed point $\mathbf{x} = g(\mathbf{y})$ is the limit of the sequence $\{\mathbf{x}_m\}_0^\infty$ defined inductively by

$$\mathbf{x}_0 = \mathbf{0}, \qquad \mathbf{x}_{m+1} = \mathbf{x}_m - f(\mathbf{x}_m) + \mathbf{y},$$

follows immediately from the contraction mapping theorem (3.1).

So it remains only to show that $g : V \to U$ is differentiable at $\mathbf{0}$, where $g(\mathbf{0}) = \mathbf{0}$. It suffices to show that

$$\lim_{|\mathbf{h}| \to 0} \frac{|g(\mathbf{h}) - \mathbf{h}|_0}{|\mathbf{h}|_0} = 0; \tag{3}$$

this will prove that g is differentiable at $\mathbf{0}$ with $dg_0 = I$. To verify (3), we apply (1) with $\mathbf{y} = \mathbf{0}$, $\mathbf{x} = g(\mathbf{h})$, $\mathbf{h} = f(\mathbf{x})$, obtaining

$$|g(\mathbf{h}) - \mathbf{h}|_0 = |f(\mathbf{x}) - \mathbf{x}|_0 \leqq \varepsilon |\mathbf{x}|_0 .$$

We then apply the left-hand inequality of (2) with $\mathbf{y} = \mathbf{0}$, obtaining

$$|g(\mathbf{h}) - \mathbf{h}|_0 \leqq \varepsilon |\mathbf{x}|_0 \leqq \frac{\varepsilon}{1-\varepsilon} |f(\mathbf{x})|_0 = \frac{\varepsilon}{1-\varepsilon} |\mathbf{h}|_0 . \tag{4}$$

This follows from the fact that

$$\|df_{\mathbf{x}} - I\| \leqq \varepsilon < 1 \qquad \text{for} \quad \mathbf{x} \in C_r .$$

Since f is $\mathscr{C}^1$ at $\mathbf{0}$ with $df_{\mathbf{0}} = I$, we can make $\varepsilon > 0$ as small as we like, simply by restricting our attention to a sufficiently small (new) cube centered at $\mathbf{0}$. Hence (4) implies (3). ∎

We now apply this lemma to establish the general inverse mapping theorem. It provides both the existence of a local inverse g under the condition $|f'(\mathbf{a})| \neq 0$, and also an explicit sequence $\{g_k\}_1^\infty$ of successive approximations to g. The definition of this sequence $\{g_k\}_1^\infty$ can be motivated precisely as in the 1-dimensional case (preceding the statement of Theorem 1.3).

Theorem 3.3 Suppose that the mapping $f: \mathscr{R}^n \to \mathscr{R}^n$ is $\mathscr{C}^1$ in a neighborhood W of the point $\mathbf{a}$, with the matrix $f'(\mathbf{a})$ being nonsingular. Then f is locally invertible—there exist neighborhoods $U \subset W$ of $\mathbf{a}$ and V of $\mathbf{b} = f(\mathbf{a})$, and a one-to-one $\mathscr{C}^1$ mapping $g : V \to W$ such that

$$g(f(\mathbf{x})) = \mathbf{x} \qquad \text{for} \quad \mathbf{x} \in U,$$

and

$$f(g(\mathbf{y})) = \mathbf{y} \qquad \text{for} \quad \mathbf{y} \in V.$$

In particular, the local inverse g is the limit of the sequence $\{g_k\}_0^\infty$ of successive approximations defined inductively by

$$g_0(\mathbf{y}) = \mathbf{a}, \qquad g_{k+1}(\mathbf{y}) = g_k(\mathbf{y}) - f'(\mathbf{a})^{-1}[f(g_k(\mathbf{y})) - \mathbf{y}] \tag{5}$$

for $\mathbf{y} \in V$.

PROOF We first "alter" the mapping f so as to make it satisfy the hypotheses of Lemma 3.2. Let $\tau_{\mathbf{a}}$ and $\tau_{\mathbf{b}}$ be the translations of $\mathscr{R}^n$ defined by

$$\tau_{\mathbf{a}}(\mathbf{x}) = \mathbf{x} + \mathbf{a}, \qquad \tau_{\mathbf{b}}(\mathbf{x}) = \mathbf{x} + \mathbf{b},$$

and let $T = df_{\mathbf{a}} : \mathscr{R}^n \to \mathscr{R}^n$. Then define $\tilde{f}: \mathscr{R}^n \to \mathscr{R}^n$ by

$$\tilde{f}(\mathbf{x}) = \tau_{\mathbf{b}}^{-1} \circ f \circ \tau_{\mathbf{a}} \circ T^{-1}(\mathbf{x}). \tag{6}$$

The relationship between f and $\tilde{f}$ is exhibited by the following diagram:

$$\begin{array}{ccc} \mathscr{R}^n & \underset{g}{\overset{f}{\rightleftarrows}} & \mathscr{R}^n \\ \tau_{\mathbf{a}} \circ T^{-1} \uparrow & & \uparrow \tau_{\mathbf{b}} \\ \mathscr{R}^n & \underset{\tilde{f}}{\overset{\tilde{g}}{\leftrightarrows}} & \mathscr{R}^n \end{array}$$

The assertion of Eq. (6) is that the same result is obtained by following the arrows around in either direction.

Note the $\tilde{f}(\mathbf{0}) = \mathbf{0}$. Since the differentials of $\tau_{\mathbf{a}}$ and $\tau_{\mathbf{b}}$ are both the identity mapping I of $\mathscr{R}^n$, an application of the chain rule yields

$$d\tilde{f}_{\mathbf{0}} = I.$$

Since $\tilde{f}$ is $\mathscr{C}^1$ in a neighborhood of $\mathbf{0}$, we have

$$\|d\tilde{f}_{\mathbf{x}} - I\| \leqq \varepsilon < 1$$

on a sufficiently small cube centered at $\mathbf{0}$. Therefore Lemma 3.2 applies to give neighborhoods $\tilde{U}$ and $\tilde{V}$ of $\mathbf{0}$, and a one-to-one mapping $\tilde{g}$ of $\tilde{V}$ onto $\tilde{U}$, differentiable at $\mathbf{0}$, such that the mappings

$$\tilde{f} : \tilde{U} \to \tilde{V} \qquad \text{and} \qquad \tilde{g} : \tilde{V} \to \tilde{U}$$

are inverses of each other. Moreover Lemma 3.2 gives a sequence $\{\tilde{g}_k\}_0^\infty$ of successive approximations to $\tilde{g}$, defined inductively by

$$\tilde{g}_0(\mathbf{y}) = \mathbf{0}, \qquad \tilde{g}_{k+1}(\mathbf{y}) = \tilde{g}_k(\mathbf{y}) - \tilde{f}(\tilde{g}_k(\mathbf{y})) + \mathbf{y}$$

for $\mathbf{y} \in \tilde{V}$.

We let $U = \tau_{\mathbf{a}} \circ T^{-1}(\tilde{U})$, $V = \tau_{\mathbf{b}}(\tilde{V})$, and define $g : V \to U$ by

$$g(\mathbf{y}) = \tau_{\mathbf{a}} \circ T^{-1} \circ \tilde{g} \circ \tau_{\mathbf{b}}^{-1}(\mathbf{y}).$$

(Now look at g and $\tilde{g}$ in the above diagram.) The facts, that $\tilde{g}$ is a local inverse to $\tilde{f}$, and that the mappings $\tau_{\mathbf{a}} \circ T^{-1}$ and $\tau_{\mathbf{b}}$ are one-to-one, imply that $g : V \to U$ is the desired local inverse to $f : U \to V$. The fact that $\tilde{g}$ is differentiable at $\mathbf{0}$ implies that g is differentiable at $\mathbf{b} = \tau_{\mathbf{b}}(\mathbf{0})$.

We obtain the sequence $\{g_k\}_0^\infty$ of successive approximations to g from the sequence $\{\tilde{g}_k\}_0^\infty$ of successive approximations to $\tilde{g}$, by defining

$$\begin{aligned} g_k(\mathbf{y}) &= \tau_{\mathbf{a}} \circ T^{-1} \circ \tilde{g}_k \circ \tau_{\mathbf{b}}^{-1}(\mathbf{y}) \\ &= \tau_{\mathbf{a}} \circ T^{-1} \circ g_k(\mathbf{y} - \mathbf{b}) \end{aligned} \tag{7}$$

for $\mathbf{y} \in V$ (replacing g by g_k and $\tilde{g}$ by $\tilde{g}_k$ in the above diagram).

To verify that the sequence $\{g_k\}_0^\infty$ may be defined inductively as in (5), note first that

$$\begin{aligned} g_0(\mathbf{y}) &= \tau_{\mathbf{a}} \circ T^{-1} \circ g_0(\mathbf{y} - \mathbf{b}) = \tau_{\mathbf{a}} \circ T^{-1}(\mathbf{0}) \\ &= \tau_{\mathbf{a}}(\mathbf{0}) = \mathbf{a}. \end{aligned}$$

Now start with the inductive relation

$$\tilde{g}_{k+1}(\mathbf{y}-\mathbf{b}) = \tilde{g}_k(\mathbf{y}-\mathbf{b}) - \tilde{f}(\tilde{g}(\mathbf{y}-\mathbf{b})) + (\mathbf{y}-\mathbf{b}).$$

Substituting from (6) and (7), we obtain

$$\begin{aligned} T \circ \tau_{\mathbf{a}}^{-1} \circ g_{k+1}(\mathbf{y}) &= T \circ \tau_{\mathbf{a}}^{-1} \circ g_k(\mathbf{y}) - \tau_{\mathbf{b}}^{-1} \circ f \circ \tau_{\mathbf{a}} \circ T^{-1}(T \circ \tau_{\mathbf{a}}^{-1} \circ g_k(\mathbf{y})) + (\mathbf{y}-\mathbf{b}) \\ &= T \circ \tau_{\mathbf{a}}^{-1} \circ g_k(\mathbf{y}) - [f(g_k(\mathbf{y})) - \mathbf{y}]. \end{aligned}$$

Applying $\tau_{\mathbf{a}} \circ T^{-1}$ to both sides of this equation, we obtain the desired inductive relation

$$g_{k+1}(\mathbf{y}) = g_k(\mathbf{y}) - T^{-1}[f(g_k(\mathbf{y})) - \mathbf{y}].$$

It remains only to show that $g : V \to U$ is a $\mathscr{C}^1$ mapping; at the moment we know only that g is differentiable at the point $\mathbf{b} = f(\mathbf{a})$. However what we have already proved can be applied at each point of U, so it follows that g is differentiable at each point of V.

To see that g is *continuously* differentiable on V, we note that, since

$$f(g(\mathbf{y})) = \mathbf{y}, \qquad \mathbf{y} \in V,$$

the chain rule gives

$$g'(\mathbf{y}) = [f'(g(\mathbf{y}))]^{-1}$$

for each $\mathbf{y} \in V$. Now $f'(g(\mathbf{y}))$ is a continuous (matrix-valued) mapping, because f is $\mathscr{C}^1$ by hypothesis, and the mapping g is continuous because it is differentiable (Exercise II.2.1). In addition it is clear from the formula for the inverse of a nonsingular matrix (Theorem I.6.3) that the entries of an inverse matrix A^{-1} are continuous functions of those of A. These facts imply that the entries of the matrix $g'(\mathbf{y}) = [f'(g(\mathbf{y}))]^{-1}$ are continuous functions of $\mathbf{y}$, so g is $\mathscr{C}^1$.

This last argument can be rephrased as follows. We can write $g' : V \to \mathscr{M}_{nn}$, where $\mathscr{M}_{nn}$ is the space of $n \times n$ matrices, as the composition

$$g' = \mathscr{I} \circ f' \circ g,$$

where $\mathscr{I}(A) = A^{-1}$ on the set of invertible $n \times n$ matrices (an open subset of $\mathscr{M}_{nn}$). $\mathscr{I}$ is continuous by the above remark, and $f' : U \to \mathscr{M}_{nn}$ is continuous because f is $\mathscr{C}^1$, so we have expressed g' as a composition of continuous mappings. Thus g' is continuous, so g is $\mathscr{C}^1$ by Proposition 2.4. ∎

The power of the inverse mapping theorem stems partly from the fact that the condition $\det f'(\mathbf{a}) \neq 0$ implies the invertibility of the $\mathscr{C}^1$ mapping f in a neighborhood of $\mathbf{a}$, even when it is difficult or impossible to find the local inverse mapping g explicitly. However Eqs. (5) enable us to approximate g arbitrarily closely [near $\mathbf{b} = f(\mathbf{a})$].

Example 2 Suppose the $\mathscr{C}^1$ mapping $f : \mathscr{R}^2_{uv} \to \mathscr{R}^2_{xy}$ is defined by the equations

$$\begin{aligned} x &= u + (v+2)^2 + 1, \\ y &= (u-1)^2 + v + 1. \end{aligned}$$

Let $\mathbf{a} = (1, -2)$, so $\mathbf{b} = f(\mathbf{a}) = (2, -1)$. The derivative matrix is

$$f'(u, v) = \begin{pmatrix} 1 & 2(v+2) \\ 2(u-1) & 1 \end{pmatrix},$$

so $f'(\mathbf{a})$ is the identity matrix with determinant 1. Therefore f is invertible near $\mathbf{a} = (1, -2)$. That is, the above equations can be solved for u and v as functions of x and y, $(u, v) = g(x, y)$, if the point (x, y) is sufficiently close to $\mathbf{b} = (2, -1)$. According to Eqs. (5), the sequence of successive approximations $\{g_k\}_1^\infty$ is defined inductively by

$$g_0(x, y) = (1, -2), \qquad g_{k+1}(x, y) = g_k(x, y) - f(g_k(x, y)) + (x, y).$$

Writing $g_k(x, y) = (u_k, v_k)$, we have

$$\begin{aligned}(u_{k+1}, v_{k+1}) &= (u_k, v_k) - (u_k + (v_k+2)^2 + 1, (u_k - 1)^2 + v_k + 1) + (x, y) \\ &= (1 + (x-2) - (v_k+2)^2, -2 + (y+1) - (u_k - 1)^2).\end{aligned}$$

The first several approximations are

$$\begin{aligned}
u_0 &= 1, & v_0 &= -2, \\
u_1 &= 1 + (x-2), & v_1 &= -2 + (y+1), \\
u_2 &= 1 + (x-2) - (y+1)^2, & v_2 &= -2 + (y+1) - (x-2)^2
\end{aligned}$$

$$\begin{aligned}
u_3 &= 1 + (x-2) - [(y+1) - (x-2)^2]^2 \\
&= 1 + (x-2) - (y+1)^2 + 2(x-2)^2(y+1) - (x-2)^4, \\
v_3 &= -2 + (y+1) - [(x-2) - (y+1)^2]^2 \\
&= -2 + (y+1) - (x-2)^2 + 2(x-2)(y+1)^2 - (y+1)^4.
\end{aligned}$$

It appears that we are generating Taylor polynomials for the component functions of g in powers of $(x-2)$ and $(y+1)$. This is true, but we will not verify it.

The inverse mapping theorem tells when we can, in principle, solve the equation $\mathbf{x} = f(\mathbf{y})$, or equivalently $\mathbf{x} - f(\mathbf{y}) = \mathbf{0}$, for $\mathbf{y}$ as a function of $\mathbf{x}$. The *implicit* mapping theorem deals with the problem of solving the general equation $G(\mathbf{x}, \mathbf{y}) = \mathbf{0}$, for $\mathbf{y}$ as a function of $\mathbf{x}$. Although the latter equation may appear considerably more general, we will find that its solution reduces quite easily to the special case considered in the inverse mapping theorem.

In order for us to reasonably expect a unique solution of $G(\mathbf{x}, \mathbf{y}) = \mathbf{0}$, there should be the same number of equations as unknowns. So if $\mathbf{x} \in \mathscr{R}^m$, $\mathbf{y} \in \mathscr{R}^n$, then G should be a mapping from $\mathscr{R}^{m+n}$ to $\mathscr{R}^n$. Writing out components of the vector equation $G(\mathbf{x}, \mathbf{y}) = \mathbf{0}$, we obtain

$$\begin{gathered}
G_1(x_1, \ldots, x_m, y_1, \ldots, y_n) = 0, \\
\vdots \\
G_n(x_1, \ldots, x_m, y_1, \ldots, y_n) = 0.
\end{gathered}$$

We want to discuss the solution of this system of equations for $y_1, \ldots, y_n$ in terms of $x_1, \ldots, x_m$.

In order to investigate this problem we need the notion of partial differential linear mappings. Given a mapping $G : \mathscr{R}^{m+n} \to \mathscr{R}^k$ wihch is differentiable at the point $\mathbf{p} = (\mathbf{a}, \mathbf{b}) \in \mathscr{R}^{m+n}$, the *partial differentials* of G are the linear mappings $d_\mathbf{x} G_\mathbf{p} : \mathscr{R}^m \to \mathscr{R}^k$ and $d_\mathbf{y} G_\mathbf{p} : \mathscr{R}^n \to \mathscr{R}^k$ defined by

$$d_\mathbf{x} G_\mathbf{p}(\mathbf{r}) = dG_\mathbf{p}(\mathbf{r}, \mathbf{0}) \qquad \text{and} \qquad d_\mathbf{y} G_\mathbf{p}(\mathbf{s}) = dG_\mathbf{p}(\mathbf{0}, \mathbf{s}),$$

respectively. The matrices of these partial differential linear mappings are called the *partial derivatives* of G with respect to $\mathbf{x}$ and $\mathbf{y}$, respectively. These partial derivative matrices are denoted by

$$D_1 G(\mathbf{a}, \mathbf{b}) \quad \text{or} \quad \frac{\partial G}{\partial \mathbf{x}} \qquad \text{and} \qquad D_2 G(\mathbf{a}, \mathbf{b}) \quad \text{or} \quad \frac{\partial G}{\partial \mathbf{y}},$$

respectively. It should be clear from the definitions that $D_1 G(\mathbf{a}, \mathbf{b})$ consists of the first m columns of $G'(\mathbf{a}, \mathbf{b})$, while $D_2 G(\mathbf{a}, \mathbf{b})$ consists of the last n columns. Thus

$$\frac{\partial G}{\partial \mathbf{x}} = \begin{pmatrix} \dfrac{\partial G_1}{\partial x_1} & \cdots & \dfrac{\partial G_1}{\partial x_m} \\ \vdots & & \vdots \\ \dfrac{\partial G_k}{\partial x_1} & \cdots & \dfrac{\partial G_k}{\partial x_m} \end{pmatrix} \qquad \text{and} \qquad \frac{\partial G}{\partial \mathbf{y}} = \begin{pmatrix} \dfrac{\partial G_1}{\partial y_1} & \cdots & \dfrac{\partial G_1}{\partial y_n} \\ \vdots & & \vdots \\ \dfrac{\partial G_k}{\partial y_1} & \cdots & \dfrac{\partial G_k}{\partial y_n} \end{pmatrix}.$$

Suppose now that $\mathbf{x} \in \mathscr{R}^m$ and $\mathbf{y} \in \mathscr{R}^n$ are both differentiable functions of $\mathbf{t} \in \mathscr{R}^l$, $\mathbf{x} = \alpha(\mathbf{t})$ and $\mathbf{y} = \beta(\mathbf{t})$, and define $\varphi : \mathscr{R}^l \to \mathscr{R}^k$ by

$$\varphi(\mathbf{t}) = G(\alpha(\mathbf{t}), \beta(\mathbf{t})).$$

Then a routine application of the chain rule gives the expected result

$$\frac{d\varphi}{dt} = \frac{\partial G}{\partial \mathbf{x}} \frac{d\mathbf{x}}{dt} + \frac{\partial G}{\partial \mathbf{y}} \frac{d\mathbf{y}}{dt} \tag{8}$$

or

$$\varphi'(\mathbf{t}) = D_1 G(\alpha(\mathbf{t}), \beta(\mathbf{t}))\alpha'(\mathbf{t}) + D_2 G(\alpha(\mathbf{t}), \beta(\mathbf{t}))\beta'(\mathbf{t})$$

in more detail. We leave this computation to the exercises (Exercise 3.16). Note that $D_1 G$ is a $k \times m$ matrix and α' is an $m \times l$ matrix, while $D_2 G$ is a $k \times n$ matrix and β' is an $n \times l$ matrix, so Eq. (8) at least makes sense.

Suppose now that the function $G : \mathscr{R}^{m+n} \to \mathscr{R}^n$ is differentiable in a neighborhood of the point $(\mathbf{a}, \mathbf{b}) \in \mathscr{R}^{m+n}$ where $G(\mathbf{a}, \mathbf{b}) = \mathbf{0}$. Suppose also that the equation $G(\mathbf{x}, \mathbf{y}) = \mathbf{0}$ implicitly defines a differentiable function $\mathbf{y} = h(\mathbf{x})$ for $\mathbf{x}$ near $\mathbf{a}$, that is, h is a differentiable mapping of a neighborhood of $\mathbf{a}$ into $\mathscr{R}^n$, with

$$h(\mathbf{a}) = \mathbf{b} \qquad \text{and} \qquad G(\mathbf{x}, h(\mathbf{x})) = \mathbf{0}.$$

We can then compute the derivative $h'(\mathbf{x})$ by differentiating the equation $G(\mathbf{x}, h(\mathbf{x})) = \mathbf{0}$. Applying Eq. (8) above with $\mathbf{x} = \mathbf{t}$, $\alpha(\mathbf{x}) = \mathbf{x}$, we obtain

$$D_1 G(\mathbf{x}, \mathbf{y}) + D_2 G(\mathbf{x}, \mathbf{y}) h'(\mathbf{x}) = 0.$$

If the $n \times n$ matrix $D_2 G(\mathbf{x}, \mathbf{y})$ is nonsingular, it follows that

$$h'(\mathbf{x}) = -D_2 G(\mathbf{x}, \mathbf{y})^{-1} D_1 G(\mathbf{x}, \mathbf{y}). \tag{9}$$

In particular, it appears that the nonvanishing of the so-called *Jacobian determinant*

$$\frac{\partial(G_1, \ldots, G_n)}{\partial(y_1, \ldots, y_n)} = \det D_2 G$$

at $(\mathbf{a}, \mathbf{b})$ is a necessary condition for us to be able to solve for $h'(\mathbf{a})$. The implicit mapping theorem asserts that, if G is $\mathscr{C}^1$ in a neighborhood of $(\mathbf{a}, \mathbf{b})$, then this condition is *sufficient* for the existence of the implicitly defined mapping h.

Given $G : \mathscr{R}^{m+n} \to \mathscr{R}^n$ and $h : U \to \mathscr{R}^n$, where $U \subset \mathscr{R}^m$, we will say that $\mathbf{y} = h(\mathbf{x})$ *solves the equation* $G(\mathbf{x}, \mathbf{y})) = \mathbf{0}$ in a neighborhood W of $(\mathbf{a}, \mathbf{b})$ if the graph of f agrees in W with the zero set of G. That is, if $(\mathbf{x}, \mathbf{y}) \in W$ and $\mathbf{x} \in U$, then

$$G(\mathbf{x}, \mathbf{y}) = \mathbf{0} \qquad \text{if and only if} \qquad \mathbf{y} = h(\mathbf{x}).$$

Note the almost verbatim analogy between the following general implicit mapping theorem and the case $m = n = 1$ considered in Section 1.

Theorem 3.4 Let the mapping $G : \mathscr{R}^{m+n} \to \mathscr{R}^n$ be $\mathscr{C}^1$ in a neighborhood of the point $(\mathbf{a}, \mathbf{b})$ where $G(\mathbf{a}, \mathbf{b}) = \mathbf{0}$. If the partial derivative matrix $D_2 G(\mathbf{a}, \mathbf{b})$ is nonsingular, then there exists a neighborhood U of $\mathbf{a}$ in $\mathscr{R}^m$, a neighborhood W of $(\mathbf{a}, \mathbf{b})$ in $\mathscr{R}^{m+n}$, and a $\mathscr{C}^1$ mapping $h : U \to \mathscr{R}^n$, such that $\mathbf{y} = h(\mathbf{x})$ solves the equation $G(\mathbf{x}, \mathbf{y}) = 0$ in W.

In particular, the implicitly defined mapping h is the limit of the sequence of successive approximations defined inductively by

$$h_0(\mathbf{x}) = \mathbf{b}, \qquad h_{k+1}(\mathbf{x}) = h_k(\mathbf{x}) - D_2 G(\mathbf{a}, \mathbf{b})^{-1} G(\mathbf{x}, h_k(\mathbf{x}))$$

for $\mathbf{x} \in U$.

PROOF We want to apply the inverse mapping theorem to the mapping $f : \mathscr{R}^{m+n} \to \mathscr{R}^{m+n}$ defined by

$$f(\mathbf{x}, \mathbf{y}) = (\mathbf{x}, G(\mathbf{x}, \mathbf{y})),$$

for which $f(\mathbf{x}, \mathbf{y}) = (\mathbf{x}, \mathbf{0})$ if and only if $G(\mathbf{x}, \mathbf{y}) = \mathbf{0}$. Note that f is $\mathscr{C}^1$ in a neighborhood of the point $(\mathbf{a}, \mathbf{b})$ where $f(\mathbf{a}, \mathbf{b}) = (\mathbf{a}, \mathbf{0})$. In order to apply the inverse mapping theorem in a neighborhood of $(\mathbf{a}, \mathbf{b})$, we must first show that the matrix $f'(\mathbf{a}, \mathbf{b})$ is nonsingular.

It is clear that

$$f'(\mathbf{a}, \mathbf{b}) = \left(\begin{array}{c|c} I & \mathbf{0} \\ \hline D_1 G(\mathbf{a}, \mathbf{b}) & D_2 G(\mathbf{a}, \mathbf{b}) \end{array}\right)$$

where I denotes the $m \times m$ identity matrix, and $\mathbf{0}$ denotes the $m \times n$ zero matrix. Consequently

$$df_{(\mathbf{a}, \mathbf{b})}(\mathbf{r}, \mathbf{s}) = (\mathbf{r}, d_{\mathbf{x}} G_{(\mathbf{a},\mathbf{b})}(\mathbf{r}) + d_{\mathbf{y}} G_{(\mathbf{a}, \mathbf{b})}(\mathbf{s}))$$

if $(\mathbf{r}, \mathbf{s}) \in \mathscr{R}^{m+n}$. In order to prove that $f'(\mathbf{a}, \mathbf{b})$ is nonsingular, it suffices to show that $df_{(\mathbf{a}, \mathbf{b})}$ is one-to-one (why?), that is, that $df_{(\mathbf{a}, \mathbf{b})}(\mathbf{r}, \mathbf{s}) = (\mathbf{0}, \mathbf{0})$ implies $(\mathbf{r}, \mathbf{s}) = (\mathbf{0}, \mathbf{0})$. But this follows immediately from the above expression for $df_{(\mathbf{a}, \mathbf{b})}(\mathbf{r}, \mathbf{s})$ and the hypothesis that $D_2 G(\mathbf{a}, \mathbf{b})$ is nonsingular, so $d_{\mathbf{y}} G_{(\mathbf{a}, \mathbf{b})}(\mathbf{s}) = \mathbf{0}$ implies $\mathbf{s} = \mathbf{0}$.

We can therefore apply the inverse mapping theorem to obtain neighborhoods W of $(\mathbf{a}, \mathbf{b})$ and V of $(\mathbf{a}, \mathbf{0})$, and a $\mathscr{C}^1$ inverse $g : V \to W$ of $f : W \to V$, such that $g(\mathbf{a}, \mathbf{0}) = (\mathbf{a}, \mathbf{b})$. Let U be the neighborhood of $\mathbf{a} \in \mathscr{R}^m$ defined by

$$U = V \cap \mathscr{R}^m,$$

identifying $\mathscr{R}^m$ with $\mathscr{R}^m \times \mathbf{0} \subset \mathscr{R}^{m+n}$ (see Fig. 3.8).

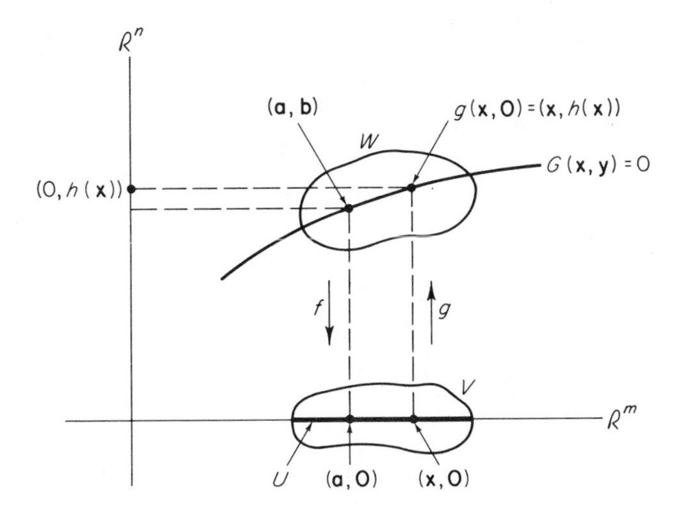

Figure 3.8

Since $f(\mathbf{x}, \mathbf{y}) = \mathbf{0}$ if and only if $G(\mathbf{x}, \mathbf{y}) = \mathbf{0}$, it is clear that g maps the set $U \times \mathbf{0} \subset \mathscr{R}^{m+n}$ one-to-one onto the intersection with W of the zero set $G(\mathbf{x}, \mathbf{y}) = \mathbf{0}$. If we now define the $\mathscr{C}^1$mapping $h : U \to \mathscr{R}^n$ by

$$g(\mathbf{x}, \mathbf{0}) = (\mathbf{x}, h(\mathbf{x})),$$

it follows that $\mathbf{y} = h(\mathbf{x})$ solves the equation $G(\mathbf{x}, \mathbf{y}) = \mathbf{0}$ in W.

It remains only to define the sequence of successive approximations $\{h_k\}_0^\infty$ to h. The inverse mapping theorem provides us with a sequence $\{g_k\}_0^\infty$ of successive approximations to g, defined inductively by

$$g_0(\mathbf{x}, \mathbf{y}) = (\mathbf{a}, \mathbf{b}), \qquad g_{k+1}(\mathbf{x}, \mathbf{y}) = g_k(\mathbf{x}, \mathbf{y}) - f'(\mathbf{a}, \mathbf{b})^{-1}[f(g_k(\mathbf{x}, \mathbf{y})) - (\mathbf{x}, \mathbf{y})].$$

We define $h_k : U \to \mathscr{R}^n$ by

$$g_k(\mathbf{x}, \mathbf{0}) = (\mathbf{x}, h_k(\mathbf{x})).$$

Then the fact, that the sequence $\{g_k\}_0^\infty$ converges to g on V, implies that the sequence $\{h_k\}_0^\infty$ converges to h on U.

Since $g_0(\mathbf{x}, \mathbf{y}) \equiv (\mathbf{a}, \mathbf{b})$, we see that $h_0(\mathbf{x}) \equiv \mathbf{b}$. Finally

$$\begin{aligned}
(\mathbf{x}, h_{k+1}(\mathbf{x})) &= g_{k+1}(\mathbf{x}, \mathbf{0}) \\
&= g_k(\mathbf{x}, \mathbf{0}) - f'(\mathbf{a}, \mathbf{b})^{-1}[f(g_k(\mathbf{x}, \mathbf{0})) - (\mathbf{x}, \mathbf{0})] \\
&= (\mathbf{x}, h_k(\mathbf{x})) - f'(\mathbf{a}, \mathbf{b})^{-1}[f(\mathbf{x}, h_k(\mathbf{x})) - (\mathbf{x}, \mathbf{0})] \\
&= (\mathbf{x}, h_k(\mathbf{x})) - f'(\mathbf{a}, \mathbf{b})^{-1}[(\mathbf{x}, G(\mathbf{x}, h_k(\mathbf{x}))) - (\mathbf{x}, \mathbf{0})],
\end{aligned}$$

so

$$\begin{pmatrix} \mathbf{x} \\ h_{k+1}(\mathbf{x}) \end{pmatrix} = \begin{pmatrix} \mathbf{x} \\ h_k(\mathbf{x}) \end{pmatrix} - \left(\begin{array}{c|c} I & \mathbf{0} \\ \hline D_1 G(\mathbf{a}, \mathbf{b}) & D_2 G(\mathbf{a}, \mathbf{b}) \end{array}\right)^{-1} \left(\begin{array}{c} \mathbf{0} \\ \hline G(\mathbf{x}, h_k(\mathbf{x})) \end{array}\right).$$

Taking second components of this equation, we obtain

$$h_{k+1}(\mathbf{x}) = h_k(\mathbf{x}) - D_2 G(\mathbf{a}, \mathbf{b})^{-1} G(\mathbf{x}, h_k(\mathbf{x}))$$

as desired. ∎

Example 3 Suppose we want to solve the equations

$$\begin{aligned}
x^2 + \tfrac{1}{2}y^2 + z^3 - z^2 - \tfrac{3}{2} &= 0, \\
x^3 + y^3 - 3y + z + 3 &= 0,
\end{aligned}$$

for y and z as functions of x in a neighborhood of $(-1, 1, 0)$. We define $G : \mathscr{R}^3 \to \mathscr{R}^2$ by

$$G(x, y, z) = (x^2 + \tfrac{1}{2}y^2 + z^3 - z^2 - \tfrac{3}{2},\ x^3 + y^3 - 3y + z + 3).$$

Since

$$\begin{aligned}
\left.\frac{\partial(G_1, G_2)}{\partial(y, z)}\right|_{(-1,1,0)} &= \left.\det\begin{pmatrix} y & 3z^2 - 2z \\ 3y^2 - 3 & 1 \end{pmatrix}\right|_{(-1,1,0)} \\
&= \det\begin{pmatrix} 1 & 0 \\ 0 & 1 \end{pmatrix} = 1 \neq 0,
\end{aligned}$$

the implicit mapping theorem assures us that the above equations do implicitly define y and z as functions of x, for x near -1. The successive approximations for $h(x) = (y(x), z(x))$ begin as follows:

$$\begin{aligned} h_0(x) &= (1, 0), \\ h_1(x) &= (1, 0) - G(x, h_0(x)) = (1, 0) - G(x, 1, 0) \\ &= (2 - x^2, -1 - x^3), \\ h_2(x) &= (2 - x^2, -1 - x^3) - G(x, 2 - x^2, -1 - x^3). \end{aligned}$$

Example 4 Let $G : \mathscr{R}^2 \times \mathscr{R}^2 = \mathscr{R}^4 \to \mathscr{R}^2$ be defined by

$$G(\mathbf{x}, \mathbf{y}) = (x_1 y_2 + x_2 y_1 - 1, x_1 x_2 - y_1 y_2),$$

where $\mathbf{x} = (x_1, x_2)$ and $\mathbf{y} = (y_1, y_2)$, and note that $G(1, 0, 0, 1) = (0, 0)$. Since

$$D_2 G(1, 0, 0, 1) = \begin{pmatrix} 0 & 1 \\ -1 & 0 \end{pmatrix}$$

is nonsingular, the equation $G(\mathbf{x}, \mathbf{y}) = \mathbf{0}$ defines $\mathbf{y}$ implicitly as a function of $\mathbf{x}$, $\mathbf{y} = h(\mathbf{x})$ for $\mathbf{x}$ near $(1, 0)$. Suppose we only want to compute the derivative matrix $h'(1, 0)$. Then from Eq. (9) we obtain

$$\begin{aligned} h'(1, 0) &= D_2 G(1, 0, 0, 1)^{-1} D_1 G(1, 0, 0, 1) \\ &= -\begin{pmatrix} 0 & 1 \\ -1 & 0 \end{pmatrix}^{-1} \begin{pmatrix} 1 & 0 \\ 0 & 1 \end{pmatrix} \\ &= -\begin{pmatrix} 0 & -1 \\ 1 & 0 \end{pmatrix} \begin{pmatrix} 1 & 0 \\ 0 & 1 \end{pmatrix} = \begin{pmatrix} 0 & 1 \\ -1 & 0 \end{pmatrix}. \end{aligned}$$

Note that, writing Eq. (9) for this example with Jacobian determinants instead of derivative matrices, we obtain the chain rule type formula

$$\frac{\partial(y_1, y_2)}{\partial(x_1, x_2)} = -\frac{\partial(G_1, G_2)/\partial(x_1, x_2)}{\partial(G_1, G_2)/\partial(y_1, y_2)}$$

or

$$\frac{\partial(G_1, G_2)}{\partial(x_1, x_2)} = -\frac{\partial(G_1, G_2)}{\partial(y_1, y_2)} \frac{\partial(y_1, y_2)}{\partial(x_1, x_2)}.$$

The partial derivatives of an implicitly defined function can be calculated using the chain rule and Cramer's rule for solving linear equations. The following example illustrates this procedure.

Example 5 Let $G_1, G_2 : \mathscr{R}^5 \to \mathscr{R}$ be $\mathscr{C}^1$ functions, with $G_1(\mathbf{a}) = G_2(\mathbf{a}) = 0$ and

$$J = \frac{\partial(G_1, G_2)}{\partial(u, v)} \neq 0$$

at the point $\mathbf{a} \in \mathscr{R}^5$. Then the two equations

$$G_1(x, y, z, u, v) = 0, \qquad G_2(x, y, z, u, v) = 0$$

implicitly define u and v as functions of x, y, z:

$$u = f(x, y, z), \qquad v = g(x, y, z).$$

Upon differentiation of the equations

$$G_i(x, y, z, f(x, y, z), g(x, y, z)) = 0, \qquad i = 1, 2,$$

with respect to x, we obtain

$$\frac{\partial G_1}{\partial x} + \frac{\partial G_1}{\partial u}\frac{\partial f}{\partial x} + \frac{\partial G_1}{\partial v}\frac{\partial g}{\partial x} = 0, \qquad \frac{\partial G_2}{\partial x} + \frac{\partial G_2}{\partial u}\frac{\partial f}{\partial x} + \frac{\partial G_2}{\partial v}\frac{\partial g}{\partial x} = 0.$$

Solving these two linear equations for $\partial f/\partial x$ and $\partial g/\partial x$, we obtain

$$\frac{\partial f}{\partial x} = -\frac{1}{J}\frac{\partial(G_1, G_2)}{\partial(x, v)} \quad \text{and} \quad \frac{\partial g}{\partial x} = -\frac{1}{J}\frac{\partial(G_1, G_2)}{\partial(u, x)}.$$

Similar formulas hold for $\partial f/\partial y$, $\partial g/\partial y$, $\partial f/\partial z$, $\partial g/\partial z$ (see Exercise 3.18).

Exercises

3.1 Show that $f(x, y) = (x/(x^2 + y^2), y/(x^2 + y^2))$ is locally invertible in a neighborhood of every point except the origin. Compute f^{-1} explicitly.

3.2 Show that the following mappings from $\mathscr{R}^2$ to itself are everywhere locally invertible.
(a) $f(x, y) = (e^x + e^y, e^x - e^y)$.
(b) $g(x, y) = (e^x \cos y, e^x \sin y)$.

3.3 Consider the mapping $f: \mathscr{R}^3 \to \mathscr{R}^3$ defined by $f(x, y, z) = (x, y^3, z^5)$. Note that f has a (global) inverse g, despite the fact that the matrix $f'(\mathbf{0})$ is singular. What does this imply about the differentiability of g at $\mathbf{0}$?

3.4 Show that the mapping $\mathscr{R}^3_{xyz} \to \mathscr{R}^3_{uvw}$, defined by $u = x + e^y$, $v = y + e^z$, $w = z + e^x$, is everywhere locally invertible.

3.5 Show that the equations

$$\sin(x + z) + \log yz^2 = 0, \qquad e^{x+z} + yz = 0$$

implicitly define z near -1, as a function of (x, y) near $(1, 1)$.

3.6 Can the surface whose equation is

$$xy - y \log z + \sin xz = 0$$

be represented in the form $z = f(x, y)$ near $(0, 2, 1)$?

3.7 Decide whether it is possible to solve the equations

$$xu^2 + yzv + x^2z = 3, \qquad xyv^3 + 2zu - u^2v^2 = 2$$

for (u, v) near $(1, 1)$ as a function of (x, y, z) near $(1, 1, 1)$.

3.8 The point $(1, -1, 1)$ lies on the surfaces

$$x^3(y^3 + z^3) = 0 \quad \text{and} \quad (x - y)^3 - z^2 = 7.$$

Show that, in a neighborhood of this point, the curve of intersection of the surfaces can be described by a pair of equations of the form $y = f(x)$, $z = g(x)$.

3.9 Determine an approximate solution of the equation

$$z^3 + 3xyz^2 - 5x^2y^2z + 14 = 0$$

for z near 2, as a function of (x, y) near $(1, -1)$.

3.10 If the equations $f(x, y, z) = 0$, $g(x, y, z) = 0$ can be solved for y and z as differentiable functions of x, show that

$$\frac{dy}{dx} = \frac{1}{J}\frac{\partial(f, g)}{\partial(z, x)}, \qquad \frac{dz}{dx} = \frac{1}{J}\frac{\partial(f, g)}{\partial(x, y)},$$

where $J = \partial(f, g)/\partial(y, z)$.

3.11 If the equations $f(x, y, u, v) = 0$, $g(x, y, u, v) = 0$ can be solved for u and v as differentiable functions of x and y, compute their first partial derivatives.

3.12 Suppose that the equation $f(x, y, z) = 0$ can be solved for each of the three variables x, y, z as a differentiable function of the other two. Then prove that

$$\frac{\partial x}{\partial y}\frac{\partial y}{\partial z}\frac{\partial z}{\partial x} = -1.,$$

Verify this in the case of the ideal gas equation $pv = RT$ (where p, v, T are the variables and R is a constant).

3.13 Let $f\colon \mathscr{R}^2_{\mathbf{x}} \to \mathscr{R}^2_{\mathbf{y}}$ and $g\colon \mathscr{R}^2_{\mathbf{y}} \to \mathscr{R}^2_{\mathbf{x}}$ be $\mathscr{C}^1$ inverse functions. Show that

$$\frac{\partial g_1}{\partial y_1} = \frac{1}{J}\frac{\partial f_2}{\partial x_2}, \qquad \frac{\partial g_1}{\partial y_2} = -\frac{1}{J}\frac{\partial f_1}{\partial x_2},$$

$$\frac{\partial g_2}{\partial y_1} = -\frac{1}{J}\frac{\partial f_2}{\partial x_1}, \qquad \frac{\partial g_2}{\partial y_2} = \frac{1}{J}\frac{\partial f_1}{\partial x_1},$$

where $J = \partial(f_1, f_2)/\partial(x_1, x_2)$.

3.14 Let $f\colon \mathscr{R}^3_{\mathbf{x}} \to \mathscr{R}^3_{\mathbf{y}}$ and $g\colon \mathscr{R}^3_{\mathbf{y}} \to \mathscr{R}^3_{\mathbf{x}}$ be $\mathscr{C}^1$ inverse functions. Show that

$$\frac{\partial g_1}{\partial y_1} = \frac{1}{J}\frac{\partial(f_2, f_3)}{\partial(x_2, x_3)}, \qquad J = \frac{\partial(f_1, f_2, f_3)}{\partial(x_1, x_2, x_3)},$$

and obtain similar formulas for the other derivatives of component functions of g.

3.15 Verify the statement of the implicit mapping theorem given in Section II.5.

3.16 Verify Eq. (8) in this section.

3.17 Suppose that the pressure p, volume v, temperature T, and internal energy u of a gas satisfy the equations

$$f(p, v, T, u) = 0, \qquad g(p, v, T, u) = 0,$$

and that these two equations can be solved for any two of the four variables as functions of the other two. Then the symbol $\partial u/\partial T$, for example, is ambiguous. We denote by $(\partial u/\partial T)_p$ the partial derivative of u with respect to T, with u and v considered as functions of p and T, and by $(\partial u/\partial T)_v$ the partial derivative of u, with u and p considered as functions of v and T. With this notation, apply the results of Exercise 3.11 to show that

$$\left(\frac{\partial u}{\partial p}\right)_v = \left(\frac{\partial u}{\partial T}\right)_v\left(\frac{\partial T}{\partial p}\right)_v = \left(\frac{\partial u}{\partial T}\right)_p\left(\frac{\partial T}{\partial p}\right)_v + \left(\frac{\partial u}{\partial p}\right)_T.$$

3.18 If $y_1, \ldots, y_n$ are implicitly defined as differentiable functions of $x_1, \ldots, x_m$ by the equations

$$f_i(\mathbf{x}, \mathbf{y}) = 0, \qquad i = 1, \ldots, n,$$

show, by generalizing the method of Example 5, that

$$\frac{\partial y_j}{\partial x_i} = -\frac{1}{J}\frac{\partial(f_1, \ldots, f_j, \ldots, f_n)}{\partial(y_1, \ldots, x_i, \ldots, y_n)},$$

where

$$J = \frac{\partial(f_1, \ldots, f_j, \ldots, f_n)}{\partial(y_1, \ldots, y_j, \ldots, y_n)}.$$

4 MANIFOLDS IN $\mathscr{R}^n$

We continue here the discussion of manifolds that was begun in Section II.5. Recall that a k-dimensional manifold in $\mathscr{R}^n$ is a set M that looks locally like the graph of a mapping from $\mathscr{R}^k$ to $\mathscr{R}^{n-k}$. That is, every point of M lies in an open subset V of $\mathscr{R}^n$ such that $P = V \cap M$ is a k-dimensional patch. Recall that this means there exists a permutation $x_{i_1}, \ldots, x_{i_n}$ of $x_1, \ldots, x_n$, and a differentiable mapping $h : U \to \mathscr{R}^{n-k}$ defined on an open set $U \subset \mathscr{R}^k$, such that

$$P = \{\mathbf{x} \in \mathscr{R}^n : (x_{i_1}, \ldots, x_{i_k}) \in U \quad \text{and} \quad (x_{i_{k+1}}, \ldots, x_{i_n}) = h(x_{i_1}, \ldots, x_{i_k})\}.$$

See Fig. 3.9. We will call P a $\mathscr{C}^1$ or *smooth* patch if the mapping h is $\mathscr{C}^1$. The manifold M will be called $\mathscr{C}^1$ or *smooth* if it is a union of smooth patches.

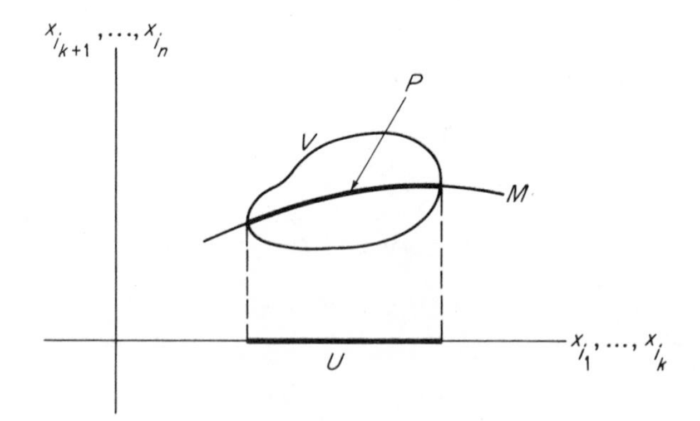

Figure 3.9

There are essentially three ways that a k-manifold in $\mathscr{R}^n$ can appear (locally):

(a) as the graph of a mapping $\mathscr{R}^k \to \mathscr{R}^{n-k}$,
(b) as the zero set of a mapping $\mathscr{R}^n \to \mathscr{R}^{n-k}$,
(c) as the image of a mapping $\mathscr{R}^k \to \mathscr{R}^n$.

The definition of a manifold is based on (a), which is actually a special case of both (b) and (c), since the graph of f: $\mathscr{R}^k \to \mathscr{R}^{n-k}$ is the same as the zero set of

$G(\mathbf{x}, \mathbf{y}) = f(\mathbf{x}) - \mathbf{y}$, and is also the image of $F : \mathscr{R}^k \to \mathscr{R}^n$, where $F(\mathbf{x}) = (\mathbf{x}, f(\mathbf{x}))$. We want to give conditions under which (b) and (c) are, in fact, equivalent to (a).

Recall that our study of the Lagrange multiplier method in Section II.5 was based on the appearance of manifolds in guise (b). The following result was stated (without proof) in Theorem II.5.7.

Theorem 4.1 Let $G : \mathscr{R}^n \to \mathscr{R}^m$ be a $\mathscr{C}^1$ mapping, where $k = n - m > 0$. If M is the set of all those points $\mathbf{x} \in G^{-1}(\mathbf{0})$ for which the derivative matrix $G'(\mathbf{x})$ has rank m, then M is a smooth k-manifold.

PROOF We need to show that each point $\mathbf{p}$ of M lies in a smooth k-dimensional patch on M. By Theorem I.5.4, the fact, that the rank of the $m \times n$ matrix $G'(\mathbf{p})$ is m, implies that some m of its column vectors are linearly independent. The column vectors of the matrix G' are simply the partial derivative vectors $\partial G/\partial x_1, \ldots, \partial G/\partial x_n$, and let us suppose we have rearranged the coordinates in $\mathscr{R}^n$ so that it is the *last m* of these partial derivative vectors that are linearly independent. If we write

$$(x_1, \ldots, x_n) = (\mathbf{x}, \mathbf{y}),$$

where $\mathbf{x} \in \mathscr{R}^k$, $\mathbf{y} \in \mathscr{R}^{n-k} = \mathscr{R}^m$, then it follows that the partial derivative matrix $D_2 G(\mathbf{p})$ is nonsingular.

Consequently the implicit mapping theorem applies to provide us with a neighborhood V of $\mathbf{p} = (\mathbf{a}, \mathbf{b})$ in $\mathscr{R}^n$, a neighborhood U of $\mathbf{a}$ in $\mathscr{R}^k$, and a $\mathscr{C}^1$ mapping $f : U \to \mathscr{R}^m$ such that $\mathbf{y} = f(\mathbf{x})$ solves the equation $G(\mathbf{x}, \mathbf{y}) = 0$ in V. Clearly the graph of f is the desired smooth k-dimensional patch. ∎

Thus conditions (a) and (b) are equivalent, subject to the rank hypothesis in Theorem 4.1.

When we study integration on manifolds in Chapter V, condition (c) will be the most important of the three. If M is a k-manifold in $\mathscr{R}^n$, and U is an open subset of $\mathscr{R}^k$, then a one-to-one mapping $\varphi : U \to M$ can be regarded as a parametrization of the subset $\varphi(U)$ of M. For example, the student is probably familiar with the spherical coordinates parametrization $\varphi : \mathscr{R}^2_{uv} \to S^2$ of the unit sphere $S^2 \subset \mathscr{R}^3$, defined by

$$\varphi(u, v) = (\sin u \cos v, \sin u \sin v, \cos u).$$

The theorem below asserts that, if the subset M of $\mathscr{R}^n$ can be suitably parametrized by means of mappings from open subsets of $\mathscr{R}^k$ to M, then M is a smooth k-manifold.

Let $\varphi : U \to \mathscr{R}^n$ be a $\mathscr{C}^1$ mapping defined on an open subset U of $\mathscr{R}^k$ $(k \leqq n)$. Then we call φ *regular* if the derivative matrix $\varphi'(\mathbf{u})$ has maximal rank k, for each $\mathbf{u} \in U$. According to the following theorem, a subset M of $\mathscr{R}^n$ is a smooth k-manifold if it is locally the image of a regular $\mathscr{C}^1$ mapping defined on an open subset of $\mathscr{R}^k$.

Theorem 4.2 Let M be a subset of $\mathscr{R}^n$. Suppose that, given $\mathbf{p} \in M$, there exists an open set $U \subset \mathscr{R}^k$ $(k < n)$ and a regular $\mathscr{C}^1$ mapping $\varphi : U \to \mathscr{R}^n$ such that $\mathbf{p} \in \varphi(U)$, with $\varphi(U')$ being an open subset of M for each open set $U' \subset U$. Then M is a smooth k-manifold.

The statement that $\varphi(U')$ is an open subset of M means that there exists an open set W' in $\mathscr{R}^n$ such that $W' \cap M = \varphi(U')$. The hypothesis that $\varphi(U')$ is open in M, for *every* open subset U' of U, and not just for U itself, is necessary if the conclusion that M is a k-manifold is to follow. That this is true may be seen by considering a figure six in the plane—although it is not a 1-manifold (why?); there obviously exists a one-to-one regular mapping $\varphi : (0, 1) \to \mathscr{R}^2$ that traces out the figure six.

PROOF Given $\mathbf{p} \in M$ and $\varphi : U \to M$ as in the statement of the theorem, we want to show that $\mathbf{p}$ has a neighborhood (in $\mathscr{R}^n$) whose intersection with M is a smooth k-dimensional patch. If $\varphi(\mathbf{a}) = \mathbf{p}$, then the $n \times k$ matrix $\varphi'(\mathbf{a})$ has rank k. After relabeling coordinates in $\mathscr{R}^n$ if necessary, we may assume that the $k \times k$ submatrix consisting of the first k rows of $\varphi'(\mathbf{a})$ is nonsingular.

Write $\mathbf{p} = (\mathbf{b}, \mathbf{c})$ with $\mathbf{b} \in \mathscr{R}^k$ and $\mathbf{c} \in \mathscr{R}^{n-k}$, and let $\pi : \mathscr{R}^n \to \mathscr{R}^k$ denote the projection onto the first k coordinates,

$$\pi(x_1, \ldots, x_n) = (x_1, \ldots, x_k).$$

If $f : U \to \mathscr{R}^k$ is defined by

$$f = \pi \circ \varphi,$$

then $f(\mathbf{a}) = \mathbf{b}$, and the derivative matrix $f'(\mathbf{a})$ is nonsingular, being simply the $k \times k$ submatrix of $\varphi'(\mathbf{a})$ referred to above.

Consequently the inverse mapping theorem applies to give neighborhoods U' of $\mathbf{a}$ and V' of $\mathbf{b}$ such that $f : U' \to V'$ is one-to-one, and the inverse $g : V' \to U'$ is $\mathscr{C}^1$. Now define $h : V' \to \mathscr{R}^{n-k}$ by

$$\varphi(g(\mathbf{x})) = (\mathbf{x}, h(\mathbf{x})).$$

Since the graph of h is $P = \varphi(U')$, and there exists (by hypothesis) an open set W' in $\mathscr{R}^n$ with $W' \cap M = \varphi(U')$, we see that $\mathbf{p}$ lies in a smooth k-dimensional patch on M, as desired. ∎

REMARK Note that, in the above notation, the mapping

$$\Phi = g \circ \pi : W' \to U'$$

is a $\mathscr{C}^1$ local inverse to φ. That is, Φ is a $\mathscr{C}^1$ mapping on the open subset W' of $\mathscr{R}^n$, and $\Phi(\mathbf{x}) = \varphi^{-1}(\mathbf{x})$ for $\mathbf{x} \in W \bigcap M$. This fact will be used in the proof of Theorem 4.3 below.

If M is a smooth k-manifold, and $\varphi : U \to M$ satisfies the hypotheses of Theorem 4.2, then the mapping φ is called a *coordinate patch* for M provided that it is one-to-one. That is, a coordinate patch for M is a one-to-one regular $\mathscr{C}^1$ mapping $\varphi : U \to \mathscr{R}^n$ defined on an open subset of $\mathscr{R}^k$, such that $\varphi(U')$ is an open subset of M, for each open subset U' of U.

Note that the "local graph" patches, on which we based the definition of a manifold, yield coordinate patches as follows. If M is a smooth k-manifold in $\mathscr{R}^n$, and W is an open set such that $W \cap M$ is the graph of the $\mathscr{C}^1$ mapping $f: U \to \mathscr{R}^{n-k}$ $(U \subset \mathscr{R}^k)$, then the mapping $\varphi : U \to \mathscr{R}^n$ defined by $\varphi(\mathbf{u}) = (\mathbf{u}, f(\mathbf{u}))$ is a coordinate patch for M. We leave it as exercise for the reader to verify this fact. At any rate, every smooth manifold M possesses an abundance of coordinate patches. In particular, every point of M lies in the image of some coordinate patch, so there exists a collection $\{\varphi_\alpha\}_{\alpha \in \Lambda}$ of coordinate patches for M such that

$$M = \bigcup_{\alpha \in \Lambda} \varphi_\alpha(U_\alpha),$$

U_α being the domain of definition of φ_α. Such a collection of coordinate patches is called an *atlas* for M.

The most important fact about coordinate patches is that they overlap differentiably, in the sense of the following theorem (see Fig. 3.10).

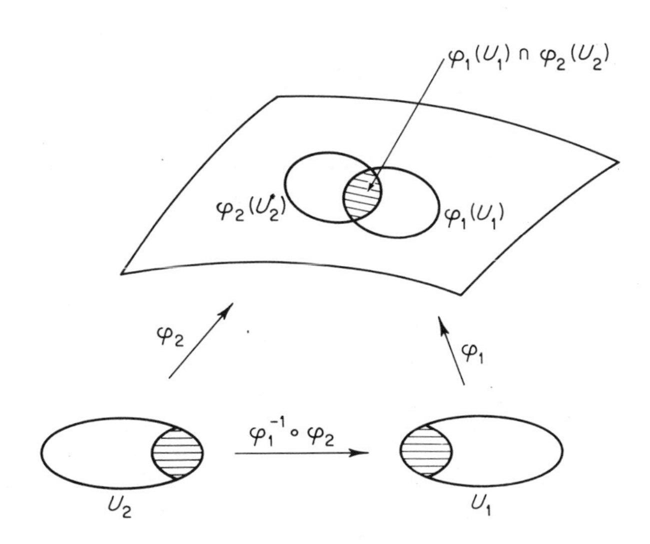

Figure 3.10

Theorem 4.3 Let M be a smooth k-manifold in $\mathscr{R}^n$, and let $\varphi_1 : U_1 \to M$ and $\varphi_2 : U_2 \to M$ be two coordinate patches with $\varphi_1(U_1) \cap \varphi_2(U_2)$ nonempty. Then the mapping

$$\varphi_1^{-1} \circ \varphi_2 : \varphi_2^{-1}(\varphi_1(U_1) \cap \varphi_2(U_2)) \to \varphi_1^{-1}(\varphi_1(U_1) \cap \varphi_2(U_2))$$

is continuously differentiable.

PROOF Given $\mathbf{p} \in \varphi_1(U_1) \cap \varphi_2(U_2)$, the remark following the proof of Theorem 4.2 provides a $\mathscr{C}^1$ local inverse Φ to φ_1, defined on a neighborhood of $\mathbf{p}$ in $\mathscr{R}^n$. Then $\varphi_1^{-1} \circ \varphi_2$ agrees, in a neighborhood of the point $\varphi_2^{-1}(\mathbf{p})$, with the composition $\Phi \circ \varphi_2$ of two $\mathscr{C}^1$ mappings. ∎

Now let $\{\varphi\}_{i \in \Lambda}$ be an atlas for the smooth k-manifold $M \subset \mathscr{R}^n$, and write

$$T_{ij} = \varphi_i^{-1} \circ \varphi_j : \varphi_j^{-1}(\varphi_i(U_i) \cap \varphi_j(U_j)) \to \varphi_i^{-1}(\varphi_i(U_i) \cap \varphi_j(U_j))$$

if $\varphi_i(U_i) \cap \varphi_j(U_j)$ is nonempty. Then T_{ij} is a $\mathscr{C}^1$ mapping by the above theorem, and has the $\mathscr{C}^1$ inverse T_{ji}. It follows from the chain rule that

$$(\det T_{ji}'(T_{ij}(\mathbf{x})))(\det T_{ij}'(\mathbf{x})) = 1,$$

so $\det T_{ij}'(\mathbf{x}) \neq 0$ wherever T_{ij} is defined.

The smooth k-manifold M is called *orientable* if there exists an atlas $\{\varphi_i\}$ for M such that each of the "change of coordinates" mappings T_{ij} defined above has *positive* Jacobian determinant,

$$\det T_{ij}' > 0,$$

wherever it is defined. The pair $(M, \{\varphi_i\})$ is then called an *oriented* k-manifold.

Not every manifold can be oriented. The classical example of a nonorientable manifold is the Möbius strip, a model of which can be made by gluing together the ends of a strip of paper after given it a half twist. We will see the importance of orientability when we study integration on manifolds in Chapter V.

Exercises

4.1 Let $\varphi : U \to \mathscr{R}^n$ and $\psi : V \to \mathscr{R}^n$ be two coordinate patches for the smooth k-manifold M. Say that φ and ψ overlap positively (respectively negatively) if $\det(\varphi^{-1} \circ \psi)'$ is positive (respectively negative) wherever defined. Now define $\rho : \mathscr{R}^k \to \mathscr{R}^k$ by

$$\rho(x_1, x_2, \ldots, x_k) = (-x_1, x_2, \ldots, x_k).$$

If φ and ψ overlap negatively, and $\tilde{\psi} = \psi \circ \rho : \rho^{-1}(V) \to \mathscr{R}^n$, prove that the coordinate neighborhoods φ and $\tilde{\psi}$ overlap positively.

4.2 Show that the unit sphere S^{n-1} has an atlas consisting of just two coordinate patches. Conclude from the preceding exercise that S^{n-1} is orientable.

4.3 Let M be a smooth k-manifold in $\mathscr{R}^n$. Given $\mathbf{p} \in M$, show that there exists an open subset W of $\mathscr{R}^n$ with $\mathbf{p} \in W$, and a one-to-one $\mathscr{C}^1$ mapping $f: W \to \mathscr{R}^n$, such that $f(W \cap M)$ is an open subset of $\mathscr{R}^k \subset \mathscr{R}^n$.

4.4 Let M be a smooth k-manifold in $\mathscr{R}^n$, and N a smooth $(k-1)$-manifold with $N \subset M$. If $\varphi : U \to M$ is a coordinate patch such that $\varphi(U) \cap N$ is nonempty, show that $\varphi^{-1}(\varphi(U) \cap N)$ is a smooth $(k-1)$-manifold in $\mathscr{R}^k$. Conclude from the preceding exercise that, given $\mathbf{p} \in N$, there exists a coordinate patch $\psi : V \to M$ with $\mathbf{p} \in \psi(V)$, such that $\psi^{-1}(\psi(V) \cap N)$ is an open subset of $\mathscr{R}^{k-1} \subset \mathscr{R}^k$.

4.5 If U is an open subset of $\mathscr{R}^2_{uv}$, and $\varphi : U \to \mathscr{R}^3$ is a $\mathscr{C}^1$ mapping, show that φ is regular if and only if $\partial\varphi/\partial u \times \partial\varphi/\partial v \neq \mathbf{0}$ at each point of U. Conclude that φ is regular if and only if, at each point of U, at least one of the three Jacobian determinants

$$\frac{\partial(\varphi_1, \varphi_2)}{\partial(u, v)}, \qquad \frac{\partial(\varphi_1, \varphi_3)}{\partial(u, v)}, \qquad \frac{\partial(\varphi_2, \varphi_3)}{\partial(u, v)}$$

is nonzero.

4.6 The 2-manifold M in $\mathscr{R}^3$ is called *two-sided* if there exists a continuous mapping $\mathbf{n}: M \to \mathscr{R}^3$ such that, for each $\mathbf{x} \in M$, the vector $\mathbf{n}(\mathbf{x})$ is perpendicular to the tangent plane $T_\mathbf{x}$ to M at $\mathbf{x}$. Show that M is two-sided if it is orientable. *Hint*: If $\varphi: U \to \mathscr{R}^3$ is a coordinate patch for M, then $\partial\varphi/\partial u(\mathbf{u}) \times \partial\varphi/\partial v(\mathbf{u})$ is perpendicular to $T_{\varphi(\mathbf{u})}$. If $\varphi: U \to \mathscr{R}^3$ and $\psi: V \to \mathscr{R}^3$ are two coordinate patches for M that overlap positively, and $\mathbf{u} \in U$ and $\mathbf{v} \in V$ are points such that $\varphi(\mathbf{u}) = \psi(\mathbf{v}) \in M$, show that the vectors

$$\frac{\partial\varphi}{\partial u}(\mathbf{u}) \times \frac{\partial\varphi}{\partial v}(\mathbf{u}) \qquad \text{and} \qquad \frac{\partial\psi}{\partial u}(\mathbf{v}) \times \frac{\partial\psi}{\partial v}(\mathbf{v})$$

are positive multiples of each other.

5 HIGHER DERIVATIVES

Thus far in this chapter our attention has been confined to $\mathscr{C}^1$ mappings, or to first derivatives. This is a brief exercise section dealing with higher derivatives.

Recall (from Section II.7) that the function $f: U \to \mathscr{R}$, where U is an open subset of $\mathscr{R}^n$, is of class $\mathscr{C}^k$ if all partial derivatives of f, of order at most k, exist and are continuous on U. Equivalently, f is of class $\mathscr{C}^k$ if and only if f and its first-order partial derivatives $D_1 f, \ldots, D_n f$ are all of class $\mathscr{C}^{k-1}$ on U. This inductive form of the definition is useful for inductive proofs.

We say that a mapping $F: U \to \mathscr{R}^m$ is of class $\mathscr{C}^k$ if each of its component functions is of class $\mathscr{C}^k$.

Exercise 5.1 If f and g are functions of class $\mathscr{C}^k$ on $\mathscr{R}^n$, and $a \in \mathscr{R}$, show that the functions $f + g$, fg, and af are also of class $\mathscr{C}^k$.

The following exercise gives the class $\mathscr{C}^k$ chain rule.

Exercise 5.2 If $f: \mathscr{R}^n \to \mathscr{R}^m$ and $g: \mathscr{R}^m \to \mathscr{R}^l$ are class $\mathscr{C}^k$ mappings, prove that the composition $g \circ f: \mathscr{R}^n \to \mathscr{R}^l$ is of class $\mathscr{C}^k$. Use the previous exercise and the matrix form of the ordinary (class $\mathscr{C}^1$) chain rule.

Recall that we introduced in Section 2 the space $\mathscr{M}_{mn}$ of $m \times n$ matrices, and showed that the differentiable mapping $f: \mathscr{R}^n \to \mathscr{R}^m$ is $\mathscr{C}^1$ if and only if its derivative $f': \mathscr{R}^n \to \mathscr{M}_{mn}$ is continuous.

Exercise 5.3 Show that the differentiable mapping $f: \mathscr{R}^n \to \mathscr{R}^m$ is of class $\mathscr{C}^k$ on the open set U if and only if $f': \mathscr{R}^n \to \mathscr{M}_{mn}$ is of class $\mathscr{C}^{k-1}$ on U.

Exercise 5.4 Denote by $\mathscr{N}_{nn}$ the open subset of $\mathscr{M}_{nn}$ consisting of all non-singular $n \times n$ matrices, and by $\mathscr{I} : \mathscr{N}_{nn} \to \mathscr{N}_{nn}$ the inversion mapping, $\mathscr{I}(A) = A^{-1}$. Show that $\mathscr{I}$ is of class $\mathscr{C}^k$ for every positive integer k. This is simply a matter of seeing that the elements of A^{-1} are $\mathscr{C}^k$ functions of those of A.

We can now establish $\mathscr{C}^k$ versions of the inverse and implicit mapping theorems.

Theorem 5.1 If, in the inverse mapping theorem, the mapping f is of class $\mathscr{C}^k$ in a neighborhood of $\mathbf{a}$, then the local inverse g is of class $\mathscr{C}^k$ in a neighborhood of $\mathbf{b} = f(\mathbf{a})$.

Theorem 5.2 If, in the implicit mapping theorem, the mapping G is of class $\mathscr{C}^k$ in a neighborhood of $(\mathbf{a}, \mathbf{b})$, then the implicitly defined mapping h is of class $\mathscr{C}^k$ in a neighborhood of $\mathbf{a}$.

It suffices to prove Theorem 5.1, since 5.2 follows from it, just as in the $\mathscr{C}^1$ case. Recall the formula

$$g' = \mathscr{I} \circ f' \circ g$$

from the last paragraph of the proof of Theorem 3.3 (the $\mathscr{C}^1$ inverse mapping theorem). We already know that g is of class $\mathscr{C}^1$. If we assume inductively that g is of class $\mathscr{C}^{k-1}$, then Exercises 5.2 and 5.4 imply that $g' : \mathscr{R}^n \to \mathscr{M}_{nn}$ is of class $\mathscr{C}^{k-1}$, so g is of class $\mathscr{C}^k$ by Exercise 5.3. ∎

With the class $\mathscr{C}^k$ inverse and implicit mapping theorems now available, the interested reader can rework Section 4 to develop the elementary theory of class $\mathscr{C}^k$ manifolds—ones for which the coordinate patches are regular class $\mathscr{C}^k$ mappings. Everything goes through with $\mathscr{C}^1$ replaced by $\mathscr{C}^k$ throughout.

IV

Multiple Integrals

Chapters II and III were devoted to multivariable differential calculus. We turn now to a study of multivariable integral calculus.

The basic problem which motivates the theory of integration is that of calculating the volumes of sets in $\mathscr{R}^n$. In particular, the integral of a continuous nonnegative, real-valued function f on an appropriate set $S \subset \mathscr{R}^n$ is supposed to equal the volume of the region in $\mathscr{R}^{n+1}$ that lies over the set S and under the graph of f. Experience and intuition therefore suggest that a thorough discussion of multiple integrals will involve the definition of a function which associates with each "appropriate" set $A \subset \mathscr{R}^n$ a nonnegative real number $v(A)$ called its *volume*, satisfying the following conditions (in which A and B are subsets of $\mathscr{R}^n$ whose volume is defined):

(a) If $A \subset B$, then $v(A) \leqq v(B)$.
(b) If A and B are nonoverlapping sets (meaning that their interiors are disjoint), then $v(A \cup B) = v(A) + v(B)$.
(c) If A and B are congruent sets (meaning that one can be carried onto the other by a rigid motion of $\mathscr{R}^n$), then $v(A) = v(B)$.
(d) If I is a *closed interval* in $\mathscr{R}^n$, that is, $I = I_1 \times I_2 \times \cdots \times I_n$ where $I_j = [a_j, b_j] \subset \mathscr{R}$, then $v(I) = (b_1 - a_1)(b_2 - a_2) \cdots (b_n - a_n)$.

Conditions (a)–(d) are simply the natural properties that should follow from any reasonable definition of volume, if it is to agree with one's intuitive notion of "size" for subsets of $\mathscr{R}^n$.

In Section 1 we discuss the concept of volume, or "area," for subsets of the plane $\mathscr{R}^2$, and apply it to the 1-dimensional integral. This will serve both as a review of introductory integral calculus, and as motivation for the treatment in

Sections 2 and 3 of volume and the integral in $\mathscr{R}^n$. Subsequent sections of the chapter are devoted to iterated integrals, change of variables in multiple integrals, and improper integrals.

1 AREA AND THE 1-DIMENSIONAL INTEGRAL

The Greeks attempted to determine areas by what is called the "method of exhaustion." Their basic idea was as follows: Given a set A whose area is to be determined, we inscribe in it a polygonal region whose area can easily be calculated (by breaking it up into nonoverlapping triangles, for instance). This should give an underestimate to the area of A. We then choose another polygonal region which more nearly fills up the set A, and therefore gives a better approximation to its area (Fig. 4.1). Continuing this process, we attempt to "exhaust"

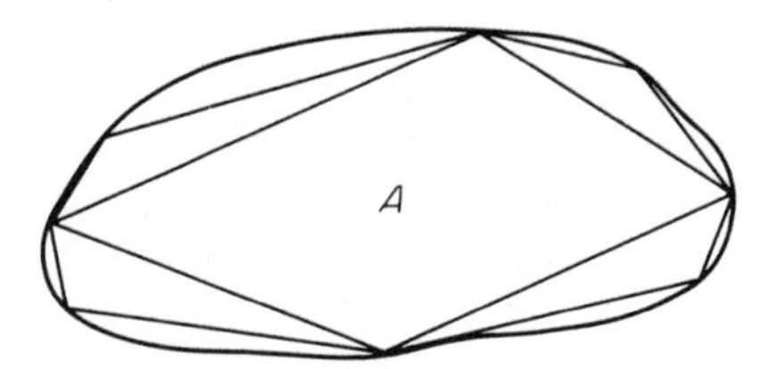

Figure 4.1

the set A with an increasing sequence of inscribed polygons, whose areas should converge to the area of the set A.

However, before attempting to determine or to measure area, we should say precisely what it is that we seek to determine. That is, we should start with a *definition* of area. The following definition is based on the idea of approximating both from within and from without, but using collections of rectangles instead of polygons. We start by *defining* the area $a(R)$ of a rectangle R to be the product of the lengths of its base and height. Then, given a *bounded* set S in the plane $\mathscr{R}^2$, we say that its *area* is α if and only if, given $\varepsilon > 0$, there exist both

(1) a finite collection $R_1', \ldots, R_k'$ of nonoverlapping rectangles, each contained in S, with $\sum_{i=1}^k a(R_i') > \alpha - \varepsilon$ (Fig. 4.2a),

and

(2) A finite collection $R_1'', \ldots, R_l''$ of rectangles which together contain S, with $\sum_{i=1}^l a(R_i'') < \alpha + \varepsilon$ (Fig. 4.2b).

If there exists no such number α, then we say that the set S does not have area, or that its area is not defined.

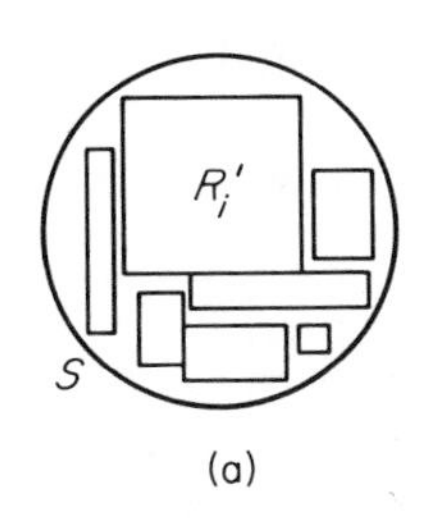

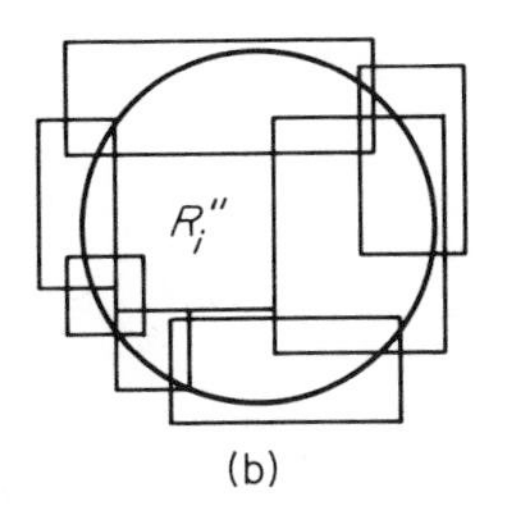

Figure 4.2

Example 1 Not every set has area. For instance, if

$$S = \{(x, y) \in \mathscr{R}^2 : x, y \in [0, 1] \text{ with both rational}\},$$

then S contains *no* (nondegenerate) rectangle. Hence $\sum_{i=1}^{k} a(R_i') = 0$ and $\sum_{i=1}^{l} a(R_i'') \geqq 1$ for any two collections as above (why?). Thus *no* number α satisfies the definition (why?).

Example 2 We verify that the area of a triangle of base b and height a is $\frac{1}{2}ab$. First subdivide the base $[0, b]$ into n equal subintervals, each of length b/n. See Fig. 4.3. Let R_k' denote the rectangle with height $(k-1)a/n$ and base $[(k-1)b/n, kb/n]$, and R_k'' the rectangle with base $[(k-1)b/n, kb/n]$ and height ka/n. Then the sum of the areas of the inscribed rectangles is

$$\sum_{k=1}^{n} a(R_k') = \sum_{k=1}^{n} \frac{b}{n} \cdot \frac{(k-1)a}{n} = \frac{ab}{n^2}(1 + 2 + \cdots + (n-1))$$

$$= \frac{ab}{n^2} \cdot \frac{n}{2}(n-1) = \frac{1}{2}ab - \frac{ab}{2n},$$

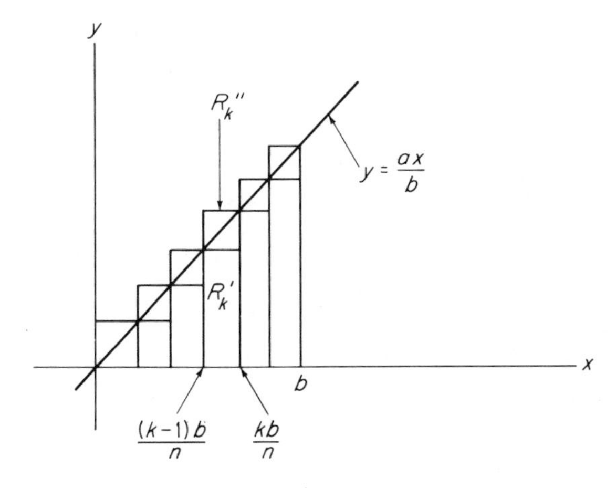

Figure 4.3

while the sum of the areas of the circumscribed rectangles is

$$\sum_{k=1}^{n} a(R_k'') = \sum_{k=1}^{n} \frac{b}{n} \cdot \frac{ka}{n} = \frac{ab}{n^2}(1 + 2 + \cdots + n)$$

$$= \frac{ab}{n^2} \cdot \frac{n}{2}(n+1) = \frac{1}{2}ab + \frac{ab}{2n}.$$

Hence let $\alpha = \frac{1}{2}ab$. Then, given $\varepsilon > 0$, we can satisfy conditions (1) and (2) of the definition by choosing n so large that $ab/2n < \varepsilon$.

We can regard area as a nonnegative valued function $a : \mathscr{A} \to \mathscr{R}$, where $\mathscr{A}$ is the collection of all those subsets of $\mathscr{R}^2$ which have area. We shall defer a systematic study of area until Section 2, where Properties A–E below will be verified. In the remainder of this section, we will employ these properties of area to develop the theory of the integral of a continuous function of one variable.

A If S and T have area and $S \subset T$, then $a(S) \leqq a(T)$ (see Exercise 1.3 below).

B If S and T are two nonoverlapping sets which have area, then so does $S \cup T$, and $a(S \cup T) = a(S) + a(T)$.

C If S and T are two congruent sets and S has area, then so does T, and $a(S) = a(T)$.

D If R is a rectangle, then $a(R) =$ the product of its base and height (obvious from the definition).

E If $S = \{(x, y) \in \mathscr{R}^2 : x \in [a, b] \text{ and } 0 \leqq y \leqq f(x)\}$ where f is a continuous nonnegative function on $[a, b]$, then S has area.

We now define the integral $\int_a^b f$ of a continuous function $f : [a, b] \to \mathscr{R}$. Suppose first that f is *nonnegative* on $[a, b]$, and consider the "ordinate set"

$$O_a^b(f) = \{(x, y) \in \mathscr{R}^2 : x \in [a, b] \quad \text{and} \quad y \in [0, f(x)]\}$$

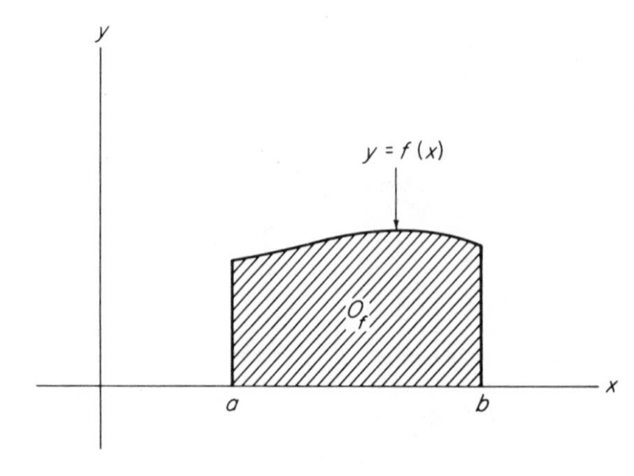

Figure 4.4

pictured in Fig. 4.4. Property E simply says that $O_a^b(f)$ has area. We define the integral of the nonnegative continuous function $f\colon [a, b] \to \mathscr{R}$ by

$$\int_a^b f = a(O_a^b(f)),$$

the area of its ordinate set. Notice that, if $0 \leqq m \leqq f(x) \leqq M$ on $[a, b]$, then

$$m(b-a) \leqq \int_a^b f \leqq M(b-a) \tag{1}$$

by Properties A and D (see Fig. 4.5a), while

$$\int_a^b f = \int_a^c f + \int_c^b f \tag{2}$$

if $c \in (a, b)$, by property B (see Fig. 4.5b).

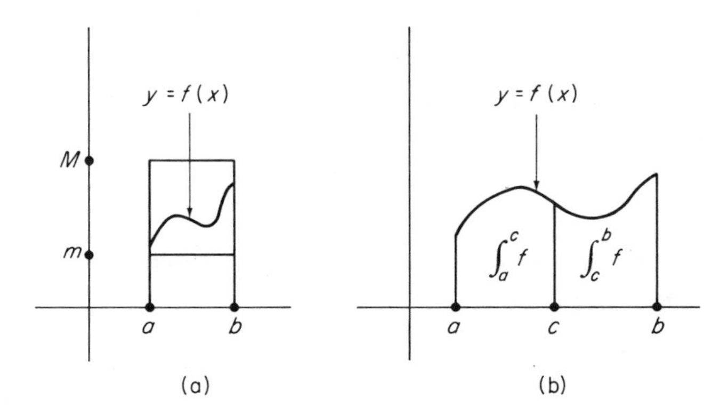

Figure 4.5

If f is an arbitrary continuous function on $[a, b]$, we consider its positive and negative parts f^+ and f^-, defined on $[a, b]$ by

$$f^+(x) = \begin{cases} f(x) & \text{if } f(x) \geqq 0, \\ 0 & \text{if } f(x) < 0, \end{cases}$$

and

$$f^-(x) = \begin{cases} -f(x) & \text{if } f(x) \leqq 0, \\ 0 & \text{if } f(x) > 0. \end{cases}$$

Notice that $f = f^+ - f^-$, and that f^+ and f^- are both nonnegative functions. Also the continuity of f implies that of f^+ and f^- (Exercise 1.4). We can therefore define the *integral* of f on $[a, b]$ by

$$\int_a^b f = \int_a^b f^+ - \int_a^b f^-.$$

In short, $\int_a^b f$ is "the area above the x-axis minus the area below the x-axis" (see Fig. 4.6). We first verify that (1) and (2) above hold for arbitrary continuous functions.

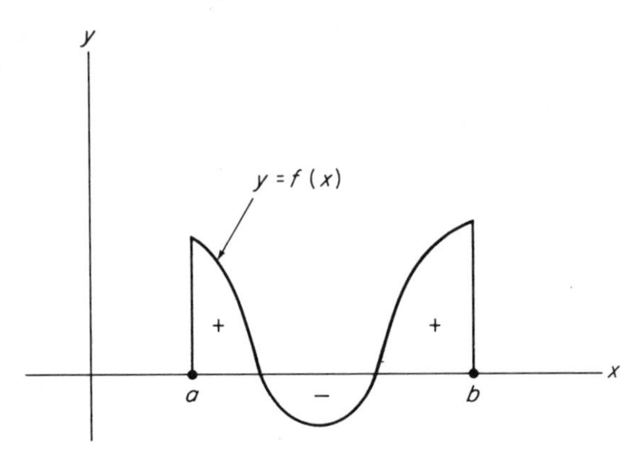

Figure 4.6

Lemma 1.1 If $c \in (a, b)$, then

$$\int_a^b f = \int_a^c f + \int_c^b f. \tag{2}$$

PROOF This is true for f because it is true for both f^+ and f^-:

$$\begin{aligned}\int_a^b f &= \int_a^b f^+ - \int_a^b f^- \\ &= \left(\int_a^c f^+ + \int_c^b f^+\right) - \left(\int_a^c f^- + \int_c^b f^-\right) \\ &= \left(\int_a^c f^+ - \int_a^c f^-\right) + \left(\int_c^b f^+ - \int_c^b f^-\right) \\ &= \int_a^c f + \int_c^b f.\end{aligned}$$

∎

Lemma 1.2 If $m \leqq f(x) \leqq M$ on $[a, b]$, then

$$m(b-a) \leqq \int_a^b f \leqq M(b-a). \tag{1}$$

PROOF We shall show that $\int_a^b f \leqq M(b-a)$; the proof that $m(b-a) \leqq \int_a^b f$ is similar and is left as an exercise.

Suppose first that $M > 0$. Then

$$\int_a^b f = \int_a^b f^+ - \int_a^b f^- \leqq \int_a^b f^+ \leqq M(b-a)$$

because $f^+(x) \leqq M$ on $[a, b]$, so that $O_a^b(f^+)$ is contained in the rectangle with base $[a, b]$ and height M (Fig. 4.7).

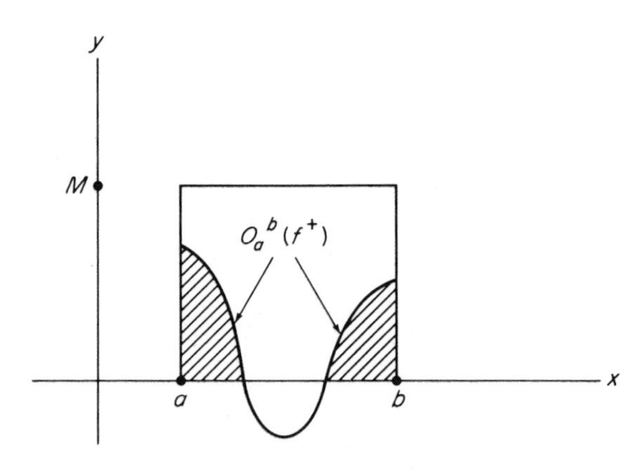

Figure 4.7

If $M \leqq 0$, then $f(x) \leqq 0$, so $f^+(x) = 0$, $f^-(x) = -f(x)$ on $[a, b]$, and $-f(x) \geqq -M$. Hence

$$\int_a^b f = -\int_a^b f^- = -\int_a^b (-f) \leqq -(-M)(b-a) = M(b-a)$$

because $O_a^b(-f)$ contains the rectangle of height $-M$. ∎

Lemmas 1.1 and 1.2 provide the only properties of the integral that are needed to prove the *fundamental theorem of calculus.*

Theorem 1.3 If $f: [a, b] \to \mathscr{R}$ is continuous, and $F: [a, b] \to \mathscr{R}$ is defined by $F(x) = \int_a^x f$, then F is differentiable, with $F' = f$.

PROOF Denote by $m(h)$ and $M(h)$ the minimum and maximum values respectively of f on $[x, x+h]$. Then

$$F(x+h) - F(x) = \int_a^{x+h} f - \int_a^x f = \int_x^{x+h} f$$

by Lemma 1.1, and

$$hm(h) \leqq \int_x^{x+h} f \leqq hM(h)$$

by Lemma 1.2, so

$$m(h) \leqq \frac{F(x+h) - F(x)}{h} \leqq M(h). \tag{3}$$

Since $\lim_{h\to 0} m(h) = \lim_{h\to 0} M(h) = f(x)$ because f is continuous at x, it follows from (3) that

$$F'(x) = \lim_{h\to 0} \frac{F(x+h) - F(x)}{h} = f(x)$$

as desired. ∎

The usual method of computing integrals follows immediately from the fundamental theorem.

Corollary 1.4 If f is continuous and $G' = f$ on $[a, b]$, then

$$\int_a^b f = G(b) - G(a).$$

PROOF If $F(x) = \int_a^x f$ on $[a, b]$, then

$$F'(x) = f(x) = G'(x)$$

by the fundamental theorem. Hence

$$F(x) = G(x) + C, \qquad C \text{ constant.}$$

Now $C = F(a) - G(a) = \int_a^a f - G(a) = -G(a)$, so

$$\int_a^b f = F(b) = G(b) + C = G(b) - G(a).$$ ∎

It also follows quickly from the fundamental theorem that integration is a linear operation.

Theorem 1.5 If the functions f and g are continuous on $[a, b]$, and $c \in \mathscr{R}$, then

$$\int_a^b (f + g) = \int_a^b f + \int_a^b g \qquad \text{and} \qquad \int_a^b (cf) = c\int_a^b f.$$

PROOF Let F and G be antiderivatives of f and g, respectively (provided by the fundamental theorem), and let $H = F + G$ on $[a, b]$. Then $H' = (F + G)' = F' + G' = f + g$, so

$$\begin{aligned}\int_a^b (f + g) &= H(b) - H(a) \qquad \text{(by Corollary 1.4)}\\ &= (F(b) + G(b)) - (F(a) + G(a))\\ &= (F(b) - F(a)) + (G(b) - G(a))\\ &= \int_a^b f + \int_a^b g \qquad \text{(by Corollary 1.4 again).}\end{aligned}$$

The proof that $\int_a^b cf = c\int_a^b f$ is similar. ∎

Theorem 1.6 If $f(x) \leqq g(x)$ on $[a, b]$, then

$$\int_a^b f \leqq \int_a^b g.$$

PROOF Since $g(x) - f(x) \geqq 0$, Theorem 1.5 and Lemma 1.2 immediately give

$$\int_a^b g - \int_a^b f = \int_a^b (g - f) \geqq 0. \qquad \blacksquare$$

Applying Theorem 1.6 with $g = |f|$, we see that

$$\left| \int_a^b f \right| \leqq M(b - a)$$

if $|f(x)| \leqq M$ on $[a, b]$.

It is often convenient to write

$$\int_a^b f = \int_a^b f(x)\, dx,$$

and we do this in the following two familiar theorems on techniques of integration.

Theorem 1.7 (Substitution) Let f have a continuous derivative on $[a, b]$, and let g be continuous on $[c, d]$, where $f([a, b]) \subset [c, d]$. Then

$$\int_a^b g(f(x))f'(x)\, dx = \int_{f(a)}^{f(b)} g(u)\, du.$$

PROOF Let G be an antiderivative of g on $[c, d]$. Then

$$DG(f(x)) = G'(f(x))f'(x) = g(f(x))f'(x)$$

by the chain rule, so

$$\begin{aligned} \int_a^b g(f(x))f'(x)\, dx &= G(f(b)) - G(f(a)) && \text{(by Corollary 1.4)} \\ &= \int_{f(a)}^{f(b)} g && \text{(Corollary 1.4 again)} \\ &= \int_{f(a)}^{f(b)} g(u)\, du. && \blacksquare \end{aligned}$$

Example 3 Using the substitution rule, we can give a quick proof that the area of a circle of radius r is indeed $A = \pi r^2$. Since π is by definition the area of the unit circle, we have (see Fig. 4.8)

$$\frac{\pi}{4} = \int_0^1 (1 - t^2)^{1/2}\, dt.$$

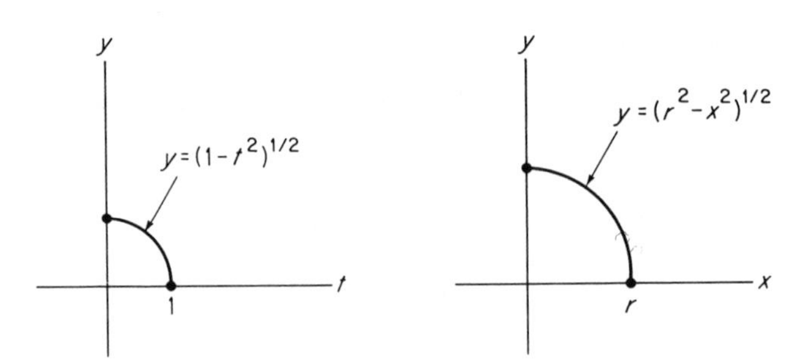

Figure 4.8

Then

$$\begin{aligned}
A &= 4\int_0^r (r^2 - x^2)^{1/2}\,dx \\
&= 4r\int_0^r \left((1 - \frac{x^2}{r^2}\right)^{1/2} dx \\
&= 4r^2 \int_0^r \left(1 - \left[\frac{x}{r}\right]^2\right)^{1/2} D\left(\frac{x}{r}\right) dx \\
&= 4r^2 \int_0^1 (1 - u^2)^{1/2}\,du \\
&= 4r^2 \cdot \frac{\pi}{4} \\
&= \pi r^2
\end{aligned}$$

Theorem 1.8 (Integration by Parts) If f and g are continuously differentiable on $[a, b]$, then

$$\int_a^b f(x)g'(x)\,dx = f(b)g(b) - f(a)g(a) - \int_a^b f'(x)g(x)\,dx.$$

PROOF

$$D[f(x)g(x)] = f(x)g'(x) + f'(x)g(x)$$

or

$$f(x)g'(x) = D[f(x)g(x)] - f'(x)g(x),$$

so

$$\begin{aligned}
\int_a^b f(x)g'(x)\,dx &= \int_a^b D(fg) - \int_a^b f'(x)g(x)\,dx \\
&= f(b)g(b) - f(a)g(a) - \int_a^b f'(x)g(x)\,dx.
\end{aligned}$$

∎

The integration by parts formula takes the familiar and memorable form

$$\int u\, dv = uv - \int v\, du$$

if we write $u = f(x)$, $v = g(x)$, and further agree to the notation $du = f'(x)\, dx$, $dv = g'(x)\, dx$ in the integrands.

Exercises

1.1 Calculate the area under $y = x^2$ over the interval $[0, 1]$ by making direct use of the definition of area.

1.2 Apply the fundamental theorem to calculate the area in the previous exercise.

1.3 Prove Property A of area. *Hint*: Given $S \subset T$, suppose to the contrary that $a(T) < a(S)$, and let $\varepsilon = \frac{1}{2}(a(S) - a(T))$. Show that you have a contradiction, assuming (as you may) that, if the rectangles $R_1'', \ldots, R_m''$ together contain the rectangles $R_1', \ldots, R_n'$, then $\sum_{i=1}^m a(R_i'') \geqq \sum_{i=1}^n a(R_i')$.

1.4 Prove that the positive and negative parts of a continuous function are continuous.

1.5 If f is continuous on $[a, b]$, and

$$G(x) = \int_x^b f \quad \text{on} \quad [a, b],$$

prove that $G'(x) = -f(x)$ on $[a, b]$. (See the proof of the fundamental theorem.)

1.6 Prove the other part of Theorem 1.5, that is, that $\int_a^b cf = c \int_a^b f$ if c is a constant.

1.7 Use integration by parts to establish the "reduction formula"

$$\int \sin^n x\, dx = -\frac{1}{n} \sin^{n-1} x \cos x + \frac{n-1}{n} \int \sin^{n-2} x\, dx.$$

In particular, $\int_0^{\pi/2} \sin^n x\, dx = [(n-1)/n] \int_0^{\pi/2} \sin^{n-2} x\, dx$.

1.8 Use the formula in Exercise 1.7 to show by mathematical induction that

$$\int_0^{\pi/2} \sin^{2n+1} x\, dx = \frac{2}{3} \cdot \frac{4}{5} \cdot \frac{6}{7} \cdot \cdots \cdot \frac{2n}{2n+1}$$

and

$$\int_0^{\pi/2} \sin^{2n} x\, dx = \frac{\pi}{2} \cdot \frac{1}{2} \cdot \frac{3}{4} \cdot \frac{5}{6} \cdot \cdots \cdot \frac{2n-1}{2n}.$$

1.9 (a) Conclude from the previous exercise that

$$\frac{\pi}{2} = \frac{2}{1} \cdot \frac{2}{3} \cdot \frac{4}{3} \cdot \frac{4}{5} \cdot \cdots \cdot \frac{2n}{2n-1} \cdot \frac{2n}{2n+1} \frac{\int_0^{\pi/2} \sin^{2n} x\, dx}{\int_0^{\pi/2} \sin^{2n+1} x\, dx}.$$

(b) Use the inequality

$$0 < \sin^{2n+1} x \leqq \sin^{2n} x \leqq \sin^{2n-1} x \qquad \text{for} \quad x \in (0, \pi/2]$$

and the first formula in Exercise 1.8 to show that

$$1 < \frac{\int_0^{\pi/2} \sin^{2n} x\, dx}{\int_0^{\pi/2} \sin^{2n+1} x\, dx} \leqq \frac{\int_0^{\pi/2} \sin^{2n-1} x\, dx}{\int_0^{\pi/2} \sin^{2n+1} x\, dx} = 1 + \frac{1}{2n}.$$

(c) Conclude from (a) and (b) that

$$\frac{\pi}{2} = \lim_{n\to\infty}\left(\frac{2}{1}\cdot\frac{2}{3}\cdot\frac{4}{3}\cdot\frac{4}{5}\cdot\cdots\cdot\frac{2n}{2n-1}\cdot\frac{2n}{2n+1}\right).$$

This result is usually written as an "infinite product," known as *Wallis' product*:

$$\frac{\pi}{2} = \frac{2}{1}\cdot\frac{2}{3}\cdot\frac{4}{3}\cdot\frac{4}{5}\cdot\frac{6}{5}\cdot\frac{6}{7}\cdot\cdots.$$

1.10 Deduce from Wallis' product that

$$\lim_{n\to\infty}\frac{(n!)^2 2^{2n}}{(2n)!\sqrt{n}} = \sqrt{\pi}.$$

Hint: Multiply and divide the right-hand side of

$$P_n = \frac{2}{1}\cdot\frac{2}{3}\cdot\frac{4}{3}\cdot\frac{4}{5}\cdot\cdots\cdot\frac{2n}{2n-1}\cdot\frac{2n}{2n+1}$$

by $2\cdot 2\cdot 4\cdot 4\cdot\cdots\cdot(2n)\cdot(2n)$. This yields

$$P_n = \frac{(n!)^4 2^{4n}}{[(2n)!]^2(2n+1)}.$$

1.11 Write $n! = a_n n^n\sqrt{n}\,e^{-n}$ for each n, thereby defining the sequence $\{a_n\}_1^\infty$. Assuming that $\lim_{n\to\infty} a_n = a \neq 0$, deduce from the previous problem that $a = (2\pi)^{1/2}$. Hence $n!$ and $(2\pi n)^{1/2}(n/e)^n$ are *asymptotic* as $n\to\infty$,

$$n! \sim (2\pi n)^{1/2}\left(\frac{n}{e}\right)^n,$$

meaning that the limit of their ratio is 1. This is a weak form of "Stirling's formula."

2 VOLUME AND THE n-DIMENSIONAL INTEGRAL

In Section 1 we used the properties of area (without verification) to study the integral of a function of one variable. This was done mainly for purposes of review and motivation. We now start anew, with definitions of volume for subsets of $\mathscr{R}^n$, and of the integral for functions on $\mathscr{R}^n$.

These definitions are the obvious generalizations to higher dimensions of those given in Section 1. Recall that a *closed interval* in $\mathscr{R}^n$ is a set $I = I_1 \times I_2 \times\cdots\times I_n$, where $I_j = [a_j, b_j] \subset \mathscr{R}$, $j = 1, \ldots, n$. Thus a closed interval I in $\mathscr{R}^n$ is simply a Cartesian product of "ordinary" closed intervals. The volume of I is, by definition, $v(I) = (b_1 - a_1)(b_2 - a_2)\cdots(b_n - a_n)$. Now let A be a *bounded* subset of $\mathscr{R}^n$. Then we say that A is a *contented* set with *volume* $v(A)$ if and only if, given $\varepsilon > 0$, there exist

(a) nonoverlapping closed intervals $I_1, \ldots, I_p \subset A$ such that $\sum_{i=1}^p v(I_i) > v(A) - \varepsilon$, and

(b) closed intervals $J_1, \ldots, J_q$ such that

$$A \subset \bigcup_{j=1}^{q} J_j \qquad \text{and} \qquad \sum_{j=1}^{q} v(J_j) < v(A) + \varepsilon.$$

We have seen (Example 1 of Section 1) that not all sets are contented. Consequently we need an effective means of detecting those that are. This will be given in terms of the boundary of a set. Recall that the *boundary* ∂A of the set $A \subset \mathscr{R}^n$ is the set of all those points of $\mathscr{R}^n$ that are limit points of both A and $\mathscr{R}^n - A$.

The contented set A is called *negligible* if and only if $v(A) = 0$. Referring to the definition of volume, we see that A is negligible if and only if, given $\varepsilon > 0$, there exist intervals $J_1, \ldots, J_q$ such that $A \subset \bigcup_{j=1}^{q} J_j$ and $\sum_{j=1}^{q} v(J_j) < \varepsilon$. It follows immediately that the union of a finite number of negligible sets is negligible, as is any subset of a negligible set.

Theorem 2.1 The bounded set A is contented if and only if its boundary is negligible.

PROOF Let R be a closed interval containing A. By a *partition* of R we shall mean a finite collection of nonoverlapping closed intervals whose union is R.

First suppose that A is contented. Given $\varepsilon > 0$, choose closed intervals $I_1, \ldots, I_p$ and $J_1, \ldots, J_q$ as in conditions (a) and (b) of the definition of $v(A)$. Let $\mathscr{P}$ be a partition of R such that each I_i and each J_j is a union of closed intervals of $\mathscr{P}$. Let $R_1, \ldots, R_k$ be those intervals of $\mathscr{P}$ which are contained in $\bigcup_{j=1}^{p} I_i$, and $R_{k+1}, \ldots, R_{k+l}$ the additional intervals of $\mathscr{P}$ which are contained in $\bigcup_{j=1}^{q} J_j$. Then

$$\partial A \subset \bigcup_{i=k+1}^{k+l} R_i,$$

and

$$\sum_{i=k+1}^{k+l} v(R_i) = \sum_{i=1}^{k+l} v(R_i) - \sum_{i=1}^{k} v(R_i) \leqq \sum_{j=1}^{q} v(J_j) - \sum_{j=1}^{p} v(I_i)$$
$$< [v(A) + \varepsilon] - [v(A) - \varepsilon] = 2\varepsilon.$$

Now suppose that ∂A is negligible. Note first that, in order to prove that A is contented, it suffices to show that, given $\varepsilon > 0$, there exist intervals $J_1, \ldots, J_q$ containing A and nonoverlapping intervals $I_1, \ldots, I_p$ contained in A such that

$$\sum_{j=1}^{q} v(J_j) - \sum_{i=1}^{p} v(I_i) < \varepsilon,$$

for then $v(A) = \operatorname{glb}\{\sum_{j=1}^{q} v(J_j)\} = \operatorname{lub}\{\sum_{i=1}^{p} v(I_i)\}$ (the greatest lower and least upper bounds for all such collections of intervals) satisfies the definition of volume for A.

To show this, let $R_1, \ldots, R_k$ be intervals covering ∂A with $\sum_{i=1}^k v(R_i) < \varepsilon$, and let $\mathscr{P}$ be a partition of R such that each R_i is a union of elements of $\mathscr{P}$. If $I_1, \ldots, I_p$ are the intervals of $\mathscr{P}$ which are contained in A, and $J_1, \ldots, J_q$ are the intervals of $\mathscr{P}$ which are contained in $A \cup \bigcup_{i=1}^k R_i$, then $A \subset \bigcup_{j=1}^q J_j$, and $\bigcup_{j=1}^q J_j - \bigcup_{i=1}^p I_i \subset \bigcup_{i=1}^k R_i$. Since the intervals $I_1, \ldots, I_p$ are included among the intervals $J_1, \ldots, J_q$, it follows that

$$\sum_{j=1}^{q} v(J_j) - \sum_{i=1}^{p} v(\mathrm{I}_i) \leqq \sum_{i=1}^{k} v(R_i) < \varepsilon$$

as desired. ∎

Corollary The intersection, union, or difference of two contented sets is contented.

This follows immediately from the theorem, the fact that a subset of a negligible set is negligible, and the fact that $\partial(A \cup B)$, $\partial(A \cap B)$, and $\partial(A - B)$ are all subsets of $\partial A \cup \partial B$.

The utility of Theorem 2.1 lies in the fact that negligible sets are often easily recognizable as such. For instance, we will show in Corollary 2.3 that, if $f\colon A \to \mathscr{R}$ is a continuous function on the contented set $A \subset \mathscr{R}^{n-1}$, then the graph of f is a negligible subset of $\mathscr{R}^n$.

Example Let B^n be the unit ball in $\mathscr{R}^n$, with $\partial B^n = S^{n-1}$, the unit $(n-1)$-sphere. Assume by induction that B^{n-1} is a contented subset of $\mathscr{R}^{n-1} \subset \mathscr{R}^n$. Then S^{n-1} is the union of the graphs of the continuous functions

$$f^+(x_1, \ldots, x_{n-1}) = \left[1 - \sum_{i=1}^{n-1} x_i^2\right]^{1/2}$$

and

$$f^-(x_1, \ldots, x_{n-1}) = -\left[1 - \sum_{i=1}^{n-1} x_i^2\right]^{1/2}.$$

Hence S^{n-1} is negligible by the above remark, so B^n is contented by Theorem 2.1.

We turn now to the definition of the integral. Given a nonnegative function $f\colon \mathscr{R}^n \to \mathscr{R}$, we consider the *ordinate set*

$$O_f = \{(x_1, \ldots, x_{n+1}) \in \mathscr{R}^{n+1} : 0 < x_{n+1} \leqq f(x_1, \ldots, x_n)\}.$$

In order to ensure that the set O_f is bounded, we must require that f be bounded and have bounded support. We say that $f\colon \mathscr{R}^n \to \mathscr{R}$ is *bounded* if there exists $M \in \mathscr{R}$ such that $|f(\mathbf{x})| \leqq M$ for all $\mathbf{x} \in \mathscr{R}^n$, and *has bounded support* if there exists a closed interval $I \subset \mathscr{R}^n$ such that $f(\mathbf{x}) = 0$ if $\mathbf{x} \notin I$ (Fig. 4.9). We will define the integral *only* for bounded functions on $\mathscr{R}^n$ that have bounded support; it will turn out that this restriction entails no loss of generality.

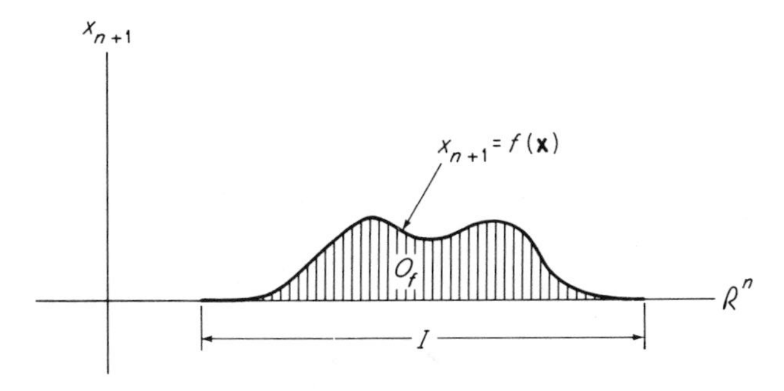

Figure 4.9

Given $f\colon R^n \to R$, we define the positive and negative parts f^+ and f^- of f by

$$f^+(\mathbf{x}) = \max\{0, f(\mathbf{x})\} \qquad \text{and} \qquad f^-(\mathbf{x}) = \max\{0, -f(\mathbf{x})\}.$$

Then $f = f^+ - f^-$ (see Fig. 4.10). That is, $f^+(\mathbf{x}) = f(\mathbf{x})$ if $f(\mathbf{x}) > 0$, and $f^-(\mathbf{x}) = -f(\mathbf{x})$ if $f(\mathbf{x}) < 0$, and each is 0 otherwise.

Suppose now that $f\colon \mathscr{R}^n \to \mathscr{R}$ is *bounded and has bounded support*. We then say that f is (Riemann) *integrable* (and write $f \in \mathscr{I}$) if and only if the ordinate sets O_{f^+} and O_{f^-} are both contented, in which case we define

$$\int f = v(O_{f^+}) - v(O_{f^-}).$$

Thus $\int f$ is, by definition, the volume of the region in $\mathscr{R}^{n+1}$ above $\mathscr{R}^n$ and below the graph of f, minus the volume of the region below $\mathscr{R}^n$ and above the graph of f (just as before, in the 1-dimensional case).

Although $f\colon \mathscr{R}^n \to \mathscr{R}$ has bounded support, we may think of $\int f$ as the "integral of f over $\mathscr{R}^n$." In order to define the integral of f over the set $A \subset \mathscr{R}^n$, thinking of the volume of the set in $\mathscr{R}^{n+1}$ which lies "under the graph of f and over the set A," we introduce the *characteristic function* φ_A of A, defined by

$$\varphi_A(\mathbf{x}) = \begin{cases} 1 & \text{if } \ \mathbf{x} \in A, \\ 0 & \text{otherwise.} \end{cases}$$

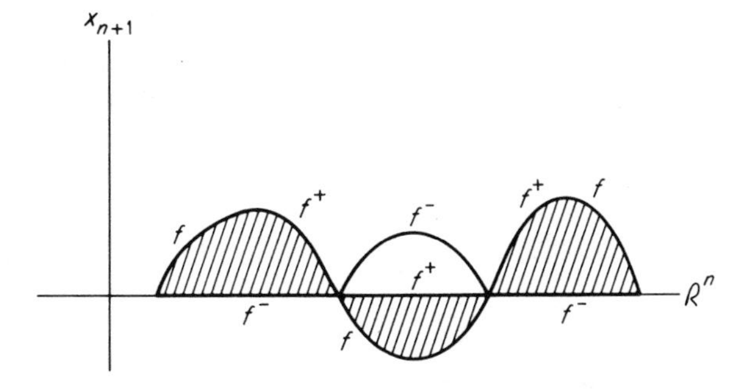

Figure 4.10

We then define

$$\int_A f = \int f\varphi_A$$

provided that the product $f\varphi_A$ is integrable; otherwise $\int_A f$ is not defined, and f is not integrable on A (see Fig. 4.11).

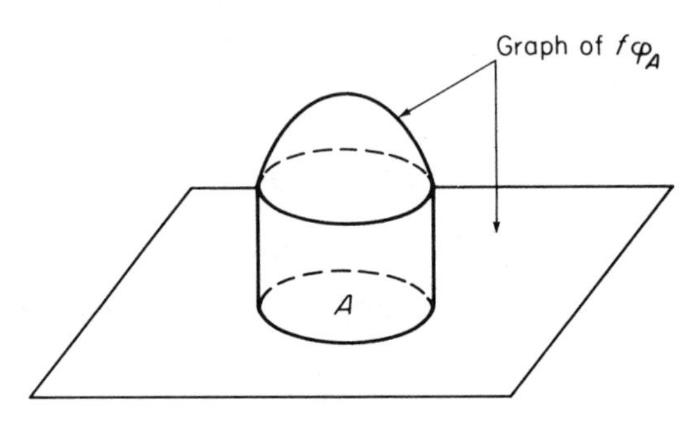

Figure 4.11

The basic properties of the integral are the following four "axioms" (which of course must be established as theorems).

Axiom I The set $\mathscr{I}$ of integrable functions is a vector space.

Axiom II The mapping $\int : \mathscr{I} \to \mathscr{R}$ is linear.

Axiom III If $f \geqq 0$ everywhere, then $\int f \geqq 0$.

Axiom IV If the set A is contented, then

$$\int \varphi_A = v(A).$$

Axiom III follows immediately from the definition of the integral. Axiom IV is almost as easy. For if $I_1, \ldots, I_q$ and $J_1, \ldots, J_p$ are intervals in $\mathscr{R}^n$ such that

$$\bigcup_{i=1}^{q} \mathrm{I}_i \subset A \subset \bigcup_{j=1}^{p} J_j, \qquad \sum_{i=1}^{q} v(\mathrm{I}_i) > v(A) - \varepsilon, \qquad \sum_{j=1}^{p} v(J_j) < v(A) + \varepsilon,$$

and

$$I_i' = I_i \times [0, 1], \qquad J_j' = J_j \times [0, 1] \subset \mathscr{R}^{n+1},$$

then

$$\bigcup_{i=1}^{q} I_i' \subset O_{\varphi_A} \subset \bigcup_{j=1}^{p} J_j', \qquad \sum_{i=1}^{q} v(\mathrm{I}_i') > v(A) - \varepsilon, \qquad \sum_{j=1}^{q} v(J_j') < v(A) + \varepsilon.$$

So it follows that

$$\int \varphi_A = v(O_{\varphi_A}) = v(A). \qquad \blacksquare$$

Axioms I and II, which assert that, if $f, g \in \mathscr{I}$ and $a, b \in \mathscr{R}$, then $af + bg \in \mathscr{I}$ and $\int (af + bg) = a\int f + b\int g$, will be verified in the following section.

In lieu of giving necessary and sufficient conditions for integrability, we next define a large class of integrable functions which includes most functions of frequent occurrence in practice.

The function f is called *admissible* if and only if

(a) f is bounded,
(b) f has bounded support, and
(c) f is continuous except on a negligible set.

Condition (c) means simply that, if D is the set of all those points at which f is *not* continuous, then D is negligible. It is easily verified that the set $\mathscr{A}$ of all admissible functions on $\mathscr{R}^n$ is a vector space (Exercise 2.1).

Theorem 2.2 Every admissible function is integrable.

PROOF Let f be an admissible function on $\mathscr{R}^n$, and R an interval outside of which f vanishes. Since f^+ and f^- are admissible (Exercise 2.2), and $\mathscr{A}$ is a vector space, we may assume that f is nonnegative.

Choose M such that $0 \leqq f(\mathbf{x}) \leqq M$ for all $\mathbf{x}$, and let D denote the negligible set of points at which f is not continuous.

Given $\varepsilon > 0$, let $Q_1, \ldots, Q_k$ be a collection of closed intervals in $\mathscr{R}^n$ such that

$$D \subset \bigcup_{i=1}^{k} Q_i \qquad \text{and} \qquad \sum_{i=1}^{k} v(Q_i) < \frac{\varepsilon}{2M}.$$

By expanding the Q_i's slightly if necessary, we may assume that $D \subset \bigcup_{i=1}^{k} \operatorname{int} Q_i$.

Then f is continuous on the set $Q = R - \bigcup_{i=1}^{k} \operatorname{int} Q_i$. Since Q is *closed* and bounded, it follows from Theorem 8.9 of Chapter I that f is *uniformly* continuous on Q, so there exists $\delta > 0$ such that

$$|f(\mathbf{x}) - f(\mathbf{y})| < \frac{\varepsilon}{2v(R)}$$

if $\mathbf{x}, \mathbf{y} \in Q$ and $|\mathbf{x} - \mathbf{y}| < \delta$.

Now let $\mathscr{P}$ be a partition of R such that each Q_i is a union of intervals of $\mathscr{P}$, and each interval of $\mathscr{P}$ has diameter $< \delta$. If $R_1, \ldots, R_q$ are the intervals of $\mathscr{P}$ that are contained in Q, and

$$a_i = \text{glb of } f(\mathbf{x}) \qquad \text{for} \quad \mathbf{x} \in R_i,$$
$$b_i = \text{lub of } f(\mathbf{x}) \qquad \text{for} \quad \mathbf{x} \in R_i,$$

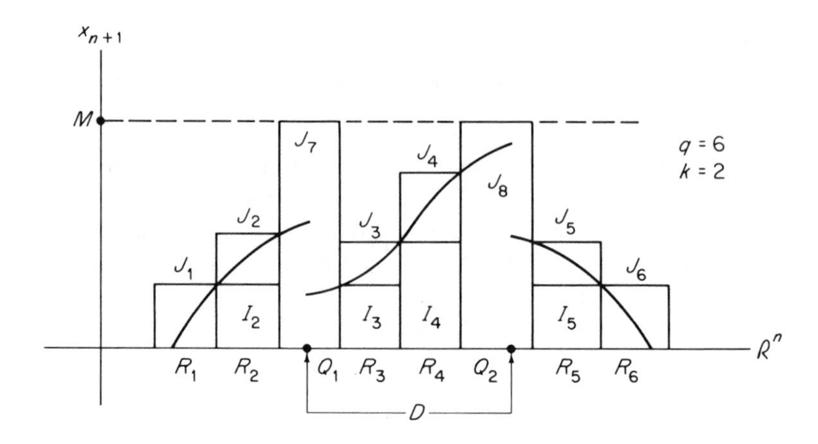

Figure 4.12

then $b_i - a_i < \varepsilon/2v(R)$ for $i = 1, \ldots, q$. Finally let, as in Fig. 4.12,

$$
\begin{aligned}
I_i &= R_i \times [0, a_i] \qquad &&\text{for} \quad i = 1, \ldots, q,\\
J_i &= R_i \times [0, b_i] \qquad &&\text{for} \quad i = 1, \ldots, q,\\
J_{q+i} &= Q_i \times [0, M] \qquad &&\text{for} \quad i = 1, \ldots, k.
\end{aligned}
$$

Let $O_f^* = O_f \cup R$. Since $O_f = O_f^* - R$, and the interval R is contented, it suffices to see that O_f^* is contented (Fig. 4.13). But $I_1, \ldots, I_q$ is a collection of

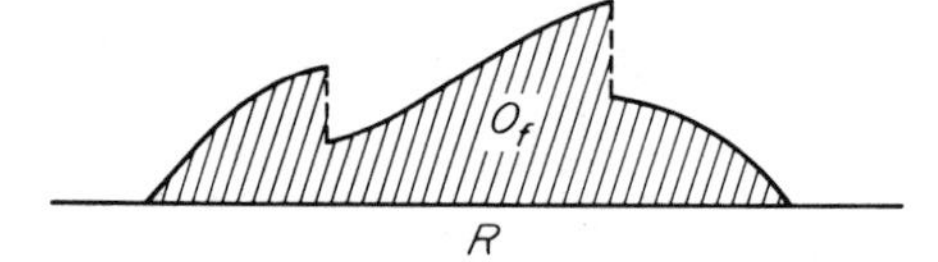

Figure 4.13

nonoverlapping intervals contained in O_f^*, and $J_1, \ldots, J_{q+k}$ is a collection of intervals containing O_f^*, while

$$
\begin{aligned}
\sum_{j=1}^{q+k} v(J_j) - \sum_{i=1}^{q} v(\mathrm{I}_i) &= \sum_{i=1}^{k} v(J_{q+i}) + \sum_{i=1}^{q} (b_i - a_i) v(R_i)\\
&< M \sum_{i=1}^{k} v(Q_i) + \frac{\varepsilon}{2v(R)} v(R_i)\\
&\leqq M \cdot \frac{\varepsilon}{2M} + \frac{\varepsilon}{2v(R)} \cdot v(R) = \varepsilon.
\end{aligned}
$$

Hence it follows that O_f^* is contented as desired. ∎

We can now establish the previous assertion about graphs of continuous functions.

Corollary 2.3 If $f: A \to \mathscr{R}$ is continuous, and $A \subset \mathscr{R}^n$ is contented, then the graph G_f of f is a negligible set in $\mathscr{R}^{n+1}$.

PROOF If we extend f to $\mathscr{R}^n$ by $f(\mathbf{x}) = 0$ for $\mathbf{x} \notin A$, then f is admissible because its discontinuities lie in ∂A, which is negligible because A is contented. Hence O_f is contented by the Theorem, so ∂O_f is negligible. But G_f is a subset of ∂O_f. ∎

It follows easily from Corollary 2.3 that, if $f: A \to \mathscr{R}$ is a nonnegative continuous function on a contented set $A \subset \mathscr{R}^n$, then the ordinate set of f is a contented subset of $\mathscr{R}^{n+1}$ (see Exercise 2.7). This is the analogue for volume of Property E in Section 1.

The following proposition shows that $\int_A f$ is defined if $f: \mathscr{R}^n \to \mathscr{R}$ is admissible and $A \subset \mathscr{R}^n$ is contented. This fact serves to justify our concentration of attention (in the study of integration) on the class of admissible functions—every function that we will have "practical" need to integrate will, in fact, be an admissible function on a contented set.

Proposition 2.4 If f is admissible and A is contented, then $f\varphi_A$ is admissible (and, therefore, is integrable by Theorem 2.2).

PROOF If D is the negligible set of points at which f is not continuous, then $f\varphi_A$ is continuous at every point not in the negligible set $D \cup \partial A$ (see Exercise 2.3). ∎

Proposition 2.5 If f and g are admissible functions on $\mathscr{R}^n$ with $f \leqq g$ everywhere, and A is a contented set, then

$$\int_A f \leqq \int_A g.$$

PROOF $g\varphi_A - f\varphi_A \geqq 0$ everywhere, so Axiom III gives $\int (g\varphi_A - f\varphi_A) \geqq 0$. Then Axiom II gives

$$\int_A f = \int f\varphi_A \leqq \int g\varphi_A = \int_A g. \qquad ∎$$

It follows easily from Proposition 2.5 that, if A and B are contented sets in $\mathscr{R}^n$, with $A \subset B$, then $v(A) \leqq v(B)$ (see Exercise 2.4). This is the analog for volume of Property A of Section 1.

Proposition 2.6 If the admissible function f satisfies $|f(\mathbf{x})| \leqq M$ for all $\mathbf{x} \in \mathscr{R}^n$, and A is contented, then

$$\left| \int_A f \right| \leqq Mv(A).$$

In particular, $\int_A f = 0$ if A is negligible.

PROOF Since $-M\varphi_A \leqq f\varphi_A \leqq M\varphi_A$, this follows immediately from Proposition 2.5 and Axioms II and IV. ∎

Proposition 2.7 If A and B are contented sets with $A \cap B$ negligible, and f is admissible, then

$$\int_{A\cup B} f = \int_A f + \int_B f.$$

PROOF First consider the special case $A \cap B = \varnothing$. Then $\varphi_{A \cup B} = \varphi_A + \varphi_B$, so

$$\int_{A \cup B} f = \int f\varphi_{A \cup B} = \int (f\varphi_A + f\varphi_B)$$

$$= \int f\varphi_A + \int f\varphi_B = \int_A f + \int_B f.$$

If $A \cap B$ is negligible, then $\int_{A \cap B} f = 0$ by Proposition 2.6. Then, applying the special case, we obtain

$$\int_{A \cup B} f = \int_{A-B} f + \int_{A \cap B} f + \int_{B-A} f$$

$$= \left(\int_{A-B} f + \int_{A \cap B} f\right) + \left(\int_{A \cap B} f + \int_{B-A} f\right)$$

$$= \int_A f + \int_B f.$$ ∎

It follows easily from Proposition 2.7 that, if A and B are contented sets with $A \cap B$ negligible, then $A \cup B$ is contented, and $v(A \cup B) = v(A) + v(B)$ (see Exercise 2.5). This is the analog for volume of Property B of Section 1.

Theorem 2.8 Let A be contented, and suppose the admissible functions f and g agree except on the negligible set D. Then

$$\int_A f = \int_A g$$

PROOF Let $h = f - g$, so $h = 0$ except on D. We need to show that $\int_A h = 0$. But

$$\int_A h = \int h\varphi_A = \int h\varphi_{A \cap D} = \int_{A \cap D} h = 0$$

by Proposition 2.6, because $A \cap D$ is negligible. ∎

This theorem indicates why sets with volume zero are called "negligible"—insofar as integration is concerned, they do not matter.

Exercises

2.1 If f and g are admissible functions on $\mathscr{R}^n$ and $c \in \mathscr{R}$, show that $f+g$ and cf are admissible.

2.2 Show that the positive and negative parts of an admissible function are admissible.

2.3 If D is the set of discontinuities of $f: \mathscr{R}^n \to \mathscr{R}$, show that the set of discontinuities of $f\varphi_A$ is contained in $D \cup \partial A$.

2.4 If A and B are contented sets with $A \subset B$, apply Proposition 2.5 with $f = \varphi_A = \varphi_B \varphi_A$, $g = \varphi_B$ to show that

$$v(A) \leqq v(B).$$

2.5 Let A and B be contented sets with $A \cap B$ negligible. Apply Proposition 2.7 with $f = \varphi_{A \cup B}$ to show that $A \cup B$ is contented with

$$v(A \cup B) = v(A) + v(B).$$

2.6 If f and g are integrable functions with $f(\mathbf{x}) \leqq g(\mathbf{x})$ for all $\mathbf{x}$, prove that $\int f \leqq \int g$ *without* using Axioms I and II. *Hint*: $f \leqq g \Rightarrow f^+ \leqq g^+$ and $f^- \geqq g^- \Rightarrow O_{f^+} \subset O_{g^+}$ and $O_{g^-} \subset O_{f^-}$.

2.7 If A is a contented subset of $\mathscr{R}^n$ and $f: A \to \mathscr{R}$ is continuous and nonnegative, apply Corollary 2.3 to show that O_f is contented. *Hint*: Note that the boundary of O_f is the union of $A \subset \mathscr{R}^n \subset \mathscr{R}^{n+1}$, the graph of f, and the ordinate set of the restriction $f: \partial A \to \mathscr{R}$. Conclude that ∂O_f is negligible, and apply Theorem 2.1.

3 STEP FUNCTIONS AND RIEMANN SUMS

As a tool for studying the integral, we introduce in this section a class of very simple functions called step functions. We shall see first that the properties of step functions are easily established, and then that an arbitrary function is integrable if and only if it is, in a certain precise sense, closely approximated by step functions.

The function $h: \mathscr{R}^n \to \mathscr{R}$ is called a *step function* if and only if h can be written as a linear combination of characteristic functions $\varphi_1, \ldots, \varphi_p$ of intervals $I_1, \ldots, I_p$ whose interiors are mutually disjoint, that is,

$$h = \sum_{i=1}^{p} a_i \varphi_i$$

with coefficients $a_i \in \mathscr{R}$. Here the intervals $I_1, I_2, \ldots, I_p$ are not necessarily closed—each is simply a product of intervals in $\mathscr{R}$ (the latter may be either open or closed or "half-open").

Theorem 3.1 If h is a step function, $h = \sum_{i=1}^{p} a_i \varphi_i$ as above, then h is integrable with

$$\int h = \sum_{i=1}^{p} a_i v(I_i).$$

PROOF In fact h is admissible, since it is clearly continuous except possibly at the points of the negligible set $\bigcup_{i=1}^{p} \partial I_i$.

Assume for simplicity that each $a_i > 0$. Then the ordinate set O_h contains the set

$$A = \bigcup_{i=1}^{p} (\text{int } I_i) \times (0, a_i]$$

whose volume is clearly $\sum_{i=1}^{p} a_i v(I_i)$, and is contained in the union of A and the negligible set

$$\left(\bigcup_{i=1}^{p} \partial I_i\right) \times [0, a_1 + \cdots + a_p].$$

It follows easily that

$$\int h = v(O_h) = \sum_{i=1}^{p} a_i v(I_i).$$

∎

It follows immediately from Theorem 3.1 that, if h is a step function and $c \in \mathscr{R}$, then $\int ch = c \int h$. The following theorem gives the other half of the linearity of the integral on step functions.

Theorem 3.2 If h and k are step functions, then so is $h + k$, and $\int (h + k) = \int h + \int k$.

PROOF In order to make clear the idea of the proof, it will suffice to consider the simple special case

$$h = a\varphi, \qquad k = b\psi,$$

where φ and ψ are the characteristic functions of intervals I and J. If I and J have disjoint interiors, then the desired result follows from the previous theorem.

Otherwise $I \cap J$ is an interval I_0, and it is easily seen that $I - I_0$ and $J - I_0$ are disjoint unions of intervals (Fig. 4.14)

$$I - I_0 = I_1' \cup \cdots \cup I_q',$$
$$J - I_0 = I_1'' \cup \cdots \cup I_p''.$$

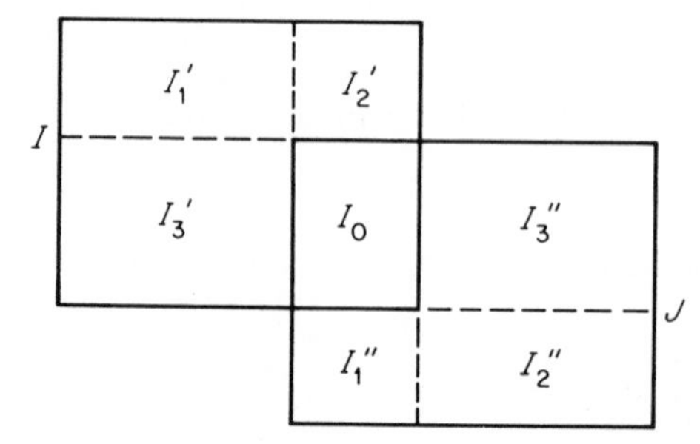

Figure 4.14

Then

$$h + k = a\varphi + b\psi$$
$$= (a + b)\varphi_{I_0} + \sum_{i=1}^{q} a\varphi_{I_i'} + \sum_{i=1}^{p} b\varphi_{I_i''},$$

so we have expressed $h + k$ as a step function. Theorem 3.1 now gives

$$\int (h + k) = (a + b)v(I_0) + a\sum_{i=1}^{q} v(I_i') + b\sum_{i=1}^{p} v(I_i'')$$
$$= a[v(I_0) + \sum_{i=1}^{q} v(I_i')] + b[v(I_0) + \sum_{i=1}^{p} v(I_i'')]$$
$$= av(I) + bv(J)$$
$$= \int h + \int k$$

as desired. The proof for general step functions h and k follows from this construction by induction on the number of intervals involved. ∎

Our main reason for interest in step functions lies in the following characterization of integrable functions.

Theorem 3.3 Let $f: \mathscr{R}^n \to \mathscr{R}$ be a bounded function with bounded support. Then f is integrable if and only if, given $\varepsilon > 0$, there exist step functions h and k such that

$$h \leqq f \leqq k \qquad \text{and} \qquad \int (k - h) < \varepsilon,$$

in which case $\int h \leqq \int f \leqq \int k$.

PROOF Suppose first that, given $\varepsilon > 0$, step functions h and k exist as prescribed. Then the set

$$S = \{(x_1, \ldots, x_{n+1}) \in \mathscr{R}^{n+1} : h(x_1, \ldots, x_n) \leqq x_{n+1} \leqq k(x_1, \ldots, x_n)\} \cup \partial O_k \cup \partial O_h$$

has volume equal to $\int (k - h) < \varepsilon$. But S contains the set $\hat{G}_f = \partial O_f - \mathscr{R}^n$ (Fig. 4.15). Thus for every $\varepsilon > 0$, $\hat{G}_f$ lies in a set of volume $< \varepsilon$. It follows easily that $\hat{G}_f$ is a negligible set. If Q is a rectangle in $\mathscr{R}^n$ such that $f = 0$ outside Q, then ∂O_{f^+} and ∂O_{f^-} are both subsets of $Q \cup \hat{G}_f$, so it follows that both are negligible. Therefore the ordinate sets O_{f^+} and O_{f^-} are both contented, so f is integrable. The fact that $\int h \leqq \int f \leqq \int k$ follows from Exercise 2.6 (without using Axioms I and II, which we have not yet proved).

Now suppose that f is integrable. Since this implies that both f^+ and f^- are integrable, we may assume without loss that $f \geqq 0$. Then the fact that f is integrable means by definition that O_f is contented. Hence, given $\varepsilon > 0$, there exist

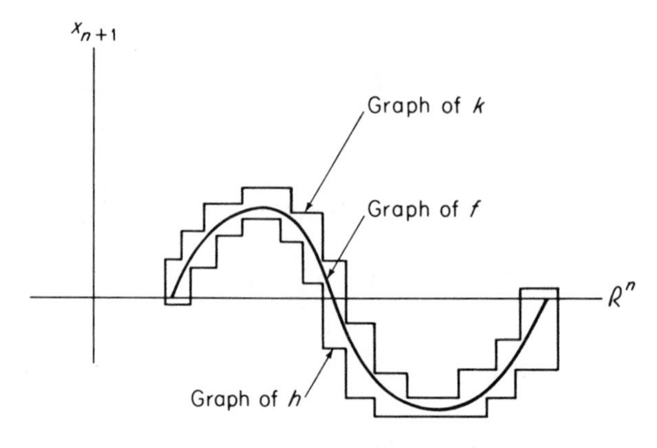

Figure 4.15

nonoverlapping intervals $I_1, \ldots, I_q$ contained in O_f, and intervals $J_1, \ldots, J_p$ with $O_f \subset \bigcup_{j=1}^{p} J_j$, such that

$$v(A) - \frac{\varepsilon}{2} < \sum_{i=1}^{q} v(I_i) \leqq \sum_{j=1}^{p} v(J_j) < v(A) + \frac{\varepsilon}{2}, \qquad A = O_f.$$

See Fig. 4.16.

Given $\mathbf{x} \in \mathscr{R}^n$, define

$$h(\mathbf{x}) = \max\{y \in \mathscr{R} : (\mathbf{x}, y) \in \bigcup_{i=1}^{q} I_i\},$$

$$k(\mathbf{x}) = \max\{y \in \mathscr{R} : (\mathbf{x}, z) \in \bigcup_{j=1}^{p} J_j \quad \text{if} \quad z \in [0, y]\},$$

if the vertical line in $\mathscr{R}^{n+1}$ through $\mathbf{x} \in \mathscr{R}^n$ intersects some I_i (respectively, some J_j), and let $h(\mathbf{x}) = 0$ (respectively, $k(\mathbf{x}) = 0$) otherwise. Then h and k are step

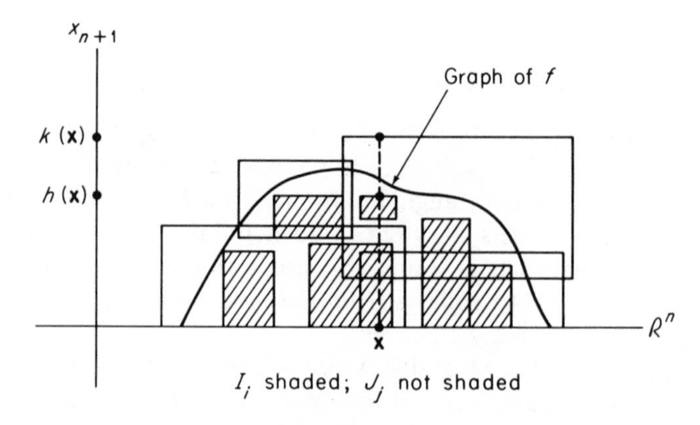

Figure 4.16

functions such that $h \leqq f \leqq k$. Since

$$O_h \supset \bigcup_{i=1}^{q} I_i \qquad \text{and} \qquad O_k \subset \bigcup_{j=1}^{p} J_j,$$

the above inequalities imply that

$$\begin{aligned}\int (k - h) &= \int k - \int h \\ &= v(O_k) - v(O_h) \\ &< \left(v(A) + \frac{\varepsilon}{2}\right) - \left(v(A) - \frac{\varepsilon}{2}\right) = \varepsilon\end{aligned}$$

as desired. ∎

Example We can apply Theorem 3.3 to prove again that continuous functions are integrable. Let $f : \mathscr{R}^n \to \mathscr{R}$ be a continuous function having compact support. Since it follows that the nonnegative functions f^+ and f^- are continuous, we may as well assume that f itself is nonnegative. Let $Q \subset \mathscr{R}^n$ be a closed interval such that $f = 0$ outside Q. Then f is uniformly continuous on Q (by Theorem 8.9 of Chapter I) so, given $\varepsilon > 0$, there exists $\delta > 0$ such that

$$|\mathbf{x} - \mathbf{y}| < \delta \Rightarrow |f(\mathbf{x}) - f(\mathbf{y})| < \frac{\varepsilon}{v(Q)}.$$

Now let $\mathscr{P} = \{Q_1, \ldots, Q_l\}$ be a partition of Q into nonoverlapping closed intervals, each of diameter less than δ. If

$$\begin{aligned} m_i &= \text{minimum value of } f \text{ on } Q_i, \\ M_i &= \text{maximum value of } f \text{ on } Q_i, \\ \varphi_i &= \text{the characteristic function of int } Q_i, \\ \psi_i &= \text{the characteristic function of } Q_i, \end{aligned}$$

and $h = \sum_{i=1}^{l} m_i \varphi_i$, $k = \sum_{i=1}^{l} M_i \psi_i$, then $h \leqq f \leqq k$ and

$$\int (k - h) = \sum_{i=1}^{l} (M_i - m_i) v(Q_i) < \frac{\varepsilon}{v(Q)} \sum_{i=1}^{l} v(Q_i) = \varepsilon,$$

so Theorem 3.3 applies.

We are now prepared to verify Axioms I and II of the previous section. Given integrable functions f_1 and f_2, and real numbers a_1 and a_2, we want to prove that $a_1 f_1 + a_2 f_2$ is integrable with

$$\int (a_1 f_1 + a_2 f_2) = a_1 \int f_1 + a_2 \int f_2 .$$

We suppose for simplicity that $a_1 > 0$, $a_2 > 0$, the proof being similar in the other cases.

Given $\varepsilon > 0$, Theorem 3.3 provides step functions h_1, h_2, k_1, k_2 such that

$$h_i \leqq f_i \leqq k_i \tag{1}$$

and $\int (k_i - h_i) < \varepsilon/2a_i$ for $i = 1, 2$. Then

$$h = a_1 h_1 + a_2 h_2 \leqq a_1 f_1 + a_2 f_2 \leqq a_1 k_1 + a_2 k_2 = k, \tag{2}$$

where h and k are step functions such that

$$\int (k - h) = a_1 \int (k_1 - h_1) + a_2 \int (k_2 - h_2) < \varepsilon.$$

So it follows from Theorem 3.3 that $a_1 f_1 + a_2 f_2$ is integrable.

At the same time it follows from (1) that

$$\int h = a_1 \int h_1 + a_2 \int h_2 \leqq a_1 \int f_1 + a_2 \int f_2 \leqq a_1 \int k_1 + a_2 \int k_2 = \int k$$

(by Exercise 2.6), and similarly from (2) that

$$\int h \leqq \int (a_1 f_1 + a_2 f_2) \leqq \int k.$$

Since $\int k - \int h < \varepsilon$, it follows that $\int (a_1 f_1 + a_2 f_2)$ and $a_1 \int f_1 + a_2 \int f_2$ differ by less than ε. This being true for every $\varepsilon > 0$, we conclude that

$$\int (a_1 f_1 + a_2 f_2) = a_1 \int f_1 + a_2 \int f_2 . \qquad \blacksquare$$

We now relate our definition of the integral to the "Riemann sum" definition of the single-variable integral which the student may have seen in introductory calculus. The motivation for this latter approach is as follows. Let $f: [a, b] \to \mathscr{R}$ be a nonnegative function. Let the points $a = x_0 < x_1 < \cdots < x_k = b$ subdivide the interval $[a, b]$ into k subintervals $[x_0, x_1]$, $[x_1, x_2]$, ..., $[x_{k-1}, x_k]$. For each $i = 1, \ldots, k$, choose a point $x_i^* \in [x_{i-1}, x_i]$. (Fig. 4.17). Then $f(x_i^*)(x_i - x_{i-1})$ is the area of the rectangle of height $f(x_i^*)$ whose base is the ith subinterval $[x_{i-1}, x_i]$, so one suspects that the sum

$$R = \sum_{i=1}^{k} f(x_i^*)(x_i - x_{i-1})$$

should be a "good approximation" to the area of O_f if the subintervals are sufficiently small. Notice that the Riemann sum R is simply the integral of the step function

$$h = \sum_{i=1}^{k} f(x_i^*)\varphi_i,$$

where φ_i denotes the characteristic function of the interval $[x_{i-1}, x_i]$.

Recall that a *partition* of the interval Q is a collection $\mathscr{P} = \{Q_1, \ldots, Q_k\}$ of closed intervals, with disjoint interiors, such that $Q = \bigcup_{i=1}^{k} Q_i$. By the *mesh* of $\mathscr{P}$

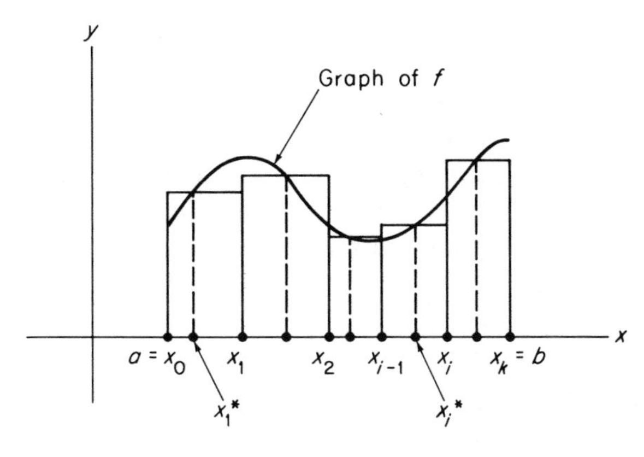

Figure 4.17

is meant the maximum of the diameters of the Q_i. A *selection* for $\mathscr{P}$ is a set $\mathscr{S} = \{\mathbf{x}_1, \ldots, \mathbf{x}_k\}$ of points such that $\mathbf{x}_i \in Q_i$ for each i. If $f: \mathscr{R}^n \to \mathscr{R}$ is a function such that $f = 0$ outside of Q, then the Riemann sum for f corresponding to the partition $\mathscr{P}$ and selection $\mathscr{S}$ is

$$R(f, \mathscr{P}, \mathscr{S}) = \sum_{i=1}^{k} f(\mathbf{x}_i) v(Q_i).$$

Notice that, by Theorem 3.1, the Riemann sum $R(f, \mathscr{P}, \mathscr{S})$ is simply the integral of the step function

$$h = \sum_{i=1}^{k} f(\mathbf{x}_i)\varphi_i ,$$

where φ_i denotes the characteristic function of Q_i.

Theorem 3.4 Suppose $f: \mathscr{R}^n \to \mathscr{R}$ is bounded and vanishes outside the interval Q. Then f is integrable with $\int f = I$ if and only if, given $\varepsilon > 0$, there exists $\delta > 0$ such that

$$|I - R(f, \mathscr{P}, \mathscr{S})| < \varepsilon$$

whenever $\mathscr{P}$ is a partition of Q with mesh $< \delta$ and $\mathscr{S}$ is a selection for $\mathscr{P}$.

PROOF If f is integrable, choose $M > 0$ such that $|f(\mathbf{x})| \leqq M$ for all $\mathbf{x}$. By Theorem 3.3 there exist step functions h and k such that $h \leqq f \leqq k$ and $\int (k - h) < \varepsilon/2$. By the construction of the proof of Theorem 3.2, we may assume that h and k are linear combinations of the characteristic functions of intervals whose closures are the same. That is, there is a partition $\mathscr{P}_0 = \{Q_1, \ldots, Q_s\}$ of Q such that

$$h = \sum_{i=1}^{s} a_i \varphi_i \qquad \text{and} \qquad k = \sum_{i=1}^{s} b_i \psi_i ,$$

where Q_i is the closure of the interval of which φ_i is the characteristic function, and likewise for ψ_i.

Now let $A = Q - \bigcup_{i=1}^{s} \operatorname{int} Q_i$. Then $v(A) = 0$, so *there exists* $\delta > 0$ such that, if $\mathscr{P}$ is a partition of Q with mesh $<\delta$, then the sum of the volumes of those intervals $P_1, \ldots, P_k$ of $\mathscr{P}$ which intersect A is less than $\varepsilon/4M$. Let $P_{k+1}, \ldots, P_l$ be the remaining intervals of $\mathscr{P}$, that is, those which lie interior to the Q_i.

If $\mathscr{S} = \{\mathbf{x}_1, \ldots, \mathbf{x}_l\}$ is a selection for $\mathscr{P}$, then $h(\mathbf{x}_i) \leqq f(\mathbf{x}_i) \leqq k(\mathbf{x}_i)$ if $i = k+1, \ldots, l$, so that

$$\sum_{i=k+1}^{l} f(\mathbf{x}_i)v(P_i) \qquad \text{and} \qquad \sum_{i=k+1}^{l} \int_{P_i} f$$

are both between $\sum_{i=k+1}^{l} \int_{P_i} h$ and $\sum_{i=k+1}^{l} \int_{P_i} k$, so it follows that

$$\left| \sum_{i=k+1}^{l} \int_{P_i} f - \sum_{i=k+1}^{l} f(\mathbf{x}_i)v(P_i) \right| < \frac{\varepsilon}{2}, \tag{3}$$

because $\int_Q (k - h) < \varepsilon/2$ by assumption.

Since $-M \leqq f(\mathbf{x}) \leqq M$ for all $\mathbf{x}$, both

$$\sum_{i=1}^{k} \int_{P_i} f \qquad \text{and} \qquad \sum_{i=1}^{k} f(\mathbf{x}_i)v(P_i)$$

lie between $-M \sum_{i=1}^{k} v(P_i) > -\varepsilon/4$ and $M \sum_{i=1}^{k} v(P_i) < \varepsilon/4$, so it follows that

$$\left| \sum_{i=1}^{k} \int_{P_i} f - \sum_{i=1}^{k} f(\mathbf{x}_i)v(P_i) \right| < \frac{\varepsilon}{2}. \tag{4}$$

Since $I = \int f = \sum_{i=1}^{l} \int_{P_i} f$, (3) and (4) finally imply by the triangle inequality that $|I - R(f, \mathscr{P}, \mathscr{S})| < \varepsilon$ as desired.

Conversely, suppose $\mathscr{P} = \{P_1, \ldots, P_p\}$ is a partition of Q such that, given any selection $\mathscr{S}$ for $\mathscr{P}$, we have

$$|I - R(f, \mathscr{P}, \mathscr{S})| < \frac{\varepsilon}{4}.$$

Let $Q_1, \ldots, Q_p$ be *disjoint* intervals (not closed) such that $\bar{Q}_i = P_i$, $i = 1, \ldots, p$, and denote by φ_i the characteristic function of Q_i. If

$$a_i = \text{glb of } f(\mathbf{x}) \quad \text{for} \quad \mathbf{x} \in P_i,$$
$$b_i = \text{lub of } f(\mathbf{x}) \quad \text{for} \quad \mathbf{x} \in P_i,$$

then

$$h = \sum_{i=1}^{p} a_i \varphi_i \qquad \text{and} \qquad k = \sum_{i=1}^{p} b_i \varphi_i$$

are step functions such that $h \leqq f \leqq k$.

Choose selections $\mathscr{S}' = \{\mathbf{x}_1', \ldots, \mathbf{x}_p'\}$ and $\mathscr{S}'' = \{\mathbf{x}_1'', \ldots, \mathbf{x}_p''\}$ for $\mathscr{P}$ such that

$$|a_i - f(\mathbf{x}_i')| < \frac{\varepsilon}{4v(Q)} \qquad \text{and} \qquad |b_i - f(\mathbf{x}_i'')| < \frac{\varepsilon}{4v(Q)}$$

for each i. Then

$$\left| R(f, \mathscr{P}, \mathscr{S}') - \int h \right| = \left| \sum_{i=1}^{p} (f(\mathbf{x}_i') - a_i) v(Q_i) \right|$$

$$< \frac{\varepsilon}{4v(Q)} \sum_{i=1}^{p} v(Q_i)$$

$$= \frac{\varepsilon}{4},$$

and similarly

$$\left| R(f, \mathscr{P}, \mathscr{S}'') - \int k \right| < \frac{\varepsilon}{4}.$$

Then

$$\int (k - h) \leqq \left| \int k - R(f, \mathscr{P}, \mathscr{S}'') \right| + |R(f, \mathscr{P}, \mathscr{S}'') - I|$$

$$+ |I - R(f, \mathscr{P}, \mathscr{S}')| + \left| R(f, \mathscr{P}, \mathscr{S}') - \int h \right|$$

$$< \frac{\varepsilon}{4} + \frac{\varepsilon}{4} + \frac{\varepsilon}{4} + \frac{\varepsilon}{4} = \varepsilon,$$

so it follows from Theorem 3.3 that f is integrable. ∎

Theorem 3.4 makes it clear that the operation of integration is a limit process. This is even more apparent in Exercise 3.2, which asserts that, if $f : \mathscr{R}^n \to \mathscr{R}$ is an integrable function which vanishes outside the interval Q, and $\{\mathscr{P}_k\}_1^\infty$ is a sequence of partitions of Q with associated selections $\{\mathscr{S}_k\}_1^\infty$ such that $\lim_{k \to \infty}$ (mesh of $\mathscr{P}_k$) $= 0$, then

$$\int f = \lim_{k \to \infty} R(f, \mathscr{P}_k, \mathscr{S}_k).$$

This observation leads to the formulation of several natural and important integration questions as *interchange of limit operations* questions.

(1) Let $A \subset \mathscr{R}^m$ and $B \subset \mathscr{R}^n$ be contented sets, and $f : A \times B \to \mathscr{R}$ a continuous function. Define $g : A \to \mathscr{R}$ by

$$g(\mathbf{x}) = \int_B f(\mathbf{x}, \mathbf{y})\, d\mathbf{y} = \int_B f_{\mathbf{x}},$$

where $f_{\mathbf{x}}(\mathbf{y}) = f(\mathbf{x}, \mathbf{y})$. Is g continuous on A? This is the question as to whether

$$\lim_{\mathbf{x} \to \mathbf{a}} \int_B f(\mathbf{x}, \mathbf{y})\, d\mathbf{y} = \int_B \lim_{\mathbf{x} \to \mathbf{a}} f(\mathbf{x}, \mathbf{y})\, d\mathbf{y}$$

for each $\mathbf{a} \in A$. According to Exercise 3.3, this is true if f is *uniformly* continuous.

(2) Let $\{f_n\}_1^\infty$ be a sequence of integrable functions on the contented set A, which converges (pointwise) to the integrable function $f: A \to \mathscr{R}$. We ask whether

$$\lim_{n\to\infty} \int_A f_n = \int \lim_{n\to\infty} f_n = \int f ?$$

According to Exercise 3.4, this is true if the sequence $\{f_n\}_1^\infty$ converges *uniformly* to f on A. This means that, given $\varepsilon > 0$, there exists N such that

$$n \geqq N \Rightarrow |f_n(\mathbf{x}) - f(\mathbf{x})| < \varepsilon$$

for all $\mathbf{x} \in A$. The following example shows that the hypothesis of *uniform* convergence is necessary. Let f_n be the function on $[0, 1]$ whose graph is pictured in Fig. 4.18. Then

$$\lim_{n\to\infty} f_n(x) = 0 \qquad \text{for all} \quad x \in [0, 1],$$

so $\int_0^1 \lim_{n\to\infty} f_n = 0$. However $\lim_{n\to\infty} \int_0^1 f_n = \lim_{n\to\infty} 1 = 1$. It is evident that the convergence of $\{f_n\}_1^\infty$ is not uniform (why?).

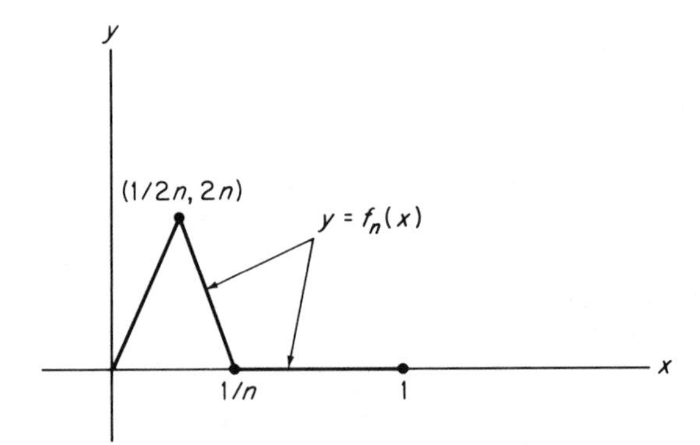

Figure 4.18

(3) *Differentiating under the integral.* Let $f: A \times J \to \mathscr{R}$ be a continuous function, where $A \subset \mathscr{R}^n$ is contented and $J \subset \mathscr{R}$ is an open interval. Define the partial derivative $D_2 f: A \times J \to \mathscr{R}$ by

$$D_2 f(\mathbf{x}, t) = \lim_{h\to 0} \frac{f(\mathbf{x}, t+h) - f(\mathbf{x}, t)}{h},$$

and the function $g: J \to \mathscr{R}$ by $g(t) = \int_A f(\mathbf{x}, t)\, d\mathbf{x}$. We ask whether

$$g'(t) = \int_A D_2 f(\mathbf{x}, t)\, d\mathbf{x},$$

that is, whether

$$\frac{\partial}{\partial t}\int_A f(\mathbf{x}, t)\,d\mathbf{x} = \int_A \frac{\partial}{\partial t} f(\mathbf{x}, t)\,d\mathbf{x}?$$

According to Exercise 3.5, this is true if $D_2 f$ is *uniformly* continuous on $A \times J$. Since differentiation is also a limit operation, this is another "interchange of limit operations" result.

Exercises

3.1 (a) If f is integrable, prove that f^2 is integrable. *Hint:* Given $\varepsilon > 0$, let h and k be step functions such that $h \leqq f \leqq k$ and $\int (k - h) < \varepsilon/M$, where M is the maximum value of $|k(\mathbf{x}) + h(\mathbf{x})|$. Then prove that h^2 and k^2 are step functions with $h^2 \leqq f^2 \leqq k^2$ (we may assume that $0 \leqq h \leqq f \leqq k$ since f is integrable if and only if $|f|$ is—why?), and that $\int (k^2 - h^2) < \varepsilon$. Then apply Theorem 3.3.
(b) If f and g are integrable, prove that fg is integrable. *Hint:* Express fg in terms of $(f+g)^2$ and $(f-g)^2$; then apply (a).

3.2 Let $f: \mathscr{R}^n \to \mathscr{R}$ be a bounded function which vanishes outside the interval Q. Show that f is integrable with $\int f = I$ if and only if

$$\lim_{k\to\infty} R(f, \mathscr{P}_k, \mathscr{S}_k) = I$$

for every sequence of partitions $\{\mathscr{P}_k\}_1^\infty$ and associated selections $\{\mathscr{S}_k\}_1^\infty$ such that $\lim_{k\to\infty}(\text{mesh of } \mathscr{P}_k) = 0$.

3.3 Let A and B be contented sets, and $f: A \times B \to \mathscr{R}$ a uniformly continuous function. If $g: A \to \mathscr{R}$ is defined by

$$g(\mathbf{x}) = \int_B f(\mathbf{x}, \mathbf{y})\,d\mathbf{y} \qquad \text{for all} \quad \mathbf{x} \in A,$$

prove that g is continuous. *Hint:* Write $g(\mathbf{x}) - g(\mathbf{a}) = \int_B [f(\mathbf{x}, \mathbf{y}) - f(\mathbf{a}, \mathbf{y})]\,d\mathbf{y}$ and apply the uniform continuity of f.

3.4 Let $\{f_n\}_1^\infty$ be a sequence of integrable functions which converges *uniformly* on the contented set A to the integrable function f. Then prove that

$$\lim_{n\to\infty} \int_A f_n = \int_A f = \int_A \lim_{n\to\infty} f_n .$$

Hint: Note that $|\int_A f_n - \int_A f| \leqq \int_A |f_n - f|$.

3.5 Let $f: A \times J \to \mathscr{R}$ be a continuous function, where $A \subset \mathscr{R}^n$ is contented and $J \subset \mathscr{R}$ is an open interval. Suppose that $D_2 f = \partial f/\partial t$ $(t \in J)$ is uniformly continuous on $A \times J$. If

$$g(t) = \int_A f(\mathbf{x}, t)\,d\mathbf{x},$$

prove that

$$g'(t) = \int_A D_2 f(\mathbf{x}, t)\,d\mathbf{x}.$$

Outline: It suffices to prove that, if $\{t_n\}_1^\infty$ is a sequence of distinct points of J converging to $a \in J$, then

$$\lim_{n\to\infty} \frac{g(t_n) - g(a)}{t_n - a} = \int_A D_2 f(\mathbf{x}, a)\, d\mathbf{x}.$$

For each fixed $\mathbf{x} \in A$, the mean value theorem gives $\tau_n(\mathbf{x}) \in (a, t_n)$ such that

$$f(\mathbf{x}, t_n) - f(\mathbf{x}, a) = (t_n - a) D_2 f(\mathbf{x}, \tau_n(\mathbf{x})).$$

Let

$$\varphi_n(\mathbf{x}) = \frac{f(\mathbf{x}, t_n) - f(\mathbf{x}, a)}{t_n - a} = D_2 f(\mathbf{x}, \tau_n(\mathbf{x})).$$

Now the hypothesis that $D_2 f$ is *uniformly* continuous implies that the sequence $\{\varphi_n\}_1^\infty$ converges *uniformly* on A to the function $D_2 f(\mathbf{x}, a)$ of $\mathbf{x}$ (explain why). Therefore, by the previous problem, we have

$$\begin{aligned}\int_A D_2 f(\mathbf{x}, a)\, d\mathbf{x} &= \lim_{n\to\infty} \int_A \varphi_n \\ &= \lim_{n\to\infty} \int_A \frac{f(\mathbf{x}, t_n) - f(\mathbf{x}, a)}{t_n - a}\, d\mathbf{x} \\ &= \lim_{n\to\infty} \frac{g(t_n) - g(a)}{t_n - a}\end{aligned}$$

as desired.

3.6 Let $f\colon [a, b] \times [c, d] \to \mathscr{R}$ be a continuous function. Prove that

$$\int_c^d \left(\int_a^b f(x, y)\, dx \right) dy = \int_a^b \left(\int_c^d f(x, y)\, dy \right) dx$$

by computing the derivatives of the two functions $g, h : [a, b] \to \mathscr{R}$ defined by

$$g(t) = \int_c^d \left(\int_a^t f(x, y)\, dx \right) dy$$

and

$$h(t) = \int_a^t \left(\int_c^d f(x, y)\, dy \right) dx,$$

using Exercise 3.5 and the fundamental theorem of calculus. This is still another interchange of limit operations.

3.7 For each positive integer n, the Bessel function $J_n(x)$ may be defined by

$$J_n(x) = \frac{x^n}{1 \cdot 3 \cdot 5 \cdot \cdots \cdot (2n-1)\pi} \int_{-1}^{+1} (1 - t^2)^{n-1/2} \cos xt\, dt.$$

Prove that $J_n(x)$ satisfies Bessel's differential equation

$$J_n'' + \frac{1}{x} J_n' + \left(1 - \frac{n^2}{x^2}\right) J_n = 0.$$

3.8 Establish the conclusion of Exercise 3.4 without the hypothesis that the limit function f is integrable. *Hint:* Let Q be a rectangle containing A, and use Theorem 3.4 as follows to prove that f must be integrable. Note first that $I = \lim \int f_n$ exists because $\{\int f_n\}_1^\infty$ is a Cauchy sequence of real numbers (why?).

Given $\varepsilon > 0$, choose N such that both $|I - \int f_N| < \varepsilon/3$ and $|f(\mathbf{x}) - f_N(\mathbf{x})| < \varepsilon/3v(Q)$ for all $\mathbf{x}$. Then choose $\delta > 0$ such that, for any partition $\mathscr{P}$ of mesh $< \delta$ and any selection $\mathscr{S}$,

$$\left| \int f_N - R(f_N, \mathscr{P}, \mathscr{S}) \right| < \frac{\varepsilon}{3}.$$

Noting that

$$|R(f_N, \mathscr{P}, \mathscr{S}) - R(f, \mathscr{P}, \mathscr{S})| < \frac{\varepsilon}{3}$$

also, conclude that

$$|I - R(f, \mathscr{P}, \mathscr{S})| < \varepsilon$$

as desired.

The following example shows that the uniform convergence of $\{f_n\}_1^\infty$ is necessary. Let $Q = \{r_k\}_1^\infty$ be the set of all rational numbers in the unit interval $[0, 1]$. If f_n denotes the characteristic function of the set $\{r_1, \ldots, r_n\}$, then $\int_0^1 f_n = 0$ for each $n = 1, 2, \ldots$, but the limit function

$$\lim_{n \to \infty} f_n(x) = \begin{cases} 1 & \text{if } x \text{ is rational,} \\ 0 & \text{otherwise,} \end{cases}$$

is not integrable. The point is that the convergence of this sequence $\{f_n\}_1^\infty$ is not uniform.

3.9 If $\varphi(t) = \int_{g(t)}^{h(t)} f(x, t)\, dx$, apply Exercise 3.5, the fundamental theorem of calculus, and the chain rule to prove Leibniz' rule:

$$\varphi'(t) = f(h(t), t)h'(t) - f(g(t), t)g'(t) + \int_{g(t)}^{h(t)} D_2 f(x, t)\, dx$$

under appropriate hypotheses (state them).

4 ITERATED INTEGRALS AND FUBINI'S THEOREM

If f is a continuous function on the rectangle $R = [a, b] \times [c, d] \subset \mathscr{R}^2$, then

$$\int_a^b \left(\int_c^d f(x, y)\, dy \right) dx = \int_c^d \left(\int_a^b f(x, y)\, dx \right) dy \tag{1}$$

by Exercise 3.6. According to the main theorem of this section, the integral $\int_R f$ is equal to the common value in (1). Thus $\int_R f$ may be computed as the result of two "iterated" single-variable integrations (which are easy, using the fundamental theorem of calculus, provided that the necessary antiderivatives can be found). In a similar way, the integral over an interval in $\mathscr{R}^n$ of a continuous function of n variables can be computed as the result of n iterated single-variable integrations (as we shall see).

In introductory calculus, the fact, that

$$\int_R f = \int_c^d \left(\int_a^b f(x, y)\, dx \right) dy, \tag{2}$$

is usually presented as an application of a "volume by cross-sections" approach. Given $t \in [c, d]$, let

$$A_t = \{(x, t, z) \in \mathscr{R}^3 : x \in [a, b] \quad \text{and} \quad z \in [0, f(x, t)]\},$$

so A_t is the "cross-section" in which the plane $y = t$ intersects the ordinate set of $f: R \to \mathscr{R}$ (see Fig. 4.19). One envisions the ordinate set of f as being swept

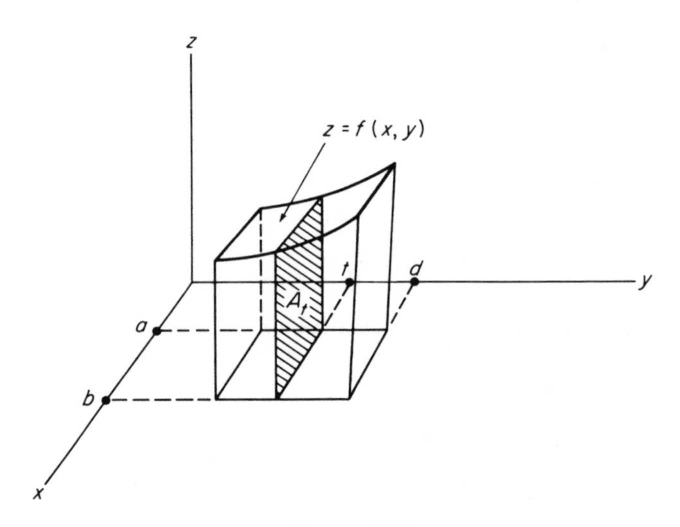

Figure 4.19

out by A_t as t increases from c to d. Let $A(t)$ denote the area of A_t, and $V(t)$ the volume over $[a, b] \times [c, t] = R_t$, that is,

$$V(t) = \int_{R_t} f.$$

If one accepts the assertion that

$$V'(t) = A(t), \tag{3}$$

then (2) follows easily. Since

$$A(t) = \int_a^b f(x, t)\, dx,$$

(3) and the fundamental theorem of calculus give

$$V(t) = \int_c^t A(y)\, dy + C,$$

so

$$V(t) = \int_c^t \left(\int_a^b f(x, y)\, dx \right) dy,$$

because $V(c) = 0$ gives $C = 0$. Then

$$\int_R f = V(d) = \int_c^d \left(\int_a^b f(x, y)\, dx \right) dy$$

as desired.

It is the fact, that $V'(t) = A(t)$, for which a heuristic argument is sometimes given at the introductory level. To prove this, assuming that f is continuous on R, let

$$\begin{aligned} \varphi(x, h) &= \text{the minimum value of } f(x, y) \quad \text{for} \quad y \in [t, t+h], \\ \psi(x, h) &= \text{the maximum value of } f(x, y) \quad \text{for} \quad y \in [t, t+h]. \end{aligned}$$

Then φ and ψ are continuous functions of (x, h) and

$$\lim_{h \to 0} \varphi(x, h) = \lim_{h \to 0} \psi(x, h) = f(x, t). \tag{4}$$

We temporarily fix t and h, and regard φ and ψ as functions of x, for $x \in [a, b]$. It is clear from the definitions of φ and ψ that the volume between A_t and A_{t+h} under $z = f(x, y)$ contains the set

$$O_\varphi \times [t, t+h] \qquad (O_\varphi = \text{ordinate set of } \varphi)$$

and is contained in the set

$$O_\psi \times [t, t+h].$$

The latter two sets are both "cylinders" of height h, with volumes

$$h \int_a^b \varphi(x, h)\, dx \qquad \text{and} \qquad h \int_a^b \psi(x, h)\, dx,$$

respectively. Therefore

$$h \int_a^b \varphi(x, h)\, dx \leqq V(t+h) - V(t) \leqq h \int_a^b \psi(x, h)\, dx.$$

We now divide through by h and take limits as $h \to 0$. Since Exercise 3.3 permits us to take the limit under the integral signs, substitution of (4) gives

$$A(t) = \int_a^b f(x, t)\, dx \leqq V'(t) \leqq \int_a^b f(x, t)\, dx = A(t),$$

so $V'(t) = A(t)$ as desired. This completes the proof of (2) if f is continuous on R.

However the hypothesis that f is continuous on R is too restrictive for most applications. For example, in order to reduce $\int_A f$ to iterated integrals, we would choose an interval R containing the contented set A, and define

$$g = \begin{cases} f & \text{on } A, \\ 0 & \text{outside } A. \end{cases}$$

Then $\int_A f = \int_R g$. But even if f is continuous on A, g will in general fail to be continuous at points of ∂A (see Fig. 4.20).

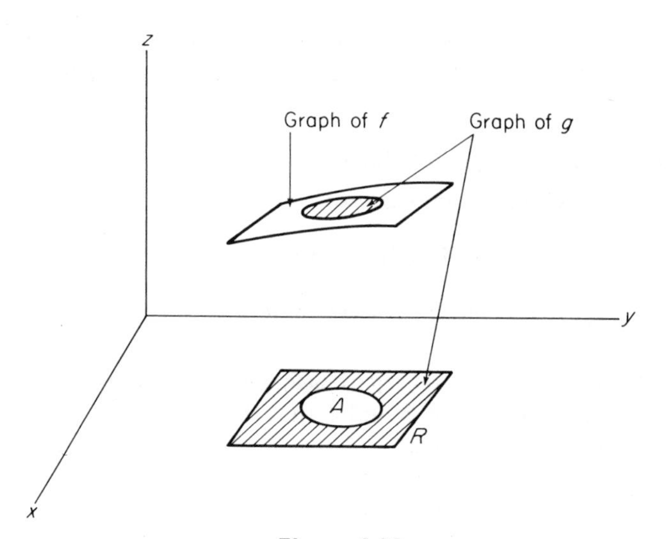

Figure 4.20

The following version of what is known as "Fubini's Theorem" is sufficiently general for our purposes. Notice that the integrals in its statement exist by the remark that, if $g : \mathscr{R}^k \to \mathscr{R}$ is integrable and $C \subset \mathscr{R}^k$ is contented, then $\int_C g$ exists, because the product of the integrable functions g and φ_C is integrable by Exercise 3.1.

Theorem 4.1 Let $f : \mathscr{R}^{m+n} = \mathscr{R}^m \times \mathscr{R}^n \to \mathscr{R}$ be an integrable function such that, for each $\mathbf{x} \in \mathscr{R}^m$, the function $f_\mathbf{x} : \mathscr{R}^n \to \mathscr{R}$, defined by $f_\mathbf{x}(\mathbf{y}) = f(\mathbf{x}, \mathbf{y})$, is integrable. Given contented sets $A \subset \mathscr{R}^m$ and $B \subset \mathscr{R}^n$, let $F : \mathscr{R}^m \to \mathscr{R}$ be defined by

$$F(\mathbf{x}) = \int_B f_\mathbf{x} = \int_B f(\mathbf{x}, \mathbf{y})\, d\mathbf{y}.$$

Then F is integrable, and

$$\int_{A \times B} f = \int_A F.$$

That is, in the usual notation,

$$\int_{A\times B} f = \int_A \left(\int_B f(\mathbf{x}, \mathbf{y})\, d\mathbf{y}\right) d\mathbf{x}.$$

REMARK The roles of $\mathbf{x}$ and $\mathbf{y}$ in this theorem can, of course, be interchanged. That is,

$$\int_{A\times B} f = \int_B \left(\int_A f(\mathbf{x}, \mathbf{y})\, d\mathbf{x}\right) d\mathbf{y}$$

under the hypothesis that, for each $\mathbf{y} \in \mathscr{R}^n$, the function $f_\mathbf{y} : \mathscr{R}^n \to \mathscr{R}$, defined by $f_\mathbf{y}(\mathbf{x}) = f(\mathbf{x}, \mathbf{y})$, is integrable.

PROOF We remark first that it suffices to prove the theorem under the assumption that $f(\mathbf{x}, \mathbf{y}) = 0$ unless $(\mathbf{x}, \mathbf{y}) \in A \times B$. For if $f^* = f\varphi_{A\times B}$, $f_\mathbf{x}^* = f_\mathbf{x}\varphi_B$, and $F^* = \int f_\mathbf{x}^*$, then

$$\int_{A\times B} f = \int_{\mathscr{R}^{m+n}} f^* \qquad \text{and} \qquad \int_A \left(\int_B f(\mathbf{x}, \mathbf{y})\, d\mathbf{y}\right) d\mathbf{x} = \int_{\mathscr{R}^m} \left(\int_{\mathscr{R}^n} f^*(\mathbf{x}, \mathbf{y})\, d\mathbf{y}\right) d\mathbf{x}.$$

In essence, therefore, we may replace A, B, and $A \times B$ by $\mathscr{R}^m$, $\mathscr{R}^n$, and $\mathscr{R}^{m\times n}$ throughout the proof.

We shall employ step functions via Theorem 3.3. First note that, if φ is the characteristic function of an interval $I \times J \subset \mathscr{R}^m \times \mathscr{R}^n$, then

$$\begin{aligned}\int \left(\int \varphi(\mathbf{x}, \mathbf{y})\, d\mathbf{y}\right) d\mathbf{x} &= \int_I \left(\int_J 1\, d\mathbf{y}\right) d\mathbf{x}\\ &= \int_I v(J)\, d\mathbf{x}\\ &= v(I)v(J)\\ &= v(I \times J)\\ &= \int \varphi.\end{aligned}$$

From this it follows that, if $h = \sum_{i=1}^k c_i \varphi_i$ is a step function, then

$$\begin{aligned}\int h = \sum_{i=1}^k c_i \int \varphi_i &= \sum_{i=1}^k c_i \int \left(\int \varphi_i(\mathbf{x}, \mathbf{y})\, d\mathbf{y}\right) d\mathbf{x}\\ &= \int \left(\int \sum_{i=1}^k c_i \varphi_i(\mathbf{x}, \mathbf{y})\, d\mathbf{y}\right) d\mathbf{x} = \int \left(\int h(\mathbf{x}, \mathbf{y})\, d\mathbf{y}\right) d\mathbf{x}.\end{aligned}$$

So the theorem holds for step functions.

Now, given $\varepsilon > 0$, let h and k be step functions such that $h \leqq f \leqq k$ and $\int (h - k) < \varepsilon$. Then for each $\mathbf{x}$ we have $h_\mathbf{x} \leqq f_\mathbf{x} \leqq k_\mathbf{x}$. Hence, if

$$H(\mathbf{x}) = \int h_\mathbf{x} \qquad \text{and} \qquad K(\mathbf{x}) = \int k_\mathbf{x},$$

then H and K are step functions on $\mathscr{R}^m$ such that $H \leqq F \leqq K$ and

$$\int_{\mathscr{R}^m} (K - H) = \int_{\mathscr{R}^{m+n}} (k - h) < \varepsilon,$$

by the fact that we have proved the theorem for step functions. Since $\varepsilon > 0$ is arbitrary, Theorem 3.3 now implies that F is integrable, with $\int H \leqq \int F \leqq \int K$. Since we now have $\int f$ and $\int F$ both between $\int h = \int H$ and $\int k = \int K$, and the latter integrals differ by $<\varepsilon$, it follows that

$$\left| \int f - \int F \right| < \varepsilon.$$

This being true for all $\varepsilon > 0$, the proof is complete. ∎

The following two applications of Fubini's theorem are generalizations of methods often used in elementary calculus courses either to define or to calculate volumes. The first one generalizes to higher dimensions the method of "volumes by cross-sections."

Theorem 4.2 (Cavalieri's Principle) Let A be a contented subset of $\mathscr{R}^{n+1}$, with $A \subset R \times [a, b]$, where $R \subset \mathscr{R}^n$ and $[a, b] \subset \mathscr{R}$ are intervals. Suppose

$$A_t = \{\mathbf{x} \in \mathscr{R}^n : (\mathbf{x}, t) \in A\} \subset \mathscr{R}^n$$

is contented for each $t \in [a, b]$, and write $A(t) = v(A_t)$. Then

$$v(A) = \int_a^b A(t)\, dt.$$

PROOF (See Fig. 4.21).

$$\begin{aligned} v(A) &= \int_A 1 \\ &= \int_{R \times [a, b]} \varphi_A \\ &= \int_{[a, b]} \left(\int_R \varphi_{A_t} \right) \qquad \text{(by Fubini's theorem)} \\ &= \int_a^b \left(\int_{A_t} 1 \right) \\ &= \int_a^b A(t)\, dt. \end{aligned}$$

∎

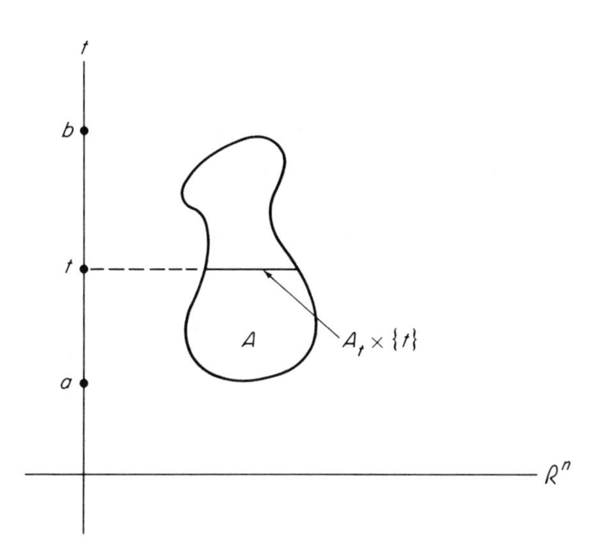

Figure 4.21

The most typical application of Cavalieri's principle is to the computation of *volumes of revolution* in $\mathscr{R}^3$. Let $f: [a, b] \to \mathscr{R}$ be a positive continuous function. Denote by A the set in $\mathscr{R}^3$ obtained by revolving about the x-axis the ordinate set of f (Fig. 4.22). It is clear that $A(t) = \pi[f(t)]^2$, so Theorem 4.2 gives

$$v(A) = \pi \int_a^b [f(x)]^2 \, dx.$$

For example, to compute the volume of the 3-dimensional ball $B_r^{\,3}$ of radius r, take $f(x) = (r^2 - x^2)^{1/2}$ on $[-r, r]$. Then we obtain

$$v(B_r^{\,3}) = \pi \int_{-r}^{r} (r^2 - x^2) \, dx = \tfrac{4}{3}\pi r^3.$$

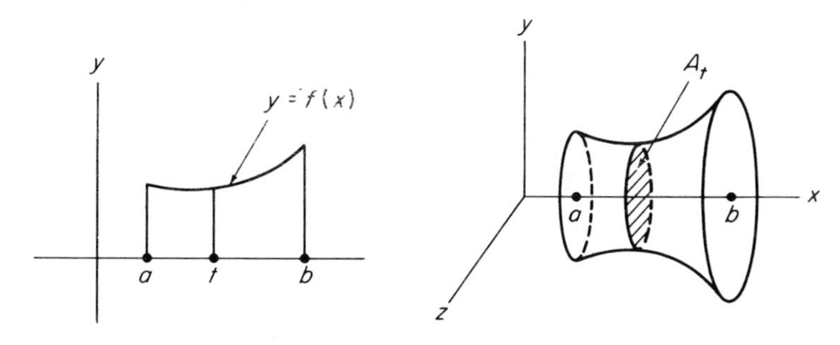

Figure 4.22

Theorem 4.3 If $A \subset \mathscr{R}^n$ is a contented set, and f_1 and f_2 are continuous functions on A such that $f_1 \leqq f_2$, then

$$C = \{(\mathbf{x}, y) \in \mathscr{R}^{n+1} : \mathbf{x} \in A \quad \text{and} \quad f_1(\mathbf{x}) \leqq y \leqq f_2(\mathbf{x})\}$$

is a contented set. If $g : C \to \mathscr{R}$ is continuous, then

$$\int_C g = \int_A \left(\int_{f_1(\mathbf{x})}^{f_2(\mathbf{x})} g(\mathbf{x}, y)\, dy \right) d\mathbf{x}.$$

PROOF The fact that C is contented follows easily from Corollary 2.3. Let B be a closed interval in $\mathscr{R}$ containing $f_1(A) \cup f_2(A)$, so that $C \subset A \times B$ (see Fig. 4.23).

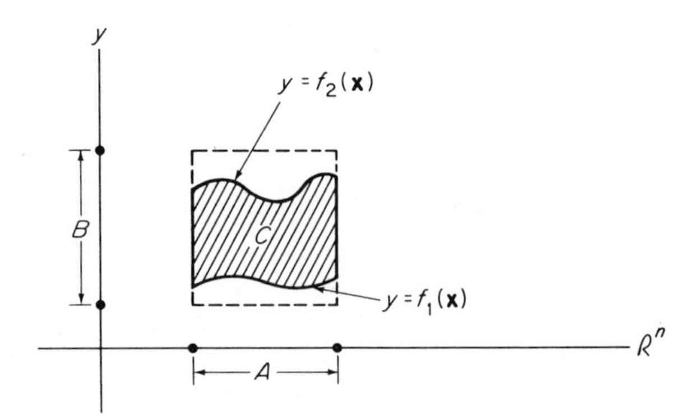

Figure 4.23

If $h = g\varphi_C$, then it is clear that the functions $h : \mathscr{R}^{n+1} \to \mathscr{R}$ and $h_{\mathbf{x}} : \mathscr{R} \to \mathscr{R}$ are admissible, so Fubini's theorem gives

$$\int_C g = \int_{A \times B} h = \int_A \left(\int_B h_{\mathbf{x}}\, dy \right) = \int_A \left(\int_{f_1(\mathbf{x})}^{f_2(\mathbf{x})} g(\mathbf{x}, y)\, dy \right) d\mathbf{x}$$

because $h_{\mathbf{x}}(y) = 1$ if $y \in [f_1(\mathbf{x}), f_2(\mathbf{x})]$, and 0 otherwise. ∎

For example, if g is a continuous function on the unit ball $B^3 \subset \mathscr{R}^3$, then

$$\int_{B^3} g = \int_{B^2} \left(\int_{-(1-x^2-y^2)^{1/2}}^{+(1-x^2-y^2)^{1/2}} g(\mathbf{x}, z)\, dz \right) d\mathbf{x},$$

where $\mathbf{x} = (x, y) \in B^2$.

The case $m = n = 1$ of Fubini's theorem, with $A = [a, b]$ and $B = [c, d]$, yields the formula

$$\int_{A \times B} f = \int_a^b \left(\int_c^d f(x, y)\, dy \right) dx,$$

alluded to at the beginning of this section. Similarly, if $f : Q \to \mathscr{R}$ is continuous, and $Q = [a_1, b_1] \times \cdots \times [a_n, b_n]$ is an *interval* in $\mathscr{R}^n$, then $n - 1$ applications of

Fubini's theorem yield the formula

$$\int_Q f = \int_{a_n}^{b_n} \left(\int_{a_{n-1}}^{b_{n-1}} \cdots \left(\int_{a_1}^{b_1} f(\mathbf{x})\, dx_1 \right) \cdots dx_{n-1} \right) dx_n .$$

This is the computational significance of Fubini's theorem—it reduces a multivariable integration over an *interval* in $\mathscr{R}^n$ to a sequence of n successive single-variable integrations, in each of which the fundamental theorem of calculus can be employed.

If Q is *not* an interval, it may be possible, by appropriate substitutions, to "transform" $\int_Q f$ to an integral $\int_R g$, where R *is* an interval (and then evaluate $\int_R g$ by use of Fubini's theorem and the fundamental theorem of calculus). This is the role of the "change of variables formula" of the next section.

Exercises

4.1 Let $A \subset \mathscr{R}^m$ and $B \subset \mathscr{R}^n$ be contented sets, and $f : \mathscr{R}^m \to \mathscr{R}$ and $g : \mathscr{R}^n \to \mathscr{R}$ integrable functions. Define $h : \mathscr{R}^{m+n} \to \mathscr{R}$ by $h(\mathbf{x}, \mathbf{y}) = f(\mathbf{x})g(\mathbf{y})$, and prove that

$$\int_{A \times B} h = \left(\int_A f \right) \left(\int_B g \right).$$

Conclude as a corollary that $v(A \times B) = v(A)v(B)$.

4.2 Use Fubini's theorem to give an easy proof that $\partial^2 f/\partial x\, \partial y = \partial^2 f/\partial y\, \partial x$ if these second derivatives are both continuous. *Hint:* If $D_1 D_2 f - D_2 D_1 f > 0$ at some point, then there is a rectangle R on which it is positive. However use Fubini's theorem to calculate $\int_R (D_1 D_2 f - D_2 D_1 f) = 0$.

4.3 Define $f : I \times I \to \mathscr{R}$, $I = \{0, 1\}$ by

$$f(x, y) = \begin{cases} 0 & \text{if either } x \text{ or } y \text{ is irrational,} \\ \dfrac{1}{q} & \text{if } x \text{ and } y \text{ are rational and } y = p/q \text{ with } p \text{ and } q \text{ relatively prime.} \end{cases}$$

Then show that

(a) $\int_{I \times I} f = 0$,

(b) $\int_0^1 f(x, y)\, dy = 0$ for all $x \in [0, 1]$,

(c) $\int_0^1 f(x, y)\, dx = 0$ if y is irrational, but does not exist if y is rational.

4.4 Let T be the solid torus in $\mathscr{R}^3$ obtained by revolving the circle $(y - a)^2 + z^2 \leq b^2$, in the yz-plane, about the z-axis. Use Cavalieri's principle to compute $v(T) = 2\pi^2 ab^2$.

4.5 Let S be the intersection of the cylinders $x^2 + z^2 \leq 1$ and $y^2 + z^2 \leq 1$. Use Cavalieri's principle to compute $v(S) = \frac{16}{3}$.

4.6 The area of the ellipse $x^2/a^2 + y^2/b^2 \leq 1$, with semiaxes a and b, is $A = \pi ab$. Use this fact and Cavalieri's principle to show that the volume enclosed by the ellipsoid

$$\frac{x^2}{a^2} + \frac{y^2}{b^2} + \frac{z^2}{c^2} = 1$$

is $v = \frac{4}{3}\pi abc$. *Hint:* What is the area $A(t)$ of the *ellipse* of intersection of the plane $x = t$ and the ellipsoid? What are the semiaxes of this ellipse?

4.7 Use the formula $V = \frac{4}{3}\pi r^3$, for the volume of a 3-dimensional ball of radius r, and Cavalieri's principle, to show that the volume of the 4-dimensional unit ball $B^4 \subset \mathscr{R}^4$ is $\pi^2/2$. *Hint:* What is the volume $A(t)$ of the 3-dimensional ball in which the hyperplane $x_4 = t$ intersects B^4?

4.8 Let C be the 4-dimensional "solid cone" in $\mathscr{R}^4$ that is bounded above by the 3-dimensional ball of radius a that is centered at $(0, 0, 0, h)$ in the hyperplane $x_4 = h$, and below by the conical "surface" $x_4 = (x_1^2 + x_2^2 + x_2^2)^{1/2}$. Show that $v(C) = \frac{1}{3}\pi a^3 h$. Note that this is one-fourth times the height of the cone C times the volume of its base.

5 CHANGE OF VARIABLES

The student has undoubtedly seen change of variables formulas such as

$$\iint f(x, y)\, dx\, dy = \iint f(r \cos \theta, r \sin \theta) r\, dr\, d\theta$$

and

$$\iiint f(x, y, z)\, dx\, dy\, dz$$
$$= \iiint f(\rho \sin \varphi \cos \theta, \rho \sin \varphi \sin \theta, \rho \cos \varphi) \rho^2 \sin \varphi\, d\rho\, d\varphi\, d\theta,$$

which result from changes from rectangular coordinates to polar and spherical coordinates respectively. The appearance of the factor r in the first formula and the factor $\rho^2 \sin \varphi$ in the second one is sometimes "explained" by mythical pictures, such as Figure 4.24, in which it is alleged that dA is an "infinitesimal"

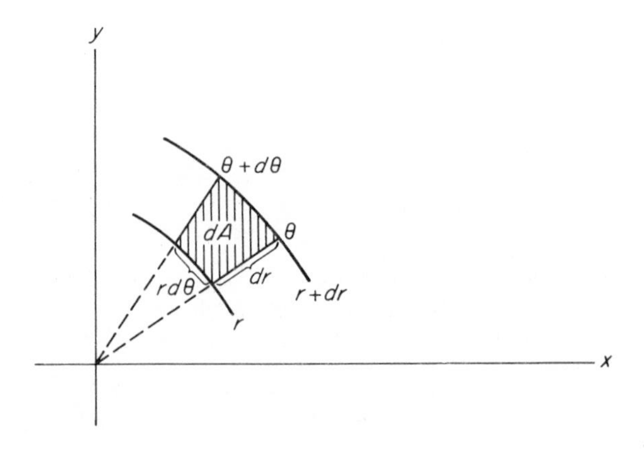

Figure 4.24

rectangle with sides dr and $r\,d\theta$, and therefore has area $r\,dr\,d\theta$. In this section we shall give a mathemtaically acceptable explanation of the origin of such factors in the transformation of multiple integrals from one coordinate system to another.

This will entail a thorough discussion of the technique of substitution for multivariable integrals. The basic problem is as follows. Let $T : \mathscr{R}^n \to \mathscr{R}^n$ be a mapping which is $\mathscr{C}^1$ (continuously differentiable) and also $\mathscr{C}^1$-*invertible* on some neighborhood U of the set $A \subset \mathscr{R}^n$, meaning that T is one-to-one on U and that $T^{-1} : T(U) \to U$ is also a $\mathscr{C}^1$ mapping. Given an integrable function $f : T(A) \to \mathscr{R}$, we would like to "transform" the integral $\int_{T(A)} f$ into an appropriate integral over A (which may be easier to compute).

For example, suppose we wish to compute $\int_Q f$, where Q is an annular sector in the plane $\mathscr{R}^2$, as pictured in Fig. 4.25. Let $T : \mathscr{R}^2 \to \mathscr{R}^2$ be the "polar coordinates" mapping defined by $T(r, \theta) = (r \cos \theta, r \sin \theta)$. Then $Q = T(A)$, where A is the rectangle $\{(r, \theta) \in \mathscr{R}^2 : a \leqq r \leqq b \text{ and } \alpha \leqq \theta \leqq \beta\}$. If $\int_Q f$ can be transformed into an integral over A then, because A *is* a rectangle, the latter integral will be amenable to computation by iterated integrals (Fubini's theorem).

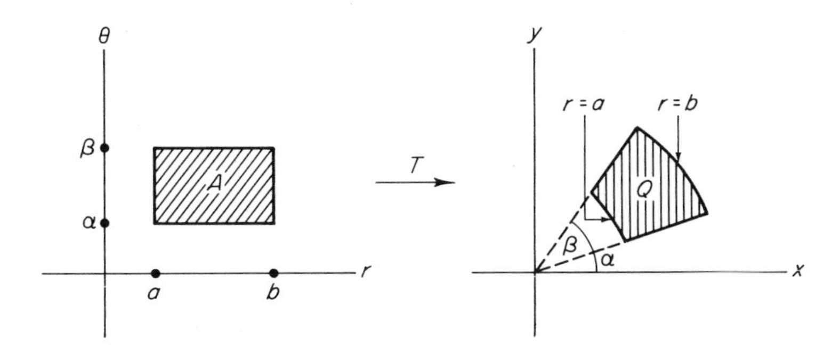

Figure 4.25

The principal idea in this discussion will be the local approximation of the transformation T by its differential linear mapping dT. Recall that $dT_{\mathbf{a}} : \mathscr{R}^n \to \mathscr{R}^n$ is that linear mapping which "best approximates" the mapping $\mathbf{h} \to T(\mathbf{a} + \mathbf{h}) - T(\mathbf{a})$ in a neighborhood of the point $\mathbf{a} \in \mathscr{R}^n$, and that the matrix of $dT_{\mathbf{a}}$ is the derivative matrix $T'(\mathbf{a}) = (D_j T_i(\mathbf{a}))$, where $T_1, \ldots, T_n$ are the component functions of $T : \mathscr{R}^n \to \mathscr{R}^n$.

We therefore discuss first the behavior of volumes under linear distortions. That is, given a linear mapping $\lambda : \mathscr{R}^n \to \mathscr{R}^n$ and a contented set $A \subset \mathscr{R}^n$, we investigate the relationship between $v(A)$ and $v(\lambda(A))$ [assuming, as we will prove, that the image $\lambda(A)$ is also contented]. If λ is represented by the $n \times n$ matrix A, that is, $\lambda(\mathbf{x}) = A\mathbf{x}$, we write

$$\det \lambda = |A| = \det A.$$

Theorem 5.1 If $\lambda : \mathscr{R}^n \to \mathscr{R}^n$ is a linear mapping, and $B \subset \mathscr{R}^n$ is contented, then $\lambda(B)$ is also contented, and

$$v(\lambda(B)) = |\det \lambda| v(B). \tag{1}$$

PROOF First suppose that $\det \lambda = 0$, so the column vectors of the matrix of λ are linearly dependent (by Theorem I.6.1). Since these column vectors are the images under λ of the standard unit vectors $\mathbf{e}_1, \ldots, \mathbf{e}_n$ in $\mathscr{R}^n$, it then follows that $\lambda(\mathscr{R}^n)$ is a *proper* subspace of $\mathscr{R}^n$. Since it is easily verified that any bounded subset of a proper subspace of $\mathscr{R}^n$ is negligible (see Exercise 5.1), we see that (1) is trivially satisfied if $\det \lambda = 0$.

If $\det \lambda \neq 0$, then a standard theorem of linear algebra asserts that the matrix A of λ can be written as a product

$$A = A_1 A_2 \cdots A_k,$$

where each A_i is either of the form

$$\begin{pmatrix} 1 & & & & O \\ & \ddots & & & \\ & & a & & \\ & & & \ddots & \\ O & & & & 1 \end{pmatrix} \quad \begin{array}{l}\text{(every diagonal}\\ \text{element but one}\\ \text{equal to 1)}\end{array} \tag{2}$$

or of the form

$$\begin{pmatrix} 1 & & & & O \\ & \ddots & & & \\ & & 1 \text{---} 1 \text{---} & & \\ O & & & \ddots & \\ & & & & 1 \end{pmatrix} \quad \begin{array}{l}\text{(a single off-}\\ \text{diagonal element).}\end{array} \tag{3}$$

Hence λ is a composition $\lambda = \lambda_1 \circ \lambda_2 \circ \cdots \circ \lambda_k$, where λ_i is the linear mapping represented by the matrix A_i.

If A_i is of the form (2), with a appearing in the pth row and column, then

$$\lambda_i(x_1, \ldots, x_n) = (x_1, \ldots, x_{p-1}, ax_p, x_{p+1}, \ldots, x_n),$$

so it is clear that, if I is an interval, then $\lambda_i(I)$ is also an interval, with $v(\lambda_i(\mathrm{I})) = |a| v(I)$. From this, and the definition of volume in terms of intervals, it follows easily that, if B is a contented set, then $\lambda_i(B)$ is contented with $v(\lambda_i(B)) = |a| v(B) = |\det \lambda_i| v(B)$.

We shall show below that, if A_i is of the form (3), then $v(\lambda_i(B)) = v(B) = |\det \lambda_i| v(B)$ for any contented set. Granting this, the theorem follows, because we then have

$$\begin{aligned} v(\lambda(B)) &= v(\lambda_1 \circ \lambda_2 \circ \cdots \circ \lambda_k(B)) \\ &= |\det \lambda_1| \cdot |\det \lambda_2| \cdot \cdots \cdot |\det \lambda_k| v(B) \\ &= |\det \lambda| v(B) \end{aligned}$$

as desired, since the determinant of a product of matrices is the product of their determinants.

So it remains only to verify that λ_i preserves volumes if the matrix A_i of λ_i is of the form (3). We consider the typical case in which the off-diagonal element of A_i is in the first row and second column, that is,

$$A_i = \begin{pmatrix} 1 & 1 & & O \\ 0 & 1 & & \\ & & \ddots & \\ O & & & 1 \end{pmatrix}$$

so

$$\lambda_i(x_1, \ldots, x_n) = (x_1 + x_2, x_2, \ldots, x_n). \tag{4}$$

We show first that λ_i preserves the volume of an interval

$$I = [\mathbf{a}, \mathbf{b}] = \{\mathbf{x} \in \mathscr{R}^n : x_i \in [a_i, b_i], \quad i = 1, \ldots, n\}.$$

If we write $I = I' \times I''$ where $I' \subset \mathscr{R}^2$ and $I'' \subset \mathscr{R}^{n-2}$, then (4) shows that

$$\lambda_i(I) = I^* \times I'',$$

where I^* is the pictured parallelogram in $\mathscr{R}^2$ (Fig. 4.26). Since it is clear that $v(I^*) = v(I')$, Exercise 4.1 gives

$$\begin{aligned} v(\lambda_i(I)) &= v(I^*)v(I'') \\ &= v(I')v(I'') = v(I). \end{aligned}$$

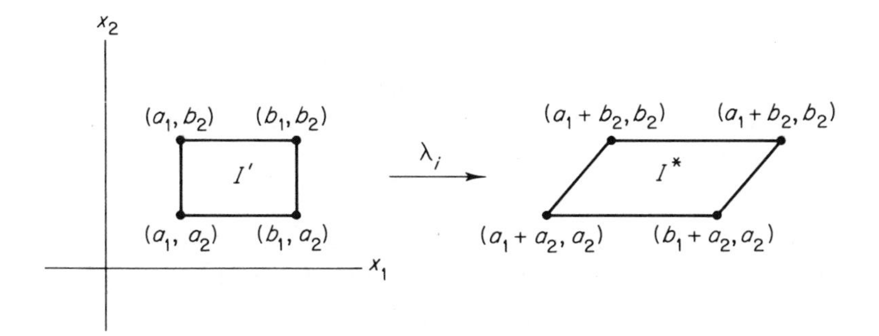

Figure 4.26

Now let B be a contented set in $\mathscr{R}^n$, and, given $\varepsilon > 0$, consider nonoverlapping intervals $I_1, \ldots, I_p$ and intervals $J_1, \ldots, J_q$ such that

$$\bigcup_{j=1}^{p} I_j \subset B \subset \bigcup_{j=1}^{q} J_j$$

and

$$\sum_{j=1}^{p} v(I_j) > v(B) - \varepsilon, \qquad \sum_{j=1}^{q} v(J_j) < v(B) + \varepsilon.$$

Since we have seen that λ_i preserves the volumes of intervals, we then have

$$\bigcup_{j=1}^{p} \lambda_i(I_j) \subset \lambda_i(B) \subset \bigcup_{j=1}^{q} \lambda_i(J_j)$$

and

$$\sum_{j=1}^{p} v(\lambda_i(I_j)) > v(B) - \varepsilon, \qquad \sum_{j=1}^{q} v(\lambda_i(J_j)) < v(B) + \varepsilon.$$

Finally it follows easily from these inequalities that $\lambda_i(B)$ is contented, with $v(\lambda_i(B)) = v(B)$ as desired (see Exercise 5.2). ∎

We are now in a position to establish the invariance of volume under rigid motions.

Corollary 5.2 If $A \subset \mathscr{R}^n$ is a contented set and $\rho : \mathscr{R}^n \to \mathscr{R}^n$ is a rigid motion, then $\rho(A)$ is contented and $v(\rho(A)) = v(A)$.

The statement that ρ is a *rigid motion* means that $\rho = \tau_{\mathbf{a}} \circ \mu$, where $\tau_{\mathbf{a}}$ is the translation by $\mathbf{a}$, $\tau_{\mathbf{a}}(\mathbf{x}) = \mathbf{x} + \mathbf{a}$, and μ is an orthogonal transformation, so that $|\det \mu| = 1$ (see Exercise I.6.10). The fact that $\rho(A)$ is contented with $v(\rho(A)) = v(A)$ therefore follows immediately from Theorem 5.1.

Example 1 Consider the ball

$$B_r^{\,n} = \{\mathbf{x} \in \mathscr{R}^n : \sum_1^n x_i^{\,2} \leqq r^2\}$$

of radius r in $\mathscr{R}^n$. Note that $B_r^{\,n}$ is the image of the unit n-dimensional ball $B_1^{\,n} \subset \mathscr{R}^n$ under the linear mapping $T(\mathbf{x}) = r\mathbf{x}$. Since $|\det T'| = |\det T| = r^n$, Theorem 5.1 gives

$$v(B_r^{\,n}) = r^n v(B_1^{\,n}).$$

Thus the volume of an n-dimensional ball is proportional to the nth power of its radius. Let α_n denote the volume of the unit n-dimensional ball $B_1^{\,n}$, so

$$v(B_r^{\,n}) = \alpha_n r^n.$$

In Exercises 5.17, 5.18, and 5.19, we shall see that

$$\alpha_{2m} = \frac{\pi^m}{m!} \qquad \text{and} \qquad \alpha_{2m+1} = \frac{2^{m+1}\pi^m}{1 \cdot 3 \cdot 5 \cdot \cdots \cdot (2m+1)}.$$

These formulas are established by induction on m, starting with the familiar $\alpha_2 = \pi$ and $\alpha_3 = 4\pi/3$.

In addition to the effect on volumes of linear distortions, discussed above, a key ingredient in the derivation of the change of variables formula is the fact that, if $F : \mathscr{R}^n \to \mathscr{R}^n$ is a $\mathscr{C}^1$ mapping such that $dF_{\mathbf{0}} = I$ (the identity transformation of $\mathscr{R}^n$), then in a sufficiently small neighborhood of $\mathbf{0}$, F differs very little from the identity mapping, and in particular does not significantly alter the volumes of sets within this neighborhood.

This is the effect of the following lemma, which was used in the proof of the inverse function theorem in Chapter III. Recall that the *norm* $\|\lambda\|$ of the linear mapping $\lambda : \mathscr{R}^n \to \mathscr{R}^n$ is by definition the maximum value of $|\lambda(\mathbf{x})|_0$ for all $\mathbf{x} \in \partial C_1$, where

$$|(x_1, \ldots, x_n)|_0 = \max\{|x_1|, \ldots, |x_n|\}$$

and $C_r = \{\text{all } \mathbf{x} \in \mathscr{R}^n \text{ such that } |\mathbf{x}|_0 \leqq r\}$. Thus C_r is a "cube" centered at $\mathbf{0}$ with $(r, r, \ldots, r)$ as one vertex, and is referred to as "the cube of radius r centered at the origin."

The lemma then asserts that, if $dF_{\mathbf{x}}$ differs (in norm) only slightly from the identity mapping I for all $\mathbf{x} \in C_r$, then the image $F(C_r)$ contains a cube slightly smaller than C_r, and is contained in one slightly larger than C_r.

Lemma 5.3 (Lemma III.3.2) Let U be an open set in $\mathscr{R}^n$ containing C_r, and $F : U \to \mathscr{R}^n$ a $\mathscr{C}^1$ mapping such that $F(\mathbf{0}) = \mathbf{0}$ and $dF_{\mathbf{0}} = I$. Suppose also that there exists $\varepsilon \in (0, 1)$ such that

$$\|dF_{\mathbf{x}} - I\| \leqq \varepsilon \tag{5}$$

for all $\mathbf{x} \in C_r$. Then

$$C_{(1-\varepsilon)r} \subset F(C_r) \subset C_{(1+\varepsilon)r}$$

(see Fig. 4.27).

Now let Q be an interval centered at the point $\mathbf{a} \in \mathscr{R}^n$, and suppose that $T : U \to \mathscr{R}^n$ is a $\mathscr{C}^1$-invertible mapping on a neighborhood U of Q. Suppose in addition that the differential $dT_{\mathbf{x}}$ is almost constant on Q, $dT_{\mathbf{x}} \approx dT_{\mathbf{a}}$, and in particular that

$$\|dT_{\mathbf{a}}^{-1} \circ dT_{\mathbf{x}} - I\| \leqq \varepsilon \tag{6}$$

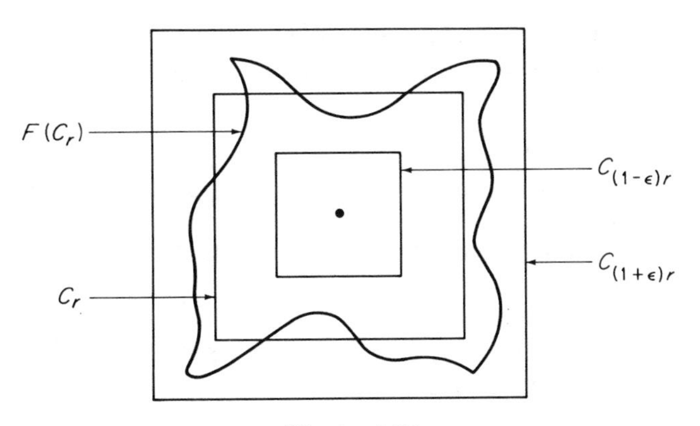

Figure 4.27

for all $\mathbf{x} \in Q$. Also let $Q_{1-\varepsilon}$ and $Q_{1+\varepsilon}$ be intervals centered at $\mathbf{a}$ and both "similar" to Q, with each edge of $Q_{1\pm\varepsilon}$ equal (in length) to $1 \pm \varepsilon$ times the corresponding edge of Q, so $v(Q_{1\pm\varepsilon}) = (1 \pm \varepsilon)^n v(Q)$. (See Fig. 4.28.)

Then the idea of the proof of Theorem 5.4 below is to use Lemma 5.3 to show that (6) implies that the image set $T(Q)$ closely approximates the "parallelepiped" $dT_{\mathbf{a}}(Q)$, in the sense that

$$dT_{\mathbf{a}}(Q_{1-\varepsilon}) \subset T(Q) \subset dT_{\mathbf{a}}(Q_{1+\varepsilon}).$$

It will then follow from Theorem 5.1 that

$$v(T(Q)) \approx |\det T'(\mathbf{a})| \, v(Q).$$

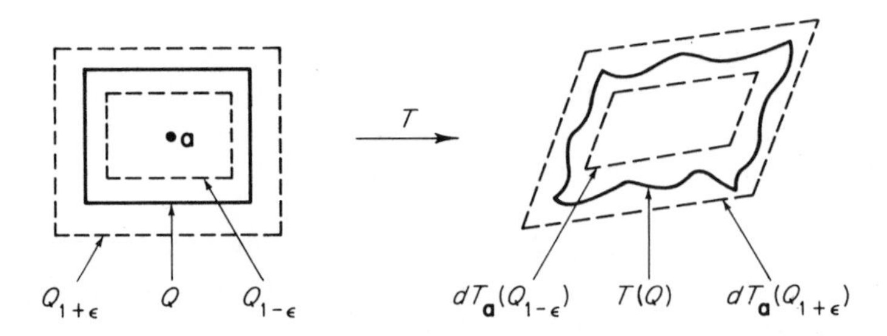

Figure 4.28

Theorem 5.4 Let Q be an interval centered at the point $\mathbf{a} \in \mathscr{R}^n$, and suppose $T: U \to \mathscr{R}^n$ is a $\mathscr{C}^1$-invertible mapping on a neighborhood U of Q. If there exists $\varepsilon \in (0, 1)$ such that

$$\|dT_{\mathbf{a}}^{-1} \circ dT_{\mathbf{x}} - I\| \leqq \varepsilon. \tag{6}$$

for all $\mathbf{x} \in Q$, then $T(Q)$ is contented with

$$(1 - \varepsilon)^n |\det T'(\mathbf{a})| \, v(Q) \leqq v(T(Q)) \leqq (1 + \varepsilon)^n \, |\det T'(a)| \, v(Q). \tag{7}$$

PROOF We leave to Exercise 5.3 the proof that $T(Q)$ is contented, and proceed to verify (7).

First suppose that Q is a cube of radius r. Let $\tau_{\mathbf{a}}$ denote translation by $\mathbf{a}$, $\tau_{\mathbf{a}}(\mathbf{x}) = \mathbf{a} + \mathbf{x}$, so $\tau_{\mathbf{a}}(C_r) = Q$. Let $\mathbf{b} = T(\mathbf{a})$, and $\lambda = dT_{\mathbf{a}}$. Let $F = \lambda^{-1} \circ \tau_{\mathbf{b}}^{-1} \circ T \circ \tau_{\mathbf{a}}$ (see Fig. 4.29).

By the chain rule we obtain

$$dF = d\lambda^{-1} \circ d\tau_{\mathbf{b}}^{-1} \circ dT \circ d\tau_{\mathbf{a}}.$$

Since the differential of a translation is the identity mapping, this gives

$$dF_{\mathbf{x}} = dT_{\mathbf{a}}^{-1} \circ dT_{\tau_{\mathbf{a}}(\mathbf{x})}$$

for all $\mathbf{x} \in C_r$. Consequently hypothesis (6) implies that $F : C_r \to \mathscr{R}^n$ satisfies (5). Consequently Lemma 5.3 applies to yield

$$C_{(1-\varepsilon)r} \subset F(C_r) \subset C_{(1+\varepsilon)r},$$

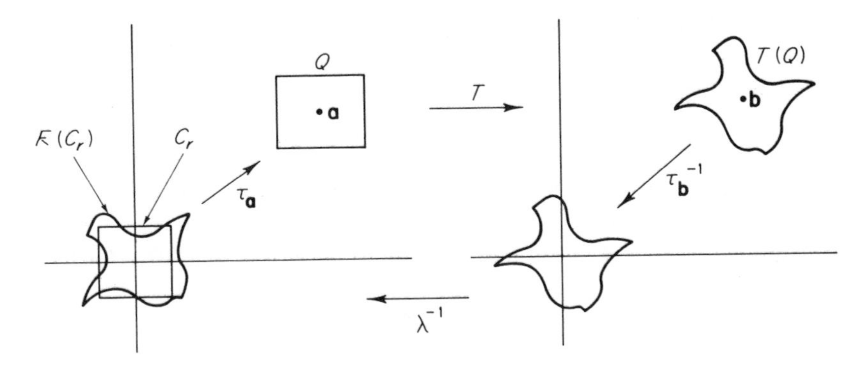

Figure 4.29

so

$$(1-\varepsilon)^n v(C_r) \leqq v(F(C_r)) \leqq (1+\varepsilon)^n v(C_r). \tag{8}$$

Since $\lambda(F(C_r)) = \tau_{\mathbf{b}}^{-1}(T(Q))$, Theorem 5.1 implies that

$$v(T(Q)) = |\det \lambda|\, v(F(C_r)) = |\det T'(\mathbf{a})|\, v(F(C_r)).$$

Since $v(Q) = v(C_r)$, we therefore obtain (7) upon multiplication of (8) by $|\det T'(\mathbf{a})|$. This completes the proof in case Q is a cube.

If Q is not a cube, let $\rho : \mathscr{R}^n \to \mathscr{R}^n$ be a linear mapping such that $\rho(C_1)$ is congruent to Q, and $\tau_{\mathbf{a}} \circ \rho(C_1) = Q$ in particular. Then $v(Q) = |\det \rho|\, v(C_1)$ by Theorem 5.1. Let

$$S = T \circ \tau_{\mathbf{a}} \circ \rho.$$

Then the chain rule gives

$$\begin{aligned} dS_{\mathbf{0}}^{-1} \circ dS_{\mathbf{x}} &= (dT_{\mathbf{a}} \circ \rho)^{-1} \circ (dT_{\mathbf{y}} \circ \rho) \\ &= \rho^{-1} \circ dT_{\mathbf{a}}^{-1} \circ dT_{\mathbf{y}} \circ \rho, \qquad \mathbf{y} = \tau_{\mathbf{a}}(\mathbf{x}) \in Q, \end{aligned}$$

because $d\rho = \rho$ since ρ is linear. Therefore

$$\|dS_{\mathbf{0}}^{-1} \circ dS_{\mathbf{x}} - I\| \leqq \|\rho\|^{-1} \cdot \|dT_{\mathbf{a}}^{-1} \circ T_{\mathbf{y}} - I\| \cdot \|\rho\| \leqq \varepsilon$$

by (6), so $S : C_1 \to \mathscr{R}^n$ itself satisfies the hypothesis of the theorem, with C_1 a cube.

What we have already proved therefore gives

$$(1-\varepsilon)^n |\det S'(\mathbf{0})|\, v(C_1) \leqq v(S(C_1)) \leqq (1+\varepsilon)^n |\det S'(\mathbf{0})|\, v(C_1).$$

Since $S(C_1) = T(Q)$ and

$$|\det S'(\mathbf{0})|\, v(C_1) = |\det T'(\mathbf{a})|\, |\det \rho|\, v(C_1) = |\det T'(\mathbf{a})|\, v(Q),$$

this last inequality is (7) as desired. ∎

Since $(1 \pm \varepsilon)^n \approx 1$ if ε is very small, the intuitive content of Theorem 5.4 is that, *if $dT_{\mathbf{x}}$ is approximately equal to $dT_{\mathbf{a}}$ for all* $\mathbf{x} \in Q$ (and this should be true if Q is sufficiently small), *then $v(T(Q))$ is approximately equal to* $|\det T'(\mathbf{a})|\, v(Q)$. Thus the factor $|\det T'(\mathbf{a})|$ seems to play the role of a "local magnification factor." With this interpretation of Theorem 5.4, we can now give a heuristic derivation of the change of variables formula for multiple integrals.

Suppose that $T: \mathscr{R}^n \to \mathscr{R}^n$ is $\mathscr{C}^1$-invertible on a neighborhood of the interval Q, and suppose that $f: \mathscr{R}^n \to \mathscr{R}$ is an integrable function such that $f \circ T$ is also integrable.

Let $\mathscr{P} = \{Q_1, \ldots, Q_k\}$ be a partition of Q into very small subintervals, and $\mathscr{S} = \{\mathbf{a}_1, \ldots, \mathbf{a}_k\}$ the selection for $\mathscr{P}$ such that $\mathbf{a}_i$ is the center-point of Q_i (Fig. 4.30). Then

$$
\begin{aligned}
\int_{T(Q)} f &= \sum_{i=1}^{k} \int_{T(Q_i)} f \\
&\approx \sum_{i=1}^{k} f(T(\mathbf{a}_i))v(T(Q_i)) \qquad \text{(because } T(Q_i) \text{ is very small)} \\
&\approx \sum_{i=1}^{k} f(T(\mathbf{a}_i))|\det T'(\mathbf{a}_i)|\, v(Q_i) \qquad \text{(by interpretation of 5.4)} \\
&= R(|\det T'|(f \circ T), \mathscr{P}, \mathscr{S}) \\
&\approx \int_Q |\det T'|(f \circ T).
\end{aligned}
$$

The proof of the main theorem is a matter of supplying enough epsilonics to convert the above discussion into a precise argument.

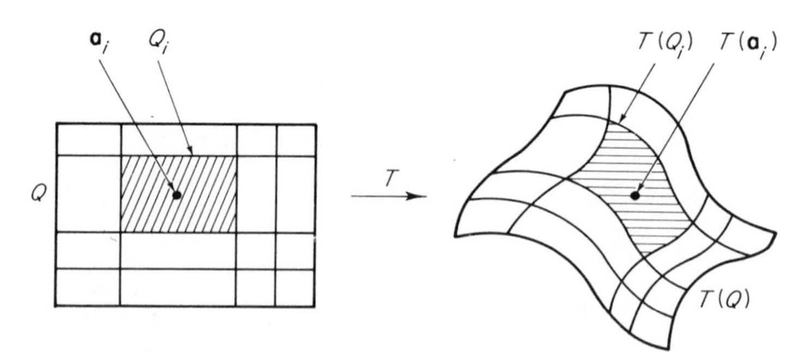

Figure 4.30

Theorem 5.5 Let Q be an interval in $\mathscr{R}^n$, and $T: \mathscr{R}^n \to \mathscr{R}^n$ a mapping which is $\mathscr{C}^1$-invertible on a neighborhood of Q. If $f: \mathscr{R}^n \to \mathscr{R}$ is an integrable function such that $f \circ T$ is also integrable, then

$$
\int_{T(Q)} f = \int_Q (f \circ T)|\det T'| \qquad \text{(change of variables formula).} \tag{9}
$$

PROOF Let $\eta > 0$ be given to start with. Given a partition $\mathscr{P} = \{Q_1, \ldots, Q_k\}$ of Q, let $\mathscr{S} = \{\mathbf{a}_1, \ldots, \mathbf{a}_k\}$ be the selection consisting of the center-points of the intervals of $\mathscr{P}$. By Theorem 3.4 there exists $\delta_1 > 0$ such that

$$\left| R - \int_Q (f \circ T) |\det T'| \right| < \frac{\eta}{2} \tag{10}$$

if mesh $\mathscr{P} < \delta_1$, where R denotes the Riemann sum

$$R((f \circ T)|\det T'|, \mathscr{P}, \mathscr{S}) = \sum_{i=1}^{k} f(T(\mathbf{a}_i))|\det T'(\mathbf{a}_i)| v(Q_i).$$

Choose $A > 0$ and $M > 0$ such that

$$|\det T'(\mathbf{x})| < A \qquad \text{and} \qquad |f(T(\mathbf{x}))| < M$$

for all $\mathbf{x} \in Q$. This is possible because $\det T'(\mathbf{x})$ is continuous (hence bounded) on Q, and $f \circ T$ is integrable (hence bounded) on Q. If

$$m_i = \text{glb of } f(T(\mathbf{x})) \qquad \text{for} \quad \mathbf{x} \in Q_i,$$
$$M_i = \text{lub of } f(T(\mathbf{x})) \qquad \text{for} \quad \mathbf{x} \in Q_i,$$

then

$$\alpha = \sum_{i=1}^{k} m_i \Delta_i v(Q_i) \leqq R \leqq \sum_{i=1}^{k} M_i \Delta_i v(Q_i) = \beta, \tag{11}$$

where

$$\Delta_i = |\det T'(\mathbf{a}_i)|.$$

By Theorem 3.4 there exists $\delta_2 > 0$ such that, if mesh $\mathscr{P} < \delta_2$, then any two Riemann sums for $f \circ T$ on $\mathscr{P}$ differ by less than $\eta/6A$, because each differs from $\int_Q f \circ T$ by less than $\eta/12A$. It follows that

$$\sum_{i=1}^{k} (M_i - m_i) v(Q_i) \leqq \frac{\eta}{6A} \tag{12}$$

because there are Riemann sums for $f \circ T$ arbitrarily close to both $\sum_{i=1}^{k} M_i v(Q_i)$ and $\sum_{i=1}^{k} m_i v(Q_i)$. Hence (11) and (12) give

$$\beta - \alpha = \sum_{i=1}^{k} (M_i - m_i) \Delta_i v(Q_i) < \frac{\eta}{6} \tag{13}$$

because each $\Delta_i < A$. Summarizing our progress thus far, we have the Riemann sum R which differs from $\int_Q (f \circ T)|\det T'|$ by less than $\eta/2$, and lies between the two numbers α and β, which in turn differ by less than $\eta/6$.

Next we want to locate $\int_{T(Q)} f$ between two numbers α' and β' which are close to α and β, respectively. Since the sets $T(Q_i)$ are contented (by Exercise 5.3) and intersect only in their boundaries,

$$\int_{T(Q)} f = \sum_{i=1}^{k} \int_{T(Q_i)} f$$

by Proposition 2.7. Therefore

$$\alpha' = \sum_{i=1}^{k} m_i v(T(Q_i)) \leqq \int_{T(Q)} f \leqq \sum_{i=1}^{k} M_i v(T(Q_i)) = \beta' \tag{14}$$

by Proposition 2.5. Our next task is to estimate $\alpha' - \alpha$ and $\beta' - \beta$, and this is where Theorem 5.4 will come into play.

Choose $\varepsilon \in (0, 1)$ such that

$$(1 + \varepsilon)^n - (1 - \varepsilon)^n < \frac{\eta}{6AMv(Q)}. \tag{15}$$

Since T is $\mathscr{C}^1$-invertible in a neighborhood of Q, we can choose an upper bound B for the value of the norm $\|dT_{\mathbf{x}}^{-1}\|$ for $\mathbf{x} \in Q$. By uniform continuity there exists $\delta_3 > 0$ such that, if mesh $\mathscr{P} < \delta_3$, then

$$\|dT_{\mathbf{x}} - dT_{\mathbf{a}_i}\| < \frac{\varepsilon}{B}$$

for all $\mathbf{x} \in Q_i$. This in turn implies that

$$\begin{aligned} \|dT_{\mathbf{a}}^{-1} \circ dT_{\mathbf{x}} - I\| &\leqq \|dT_{\mathbf{a}}^{-1}\| \cdot \|dT_{\mathbf{x}} - dT_{\mathbf{a}}\| \\ &< B \cdot \frac{\varepsilon}{B} = \varepsilon \end{aligned}$$

for all $\mathbf{x} \in Q_i$. Consequently Theorem 5.4 applies to show that $v(T(Q_i))$ lies between

$$(1 - \varepsilon)^n \, \Delta_i v(Q_i) \qquad \text{and} \qquad (1 + \varepsilon)^n \, \Delta_i v(Q_i),$$

as does $\Delta_i v(Q_i)$.

It follows that

$$|v(T(Q_i)) - \Delta_i v(Q_i)| \leqq [(1 + \varepsilon)^n - (1 - \varepsilon)^n] \, \Delta_i v(Q_i) \tag{16}$$

if the mesh of $\mathscr{P}$ is less than δ_3. We now fix the partition $\mathscr{P}$ with mesh less than $\delta = \min(\delta_1, \delta_2, \delta_3)$. Then

$$\begin{aligned} |\beta' - \beta| &\leqq \sum_{i=1}^{k} |M_i| \, |v(T(Q_i)) - \Delta_i v(Q_i)| \\ &\leqq M \sum_{i=1}^{k} [(1 + \varepsilon)^n - (1 - \varepsilon)^n] \, \Delta_i v(Q_i) \\ &\leqq AMv(Q)[(1 + \varepsilon)^n - (1 - \varepsilon)^n] < \frac{\eta}{6} \end{aligned} \tag{17}$$

by (15) and (16). Similarly

$$|\alpha' - \alpha| < \frac{\eta}{6}. \tag{18}$$

We are finally ready to move in for the kill. If $\alpha^* = \min(\alpha, \alpha')$ and $\beta^* = \max(\beta, \beta')$ then, by (11) and (14), both R and $\int_{T(Q)} f$ lie between α^* and β^*.

Since (13), (17), and (18) imply that $\beta^* - \alpha^* < \eta/2$, it follows that

$$|R - \textstyle\int_{T(Q)} f| < \frac{\eta}{2}. \tag{19}$$

Thus $\int_{T(Q)} f$ and $\int_Q (f \circ T)|\det T'|$ both differ from R by less than $\eta/2$, so

$$\left| \textstyle\int_{T(Q)} f - \int_Q (f \circ T)|\det T'| \right| < \eta.$$

This being true for every $\eta > 0$, we conclude that

$$\int_{T(Q)} f = \int_Q (f \circ T)|\det T'|. \quad \blacksquare$$

Theorem 5.5 is not quite general enough for some of the standard applications of the change of variables formula. However, the needed generalizations are usually quite easy to come by, the real work having already been done in the proof of Theorem 5.5. For example, the following simple extension will suffice for most purposes.

Addendum 5.6 The conclusion of Theorem 5.5 still holds if the $\mathscr{C}^1$ mapping $T: \mathscr{R}^n \to \mathscr{R}^n$ is only assumed to be $\mathscr{C}^1$-invertible on the interior of the interval Q (instead of on a neighborhood of Q).

PROOF Let Q^* be an interval interior to Q, and write $K = Q - \operatorname{int} Q^*$ (Fig. 4.31). Then Theorem 5.5 as stated gives

$$\int_{T(Q^*)} f = \int_{Q^*} (f \circ T)|\det T'|.$$

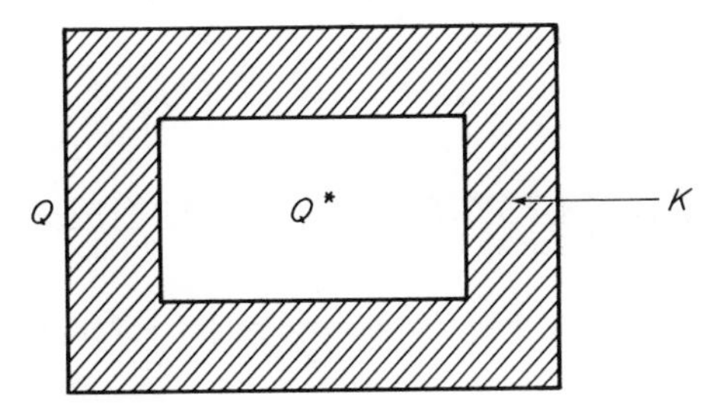

Figure 4.31

But

$$\int_{T(Q)} f = \int_{T(Q^*)} f + \int_{T(K)} f$$

and

$$\int_Q (f \circ T)\, |\det T'| = \int_{Q^*} (f \circ T)|\det T'| + \int_K (f \circ T)|\det T'|,$$

and the integrals over K and $T(K)$ can be made as small as desired by making the volume of K, and hence that of $T(K)$, sufficiently small. The easy details are left to the student (Exercise 5.4). ∎

Example 2 Let A denote the annular region in the plane bounded by the circles of radii a and b centered at the origin (Fig. 4.32). If $T : \mathscr{R}^2 \to \mathscr{R}^2$ is the "polar coordinates mapping" defined by

$$T(r, \theta) = (r \cos \theta, r \sin \theta)$$

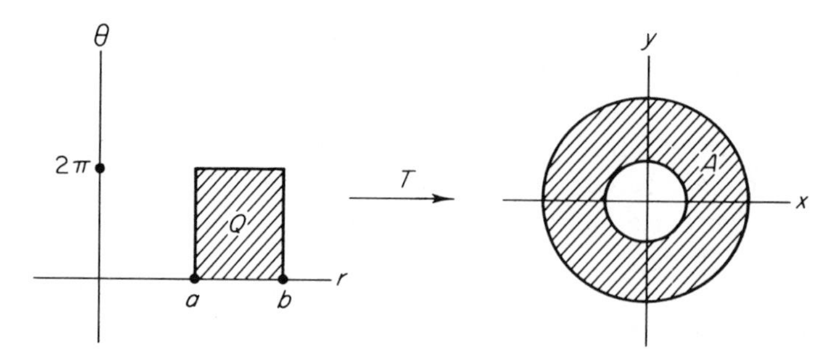

Figure 4.32

(for clarity we use r and θ as coordinates in the domain, and x and y as coordinates in the image), then A is the image under T of the rectangle

$$Q = \{(r, \theta) \in \mathscr{R}^2 : r \in [a, b] \text{ and } \theta \in [0, 2\pi]\}.$$

T is not one-to-one on Q, because $T(r, 0) = T(r, 2\pi)$, but *is* $\mathscr{C}^1$-invertible on the interior of Q, so Addendum 5.6 gives

$$\int_A f = \int_Q (f \circ T)\, |\det T'|.$$

Since $|\det T'(r, \theta)| = r$, Fubini's theorem then gives

$$\int_A f = \int_0^{2\pi} \int_a^b f(r \cos \theta, r \sin \theta) r\, dr\, d\theta.$$

Example 3 Let A be an "ice cream cone" bounded above by the sphere of radius 1 and below by a cone with vertex angle $\pi/6$ (Fig. 4.33). If $T : \mathscr{R}^3 \to \mathscr{R}^3$ is the "spherical coordinates mapping" defined by

$$T(\rho, \varphi, \theta) = (\rho \sin \varphi \cos \theta, \rho \sin \varphi \sin \theta, \rho \cos \varphi),$$

then A is the image under T of the interval

$$Q = \{(\rho, \varphi, \theta) \in \mathscr{R}^3 : \rho \in [0, 1],\ \varphi \in [0, \pi/12],\ \theta \in [0, 2\pi]\}.$$

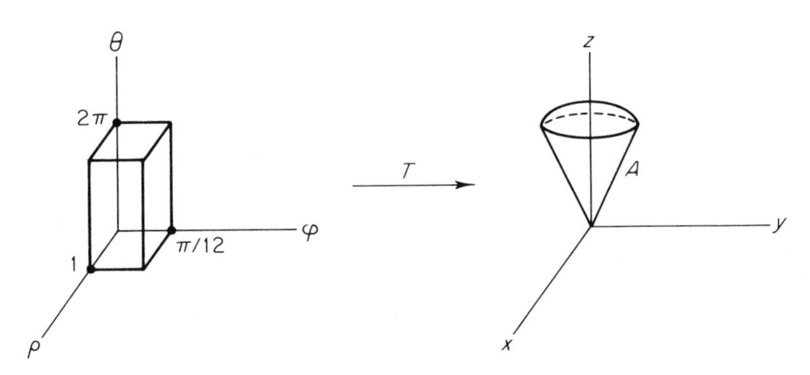

Figure 4.33

T is not one-to-one on Q (for instance, the image under T of the back face $\rho = 0$ of Q is the origin), but is $\mathscr{C}^1$-invertible on the interior of Q, so Addendum 5.6 together with Fubini's theorem gives

$$\int_A f = \int_0^{2\pi} \int_0^{\pi/12} \int_0^1 f(\rho \sin \varphi \cos \theta, \rho \sin \varphi \sin \theta, \rho \cos \varphi)\, \rho^2 \sin \varphi \, d\rho \, d\varphi \, d\theta$$

since $|\det T'(\rho, \varphi, \theta)| = \rho^2 \sin \varphi$.

Example 4 To compute the volume of the unit ball $B^3 \subset \mathscr{R}^3$, take $f(\mathbf{x}) \equiv 1$, T the spherical coordinates mapping of Example 2, and

$$Q = \{(\rho, \varphi, \theta) \in \mathscr{R}^3 : \rho \in [0, 1], \ \varphi \in [0, \pi], \ \theta \in [0, 2\pi]\}.$$

Then $T(Q) = B^3$, and T is $\mathscr{C}^1$-invertible on the interior of Q. Therefore Addendum 5.6 yields

$$\begin{aligned} v(B^3) &= \int_{B^3} 1 = \int_{B^3} f \\ &= \int_Q (f \circ T)|\det T'| \\ &= \int_0^{2\pi} \int_0^{\pi} \int_0^1 \rho^2 \sin \varphi \, d\rho \, d\varphi \, d\theta = \tfrac{4}{3}\pi, \end{aligned}$$

the familiar answer.

In integration problems that display an obvious cylindrical or spherical symmetry, the use of polar or spherical coordinates is clearly indicated. In general, in order to evaluate $\int_R f$ by use of the change of variables formula, one chooses the transformation T (often in an ad hoc manner) so as to simplify either the function f or the region R (or both). The following two examples illustrate this.

Example 5 Suppose we want to compute

$$\iint_R (\sqrt{x} + \sqrt{y})^{1/2}\, dx\, dy,$$

where R is the region in the first quadrant which is bounded by the coordinate axes and the curve $\sqrt{x} + \sqrt{y} = 1$ (a parabola). The substitution $u = \sqrt{x}$, $v = \sqrt{y}$ seems promising (it simplifies the integrand considerably), so we consider the mapping T from uv-space to xy-space defined by $x = u^2$, $y = v^2$ (see Fig. 4.34). It maps onto R the triangle Q in the uv-plane bounded by the co-

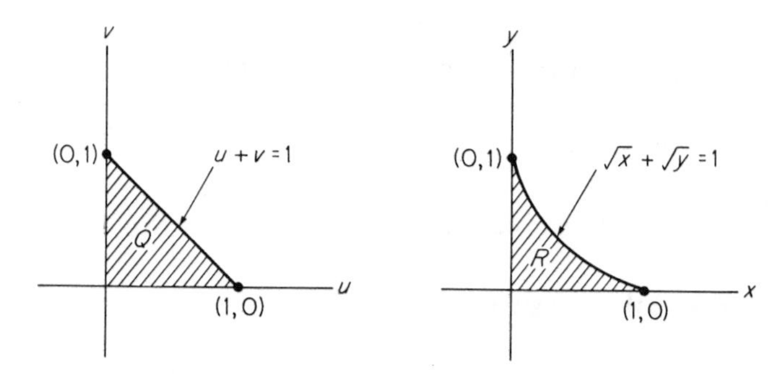

Figure 4.34

ordinate axes and the line $u + v = 1$. Now $\det T' = 4uv$, so the change of variables formula (which applies here, by Exercise 5.4) gives

$$\iint_R (\sqrt{x} + \sqrt{y})^{1/2}\, dx\, dy = 4 \iint_Q uv(u + v)^{1/2}\, du\, dv$$

$$= 4 \int_0^1 \left(\int_0^{1-v} uv(u + v)^{1/2}\, du \right) dv.$$

This iterated integral (obtained from Theorem 4.3) is now easily evaluated using integration by substitution (we omit the details).

Example 6 Consider $\iint_R (x^2 + y^2)\, dx\, dy$, where R is the region in the first quadrant that is bounded by the hyperbolas $xy = 1$, $xy = 3$, $x^2 - y^2 = 1$, and $x^2 - y^2 = 4$. We make the substitution $u = xy$, $v = x^2 - y^2$ so as to simplify the region; this transformation S from xy-space to uv-space maps the region R onto the rectangle Q bounded by the lines $u = 1$, $u = 3$, $v = 1$, $v = 4$ (see Fig. 4.35). From $u = xy$, $v = x^2 - y^2$ we obtain

$$x^2 + y^2 = (4u^2 + v^2)^{1/2}, \qquad 2x^2 = (4u^2 + v^2)^{1/2} + v, \qquad 2y^2 = (4u^2 + v^2)^{1/2} - v.$$

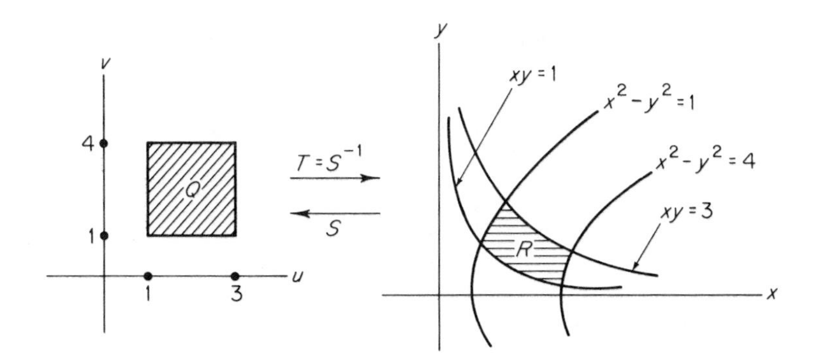

Figure 4.35

The last two equations imply that S is one-to-one on R; we are actually interested in its inverse $T = S^{-1}$. Since $S \circ T = I$, the chain rule gives $S'(T(u, v)) \circ T'(u, v) = I$, so

$$|\det T'| = \frac{1}{|\det S'|} = \frac{1}{2(x^2 + y^2)} = \frac{1}{2(4u^2 + v^2)^{1/2}}.$$

Consequently the change of variables formula gives

$$\iint_R (x^2 + y^2)\, dx\, dy = \iint_Q (4u^2 + v^2)^{1/2} \cdot \frac{1}{2(4u^2 + v^2)^{1/2}}\, du\, dv$$
$$= \tfrac{1}{2} v(Q) = 3.$$

Multiple integrals are often used to compute mass. A *body* with mass consists of a contented set A and an integrable *density* function $\mu : A \to \mathscr{R}$; its *mass* $M(A)$ is then defined by

$$M(A) = \int_A \mu.$$

Example 7 Let A be the ellipsoidal ball $x^2/a^2 + y^2/b^2 + z^2/c^2 \leqq 1$, with its density function μ being constant on concentric ellipsoids. That is, there exists a function $g : [0, 1] \to \mathscr{R}$ such that $\mu(\mathbf{x}) = g(\rho)$ at each point $\mathbf{x}$ of the ellipsoid

$$\frac{x^2}{a^2} + \frac{y^2}{b^2} + \frac{z^2}{c^2} = \rho^2.$$

In order to compute $M(A)$, we intoduce ellipsoidal coordinates by

$$x = a\rho \sin\varphi \cos\theta, \qquad y = b\rho \sin\varphi \sin\theta, \qquad z = c\rho \cos\varphi.$$

The transformation T defined by these equations maps onto A the interval

$$Q = \{(\rho, \varphi, \theta) : \rho \in [0, 1],\ \varphi \in [0, \pi],\ \theta \in [0, 2\pi]\},$$

and is invertible on the interior of Q. Since det $T' = abc\rho^2 \sin \varphi$, Addendum 5.6 gives

$$M(A) = abc \int_0^1 \int_0^\pi \int_0^{2\pi} g(\rho)\rho^2 \sin \varphi \, d\theta \, d\varphi \, d\rho$$
$$= 4\pi \, abc \int_0^1 \rho^2 g(\rho) \, d\rho.$$

For instance, taking $g(\rho) \equiv 1$ (uniform unit density), we see that the volume of the ellipsoid $x^2/a^2 + y^2/b^2 + z^2/c^2 \leqq 1$ is $\frac{4}{3}\pi abc$ (as seen previously in Exercise 4.6).

A related application of multiple integrals is to the computation of force. A force field is a vector-valued function, whereas we have thus far integrated only scalar-valued functions. The function $G : \mathscr{R}^n \to \mathscr{R}^m$ is said to be integrable on $A \subset \mathscr{R}^n$ if and only if each of its component functions $g_1, \ldots, g_m$ is, in which case we define

$$\int_A G = \left(\int_A g_1, \ldots, \int_A g_m\right) \in \mathscr{R}^m.$$

Thus vector-valued functions are simply integrated componentwise.

As an example, consider a body A with density function $\mu : A \to \mathscr{R}$, and let $\boldsymbol{\xi}$ be a point not in A. Motivated by Newton's law of gravitation, we define the *gravitational force* exerted by A, on a particle of mass m situated at $\boldsymbol{\xi}$, to be the vector

$$\mathbf{F} = \int_A \frac{\gamma m \, \mu(\mathbf{x})(\mathbf{x} - \boldsymbol{\xi})}{|\mathbf{x} - \boldsymbol{\xi}|^3} \, d\mathbf{x},$$

where γ is the "gravitational constant."

For example, suppose that A is a uniform rod (μ = constant) coincident with the unit interval $[0, 1]$ on the x-axis, and $\boldsymbol{\xi} = (0, 1)$. (See Fig. 4.36.) Then the gravitational force exerted by the rod A on a unit mass at $\boldsymbol{\xi}$ is

$$\mathbf{F} = \left(\int_0^1 \frac{\gamma\mu x}{(x^2 + 1)^{3/2}} \, dx, -\int_0^1 \frac{\gamma\mu}{(x^2 + 1)^{3/2}} \, dx\right)$$
$$= \left(\int_0^1 \frac{\gamma\mu \sin \alpha}{x^2 + 1} \, dx, -\int_0^1 \frac{\gamma\mu \cos \alpha}{x^2 + 1} \, dx\right).$$

For a more interesting example, see Exercises 5.13 and 5.14.

We close this section with a discussion of some of the classical notation and terminology associated with the change of variables formula. Given a differentiable mapping $T : \mathscr{R}^n \to \mathscr{R}^n$, recall that the determinant det $T'(\mathbf{a})$ of the deriva-

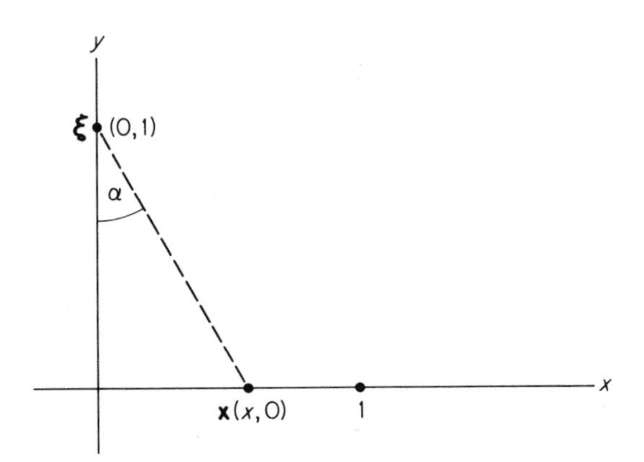

Figure 4.36

tive matrix $T'(\mathbf{a})$ is called the *Jacobian* (determinant) of T at $\mathbf{a}$. A standard notation (which we have used in Chapter III) for the Jacobian is

$$\det T' = \frac{\partial(T_1, \ldots, T_n)}{\partial(u_1, \ldots, u_n)},$$

where $T_1, \ldots, T_n$ are as usual the component functions of T, thought of as a mapping from $\mathbf{u}$-space (with coordinates $u_1, \ldots, u_n$) to $\mathbf{x}$-space (with coordinates $x_1, \ldots, x_n$). If we replace the component functions $T_1, \ldots, T_n$ in the Jacobian by the variables $x_1, \ldots, x_n$ which they give,

$$\det T' = \frac{\partial(x_1, \ldots, x_n)}{\partial(u_1, \ldots, u_n)},$$

then the change of variables formula takes the appealing form

$$\int_{T(Q)} f(\mathbf{x})\, dx_1 \cdots dx_n = \int_Q f(T(\mathbf{u})) \left| \frac{\partial(x_1, \ldots, x_n)}{\partial(u_1, \ldots, u_n)} \right| du_1 \cdots du_n. \tag{20}$$

With the abbreviations

$$T(\mathbf{u}) = \mathbf{x}(\mathbf{u}), \qquad d\mathbf{x} = dx_1 \cdots dx_n, \qquad d\mathbf{u} = du_1 \cdots du_n,$$

and

$$\frac{d\mathbf{x}}{d\mathbf{u}} = \frac{\partial(x_1, \ldots, x_n)}{\partial(u_1, \ldots, u_n)},$$

the above formula takes the simple form

$$\int_{T(Q)} f(\mathbf{x})\, d\mathbf{x} = \int_Q f(\mathbf{x}(\mathbf{u})) \left| \frac{d\mathbf{x}}{d\mathbf{u}} \right| d\mathbf{u}.$$

This form of the change of variables formula is particularly pleasant because it reads the same as the familar substitution rule for single-variable integrals (Theorem 1.7), taking into account the change of sign in a single-variable integral which results from the interchange of its limits of integration.

This observation leads to an interpretation of the change of variables formula as the result of a "mechanical" substitution procedure. Suppose T is a differentiable mapping from uv-space to xy-space. In the integral $\int_{T(Q)} f(x, y)\, dx\, dy$ we want to make the substitutions

$$dx = \frac{\partial x}{\partial u}\, du + \frac{\partial x}{\partial v}\, dv, \qquad dy = \frac{\partial y}{\partial u}\, du + \frac{\partial y}{\partial v}\, dv \tag{21}$$

suggested by the chain rule. In formally multiplying together these two "differential forms," we agree to the conventions

$$du\, du = dv\, dv = 0 \qquad \text{and} \qquad dv\, du = -du\, dv, \tag{22}$$

for *no* other reason than that they are necessary if we are to get the "right" answer. We then obtain

$$\begin{aligned} dx\, dy &= \left(\frac{\partial x}{\partial u}\, du + \frac{\partial x}{\partial v}\, dv\right)\left(\frac{\partial y}{\partial u}\, du + \frac{\partial y}{\partial v}\, dv\right) \\ &= \frac{\partial x}{\partial u}\frac{\partial y}{\partial u}\, du\, du + \frac{\partial x}{\partial u}\frac{\partial y}{\partial v}\, du\, dv \\ &\quad + \frac{\partial x}{\partial v}\frac{\partial y}{\partial u}\, dv\, du + \frac{\partial x}{\partial v}\frac{\partial y}{\partial v}\, dv\, dv \\ &= \left(\frac{\partial x}{\partial u}\frac{\partial y}{\partial v} - \frac{\partial x}{\partial v}\frac{\partial y}{\partial u}\right) du\, dv, \\ dx\, dy &= \frac{\partial(x, y)}{\partial(u, v)}\, du\, dv. \end{aligned}$$

Note that, to within sign, this result agrees with formula (20), according to which $dx\, dy$ should be replaced by $|\partial(x, y)/\partial(u, v)|\, du\, dv$ when the integral is transformed.

Thus the change of variables. $(x, y) = T(u, v)$ in a 2-dimensional integral $\int_{T(Q)} f(x, y)\, dx\, dy$ may be carried out in the following way. First replace x and y in $f(x, y)$ by their expressions in terms of u and v, then multiply out $dx\, dy$ as above using (21) and (22), and finally replace the coefficient of $du\, dv$, which is obtained, by its absolute value,

For example, in a change from rectangular to polar coordinates, we have $x = r \cos \theta$ and $y = r \sin \theta$, so

$$\begin{aligned} dx\, dy &= (\cos \theta\, dr - r \sin \theta\, d\theta)(\sin \theta\, dr + r \cos \theta\, d\theta) \\ &= \cos \theta \sin \theta\, dr\, dr + r \cos^2 \theta\, dr\, d\theta \\ &\quad - r \sin^2 \theta\, d\theta\, dr - r^2 \sin \theta \cos \theta\, d\theta\, d\theta \\ &= r\, (\cos^2 \theta + \sin^2 \theta)\, dr\, d\theta, \\ dx\, dy &= r\, dr\, d\theta, \end{aligned}$$

where the sign is correct because $r \geqq 0$.

The relation $dv\,du = -du\,dv$ requires a further comment. Once having agreed upon u as the first coordinate and v as the second one, we agree that

$$\int_Q g(u, v)\,du\,dv = \int_Q g,$$

while

$$\int_Q g(u, v)\,dv\,du = -\int_Q g.$$

So, in writing multidimensional integrals with differential notation, the order of the differentials in the integrand makes a difference (of sign). This should be contrasted with the situation in regards to iterated integrals, such as

$$\int_a^b \left(\int_c^d g(u, v)\,dv\right) du \qquad \text{or} \qquad \int_c^d \left(\int_a^b g(u, v)\,du\right) dv,$$

in which the order of the differentials indicates the order of integration. If parentheses are consistently used to indicate iterated integrals, no confusion between these two different situations should arise.

The substitution process for higher-dimensional integrals is similar. For example, the student will find it instructive to verify that, with the substitution

$$dx = \frac{\partial x}{\partial u}\,du + \frac{\partial x}{\partial v}\,dv + \frac{\partial x}{\partial w}\,dw,$$

$$dy = \frac{\partial y}{\partial u}\,du + \frac{\partial y}{\partial v}\,dv + \frac{\partial y}{\partial w}\,dw,$$

$$dz = \frac{\partial z}{\partial u}\,du + \frac{\partial z}{\partial v}\,dv + \frac{\partial z}{\partial w}\,dw,$$

formal multiplication, using the relations

$$du\,du = dv\,dv = dw\,dw = 0, \qquad dv\,du = -du\,dv, \qquad dw\,dv = -dv\,dw,$$
$$dw\,dx = -dx\,dw,$$

yields

$$dx\,dy\,dz = \frac{\partial(x, y, z)}{\partial(u, v, w)}\,du\,dv\,dw \tag{23}$$

(Exercise 5.20). These matters will be discussed more fully in Chapter V.

Exercises

5.1 Let A be a bounded subset of $\mathscr{R}^k$, with $k < n$. If $f: \mathscr{R}^k \to \mathscr{R}^n$ is a $\mathscr{C}^1$ mapping, show that $f(A)$ is negligible. *Hint:* Let Q be a cube in $\mathscr{R}^k$ containing A, and $\mathscr{P} = \{Q_1, \ldots, Q_{N^k}\}$ a partition of Q into N^k cubes with edgelength r/N, where r is the length of the edges of Q. Since f is $\mathscr{C}^1$ on Q, there exists c such that $|f(\mathbf{x}) - f(\mathbf{y})| \leqq c|\mathbf{x} - \mathbf{y}|$ for all $\mathbf{x}, \mathbf{y} \in Q$. It follows that $v(f(Q_i)) \leqq (cr/N)^n$.

5.2 Show that A is contented if and only if, given $\varepsilon > 0$, there exist contented sets A_1 and A_2 such that $A_1 \subset A \subset A_2$ and $v(A_2 - A_1) < \varepsilon$.

5.3 (a) If $A \subset \mathscr{R}^n$ is negligible and $T: \mathscr{R}^n \to \mathscr{R}^n$ is a $\mathscr{C}^1$ mapping, show that $T(A)$ is negligible. *Hint:* Apply the fact that, given an interval Q containing A, there exists $c > 0$ such that $|T(\mathbf{x}) - T(\mathbf{y})| < c|\mathbf{x} - \mathbf{y}|$ for all $\mathbf{x}, \mathbf{y} \in Q$.
(b) If A is a contented set in $\mathscr{R}^n$ and $T: \mathscr{R}^n \to \mathscr{R}^n$ is a $\mathscr{C}^1$ mapping which is $\mathscr{C}^1$-invertible on the interior of A, show that $T(A)$ is contented. *Hint:* Show that $\partial(T(A)) \subset T(\partial A)$, and then apply (a).

5.4 Establish the conclusion of Theorem 5.5 under the hypothesis that Q is a contented set such that T is $\mathscr{C}^1$-invertible on the interior of Q. *Hint:* Let $Q_1, \ldots, Q_k$ be nonoverlapping intervals interior to Q, and $K = Q - \bigcup_{i=1}^{k} Q_i$. Then

$$\sum_{i=1}^{k} \int_{T(Q_i)} f = \sum_{i=1}^{k} \int_{Q_i} (f \circ T)|\det T'|$$

by Theorem 5.5. Use the hint for Exercise 5.3(a) to show that $v(T(K))$ can be made arbitrarily small by making $v(K)$ sufficiently small.

5.5 Consider the n-dimensional solid ellipsoid

$$E = \left\{\mathbf{x} \in \mathscr{R}^n : \sum_{i=1}^{k} \frac{x_i^2}{a_i^2} \leq 1\right\}.$$

Note that E is the image of the unit ball B^n under the mapping $T: \mathscr{R}^n \to \mathscr{R}^n$ defined by

$$T(x_1, \ldots, x_n) = (a_1 x_1, \ldots, a_n x_n).$$

Apply Theorem 5.1 to show that $v(E) = a_1 a_2 \cdots a_n v(B^n)$. For example, the volume of the 3-dimensional ellipsoid $x^2/a^2 + y^2/b^2 + z^2/c^2 \leq 1$ is $\frac{4}{3}\pi abc$.

5.6 Let R be the solid torus in $\mathscr{R}^3$ obtained by revolving the circle $(y - a)^2 + z^2 \leq b^2$, in the yz-plane, about the z-axis. Note that the mapping $T: \mathscr{R}^3 \to \mathscr{R}^3$ defined by

$$x = (a + w \cos v)\cos u,$$
$$y = (a + w \cos v)\sin u,$$
$$z = w \sin v,$$

maps the interval $Q = \{(u, v, w) : u, v \in [0, 2\pi] \text{ and } w \in [0, b]\}$ onto this torus. Apply the change of variables formula to calculate its volume.

5.7 Use the substitution $u = x - y$, $v = x + y$ to evaluate

$$\iint_R e^{(x-y)/(x+y)}\, dx\, dy,$$

where R is the first quadrant region bounded by the coordinate axes and the line $x + y = 1$.

5.8 Calculate the volume of the region in $\mathscr{R}^3$ which is above the xy-plane, under the paraboloid $z = x^2 + y^2$, and inside the elliptic cylinder $x^2/9 + y^2/4 = 1$. Use elliptical coordinates $x = 3r \cos \theta$, $y = 2r \sin \theta$.

5.9 Consider the solid elliptical rod bounded by the xy-plane, the plane $z = \alpha x + \beta y + h$ through the point $(0, 0, h)$ on the positive z-axis, and the elliptical cylinder $x^2/a^2 + y^2/b^2 = 1$. Show that its volume is πabh (independent of α and β). *Hint:* Use elliptical coordinates $x = ar \cos \theta$, $y = br \sin \theta$.

5.10 Use "double polar" coordinates defined by the equations

$$x_1 = r \cos \theta, \qquad x_2 = r \sin \theta,$$
$$x_3 = \rho \cos \varphi, \qquad x_4 = \rho \sin \varphi,$$

to show that the volume of the unit ball $B^4 \subset \mathscr{R}^4$ is $\frac{1}{2}\pi^2$.

5.11 Use the transformation defined by

$$u = \frac{2x}{x^2 + y^2}, \qquad v = \frac{2y}{x^2 + y^2}$$

to evaluate

$$\iint_R \frac{dx\, dy}{(x^2 + y^2)^2},$$

where R is the region in the first quadrant of the xy-plane which is bounded by the circles $x^2 + y^2 = 6x$, $x^2 + y^2 = 4x$, $x^2 + y^2 = 8y$, $x^2 + y^2 = 2y$.

5.12 Use the transformation T from $r\theta t$-space to xyz-space, defined by

$$x = \frac{r}{t} \cos \theta, \qquad y = \frac{r}{t} \sin \theta, \qquad z = r^2,$$

to find the volume of the region R which lies between the paraboloids $z = x^2 + y^2$ and $z = 4(x^2 + y^2)$, and also between the planes $z = 1$ and $z = 4$.

5.13 Consider a spherical ball of radius a centered at the origin, with a spherically symmetric density function $d(\mathbf{x}) = g(\rho)$.

(a) Use spherical coordinates to show that its mass is

$$M = 4\pi \int_0^a \rho^2 g(\rho)\, d\rho.$$

(b) Show that the gravitational force exerted by this ball on a point mass m located at the point $\boldsymbol{\xi} = (0, 0, c)$, with $c > a$, is the same as if its mass were all concentrated at the origin, that is,

$$F = \frac{\gamma M m}{c^2}.$$

Hint: Because of symmetry considerations, you may assume that the force is directed toward $\mathbf{0}$, so only its z-component need be explicitly computed. In spherical coordinates (see Fig. 4.37),

$$F_z = -\int_0^a \int_0^\pi \int_0^{2\pi} \frac{\gamma m g(\rho) \cos \alpha}{w^2} \rho^2 \sin \varphi\, d\theta\, d\varphi\, d\rho \qquad \text{(why?)}.$$

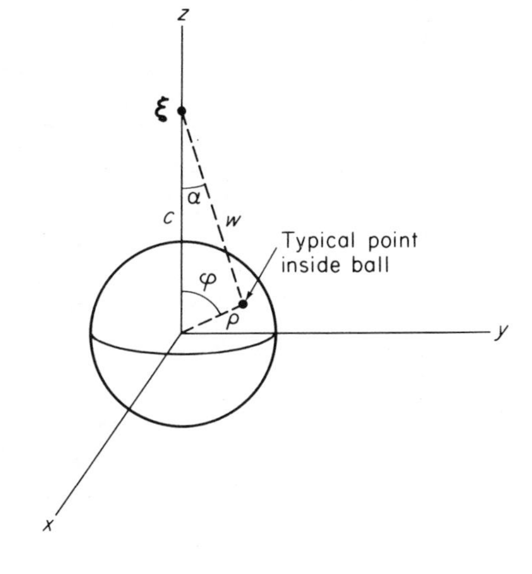

Figure 4.37

Change the second variable of integration from φ to w, using the relations

$$w^2 = \rho^2 + c^2 - 2\rho c \cos \varphi \qquad \text{(law of cosines)},$$

so $2w\,dw = 2\rho c \sin \varphi \, d\varphi$, and $w \cos \alpha + \rho \cos \varphi = c$ (why?).

5.14 Consider now a spherical shell, centered at the origin and defined by $a \leqq \rho \leqq b$, with a spherically symmetric density function. Show that this shell exerts *no* gravitational force on a point mass m located at the point $(0, 0, c)$ *inside* it (that is, $c < a$). *Hint:* The computation is the same as in the previous problem, except for the limits of integration on ρ and w.

5.15 *Volume of Revolution.* Suppose $g : [a, b] \to \mathscr{R}$ is a positive-valued continuous function, and let

$$A = \{(x, z) \in \mathscr{R}^2 : 0 \leqq x \leqq g(z),\ z \in [a, b]\}.$$

Denote by C the set generated by revolving A about the z-axis (Fig. 4.38); that is,

$$C = \{(x, y, z) \in \mathscr{R}^3 : (x^2 + y^2)^{1/2} \leqq g(z),\ z \in [a, b]\}.$$

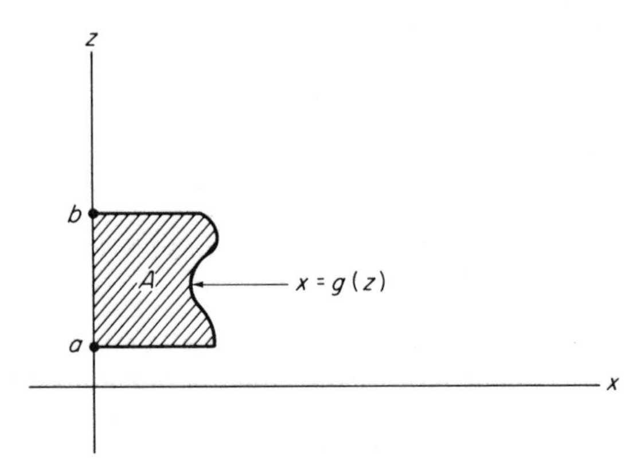

Figure 4.38

We wish to calculate $v(C) = \int_C 1$. Noting that C is the image of the contented set

$$B = \{(r, \theta, z) \in \mathscr{R}^3 : 0 \leqq r \leqq g(z),\ \theta \in [0, 2\pi],\ z \in [a, b]\}$$

under the *cylindrical coordinates mapping* $T: \mathscr{R}^3 \to \mathscr{R}^3$ defined by $T(r, \theta, z) = (r \cos \theta, r \sin \theta, z)$, show that the change of variable and repeated integrals theorems apply to give

$$v(C) = \pi \int_a^b [g(z)]^2 \, dz.$$

5.16 Let A be a contented set in the right half xz-plane $x > 0$. Define $\hat{x}$, the x-coordinate of the *centroid* of A, by $\hat{x} = [1/v(A)] \iint_A x \, dx \, dz$. If C is the set obtained by revolving A about the z-axis, that is,

$$C = \{(x, y, z) \in \mathscr{R}^3 : ((x^2 + y^2)^{1/2}, z) \in A\},$$

then *Pappus' theorem* asserts that

$$v(C) = 2\pi\hat{x}v(A),$$

that is, that the volume of C is the volume of A multiplied by the distance traveled by the centroid of A. Note that C is the image under the cylindrical coordinates map of the set

$$B = \{(r, \theta, z) \in \mathscr{R}^3 : (r, z) \in A \text{ and } \theta \in [0, 2\pi]\}.$$

Apply the change of variables and iterated integrals theorems to deduce Pappus' theorem.

The purpose of each of the next three problems is to verify the formulas

$$\alpha_{2m} = \frac{\pi^m}{m!} \quad \text{and} \quad \alpha_{2m+1} = \frac{2^{m+1}\pi^m}{1 \cdot 3 \cdot 5 \cdot \cdots \cdot (2m+1)} \tag{$*$}$$

given in this section, for the volume α_n of the unit n-ball $B_1{}^n \subset \mathscr{R}^n$.

5.17 (a) Apply Cavalieri's principle and appropriate substitutions to obtain

$$\alpha_n = \int_{B^n} 1 = \int_{-1}^{1} v(B^{n-1}_{(1-t^2)^{1/2}})\,dt$$

$$= 2\alpha_{n-1}\int_0^1 (1-t^2)^{(n-1)/2}\,dt = 2\alpha_{n-1}\int_0^{\pi/2} \sin^n\theta\,d\theta,$$

$$\alpha_n = 2\alpha_{n-1}I_n.$$

Conclude that $\alpha_n = 4\alpha_{n-2}I_nI_{n-1}$.
(b) Deduce from Exercise 1.8 that $I_nI_{n-1} = \pi/2n$ if $n \geqq 1$. Hence

$$\alpha_n = \frac{2\pi}{n}\alpha_{n-2} \quad \text{if} \quad n \geqq 2. \tag{$\#$}$$

(c) Use the recursion formula ($\#$) to establish, by separate inductions on m, formulas ($*$), starting with $\alpha_2 = \pi$ and $\alpha_3 = 4\pi/3$.

5.18 In this problem we obtain the same recursion formula without using the formulas for I_{2n} and I_{2n+1}. Let

$$B^2 = \{(x_1, x_2) \in \mathscr{R}^2 : x_1{}^2 + x_2{}^2 \leqq 1\}$$

and $Q = \{(x_3, \ldots, x_n) \in \mathscr{R}^{n-2} : \text{each } |x_i| \leqq 1\}$. Then $B^n \subset B^2 \times Q$. Let $\varphi : B^2 \times Q \to \mathscr{R}$ be the characteristic function of B^n. Then

$$\alpha_n = \int_{B^2}\left(\int_Q \varphi(x_1, \ldots, x_n)\,dx_3 \ldots dx_n\right)dx_1\,dx_2.$$

Note that, if $(x_1, x_2) \in B^2$ is a fixed point, then φ, as a function of $(x_3, \ldots, x_n)$, is the characteristic function of $B^{n-2}_{(1-x_1{}^2-x_2{}^2)^{1/2}}$. Hence

$$\int_Q \varphi(x_1, \ldots, x_n)\,dx_3 \ldots dx_n = (1 - x_1{}^2 - x_2{}^2)^{(n-2)/2}\alpha_{n-2}.$$

Now introduce polar coordinates in $\mathscr{R}^2_{x_1x_2}$ to show that

$$\int_{B^2}(1 - x_1{}^2 - x_2)^{(n-2)/2}\,dx_1\,dx_2 = \frac{2\pi}{n}$$

so that $\alpha_n = (2\pi/n)\alpha_{n-2}$ as before.

5.19 The n-dimensional spherical coordinates mapping $T: \mathscr{R}^n \to \mathscr{R}^n$ is defined by

$$\begin{aligned} x_1 &= \rho \cos \varphi_1, \\ x_2 &= \rho \sin \varphi_1 \cos \varphi_2, \\ x_3 &= \rho \sin \varphi_1 \sin \varphi_2 \cos \varphi_3, \\ &\vdots \\ x_{n-1} &= \rho \sin \varphi_1 \cdots \sin \varphi_{n-2} \cos \theta, \\ x_n &= \rho \sin \varphi_1 \cdots \sin \varphi_{n-2} \sin \theta, \end{aligned}$$

and maps the interval

$$Q = \{(\rho, \varphi_1, \ldots, \varphi_{n-2}, \theta) \in \mathscr{R}^n : \rho \in [0, 1],\ \varphi_i \in [0, \pi],\ \theta \in [0, 2\pi]\}$$

onto the unit ball B^n.

(a) Prove by induction on n that

$$|\det T'| = \rho^{n-1} \sin^{n-2} \varphi_1 \sin^{n-3} \varphi_2 \cdots \sin^2 \varphi_{n-3} \sin \varphi_{n-2}.$$

(b) Then

$$\begin{aligned} \alpha_n &= \int_{B^n} 1 = \int_Q |\det T'| \\ &= \frac{2\pi}{n} \prod_{k=1}^{n-2} \left[\int_0^\pi \sin^k \varphi \, d\varphi\right], \end{aligned}$$

$$\alpha_n = \frac{2\pi}{n} I_1' I_2' \cdots I_{n-2}',$$

where $I_k' = \int_0^\pi \sin^k \varphi \, d\varphi$. Now use the fact that

$$I_{k-1}' I_k' = \frac{2\pi}{k} \qquad \text{and} \qquad I_{2m-1}' = \frac{2 \cdot 4 \cdot \cdots \cdot (2m-2)}{3 \cdot 5 \cdot \cdots \cdot (2m-1)}$$

(by Exercise 1.8) to establish formulas (*).

5.20 Verify formulas (23) of this section.

6 IMPROPER INTEGRALS AND ABSOLUTELY INTEGRABLE FUNCTIONS

Thus far in this chapter on integration, we have confined our attention to bounded functions that have bounded support. This was required by our definition of the integral of a nonnegative function as the volume of its ordinate set (if contented)—if $f: \mathscr{R}^n \to \mathscr{R}$ is to be integrable, then its ordinate set must be contented. But every contented set is a priori bounded, so f must be bounded and have bounded support.

However there are certain types of unbounded subsets of $\mathscr{R}^n$ with which it is natural and useful to associate a notion of volume, and analogously functions which are either unbounded or have unbounded support, but for which it is nevertheless natural and useful to define an integral. The following two examples illustrate this.

Example 1 Let $f: [1, \infty) \to \mathscr{R}$ be defined by $f(x) = 1/x^2$. Let A denote the ordinate set of f, and A_n the part of A lying above the closed interval $[1, n]$ (Fig. 4.39). Then

$$v(A_n) = \int_1^n \frac{dx}{x^2} = 1 - \frac{1}{n}.$$

Since $A = \bigcup_{n=1}^{\infty} A_n$ and $\lim_{n\to\infty} v(A_n) = 1$, it would seem natural to say that

$$v(A) = \int_1^{\infty} \frac{dx}{x^2} = 1$$

despite the fact that f does not have bounded support. (Of course what we really mean is that our definitions of volume and/or the integral ought to be extended so as to make this true.)

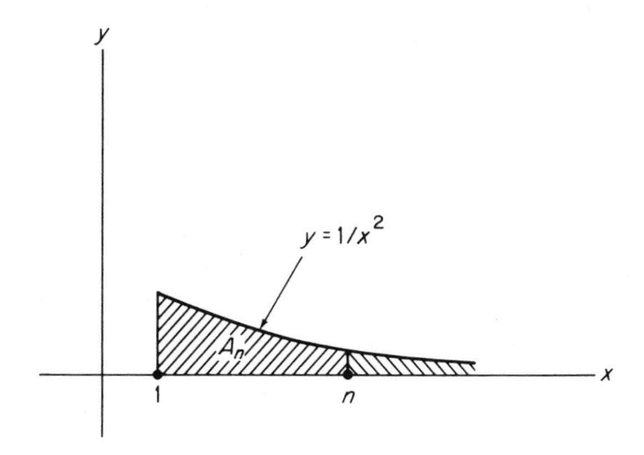

Figure 4.39

Example 2 Let f be the unbounded function defined on $(0, 1]$ by $f(x) = 1/\sqrt{x}$. Let A denote the ordinate set of f, and A_n the part of A lying above the closed interval $[1/n, 1]$ (Fig. 4.40). Then

$$v(A_n) = \int_{1/n}^{1} \frac{dx}{x^{1/2}} = 2 - \frac{2}{n^{1/2}}.$$

Since $A = \bigcup_{n=1}^{\infty} A_n$ and $\lim_{n\to\infty} v(A_n) = 2$, it would seem natural to say that

$$v(A) = \int_0^1 \frac{dx}{x^{1/2}} = 2.$$

These two examples indicate both the need for some extension (to unbounded situations) of our previous definitions, and a possible method of making this extension. Given a function $f: U \to \mathscr{R}$, with either f or U (or both) unbounded,

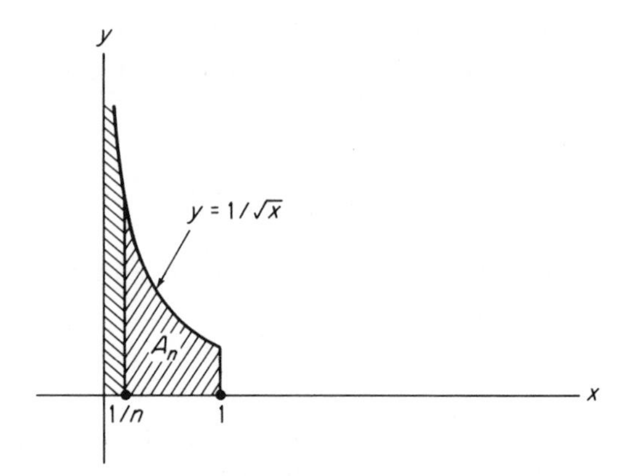

Figure 4.40

we might choose a sequence $\{A_n\}_{n=1}^{\infty}$ of subsets of U which "fill up" U in some appropriate sense, on each of which f is integrable, and then define

$$\int_U f = \lim_{n\to\infty} \int_{A_n} f$$

provided that this limit exists. Of course we would have to verify that $\int_U f$ is thereby well-defined, that is, that its value is independent of the particular chosen sequence $\{A_n\}_1^{\infty}$. Partly because of this problem of well-definition, we shall formulate our initial definition of $\int_U f$ in a slightly different manner, and then return later to the interpretation of $\int_U f$ as a sequential limit (as above).

We deal first with the necessity for $f: U \to \mathscr{R}$ to be integrable on enough subsets of U to "fill up" U. We say that f is *locally integrable* on U if and only if f is integrable on every compact (closed and bounded) contented subset of U. For instance the function $f(x) = 1/x^2$ of Example 1 is not integrable on $(1, \infty)$, but is locally integrable there. Similarly the function $f(x) = 1/x^{1/2}$ of Example 2 is not integrable on $(0, 1)$, but is locally integrable there.

It will be convenient to confine our attention to open domains of definition. So let $f: U \to \mathscr{R}$ be a locally integrable function on the *open* set $U \subset \mathscr{R}^n$. We then say that f is *absolutely integrable* on U if and only if, given $\varepsilon > 0$, there exists a compact contented subset B_ε of U such that

$$\left| \int_A f - \int_{B_\varepsilon} f \right| < \varepsilon$$

for every compact contented set $A \subset U$ which contains B_ε.

The student should verify that the sum of two absolutely integrable functions on U is absolutely integrable, as is a constant multiple of one, so the set of all absolutely integrable functions on U is a vector space. This elementary fact will be used in the proof of the following proposition, which provides the reason for our terminology.

Proposition 6.1 If the function $f: U \to \mathscr{R}$ is absolutely integrable, then so is its absolute value $|f|$.

We will later prove the converse, so f is absolutely integrable if and only if $|f|$ is.

PROOF Writing $f = f^+ - f^-$, it suffices by the above remark to show that f^+ and f^- are both absolutely integrable, since $|f| = f^+ + f^-$. We consider f^+. Given $\varepsilon > 0$, choose B_ε as in the definition, such that $A \supset B_\varepsilon$ implies

$$\left| \int_A f - \int_{B_\varepsilon} f \right| < \varepsilon$$

if A is a compact contented subset of U. It will suffice to show that $A \supset B_\varepsilon$ also implies that

$$\left| \int_A f^+ - \int_{B_\varepsilon} f^+ \right| < \varepsilon.$$

Given a compact contented set A with $B_\varepsilon \subset A \subset U$, define

$$A^+ = B_\varepsilon \cup \{\mathbf{x} \in A : f(\mathbf{x}) \geqq 0\},$$

so that $f(\mathbf{x}) = f^+(\mathbf{x})$ for $\mathbf{x} \in A^+ - B_\varepsilon$, and

$$\int_{A - B_\varepsilon} f^+ = \int_{A^+ - B_\varepsilon} f^+$$

because $f^+(\mathbf{x}) = 0$ if $\mathbf{x} \in A - A^+$. Then

$$\begin{aligned} \left| \int_A f^+ - \int_{B_\varepsilon} f^+ \right| &= \left| \int_{A - B_\varepsilon} f^+ \right| \\ &= \left| \int_{A^+ - B_\varepsilon} f^+ \right| \\ &= \left| \int_{A^+ - B_\varepsilon} f \right| \\ &= \left| \int_{A^+} f - \int_{B_\varepsilon} f \right| < \varepsilon \end{aligned}$$

as desired, because A^+ is a compact contented set such that $B_\varepsilon \subset A^+ \subset U$. ∎

The number I, which the following theorem associates with the absolutely integrable function f on the open set $U \subset \mathscr{R}^n$, will be called the *improper* (Riemann) integral of f on U. It will be temporarily denoted by $I_U f$ (until we have verified that $I_U f = \int_U f$ in case f is integrable on U).

The following definition will simplify the interpretation of $I_U f$ as a sequential limit of integrals of f on compact contented subsets of U. The sequence $\{A_k\}_1^\infty$

of compact contented subsets of U is called an *approximating sequence* for U if (a) $A_{k+1} \supset A_k$ for each $k \geqq 1$, and (b) $U = \bigcup_{k=1}^{\infty} \operatorname{int} A_k$.

Theorem 6.2 Suppose f is absolutely integrable on the open set $U \subset \mathscr{R}^n$. Then there exists a number $I = I_U f$ with the property that, given $\varepsilon > 0$, there exists a compact contented set C_ε such that

$$\left| I - \int_A f \right| < \varepsilon$$

for every compact contented subset A of U containing C_ε. Furthermore

$$I_U f = \lim_{k \to \infty} \int_{A_k} f$$

for every approximating sequence $\{A_k\}_1^\infty$ for U.

PROOF To start with, we choose a (fixed) approximating sequence $\{C_k\}_1^\infty$ for U (whose existence is provided by Exercise 6.16). Given $\eta > 0$, the Heine–Borel theorem (see the Appendix) implies that $C_k \supset B_{\eta/2}$ if k is sufficiently large. Here $B_{\eta/2}$ is the compact contented subset of U (provided by the definition) such that $B_{\eta/2} \subset A \subset U$ implies that

$$\left| \int_A f - \int_{B_{\eta/2}} f \right| < \frac{\eta}{2}.$$

Consequently, if k and l are sufficiently large, it follows that

$$\left| \int_{C_k} f - \int_{C_l} f \right| \leqq \left| \int_{C_k} f - \int_{B_{\eta/2}} f \right| + \left| \int_{C_l} f - \int_{B_{\eta/2}} f \right| < \eta.$$

Thus $\{\int_{C_k} f\}_1^\infty$ is a Cauchy sequence of numbers, and therefore has a limit I (see the Appendix).

Now let $C_\varepsilon = B_{\varepsilon/3}$. Given a compact contented subset A of U which contains C_ε, choose a fixed k sufficiently large that $C_k \supset C_\varepsilon$ and $|I - \int_{C_k} f| < \varepsilon/3$. Then

$$\left| I - \int_A f \right| \leqq \left| I - \int_{C_k} f \right| + \left| \int_{C_k} f - \int_{C_\varepsilon} f \right| + \left| \int_{C_\varepsilon} f - \int_A f \right|$$

$$< \frac{\varepsilon}{3} + \frac{\varepsilon}{3} + \frac{\varepsilon}{3} = \varepsilon$$

as desired. This completes the proof of the first assertion.

Now let $\{A_k\}_1^\infty$ be an arbitrary approximating sequence for U. Given $\delta > 0$, choose K sufficiently large that $C_k \supset C_\varepsilon$, for all $k \geqq K$. Then what we have just proved gives $|I - \int_{A_k} f| < \varepsilon$ for all $k \geqq K$, so it follows that

$$I = \lim_{k \to \infty} \int_{A_k} f. \qquad \blacksquare$$

We can now show that $I_U f$ and $\int_U f$ are equal when both are defined. Note that, if f is integrable on the open set $U \subset \mathscr{R}^n$, then it follows from Exercise 3.1 that f is locally integrable on U (why?).

Theorem 6.3 If $U \subset \mathscr{R}^n$ is open and $f: U \to \mathscr{R}$ is integrable, then f is absolutely integrable, and

$$I_U f = \int_U f.$$

PROOF Let M be an upper bound for the values of $|f(\mathbf{x})|$ on U. Given $\varepsilon > 0$, let B_ε be a compact contented subset of U such that $v(U - B_\varepsilon) < \varepsilon/M$. If A is a compact contented subset of U which contains B_ε, then

$$\left|\int_A f - \int_{B_\varepsilon} f\right| = \left|\int_{A-B_\varepsilon} f\right| \leqq \int_{A-B_\varepsilon} |f|$$
$$\leqq \int_{U-B_\varepsilon} f \leqq Mv(U - A_\varepsilon) < \varepsilon,$$

so it follows that f is absolutely integrable on U.

If in addition A contains the set C_ε of Theorem 6.2, then

$$\left|I_U f - \int_U f\right| \leqq \left|I_U f - \int_A f\right| + \left|\int_U f - \int_A f\right|$$
$$< \varepsilon + \int_{U-A} |f|$$
$$\leqq \varepsilon + \int_{U-B_\varepsilon} |f| < 2\varepsilon.$$

This being true for all $\varepsilon > 0$, it follows that $I_U f = \int_U f$. ∎

In order to proceed to compute improper integrals by taking limits over approximating sequences as in Theorem 6.2, we need an effective test for absolute integrability. For *nonnegative* functions, this is easy to provide. We say that the improper integral $\int_U f$ is *bounded* (whether or not it exists, that is, whether or not f is absolutely integrable on U) if and only if there exists $M > 0$ such that

$$\left|\int_A f\right| \leqq M \tag{1}$$

for every compact contented subset A of U.

Theorem 6.4 Suppose that the nonnegative function f is locally integrable on the open set U. Then f is absolutely integrable on U if and only if $\int_U f$ is bounded, in which case $\int_U f$ is the least upper bound of the values $\int_A f$, for all compact contented sets $A \subset U$.

PROOF Suppose first that (1) holds, and denote by I the least upper bound of $\{\int_A f\}$ for all compact contented $A \subset U$. Given $\varepsilon > 0$, $I - \varepsilon$ is not an upper bound for the numbers $\{\int_A f\}$, so there exists a compact contented set $B_\varepsilon \subset U$ such that

$$\int_{B_\varepsilon} f > I - \varepsilon.$$

If A is any compact contented set such that $B_\varepsilon \subset A \subset U$ then, since $f \geqq 0$, we have

$$I \geqq \int_A f \geqq \int_{B_\varepsilon} f > I - \varepsilon,$$

so it follows that both

$$\left| \int_A f - \int_{B_\varepsilon} f \right| < \varepsilon \qquad \text{and} \qquad \left| I - \int_A f \right| < \varepsilon.$$

The first inequality implies that f is absolutely integrable, and then the second implies that $\int_U f = I$ as desired (using Theorem 6.2).

Conversely, if f is absolutely integrable on U, then it is easily seen that $\int_A f \leqq \int_U f$ for every compact contented subset U of A, so we can take $M = \int_U f$ in (1). ∎

Corollary 6.5 Suppose the nonnegative function f is locally integrable on the open set U, and let $\{A_k\}_1^\infty$ be an approximating sequence for U. Then f is absolutely integrable with

$$\int_U f = \lim_{k \to \infty} \int_{A_k} f, \tag{2}$$

provided that this limit exists (and is finite).

PROOF The fact that the monotone increasing sequence of numbers $\{\int_{A_k} f\}_1^\infty$ is bounded (because it has a finite limit) implies easily that $\int_U f$ is bounded, so f is absolutely integrable by Theorem 6.4. Hence (2) follows from Theorem 6.2. ∎

As an application, we will now generalize Examples 1 and 2 to $\mathscr{R}^n$. Let A denote the solid annular region $B_b^n - \operatorname{int} B_a^n \subset \mathscr{R}^n$ (see Fig. 4.41), and let f be a spherically symmetric continuous function on A. That is,

$$f \circ T(\rho, \varphi_1, \ldots, \varphi_{n-2}, \theta) = g(\rho)$$

for some function g, where $T : \mathscr{R}^n \to \mathscr{R}^n$ is the n-dimensional spherical coordinates mapping of Exercise 5.19. By part (a) of that exercise, and the change of variables theorem, we have

$$\begin{aligned} \int_A f &= \int_a^b \int_0^\pi \cdots \int_0^\pi \int_0^{2\pi} g(\rho)\rho^{n-1} \sin^{n-2} \varphi_1 \cdots \sin \varphi_{n-2} \, d\theta \, d\varphi_1 \cdots d\varphi_{n-2} \, d\rho \\ &= \sigma_n \int_a^b g(\rho)\rho^{n-1} \, d\rho, \end{aligned} \tag{3}$$

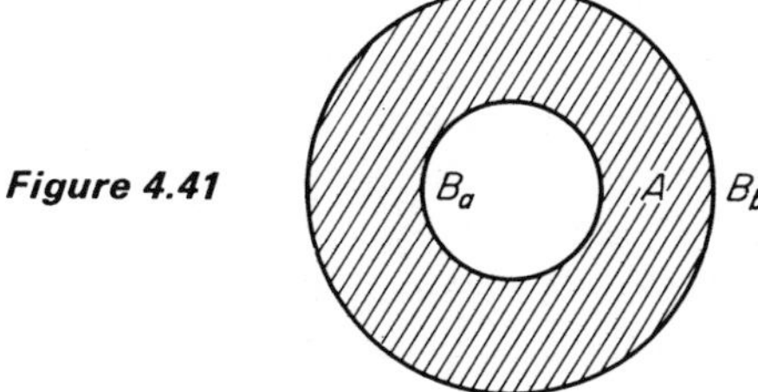

Figure 4.41

where

$$\sigma_n = \int_0^\pi \cdots \int_0^\pi \int_0^{2\pi} \sin^{n-2} \varphi_1 \cdots \sin \varphi_{n-2} \, d\theta \, d\varphi_1 \cdots d\varphi_{n-2}.$$

Example 3 Let $U = \mathscr{R}^n - B_1{}^n$ and $f(\mathbf{x}) = 1/|\mathbf{x}|^p$ where $p > n$. Writing $A_k = B_k{}^n - B_1{}^n$, Corollary 6.5 and (3) above give

$$\begin{aligned} \int_U f &= \lim_{k\to\infty} \int_{A_k} f \\ &= \lim_{k\to\infty} \sigma_n \int_1^k \rho^{n-p-1} \, d\rho \\ &= \lim_{k\to\infty} \frac{\sigma_n}{n-p} (k^{n-p} - 1) \\ &= \frac{\sigma_n}{p-n}, \end{aligned}$$

because $k^{n-p} \to 0$ since $n - p < 0$.

Example 4 Let U denote the interior of the unit ball with the origin deleted, and $f(\mathbf{x}) = 1/|\mathbf{x}|^p$ with $p < n$. So now the problem is that f, rather than U, is unbounded. Writing $A_k = B_1{}^n - B_{1/k}^n$, Corollary 6.5 and (3) give

$$\begin{aligned} \int_U f &= \lim_{k\to\infty} \int_{A_k} f \\ &= \lim_{k\to\infty} \sigma_n \int_{1/k}^1 \rho^{n-p-1} \, d\rho \\ &= \lim_{k\to\infty} \frac{\sigma_n}{n-p} [1 - (1/k)^{n-p}] \\ &= \frac{\rho_n}{n-p} \end{aligned}$$

because $p < n$.

For *nonnegative* locally integrable functions, Corollary 6.5 plays in practice the role of a "working definition." One need not worry in advance about whether the improper integral $\int_U f$ actually exists (that is, whether f is absolutely integrable on U). Simply choose a convenient approximating sequence $\{A_k\}_1^\infty$ for U, and compute $\lim_{k\to\infty} \int_{A_k} f$. If this limit is finite, then $\int_U f$ *does* exist, and its value is the obtained limit.

In the case of an arbitrary (not necessarily nonnegative) function, one must know in advance that f is absolutely integrable on U. Once this has been established, Theorem 6.2 enables us to proceed as above—choose an approximating sequence $\{A_k\}_1^\infty$ for U, and then compute $\lim_{k\to\infty} \int_{A_k} f$.

The simplest way to show that a given locally integrable function $f: U \to \mathscr{R}$ is absolutely integrable is to compare it with a function $g: U \to \mathscr{R}$ which is already known to be absolutely integrable, using the comparison test stated below. First we need the converse of Proposition 6.1.

Corollary 6.6 Let $f: U \to \mathscr{R}$ be locally integrable. If $|f|$ is absolutely integrable on U, then so is f.

PROOF Write $f = f^+ - f^-$, so $|f| = f^+ + f^-$. Since $|f|$ is absolutely integrable, $\int_U |f|$ is bounded (by Theorem 6.4). It follows immediately that $\int_U f^+$ and $\int_U f^-$ are bounded. Hence Theorem 6.4 implies that f^+ and f^- are both absolutely integrable on U, so $f = f^+ - f^-$ is also. ∎

The import of Corollary 6.6 is that, in practice, we need only to test *nonnegative* functions for absolute integrability.

Corollary 6.7 (Comparison Test) Suppose that f and g are locally integrable on U with $0 \leqq f \leqq g$. If g is absolutely integrable on U, then so is f.

PROOF Since g is absolutely integrable, $\int_U g$ is bounded (by Theorem 6.4). But then the fact that $0 \leqq f \leqq g$ implies immediately that $\int_U f$ is also bounded, so f is also absolutely integrable. ∎

Example 5 The functions

$$f(x) = \frac{1}{x^2+1}, \qquad f(x) = \frac{x}{x^3+1}, \qquad f(x) = e^{-x}, \qquad \text{and} \qquad f(x) = \frac{\sin x}{x^2}$$

are all absolutely integrable on $U = (1, \infty)$, by comparison with the absolutely integrable function $g(x) = 1/x^2$ of Examples 1 and 3. For the latter, we note that

$$\left|\frac{\sin x}{x^2}\right| \leqq \frac{1}{x^2}$$

so $|\sin x/x^2|$ is absolutely integrable by Corollary 6.7. But then it follows from Corollary 6.6 that $(\sin x)/x^2$ itself is absolutely integrable on $(1, \infty)$.

Similarly the functions

$$f(x) = \frac{1}{(x+1)^{1/2}} \qquad \text{and} \qquad f(x) = \frac{\cos x}{x^{1/2}}$$

are absolutely integrable on $U = (0, 1)$, by comparison with the function $g(x) = 1/x^{1/2}$ of Examples 2 and 4.

Next we want to define a different type of improper integral for the case of functions of a single variable. Suppose that f is locally integrable on the open interval (a, b), where possibly $a = -\infty$ or $b = +\infty$ or both. We want to define the improper integral denoted by

$$\int_a^b f(x)\, dx,$$

to be contrasted with the improper integral $\int_{(a,b)} f(x)\, dx$ which is defined if f is absolutely integrable on (a, b). It will turn out that $\int_a^b f(x)\, dx$ exists in some cases where f is not absolutely integrable on (a, b), but that $\int_a^b f(x)\, dx = \int_{(a,b)} f(x)\, dx$ if both exist.

The new improper integral $\int_a^b f(x)\, dx$ is obtained by restricting our attention to a single special approximating sequence for (a, b), instead of requiring that the same limit be obtained for all possible choices of the approximating sequence (as, by Theorem 6.2, is the case if f is absolutely integrable). Separating the four possible cases, we define

$$\int_a^b f = \lim_{n\to\infty} \int_{a+1/n}^{b-1/n} f, \qquad \int_{-\infty}^{\infty} f = \lim_{n\to\infty} \int_{-n}^{n} f,$$

$$\int_a^{\infty} f = \lim_{n\to\infty} \int_{a+1/n}^{n} f, \qquad \int_{-\infty}^{b} f = \lim_{n\to\infty} \int_{-n}^{b-1/n} f$$

(where a and b are finite), provided that the appropriate limit exists (and is finite) and say that the integral *converges*; otherwise it *diverges*.

Example 6 The integral $\int_{-\infty}^{\infty} [(1+x)/(1+x^2)]\, dx$ converges, because

$$\begin{aligned}
\int_{-\infty}^{\infty} \frac{1+x}{1+x^2}\, dx &= \lim_{n\to\infty} \int_{-n}^{n} \frac{1+x}{1+x^2}\, dx \\
&= \lim_{n\to\infty} \left[\arctan x + \tfrac{1}{2}\log(1+x^2)\right]_{-n}^{n} \\
&= 2 \lim_{n\to\infty} \arctan n \\
&= \pi.
\end{aligned}$$

However $f(x) = (1 + x)/(1 + x^2)$ is *not* absolutely integrable on $(-\infty, \infty) = \mathscr{R}$. If it were, we would have to obtain the same limit π for any approximating sequence $\{A_n\}_1^\infty$ for $\mathscr{R}$. But, taking $A_n = [-n, 2n]$, we obtain

$$\begin{aligned}\lim_{n\to\infty}\int_{A_n} f &= \lim_{n\to\infty}\int_{-n}^{2n}\frac{1+x}{1+x^2}\,dx\\ &= \lim_{n\to\infty}\left[\arctan x + \tfrac{1}{2}\log(1+x^2)\right]_{-n}^{2n}\\ &= \lim_{n\to\infty}\arctan 2n - \lim_{n\to\infty}\arctan(-n)\\ &\quad + \lim_{n\to\infty}\frac{1}{2}\log\frac{1+4n^2}{1+n^2}\\ &= \pi + \log 2.\end{aligned}$$

Example 7 Consider $\int_0^\infty [(\sin x)/x]\,dx$. Since $\lim_{x\to 0}[(\sin x)/x] = 1$, $\sin x/x$ is continuous on $[0, 1]$. So we need only consider the convergence of

$$\begin{aligned}\int_1^\infty \frac{\sin x}{x}\,dx &= \lim_{n\to\infty}\int_1^n \frac{\sin x}{x}\,dx\\ &= \lim_{n\to\infty}\left\{\left[-\frac{\cos x}{x}\right]_1^n - \int_1^n \frac{\cos x}{x^2}\,dx\right\}\\ &\qquad \text{(integration by parts)}\\ &= \cos 1 - \lim_{n\to\infty}\int_1^n \frac{\cos x}{x^2}\,dx.\end{aligned}$$

But $(\cos x)/x^2$ is absolutely convergent on $(1, \infty)$, by comparison with $1/x^2$.

Therefore $\int_0^\infty [(\sin x)/x]\,dx$ converges; its actual value is $\pi/2$ (see Exercise 6.14). However the function $f(x) = (\sin x)/x$ is *not* absolutely integrable on $(0, \infty)$ (see Exercise 6.15).

The phenomenon illustrated by Examples 6 and 7, of $\int_a^b f$ converging despite the fact that f is not absolutely convergent on (a, b), does not occur when f is nonnegative.

Theorem 6.8 Suppose $f\colon (a, b) \to \mathscr{R}$ is locally integrable with $f \geqq 0$. Then f is absolutely integrable if and only if $\int_a^b f$ converges, in which case

$$\int_{(a,b)} f = \int_a^b f.$$

PROOF This follows immediately from Theorem 6.2 and Corollary 6.5. ∎

Example 8 We want to compute $\int_0^\infty e^{-x^2}\,dx$, which converges because $e^{-x^2} \leqq e^{-x}$ if $x \geqq 1$, while

$$\int_1^\infty e^{-x}\,dx = \lim_{n\to\infty} \int_1^n e^{-x}\,dx = \frac{1}{e}.$$

To obtain the value, we must resort to a standard subterfuge.

Consider $f: \mathscr{R}^2 \to \mathscr{R}$ defined by $f(x, y) = e^{-x^2-y^2}$. Since

$$e^{-x^2-y^2} \leqq \frac{1}{(x^2+y^2)^2}$$

when $x^2 + y^2$ is sufficiently large, it follows from Example 3 and the comparison test that $e^{-x^2-y^2}$ is absolutely integrable on $\mathscr{R}^2$. If D_n denotes the disk of radius n in $\mathscr{R}^2$, then

$$\begin{aligned}\int_{\mathscr{R}^2} f &= \lim_{n\to\infty} \int_{D^n} f = \lim_{n\to\infty} \int_0^{2\pi} \int_0^n e^{-r^2} r\,dr\,d\theta \\ &= 2\pi \lim_{n\to\infty} [-\tfrac{1}{2}e^{-r^2}]_0^n = \pi.\end{aligned}$$

On the other hand, if S_n denotes the square with edgelength $2n$ centered at the origin, then

$$\begin{aligned}\int_{\mathscr{R}^2} f &= \lim_{n\to\infty} \int_{S^n} f = \lim_{n\to\infty} \left(\int_{-n}^n \int_{-n}^n e^{-x^2-y^2}\,dx\,dy\right) \\ &= \lim_{n\to\infty} \left(\int_{-n}^n e^{-x^2}\,dx\right)^2 = \left(\int_{-\infty}^\infty e^{-x^2}\,dx\right)^2.\end{aligned}$$

Comparing the two results, we see that

$$\int_{-\infty}^\infty e^{-x^2}\,dx = \sqrt{\pi}.$$

Since e^{-x^2} is an even function, it follows that

$$\int_0^\infty e^{-x^2}\,dx = \frac{\sqrt{\pi}}{2}. \tag{4}$$

Example 9 The important *gamma function* $\Gamma : (0, \infty) \to \mathscr{R}$ is defined by

$$\Gamma(x) = \int_0^\infty t^{x-1} e^{-t}\,dt. \tag{5}$$

We must show that this improper integral converges for all $x > 0$. Now

$$\int_0^\infty t^{x-1} e^{-t}\,dt = \int_0^1 t^{x-1} e^{-t}\,dt + \int_1^\infty t^{x-1} e^{-t}\,dt.$$

If $x \geqq 1$, then $f(t) = t^{x-1}e^{-t}$ is continuous on $[0, 1]$. If $x \in (0, 1)$, then $t^{x-1}e^{-t} \leqq t^{x-1}$, so the first integral on the right converges, because $g(t) = t^{x-1}$ is absolutely convergent on $(0, 1)$ by Example 4, since $1 - x < 1$.

To consider the second integral, note first that

$$\lim_{t\to\infty} \frac{t^{x-1}e^{-t}}{1/t^2} = \lim_{t\to\infty} \frac{t^{x+1}}{e^t} = 0,$$

so

$$t^{x-1}e^{-t} \leqq \frac{1}{t^2}$$

for t sufficiently large. Since $1/t^2$ is absolutely convergent on $(1, \infty)$, it follows that $\int_1^\infty t^{x-1}e^{-t}\,dt$ converges.

The fundamental property of the gamma function is that $\Gamma(x+1) = x\Gamma(x)$ if $x > 0$:

$$\begin{aligned}\Gamma(x+1) &= \lim_{n\to\infty} \int_{1/n}^{n} t^x e^{-t}\,dt,\\ &= \lim_{n\to\infty} \left(\left[-\frac{t^x}{e^t} \right]_{t=1/n}^{n} + \int_{1/n}^{n} x t^{x-1} e^{-t}\,dt \right)\\ &= \lim_{n\to\infty} x \int_{1/n}^{n} t^{x-1} e^{-t}\,dt,\\ \Gamma(x+1) &= x\Gamma(x).\end{aligned}$$

Since $\Gamma(1) = 1$, it follows easily that

$$\Gamma(n+1) = n!$$

if n is a positive integer (Exercise 6.1).

Example 10 The *beta function* is a function of two variables, defined on the open first quadrant Q of the xy-plane by

$$\mathrm{B}(x, y) = \int_0^1 t^{x-1}(1-t)^{y-1}\,dt. \tag{6}$$

If either $x < 1$ or $y < 1$ or both, this integral is improper, but can be shown to converge by methods similar to those used with the gamma function (Exercise 6.9).

Substituting $t = \sin^2\theta$ in (6) (before taking the limit in the improper cases), we obtain the alternative form

$$\mathrm{B}(x, y) = 2\int_0^{\pi/2} \sin^{2x-1}\theta \cos^{2y-1}\theta\,d\theta. \tag{7}$$

The beta function has an interesting relation with the gamma function. To see this, we first express the gamma function in the form

$$\Gamma(x) = 2\int_0^\infty u^{2x-1}e^{-u^2}\,du \tag{8}$$

by substituting $t = u^2$ in (5). From this it follows that

$$\Gamma(x)\Gamma(y) = 4\left(\int_0^\infty u^{2x-1}\, e^{-u^2}\, du\right)\left(\int_0^\infty v^{2y-1}\, e^{-v^2}\, dv\right),$$

$$\Gamma(x)\Gamma(y) = 4\iint_Q e^{-u^2-v^2}\, u^{2x-1}\, v^{2y-1}\, du\, dv. \tag{9}$$

This last equality is easily verified by use of the approximating sequence $\{A^n\}_1^\infty$ for Q, where $A_n = [1/n, n] \times [1/n, n]$.

It follows from Corollary 6.5 that the integrand function in (9) is absolutely convergent, so we may use any other approximating sequence we like. Let us try $\{B_n\}_1^\infty$, where B_n is the set of all those points with polar coordinates (r, θ) such that $1/n \leqq r \leqq n$ and $1/n \leqq \theta \leqq (\pi/2) - (1/n)$ (Fig. 4.42). Then

$$\begin{aligned}\Gamma(x)\Gamma(y) &= 4 \lim_{n\to\infty} \iint_{B_n} e^{-u^2-v^2}\, u^{2x-1} v^{2y-1}\, du\, dv \\ &= 4 \lim_{n\to\infty} \int_{1/n}^{\pi/2-1/n} \left(\int_{1/n}^{n} e^{-r^2}\, (r\cos\theta)^{2x-1}\, (r\sin\theta)^{2y-1}\, r\, dr\right) d\theta \\ &= 4 \lim_{n\to\infty} \left(\int_{1/n}^{\pi/2-1/n} \cos^{2x-1}\theta\, \sin^{2y-1}\theta\, d\theta\right)\left(\int_{1/n}^{n} r^{2(x+y)-1} e^{-r^2}\, dr\right), \\ \Gamma(x)\Gamma(y) &= \mathrm{B}(x, y)\Gamma(x+y)\end{aligned}$$

using (7) and (8) and the obvious symmetry of the beta function. Hence

$$\mathrm{B}(x, y) = \frac{\Gamma(x)\Gamma(y)}{\Gamma(x+y)}. \tag{10}$$

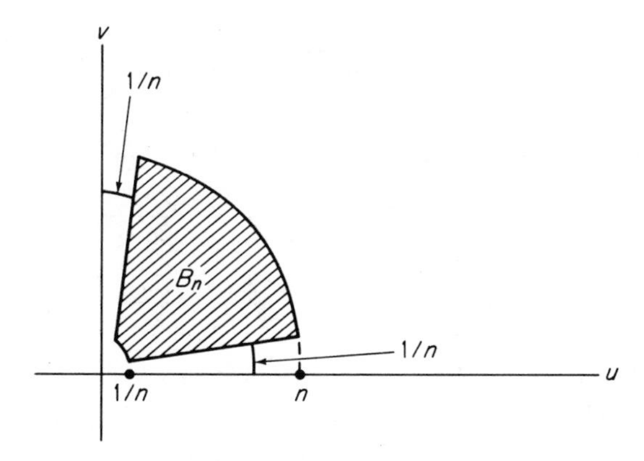

Figure 4.42

As a typical application of (7) and (10), we obtain

$$\begin{aligned}\int_0^{\pi/2} \sin^5\theta\cos^7\theta\, d\theta &= \frac{1}{2}\,\mathrm{B}(3,4)\\ &= \frac{1}{2}\frac{\Gamma(3)\Gamma(4)}{\Gamma(7)}\\ &= \frac{1}{2}\frac{2!\,3!}{6!}\\ &= \frac{1}{120}.\end{aligned}$$

Another typical example of the use of gamma functions to evaluate integrals is

$$\begin{aligned}\int_0^\infty x^{1/2}e^{-x^3}\,dx &= \tfrac{1}{3}\int_0^\infty u^{-1/2}e^{-u}\,du\\ &= \tfrac{1}{3}\Gamma(\tfrac{1}{2})\\ &= \tfrac{1}{3}\sqrt{\pi},\end{aligned}$$

making the substitution $u = x^3$ and using the fact that $\Gamma(\frac{1}{2}) = \sqrt{\pi}$ (Exercise 6.2).

Exercises

6.1 Show that $\Gamma(1) = 1$, and then deduce from $\Gamma(x+1) = x\Gamma(x)$ that $\Gamma(n+1) = n!$.

6.2 Show that $\Gamma(\frac{1}{2}) = \sqrt{\pi}$. *Hint:* Substitute $t = u^2$ in the integral defining $\Gamma(\frac{1}{2})$, and refer to Example 8. Then prove by induction on the positive integer n that

$$\Gamma(n+\tfrac{1}{2}) = \frac{1\cdot 3\cdot\cdots\cdot(2n-1)}{2^n}\sqrt{\pi}.$$

6.3 Recalling from the previous exercise section the formulas for the volume α_n of the unit n-dimensional ball $B^n \subset \mathscr{R}^n$, deduce from the previous two exercises that

$$\alpha_n = \frac{\pi^{n/2}}{\Gamma((n/2)+1)} = \frac{2\pi^{n/2}}{n\Gamma(n/2)}$$

for all $n \geqq 1$.

6.4 Apply Exercises 6.1 and 6.2, and formulas (7) and (10), to obtain a new derivation of

$$\int_0^{\pi/2} \sin^{2n-1}\theta\,d\theta = \frac{2\cdot 4\cdot\cdots\cdot(2n-2)}{1\cdot 3\cdot\cdots\cdot(2n-1)},$$

$$\int_0^{\pi/2} \sin^{2n}\theta\,d\theta = \frac{1\cdot 3\cdot\cdots\cdot(2n-1)}{2\cdot 4\cdot\cdots\cdot(2n)}\frac{\pi}{2}.$$

6.5 Show that $\int_0^1 x^4(1-x^2)^{1/2}\,dx = \pi/32$ by subtituting $x = t^{1/2}$.

6.6 Use the substitution $t = x^n$ to show that

$$\int_0^\infty x^m e^{-x^n}\,dx = \frac{1}{n}\Gamma\left(\frac{m+1}{n}\right).$$

6.7 Use the substitution $x = e^{-u}$ to show that

$$\int_0^1 x^m(\log x)^n\,dx = \frac{(-1)^n n!}{(m+1)^{n+1}}.$$

6.8 Show that

$$\int_0^{\pi/2} \frac{dx}{(\cos x)^{1/2}} = \frac{[\Gamma(\frac{1}{4})]^2}{2(2\pi)^{1/2}}.$$

6.9 Prove that the integral defining the beta function converges.

6.10 Show that the mass of a spherical ball of radius a, with density function $d(x, y, z) = x^2y^2z^2$, is $M = 4\pi a^9/945$. *Hint:* Calculate the mass of the first octant. Introduce spherical coordinates, and use formulas (7) and (10).

6.11 Use ellipsoidal coordinates to show that the mass of the ellipsoidal ball $x^2/a^2 + y^2/b^2 + z^2/c^2 \leqq 1$, with density function $d(x, y, z) = x^2 + y^2 + z^2$, is

$$M = \frac{4\pi abc}{15}(a^2 + b^2 + c^2).$$

6.12 By integration by parts show that

$$\int_0^\infty e^{-x^2}\,x^k\,dx = \frac{k-1}{2}\int_0^\infty e^{-x^2}\,x^{k-2}\,dx.$$

Deduce from this recursion formula that

$$\int_0^\infty e^{-x^2}\,x^{2m}\,dx = \frac{1\cdot 3\cdot\cdots\cdot(2m-1)}{2^{m+1}}\sqrt{\pi}$$

and

$$\int_0^\infty e^{-x^2}\,x^{2m-1}\,dx = \frac{1}{2}(m-1)!.$$

Apply Exercises 6.1 and 6.2 to conclude that

$$\int_0^\infty e^{-x^2}\,x^{n-1}\,dx = \frac{\Gamma(n/2)}{2}.$$

for all integers $n \geqq 1$.

6.13 The purpose of this exercise is to give a final computation of the volume α_n of the unit ball $B^n \subset \mathscr{R}^n$.

(a) Let $T: \mathscr{R}^n \to \mathscr{R}^n$ be the n-dimensional spherical coordinates mapping of Exercise 5.19. Note by inspection of the definition of T, without computing it explicitly, that the Jacobian of T is of the form

$$|\det T'| = \rho^{n-1}\gamma(\varphi_1, \ldots, \varphi_{n-2}, \theta)$$

for some function γ.

(b) Let $f: \mathscr{R}^n \to \mathscr{R}$ be an absolutely integrable function such that $g = f \circ T$ is a function of ρ alone. Then show that

$$\int_{B_a{}^n} f = C_n \int_0^a g(\rho)\rho^{n-1}\,d\rho \qquad (*)$$

for some constant C_n (independent of f). Setting $f = g = 1$ on B^n, $a = 1$, we see that $\alpha_n = C_n/n$, so it suffices to compute C_n.

(c) Taking limits in (*) as $a \to \infty$, we obtain

$$\int_{\mathscr{R}^n} f = C_n \int_0^\infty g(\rho)\rho^{n-1}\, d\rho,$$

so it suffices to find an absolutely integrable function f for which we can evaluate both of these improper integrals. For this purpose take

$$f(x_1, \ldots, x_n) = e^{-x_1^2 - \cdots - x_n^2}$$

so $g(\rho) = e^{-\rho^2}$. Conclude from Example 8 that $\int_{\mathscr{R}^n} f = \pi^{n/2}$, and then apply the previous exercise to compute C_n, and thereby α_n.

6.14 The purpose of the problem is to evaluate

$$\int_0^\infty \frac{\sin x}{x}\, dx.$$

(a) First show that $\int_0^\infty e^{-xy}\, dy = 1/x$ if $x > 0$.
(b) Then use integration by parts to show that

$$\int_0^\infty e^{-xy} \sin x\, dx = \frac{1}{1+y^2} \qquad \text{if} \quad y > 0.$$

(c) Hence

$$\begin{aligned}\int_0^\infty \frac{\sin x}{x}\, dx &= \int_0^\infty \left(\int_0^\infty e^{-xy} \sin x\, dy\right) dx \\ &= \int_0^\infty \left(\int_0^\infty e^{-xy} \sin x\, dx\right) dy \\ &= \int_0^\infty \frac{dy}{1+y^2},\end{aligned}$$

$$\int_0^\infty \frac{\sin x}{x}\, dx = \frac{\pi}{2}$$

provided that the interchange of limit operations can be justified.

6.15 The object of this problem is to show that $f(x) = (\sin x)/x$ is not absolutely integrable on $(0, \infty)$. Show first that

$$\int_{2k\pi}^{(2k+1)\pi} \frac{\sin x}{x}\, dx \geq \frac{2}{(2k+1)\pi}.$$

Given any compact contented set $A \subset (0, \infty)$, pick m such that $A \subset [0, 2m\pi]$, and then define

$$B_n = [0, 2m\pi] \cap \bigcup_{k=m}^{n} [2k\pi, (2k+1)\pi]$$

for all $n > m$. Now conclude, from the fact that $\sum_{k=1}^\infty 1/(2k+1)$ diverges, that

$$\lim_{n\to\infty} \int_{B_n} \frac{\sin x}{x}\, dx = \infty.$$

Why does this imply that $\int_{(0,\infty)} [(\sin x)/x]\, dx$ does not exist?

6.16 If U is an open subset of $\mathscr{R}^n$, show that there exists an increasing sequence $\{A_k\}_1^\infty$ of compact contented sets such that $U = \bigcup_{k=1}^\infty \operatorname{int} A_k$. *Hint:* Each point of U is contained in some closed ball which lies in U. Pick the sequence in such a way that A_k is the union of k closed balls.

6.17 Let $q(\mathbf{x}) = \mathbf{x}^t A \mathbf{x}$ be a positive definite quadratic form on $\mathscr{R}^n$. Then show that

$$\int_{\mathscr{R}^n} e^{-q(\mathbf{x})}\, d\mathbf{x} = \frac{\pi^{n/2}}{(\det A)^{1/2}}.$$

Outline: Let P be the orthogonal matrix provided by Exercise II.8.7, such that $P^t A P = P^{-1} A P$ is the diagonal matrix whose diagonal elements are the (positive) eigenvalues $\lambda_1, \ldots, \lambda_n$ of the symmetric matrix A. Use the linear mapping $L : \mathscr{R}_{\mathbf{y}}^n \to \mathscr{R}_{\mathbf{x}}^n$ defined by $\mathbf{x} = P\mathbf{y}$ to transform the above integral to

$$\int_{\mathscr{R}^n} \exp(-\lambda_1 y_1^2 - \cdots - \lambda_n y_n^2)\, d\mathbf{y}.$$

Then apply the fact that $\int_{-\infty}^{\infty} e^{-t^2}\, dt = \sqrt{\pi}$ by Example 8, and that fact that $\lambda_1 \lambda_2 \cdots \lambda_n = \det A$ by Lemma II.8.7.

V

Line and Surface Integrals; Differential Forms and Stokes' Theorem

This chapter is an exposition of the machinery that is necessary for the statement, proof, and application of Stokes' theorem. Stokes' theorem is a multi-dimensional generalization of the fundamental theorem of (single-variable) calculus, and may accurately be called the "fundamental theorem of multi-variable calculus." Among its numerous applications are the classical theorems of vector analysis (see Section 7).

We will be concerned with the integration of appropriate kinds of functions over surfaces (or manifolds) in $\mathscr{R}^n$. It turns out that the sort of "object," which can be integrated over a smooth k-manifold in $\mathscr{R}^n$, is what is called a differential k-form (defined in Section 5). It happens that to every differential k-form α there corresponds a differential $(k+1)$-form $d\alpha$, the differential of α, and Stokes' theorem is the formula

$$\int_V d\alpha = \int_{\partial V} \alpha,$$

where V is a compact $(k+1)$-dimensional manifold with nonempty boundary ∂V (a k-manifold). In the course of our discussion we will make clear the way in which this is a proper generalization of the fundamental theorem of calculus in the form

$$\int_a^b f'(t)\,dt = f(b) - f(a).$$

We start in Section 1 with the simplest case—integrals over curves in space, traditionally called "line integrals." Section 2 is a leisurely treatment of the 2-dimensional special case of Stokes' theorem, known as Green's theorem;

this will serve to prepare the student for the proof of the general Stokes' theorem in Section 6.

Section 3 includes the multilinear algebra that is needed for the subsequent discussions of surface area (Section 4) and differential forms (Section 5). The student may prefer (at least in the first reading) to omit the proofs in Section 3—only the definitions and statements of the theorems of this section are needed in the sequel.

1 PATHLENGTH AND LINE INTEGRALS

In this section we generalize the familiar single-variable integral (on a closed interval in $\mathscr{R}$) so as to obtain a type of integral that is associated with paths in $\mathscr{R}^n$. By a $\mathscr{C}^1$ *path* in $\mathscr{R}^n$ is meant a continuously differentiable function $\gamma : [a, b] \to \mathscr{R}^n$. The $\mathscr{C}^1$ path $\gamma : [a, b] \to \mathscr{R}^n$ is said to be *smooth* if $\gamma'(t) \neq \mathbf{0}$ for all $t \in [a, b]$. The significance of the condition $\gamma'(t) \neq \mathbf{0}$—that the direction of a $\mathscr{C}^1$ path cannot change abruptly if its velocity vector never vanishes—is indicated by the following example.

Example 1 Consider the $\mathscr{C}^1$ path $\gamma = (\gamma_1, \gamma_2) : \mathscr{R} \to \mathscr{R}^2$ defined by

$$x = \gamma_1(t) = t^3, \qquad y = \gamma_2(t) = \begin{cases} t^3 & \text{if } t \geqq 0, \\ -t^3 & \text{if } t \leqq 0. \end{cases}$$

The image of γ is the graph $y = |x|$ (Fig. 5.1). The only "corner" occurs at the origin, which is the image of the single point $t = 0$ at which $\gamma'(t) = \mathbf{0}$.

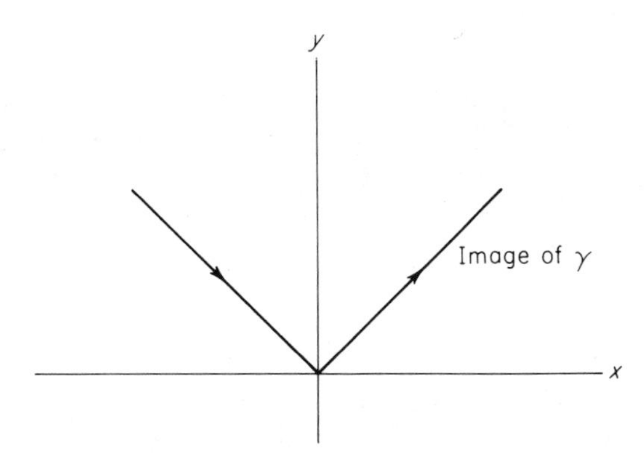

Figure 5.1

We discuss first the concept of length for paths. It is natural to approach the definition of the length $s(\gamma)$ of the path $\gamma : [a, b] \to \mathscr{R}^n$ by means of polygonal approximations to γ. Choose a partition $\mathscr{P}$ of $[a, b]$,

$$\mathscr{P} = \{a = t_0 < t_1 < t_2 < \cdots < t_{k-1} < t_k = b\},$$

and recall that the mesh $|\mathscr{P}|$ is the maximum length $t_i - t_{i-1}$ of a subinterval of $\mathscr{P}$. Then $\mathscr{P}$ defines a polygonal approximation to γ, namely the polygonal arc from $\gamma(a)$ to $\gamma(b)$ having successive vertices $\gamma(t_0), \gamma(t_1), \ldots, \gamma(t_k)$ (Fig. 5.2), and we regard the length

$$s(\gamma, \mathscr{P}) = \sum_{i=1}^{k} |\gamma(t_i) - \gamma(t_{i-1})|$$

of this polygonal arc as an approximation to $s(\gamma)$.

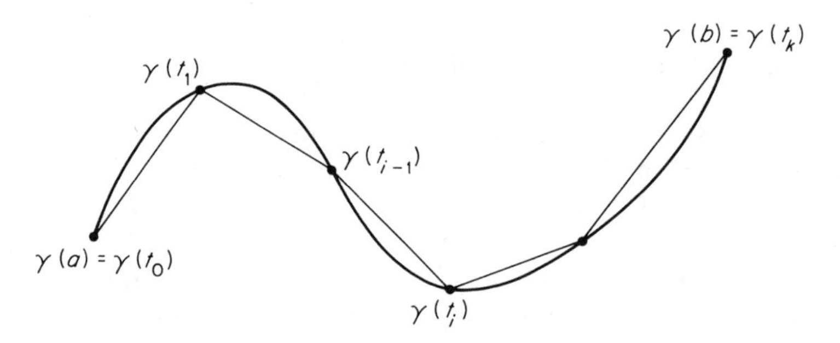

Figure 5.2

This motivates us to *define* the *length* $s(\gamma)$ of the path $\gamma : [a, b] \to \mathscr{R}^n$ by

$$s(\gamma) = \lim_{|\mathscr{P}| \to 0} s(\gamma, \mathscr{P}), \tag{1}$$

provided that this limit exists, in the sense that, given $\varepsilon > 0$, there exists $\delta > 0$ such that

$$|\mathscr{P}| < \delta \text{ implies } |s(\gamma) - s(\gamma, \mathscr{P})| < \varepsilon.$$

It turns out that the limit in (1) may not exist if γ is only assumed to be continuous, but that it does exist if γ is a $\mathscr{C}^1$ path, and moreover that there is then a pleasant means of computing it.

Theorem 1.1 If $\gamma : [a, b] \to \mathscr{R}^n$ is a $\mathscr{C}^1$ path, then $s(\gamma)$ exists, and

$$s(\gamma) = \int_a^b |\gamma'(t)| \, dt. \tag{2}$$

REMARK If we think of a moving particle in $\mathscr{R}^n$, whose position vector at time t is $\gamma(t)$, then $|\gamma'(t)|$ is its speed. Thus (2) simply asserts that the distance traveled by the particle is equal to the time integral of its speed—a result whose 1-dimensional case is probably familiar from elementary calculus.

PROOF It suffices to show that, given $\varepsilon > 0$, there exists $\delta > 0$ such that

$$\left| \int_a^b |\gamma'(t)| \, dt - s(\gamma, \mathscr{P}) \right| < \varepsilon$$

for every partition

$$\mathscr{P} = \{a = t_0 < t_1 < \cdots < t_{k-1} < t_k = b\}$$

of mesh less than δ. Recall that

$$\begin{aligned} s(\gamma, \mathscr{P}) &= \sum_{i=1}^{k} |\gamma(t_i) - \gamma(t_{i-1})| \\ &= \sum_{i=1}^{k} \left[\sum_{r=1}^{n} (\gamma_r(t_i) - \gamma_r(t_{i-1}))^2 \right]^{1/2}, \end{aligned} \tag{3}$$

where $\gamma_1, \ldots, \gamma_n$ are the component functions of γ.

An application of the mean value theorem to the function γ_r on the ith subinterval $[t_{i-1}, t_i]$ yields a point $t_i^r \in (t_{i-1}, t_i)$ such that

$$\gamma_r(t_i) - \gamma_r(t_{i-1}) = \gamma_r'(t_i^r)(t_i - t_{i-1}).$$

Consequently Eq. (3) becomes

$$s(\gamma, \mathscr{P}) = \sum_{i=1}^{k} \left[\sum_{r=1}^{n} (\gamma_r'(t_i^r))^2 \right]^{1/2} (t_i - t_{i-1}). \tag{4}$$

If the points $t_i^1, t_i^2, \ldots, t_i^n$ just happened to all be the same point $t_i^* \in [t_{i-1}, t_i]$, then (4) would take the form

$$\sum_{i=1}^{k} \left[\sum_{r=1}^{n} (\gamma_r'(t_i^*))^2 \right]^{1/2} (t_i - t_{i-1}) = \sum_{i=1}^{k} |\gamma'(t_i^*)| (t_i - t_{i-1}),$$

which is a Riemann sum for $\int_a^b |\gamma'(t)|\, dt$. This is the real reason for the validity of formula (2); the remainder of the proof consists in showing that the difference between the sum (4) and an actual Riemann sum is "negligible."

For this purpose we introduce the auxiliary function $F : I = [a, b]^n \to \mathscr{R}$ defined by

$$F(x_1, x_2, \ldots, x_n) = \left[\sum_{r=1}^{n} \gamma_r'(x_r))^2 \right]^{1/2}.$$

Notice that $F(t, t, \ldots, t) = |\gamma'(t)|$, and that F is continuous, and therefore uniformly continuous, on the n-dimensional interval I, since γ is a $\mathscr{C}^1$ path. Consequently there exists a $\delta_1 > 0$ such that

$$|F(\mathbf{x}) - F(\mathbf{y})| < \frac{\varepsilon}{2(b - a)} \tag{5}$$

if each $|x_r - y_r| < \delta_1$.

We now want to compare the approximation

$$s(\gamma, \mathscr{P}) = \sum_{i=1}^{k} F(t_i^1, t_i^2, \ldots, t_i^n)(t_i - t_{i-1})$$

with the Riemann sum

$$R(\gamma, \mathscr{P}) = \sum_{i=1}^{k} |\gamma'(t_i)|(t_i - t_{i-1})$$

$$= \sum_{i=1}^{k} F(t_i, t_i, \ldots, t_i)(t_i - t_{i-1}).$$

Since the points $t_i^1, t_i^2, \ldots, t_i^n$ all lie in the interval $[t_{i-1}, t_i]$, whose length $\leqq |\mathscr{P}|$, it follows from (5) that

$$|s(\gamma, \mathscr{P}) - R(\gamma, \mathscr{P})| \leqq \sum_{i=1}^{k} |F(t_i^1, \ldots, t_i^n) - F(t_i, \ldots, t_i)|(t_i - t_{i-1})$$

$$< \frac{\varepsilon}{2(b-a)} \sum_{i=1}^{k} (t_i - t_{i-1}) = \frac{\varepsilon}{2}$$

if $|\mathscr{P}| < \delta_1$.

On the other hand, there exists (by Theorem IV.3.4) a $\delta_2 > 0$ such that

$$\left| R(\gamma, \mathscr{P}) - \int_a^b |\gamma'(t)|\, dt \right| < \frac{\varepsilon}{2}$$

if $|\mathscr{P}| < \delta_2$.

If, finally, $\delta = \min(\delta_1, \delta_2)$, then $|\mathscr{P}| < \delta$ implies that

$$\left| s(\gamma, \mathscr{P}) - \int_a^b |\gamma'(t)|\, dt \right| \leqq |s(\gamma, \mathscr{P}) - R(\gamma, \mathscr{P})| + \left| R(\gamma, \mathscr{P}) - \int_a^b |\gamma'(t)|\, dt \right|$$

$$< \frac{\varepsilon}{2} + \frac{\varepsilon}{2} = \varepsilon$$

as desired. ∎

Example 2 Writing $\gamma(t) = (x_1(t), \ldots, x_n(t))$, and using Leibniz notation, formula (2) becomes

$$s(\gamma) = \int_a^b \left[\left(\frac{dx_1}{dt}\right)^2 + \cdots + \left(\frac{dx_n}{dt}\right)^2 \right]^{1/2} dt.$$

Given a $\mathscr{C}^1$ function $f: [a, b] \to \mathscr{R}$, the graph of f is the image of the $\mathscr{C}^1$ path $\gamma : [a, b] \to \mathscr{R}^2$ defined by $\gamma(x) = (x, f(x))$. Substituting $x_1 = t = x$ and $x_2 = y$ in the above formula with $n = 2$, we obtain the familiar formula

$$s = \int_a^b \left[1 + \left(\frac{dy}{dx}\right)^2 \right]^{1/2} dx$$

for the length of the graph of a function.

Having defined pathlength and established its existence for $\mathscr{C}^1$ paths, the following question presents itself. Suppose that $\alpha : [a, b] \to \mathscr{R}^n$ and $\beta : [c, d] \to \mathscr{R}^n$

are two $\mathscr{C}^1$ paths that are "geometrically equivalent" in the sense that they have the same *initial point* $\alpha(a) = \beta(c)$, the same *terminal point* $\alpha(b) = \beta(d)$, and trace through the same intermediate points in the same order. Do α and β then have the same length, $s(\alpha) = s(\beta)$?

Before providing the expected affirmative answer to this question, we must make precise the notion of equivalence of paths. We say that the path $\alpha : [a, b] \to \mathscr{R}^n$ is *equivalent* to the path $\beta : [c, d] \to \mathscr{R}^n$ if and only if there exists a $\mathscr{C}^1$ function

$$\varphi : [a, b] \to [c, d]$$

such that $\varphi([a, b]) = [c, d]$, $\alpha = \beta \circ \varphi$, and $\varphi'(t) > 0$ for all $t \in [a, b]$ (see Fig. 5.3). The student should show that this relation is symmetric (if α is equivalent to β, then β is equivalent to α) and transitive (if α is equivalent to β, and β is equivalent to γ, then α is equivalent to γ), and is therefore an equivalence relation (Exercise 1.1). He should also show that, if the $\mathscr{C}^1$ paths α and β are equivalent, then α is smooth if and only if β is smooth (Exercise 1.2).

The fact that $s(\alpha) = s(\beta)$ if α and β are equivalent is seen by taking $f(\mathbf{x}) \equiv 1$ in the following theorem.

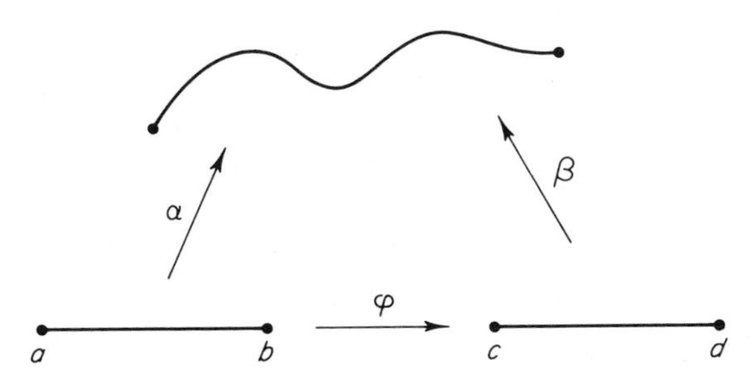

Figure 5.3

Theorem 1.2 Suppose that $\alpha : [a, b] \to \mathscr{R}^n$ and $\beta : [c, d] \to \mathscr{R}^n$ are equivalent $\mathscr{C}^1$ paths, and that f is a continuous real-valued function whose domain of definition in $\mathscr{R}^n$ contains the common image of α and β. Then

$$\int_a^b f(\alpha(t))\,|\alpha'(t)|\,dt = \int_c^d f(\beta(t))\,|\beta'(t)|\,dt.$$

PROOF If $\varphi : [a, b] \to [c, d]$ is a $\mathscr{C}^1$ function such that $\alpha = \beta \circ \varphi$ and $\varphi'(t) > 0$ for all $t \in [a, b]$, then

$$\begin{aligned}
\int_a^b f(\alpha(t))\,|\alpha'(t)|\,dt &= \int_a^b f(\beta(\varphi(t)))\,|\beta'(\varphi(t))\varphi'(t)|\,dt \qquad \text{(chain rule)} \\
&= \int_a^b f(\beta(\varphi(t)))\,|\beta'(\varphi(t))|\,\varphi'(t)\,dt \qquad (\text{because } \varphi'(t) > 0) \\
&= \int_c^d f(\beta(u))\,|\beta'(u)|\,du \qquad (\text{substitution } u = \varphi(t)).
\end{aligned}$$

∎

To provide a physical interpretation of integrals such as those of the previous theorem, let us think of a wire which coincides with the image of the $\mathscr{C}^1$ path $\gamma : [a, b] \to \mathscr{R}^n$, and whose density at the point $\gamma(t)$ is $f(\gamma(t))$. Given a partition $\mathscr{P} = \{a = t_0 < t_1 < \cdots < t_k = b\}$ of $[a, b]$, the sum

$$m(\gamma, f, \mathscr{P}) = \sum_{i=1}^{k} f(\gamma(t_i)) | \gamma(t_i) - \gamma(t_{i-1}) |$$

may be regarded as an approximation to the mass of the wire. By an argument essentially identical to that of the proof of Theorem 1.1, it can be proved that

$$\lim_{|\mathscr{P}| \to 0} m(\gamma, f, \mathscr{P}) = \int_a^b f(\gamma(t)) | \gamma'(t) | \, dt. \tag{6}$$

Consequently this integral is taken as the definition of the mass of the wire.

If γ is a smooth path, the integral (6) can be interpreted as an integral with respect to pathlength. For this we need an equivalent *unit-speed* path, that is an equivalent smooth path $\hat{\gamma}$ such that $| \hat{\gamma}'(t) | \equiv 1$.

Proposition 1.3 Every smooth path $\gamma : [a, b] \to \mathscr{R}^n$ is equivalent to a smooth unit-speed path.

PROOF Write $L = s(\gamma)$ for the length of γ, and define $\sigma : [a, b] \to [0, L]$ by

$$\sigma(t) = \int_a^t | \gamma'(u) | \, du,$$

so $\sigma(t)$ is simply the length of the path $\gamma : [a, t] \to \mathscr{R}^n$ (Fig. 5.4). Then from the fundamental theorem of calculus we see that σ is a $\mathscr{C}^1$ function with

$$\sigma'(t) = | \gamma'(t) | > 0.$$

Therefore the inverse function $\tau : [0, L] \to [a, b]$ exists, with

$$\tau'(s) = \frac{1}{\sigma'(\tau(s))} > 0.$$

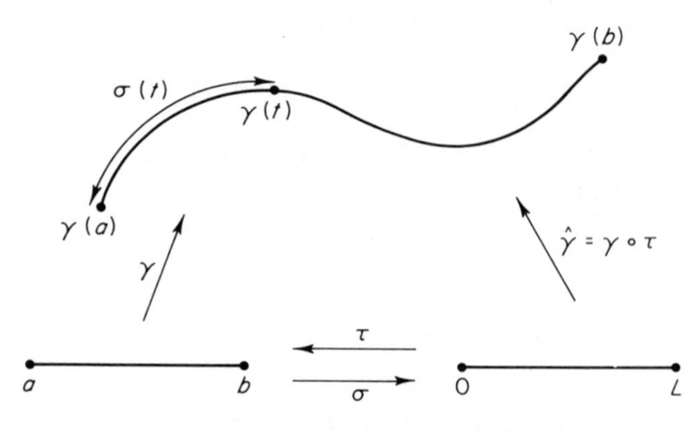

Figure 5.4

If $\hat{\gamma} : [0, L] \to \mathscr{R}^n$ is the equivalent smooth path defined by $\hat{\gamma} = \gamma \circ \tau$, the chain rule gives

$$|\hat{\gamma}'(s)| = |\gamma'(\tau(s))\tau'(s)| = \frac{|\gamma'(\tau(s))|}{|\sigma'(\tau(s))|} = 1,$$

since $\sigma'(t) = |\gamma'(t)|$. ∎

With the notation of this proposition and its proof, we have

$$\begin{aligned}\int_a^b f(\gamma(t))|\gamma'(t)|\,dt &= \int_a^b f(\hat{\gamma}(\sigma(t)))|\hat{\gamma}'(\sigma(t))\sigma'(t)|\,dt\\ &= \int_a^b f(\hat{\gamma}(\sigma(t)))\sigma'(t)\,dt\end{aligned}$$

since $|\hat{\gamma}'(\sigma(t))| = 1$ and $\sigma'(t) > 0$. Substituting $s = \sigma(t)$, we then obtain

$$\int_a^b f(\gamma(t))|\gamma'(t)|\,dt = \int_0^L f(\hat{\gamma}(s))\,ds. \tag{7}$$

(This result also follows immediately from Theorem 1.2.) Since $\hat{\gamma}(s)$ simply denotes that point whose "distance along the path" from the initial point $\gamma(a)$ is s, Eq. (7) provides the promised integral with respect to pathlength.

It also serves to motivate the common abbreviation

$$\int_a^b f(\gamma(t))|\gamma'(t)|\,dt = \int_\gamma f\,ds. \tag{8}$$

Historically this notation resulted from the expression

$$ds = |\gamma'(t)|\,dt$$

for the length of the "infinitesimal" piece of path corresponding to the "infinitesimal" time interval $[t, t + dt]$. Later in this section we will provide the expression $ds = |\gamma'(t)|\,dt$ with an actual mathematical meaning (in contrast to its "mythical" meaning here).

Now let $\gamma : [a, b] \to \mathscr{R}^n$ be a smooth path, and think of $\gamma(t)$ as the position vector of a particle moving in $\mathscr{R}^n$ under the influence of the continuous force field $\mathbf{F} : \mathscr{R}^n \to \mathscr{R}^n$ [so $\mathbf{F}(\gamma(t))$ is the force acting on the particle at time t]. We inquire as to the work W done by the force field in moving the particle along the path from $\gamma(a)$ to $\gamma(b)$. Let $\mathscr{P}$ be the usual partition of $[a, b]$. If $\mathbf{F}$ were a constant force field, then the work done by $\mathbf{F}$ in moving the particle along the straight line segment from $\gamma(t_{i-1})$ to $\gamma(t_i)$ would by definition be $\mathbf{F} \cdot (\gamma(t_i) - \gamma(t_{i-1}))$ (in the constant case, work is simply the product of force and distance). We therefore regard the sum

$$W(\gamma, \mathbf{F}, \mathscr{P}) = \sum_{i=1}^{k} \mathbf{F}(\gamma(t_i)) \cdot (\gamma(t_i) - \gamma(t_{i-1}))$$

as an approximation to W. By an argument similar to the proof of Theorem 1.1, it can be proved that

$$\lim_{|\mathscr{P}|\to 0} W(\gamma, \mathbf{F}, \mathscr{P}) = \int_a^b \mathbf{F}(\gamma(t)) \cdot \gamma'(t)\, dt,$$

so we *define* the work done by the force field $\mathbf{F}$ in moving a particle along the path γ by

$$W = \int_a^b \mathbf{F}(\gamma(t)) \cdot \gamma'(t)\, dt. \tag{9}$$

Rewriting (9) in terms of the *unit tangent vector*

$$\mathbf{T}(t) = \frac{\gamma'(t)}{|\gamma'(t)|}$$

at $\gamma(t)$ (Fig. 5.5), we obtain

$$\begin{aligned} W &= \int_a^b \mathbf{F}(\gamma(t)) \cdot \frac{\gamma'(t)}{|\gamma'(t)|}\, |\gamma'(t)|\, dt \\ &= \int_a^b \mathbf{F}(\gamma(t)) \cdot \mathbf{T}(t) |\gamma'(t)|\, dt, \end{aligned}$$

or

$$W = \int_\gamma \mathbf{F} \cdot \mathbf{T}\, ds \tag{10}$$

in the notation of (8). Thus the work done by the force field is the integral with respect to pathlength of its "tangential component" $\mathbf{F} \cdot \mathbf{T}$.

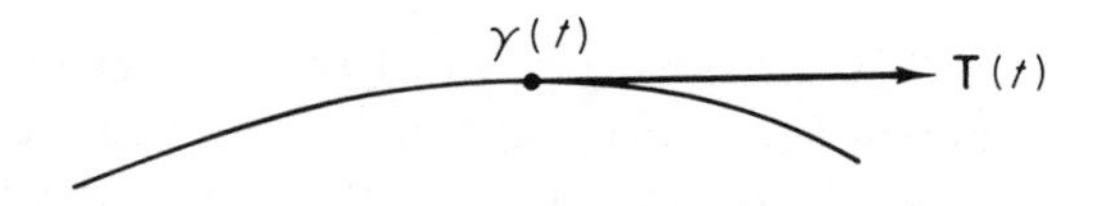

Figure 5.5

In terms of components, (9) becomes

$$W = \int_a^b [F_1(\gamma(t))\gamma_1'(t) + \cdots + F_n(\gamma(t))\gamma_n'(t)]\, dt,$$

or

$$W = \int_a^b \left(F_1 \frac{dx_1}{dt} + \cdots + F_n \frac{dx_n}{dt}\right) dt$$

in Leibniz notation. A classical abbreviation for this last formula is

$$W = \int_\gamma F_1\, dx_1 + \cdots + F_n\, dx_n.$$

This type of integral, called a line integral (or curvilinear integral), can of course be defined without reference to the above discussion of work, and indeed line integrals have a much wider range of applications than this. Given a $\mathscr{C}^1$ path $\gamma : [a, b] \to \mathscr{R}^n$ and n continuous functions $f_1, \ldots, f_n$ whose domains of definition in $\mathscr{R}^n$ all contain the image of γ, the *line integral* $\int_\gamma f_1\,dx_1 + \cdots + f_n\,dx_n$ is defined by

$$\int_\gamma f_1\,dx_1 + \cdots + f_n\,dx_n = \int_a^b [f_1(\gamma(t))\gamma_1'(t) + \cdots + f_n(\gamma(t))\gamma_n'(t)]\,dt. \tag{11}$$

Formally, we simply substitute $x_i = \gamma_i(t)$, $dx_i = \gamma_i'(t)\,dt$ into the left-hand side of (11), and then integrate from a to b.

We now provide an interpretation of (11) whose viewpoint is basic to subsequent sections. By a (linear) *differential form* on the set $U \subset \mathscr{R}^n$ is meant a mapping ω which associates with each point $\mathbf{x} \in U$ a linear function $\omega(\mathbf{x}) : \mathscr{R}^n \to \mathscr{R}$. Thus

$$\omega : U \to \mathscr{L}(\mathscr{R}^n, \mathscr{R}),$$

where $\mathscr{L}(\mathscr{R}^n, \mathscr{R})$ is the vector space of all linear (real-valued) functions on $\mathscr{R}^n$. We will frequently find it convenient to write

$$\omega(\mathbf{x}) = \omega_\mathbf{x}.$$

Recall that every linear function $L : \mathscr{R}^n \to \mathscr{R}$ is of the form

$$\begin{aligned} L(v_1, \ldots, v_n) &= a_1 v_1 + \cdots a_n + v_n \\ &= a_1 \lambda_1(\mathbf{v}) + \cdots + a_n \lambda_n(\mathbf{v}), \end{aligned} \tag{12}$$

where $\lambda_i : \mathscr{R}^n \to \mathscr{R}$ is the *ith projection function* defined by

$$\lambda_i(v_1, \ldots, v_n) = v_i, \qquad i = 1, \ldots, n.$$

If we use the customary notation $\lambda_i = dx_i$, then (12) becomes

$$L = a_1\,dx_1 + \cdots + a_n\,dx_n.$$

If $L = \omega(\mathbf{x})$ depends upon the point $\mathbf{x} \in U$, then so do the coefficients $a_1, \ldots, a_n$; this gives the following result.

Proposition 1.4 If ω is a differential form on $U \subset \mathscr{R}^n$, then there exist unique real-valued functions $a_1, \ldots, a_n$ on U such that

$$\omega(\mathbf{x}) = \omega_\mathbf{x} = a_1(\mathbf{x})\,dx_1 + \cdots + a_n(\mathbf{x})\,dx_n \tag{13}$$

for each $\mathbf{x} \in U$.

Thus we may regard a differential form as simply an "expression of the form" (13), remembering that, for each $\mathbf{x} \in U$, this expression denotes the linear function whose value at $\mathbf{v} = (v_1, \ldots, v_n) \in \mathscr{R}^n$ is

$$\omega_\mathbf{x}(\mathbf{v}) = a_1(\mathbf{x})v_1 + \cdots + a_n(\mathbf{x})v_n.$$

The differential form ω is said to be *continuous* (or differentiable, or $\mathscr{C}^1$) provided that its coefficient functions $a_1, \ldots, a_n$ are continuous (or differentiable, or $\mathscr{C}^1$).

Now let ω be a continuous differential form on $U \subset \mathscr{R}^n$, and $\gamma : [a, b] \to U$ a $\mathscr{C}^1$ path. We define the integral of ω over the path γ by

$$\int_\gamma \omega = \int_a^b \omega_{\gamma(t)}(\gamma'(t))\, dt. \tag{14}$$

In other words, if $\omega = a_1\, dx_1 + \cdots + a_n\, dx_n$, then

$$\int_\gamma \omega = \int_a^b [a_1(\gamma(t))\gamma_1'(t) + \cdots + a_n(\gamma(t))\gamma_n'(t)]\, dt.$$

Note the agreement between this and formula (11); a line integral is simply the integral of the differential form appearing as its "integrand."

Example 3 Let ω be the differential form defined on $\mathscr{R}^2$ minus the origin by

$$\omega = \frac{-y\, dx + x\, dy}{x^2 + y^2}.$$

If $\gamma_1 : [0, 1] \to \mathscr{R}^2 - \{\mathbf{0}\}$ is defined by $\gamma_1(t) = (\cos \pi t, \sin \pi t)$, then the image of γ_1 is the upper half of the unit circle, and

$$\int_{\gamma_1} \omega = \int_0^1 \frac{-(\sin \pi t)(-\pi \sin \pi t) + (\cos \pi t)(\pi \cos \pi t)}{\cos^2 \pi t + \sin^2 \pi t}\, dt = \pi.$$

If $\gamma_2 : [0, 1] \to \mathscr{R}^2 - \{\mathbf{0}\}$ is defined by $\gamma_2(t) = (\cos \pi t, -\sin \pi t)$, then the image of γ_2 is the lower half of the unit circle (Fig. 5.6), and

$$\int_{\gamma_2} \omega = \int_0^1 \frac{-(-\sin \pi t)(-\pi \sin \pi t) + (\cos \pi t)(-\pi \cos \pi t)}{\cos^2 \pi t + \sin^2 \pi t}\, dt = -\pi.$$

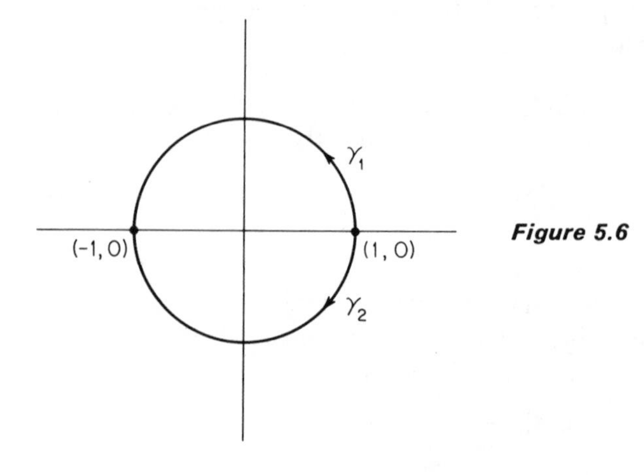

Figure 5.6

In a moment we will see the significance of the fact that $\int_{\gamma_1} \omega \neq \int_{\gamma_2} \omega$.

Recall that, if f is a differentiable real-valued function on the open set $U \subset \mathscr{R}^n$, then its differential $df_{\mathbf{x}}$ at $\mathbf{x} \in U$ is the linear function on $\mathscr{R}^n$ defined by

$$df_{\mathbf{x}}(\mathbf{v}) = D_1 f(\mathbf{x})v_1 + \cdots + D_n f(\mathbf{x})v_n.$$

Consequently we see that the differential of a differentiable function is a differential form

$$df_{\mathbf{x}} = D_1 f(\mathbf{x})\, dx_1 + \cdots + D_n f(\mathbf{x})\, dx_n,$$

or

$$df = \frac{\partial f}{\partial x_1} dx_1 + \cdots + \frac{\partial f}{\partial x_n} dx_n.$$

Example 4 Let U denote $\mathscr{R}^2$ minus the nonnegative x-axis; that is, $(x, y) \in U$ unless $x \geqq 0$ and $y = 0$. Let $\theta : U \to \mathscr{R}$ be the polar angle function defined in the obvious way (Fig. 5.7). In particular,

$$\theta(x, y) = \arctan \frac{y}{x}$$

if $x \neq 0$, so

$$D_1\theta(x, y) = \frac{-y}{x^2 + y^2} \qquad \text{and} \qquad D_2\,\theta(x, y) = \frac{x}{x^2 + y^2}.$$

Therefore

$$d\theta = \frac{-y\, dx + x\, dy}{x^2 + y^2}$$

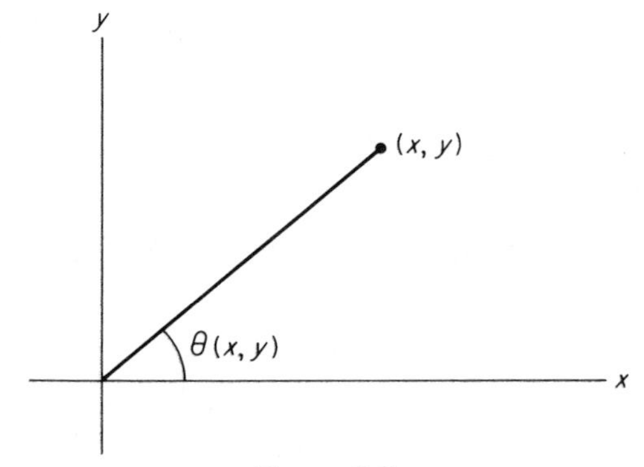

Figure 5.7

on U; the differential of θ agrees on its domain of definition U with the differential form ω of Example 3. Although

$$\omega = \frac{-y\,dx + x\,dy}{x^2 + y^2}$$

is defined on $\mathscr{R}^2 - \{\mathbf{0}\}$, it is clear that the angle function cannot be extended continuously to $\mathscr{R}^2 - \{\mathbf{0}\}$. As a consequence of the next theorem, we will see that ω is *not* the differential of any differentiable function that is defined on all of $\mathscr{R}^2 - \{\mathbf{0}\}$.

Recall the fundamental theorem of calculus in the form

$$\int_a^b f'(t)\,dt = f(b) - f(a).$$

Now $f'(t)\,dt$ is simply the differential at t of $f : \mathscr{R} \to \mathscr{R}$; $dt : \mathscr{R} \to \mathscr{R}$ is the identity mapping. If $\gamma : [a, b] \to [a, b]$ is also the identity, then

$$\int_\gamma df = \int_a^b df_t(\gamma'(t))\,dt = \int_a^b df_t(1)\,dt = \int_a^b f'(t)\,dt,$$

and $f(b) - f(a) = f(\gamma(b)) - f(\gamma(a))$, so the fundamental theorem of calculus takes the form

$$\int_\gamma df = f(\gamma(b)) - f(\gamma(a)).$$

The following theorem generalizes this formula; it is the "fundamental theorem of calculus for paths in $\mathscr{R}^n$."

Theorem 1.5 If f is a real-valued $\mathscr{C}^1$ function on the open set $U \subset \mathscr{R}^n$, and $\gamma : [a, b] \to U$ is a $\mathscr{C}^1$ path, then

$$\int_\gamma df = f(\gamma(b)) - f(\gamma(a)). \tag{15}$$

PROOF Define $g : [a, b] \to \mathscr{R}$ by $g = f \circ \gamma$. Then $g'(t) = \nabla f(\gamma(t)) \cdot \gamma'(t)$ by the chain rule, so

$$\begin{aligned}
\int_\gamma df &= \int_a^b df_{\gamma(t)}(\gamma'(t))\,dt \\
&= \int_a^b [D_1 f(\gamma(t))\gamma_1'(t) + \cdots + D_n f(\gamma(t))\gamma_n'(t)]\,dt \\
&= \int_a^b \nabla f(\gamma(t)) \cdot \gamma'(t)\,dt \\
&= \int_a^b g'(t)\,dt
\end{aligned}$$

(equation continues)

$$= g(b) - g(a)$$
$$= f(\gamma(b)) - f(\gamma(a)). \qquad \blacksquare$$

It is an immediate corollary to this theorem that, if the differential form ω on U is the differential of some $\mathscr{C}^1$ function on the open set U, then the integral $\int_\gamma \omega$ is *independent of the path* γ, to the extent that it depends only on the initial and terminal points of γ.

Corollary 1.6 If $\omega = df$, where f is a $\mathscr{C}^1$ function on U, and α and β are two $\mathscr{C}^1$ paths in U with the same initial and terminal points, then

$$\int_\alpha \omega = \int_\beta \omega.$$

We now see that the result $\int_{\gamma_1} \omega \neq \int_{\gamma_2} \omega$ of Example 3, where γ_1 and γ_2 were two different paths in $\mathscr{R}^2 - \{\mathbf{0}\}$ from $(1, 0)$ to $(-1, 0)$, implies that the differential form

$$\omega = \frac{-y\,dx + x\,dy}{x^2 + y^2}$$

is not the differential of any $\mathscr{C}^1$ function on $\mathscr{R}^2 - \{\mathbf{0}\}$. Despite this fact, it is customarily denoted by $d\theta$, because its integral over a path γ measures the polar angle between the endpoints of γ.

Recall that, if $\mathbf{F} : U \to \mathscr{R}^n$ is a force field on $U \subset \mathscr{R}^n$ and $\gamma : [a, b] \to \mathscr{R}^n$ is a $\mathscr{C}^1$ path in U, then the work done by the field $\mathbf{F}$ in transporting a particle along the path γ is defined by

$$W = \int_\gamma \omega,$$

where $\omega = F_1\,dx_1 + \cdots + F_n\,dx_n$. Suppose now that the force field $\mathbf{F}$ is *conservative* (see Exercise II.1.3), that is, there exists a $\mathscr{C}^1$ function $g : U \to \mathscr{R}$ such that $\mathbf{F} = \nabla g$. Since this means that $\omega = dg$, Corollary 1.6 implies that W depends only on the initial and terminal points of γ, and in particular that

$$W = g(\gamma(b)) - g(\gamma(a)).$$

This is the statement that the work done by a conservative force field, in moving a particle from one point to another, is equal to the "difference in potential" of the two points.

We close this section with a discussion of the arclength form ds of an oriented curve in $\mathscr{R}^n$. This will provide an explication of the notation of formulas (8) and (10).

The set C in $\mathscr{R}^n$ is called a *curve* if and only if it is the image of a smooth path γ which is *one-to-one*. Any one-to-one smooth path which is equivalent to γ is then called a *parametrization* of C.

If $\mathbf{x} = \gamma(t) \in C$, then

$$\mathbf{T}(\mathbf{x}) = \frac{\gamma'(t)}{|\gamma'(t)|}$$

is a unit tangent vector to C at $\mathbf{x}$ (Fig. 5.8), and it is easily verified that $\mathbf{T}(\mathbf{x})$ is independent of the chosen parametrization γ of C (Exercise 1.3). Such a continuous mapping $\mathbf{T}: C \to \mathscr{R}^n$, such that $\mathbf{T}(\mathbf{x})$ is a unit tangent vector to C at $\mathbf{x}$, is called an *orientation* for C. An *oriented curve* is then a pair $(C, \mathbf{T})$, where $\mathbf{T}$ is an orientation for C. However we will ordinarily abbreviate $(C, \mathbf{T})$ to C, and write $-C$ for $(C, -\mathbf{T})$, the same geometric curve with the opposite orientation.

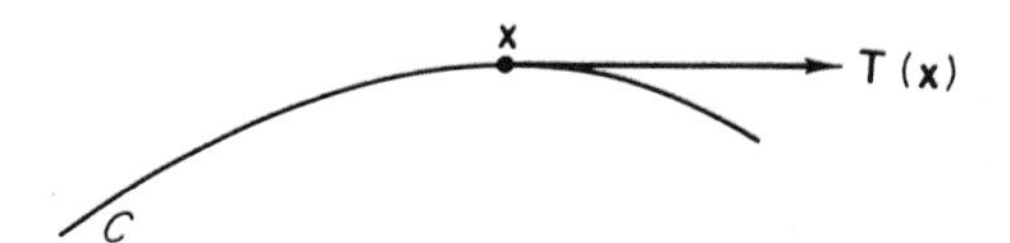

Figure 5.8

Given an oriented curve C in $\mathscr{R}^n$, its *arclength form* ds is defined for $\mathbf{x} \in C$ by

$$ds_{\mathbf{x}}(\mathbf{v}) = T(\mathbf{x}) \cdot \mathbf{v}. \tag{16}$$

Thus $ds_{\mathbf{x}}(\mathbf{v})$ is simply the component of $\mathbf{v}$ in the direction of the unit tangent vector $T(\mathbf{x})$. It is clear that $ds_{\mathbf{x}}(\mathbf{v})$ is a linear function of $\mathbf{v} \in \mathscr{R}^n$, so *ds is* a differential form on C.

The following theorem expresses in terms of ds the earlier integrals of this section [compare with formulas (8) and (10)].

Theorem 1.7 Let γ be a parametrization of the oriented curve $(C, \mathbf{T})$, and let ds be the arclength form of C. If $f: \mathscr{R}^n \to \mathscr{R}$ and $\mathbf{F}: \mathscr{R}^n \to \mathscr{R}^n$ are continuous mappings, then

(a) $\displaystyle\int_a^b f(\gamma(t))|\gamma'(t)|\, dt = \int_\gamma f\, ds,$

(so in particular $s(\gamma) = \int_\gamma ds$), and

(b) $\displaystyle\int_a^b \mathbf{F}(\gamma(t)) \cdot \gamma'(t)\, dt = \int_\gamma \mathbf{F} \cdot \mathbf{T}\, ds.$

PROOF We verify (a), and leave the proof of (*b*) as an exercise. By routine application of the definitions, we obtain

$$\begin{aligned} \textstyle\int_\gamma f\, ds &= \int_a^b f(\gamma(t))\, ds_{\gamma(t)}(\gamma'(t))\, dt \\ &= \int_a^b f(\gamma(t))\mathbf{T}(\gamma(t)) \cdot \gamma'(t)\, dt \end{aligned}$$

(equation continues)

$$= \int_a^b f(\gamma(t)) \frac{\gamma'(t)}{|\gamma'(t)|} \cdot \gamma'(t)\, dt$$

$$= \int_a^b f(\gamma(t))|\gamma'(t)|\, dt. \qquad \blacksquare$$

Finally notice that, if the differential form $\omega = F_1\, dx_1 + \cdots + F_n\, dx_n$ is defined on the oriented smooth curve $(C, \mathbf{T})$ with parametrization γ, then

$$\begin{aligned}\int_\gamma \omega &= \int_\gamma F_1\, dx_1 + \cdots + F_n\, dx_n \\ &= \int_a^b [F_1(\gamma(t))\gamma_1'(t) + \cdots + F_n(\gamma(t))\gamma_n'(t)]\, dt \\ &= \int_a^b \mathbf{F}(\gamma(t)) \cdot \gamma'(t)\, dt \\ &= \int_\gamma \mathbf{F} \cdot \mathbf{T}\, ds\end{aligned}$$

by Theorem 1.7(b). Thus every integral of a differential form, over a parametrization of an oriented smooth curve, is an "arclength integral."

Exercises

1.1 Show that the relation of equivalence of paths in $\mathscr{R}^n$ is symmetric and transitive.

1.2 If α and β are equivalent paths in $\mathscr{R}^n$, and α is smooth, show that β is also smooth.

1.3 Show that any two equivalent parametrizations of the smooth curve $\mathscr{C}$ induce the same orientation of $\mathscr{C}$.

1.4 If α and β are equivalent $\mathscr{C}^1$ paths in $\mathscr{R}^n$, and ω is a continuous differential form, show that $\int_\alpha \omega = \int_\beta \omega$.

1.5 If $\alpha : [0,1] \to \mathscr{R}^n$ is a $\mathscr{C}^1$ path and ω is a continuous differential form, define $\beta(t) = \alpha(1-t)$ for $t \in [0,1]$. Then show that $\int_\alpha \omega = -\int_\beta \omega$.

1.6 If $\alpha : [a, b] \to \mathscr{R}^n$ and $\beta : [c, d] \to \mathscr{R}^n$ are smooth one-to-one paths with the same image, and $\alpha(a) = \beta(c)$, $\alpha(b) = \beta(d)$, show that α and β are equivalent.

1.7 Show that the circumference of the ellipse $x^2/a^2 + y^2/b^2 = 1$ is

$$s = 4aE\left(\frac{1}{a}(a^2 - b^2)^{1/2}, \frac{\pi}{2}\right),$$

where $E(k, \varphi) = \int_0^\varphi (1 - k^2 \sin^2 t)^{1/2}\, dt$ denotes the standard "elliptic integral of the second kind."

1.8 (a) Given $\mathscr{C}^1$ mappings $[a, b] \xrightarrow{c} \mathscr{R}^2_{uv} \xrightarrow{T} \mathscr{R}^3$, define $\gamma = T \circ c$. Show that

$$\gamma'(t) = c_1'(t) D_1 T(c(t)) + c_2'(t) D_2 T(c(t))$$

and conclude that the length of γ is

$$s(\gamma) = \int_a^b \left[E\left(\frac{du}{dt}\right)^2 + 2F\frac{du}{dt}\frac{dv}{dt} + G\left(\frac{dv}{dt}\right)^2\right]^{1/2} dt,$$

where

$$E = \frac{\partial T}{\partial u} \cdot \frac{\partial T}{\partial u}, \qquad F = \frac{\partial T}{\partial u} \cdot \frac{\partial T}{\partial v}, \qquad G = \frac{\partial T}{\partial v} \cdot \frac{\partial T}{\partial v}.$$

(b) Let $T : \mathscr{R}^2_{\varphi\theta} \to \mathscr{R}^3_{xyz}$ be the spherical coordinates mapping given by

$$x = \sin \varphi \cos \theta, \qquad y = \sin \varphi \sin \theta, \qquad z = \cos \varphi.$$

Let $\gamma : [a, b] \to \mathscr{R}^3$ be a path on the unit sphere described by giving φ and θ as functions of t, that is, $c(t) = (\varphi(t), \theta(t))$. Deduce from (a) that

$$s(\gamma) = \int_a^b \left[\left(\frac{d\varphi}{dt} \right)^2 + \sin^2 \varphi \left(\frac{d\theta}{dt} \right)^2 \right]^{1/2} dt.$$

1.9 (a) Given $\mathscr{C}^1$ mappings $[a, b] \xrightarrow{c} \mathscr{R}^n_{\mathbf{u}} \xrightarrow{F} \mathscr{R}^n_{\mathbf{x}}$, define $\gamma = F \circ c$. Show that

$$\gamma'(t) = \sum_{i=1}^{n} c_i'(t) D_i F(c(t))$$

and conclude that the length of γ is

$$s(\gamma) = \int_a^b \left[\sum_{i,j=1}^{n} g_{ij} \frac{du_i}{dt} \frac{du_j}{dt} \right]^{1/2} dt,$$

where $g_{ij} = (\partial F/\partial u_i) \cdot (\partial F/\partial u_j)$.
(b) Let $F : \mathscr{R}^3_{r\theta z} \to \mathscr{R}^3_{xyz}$ be the cylindrical coordinates mapping given by

$$x = r \cos \theta, \qquad y = r \sin \theta, \qquad z = z.$$

Let $\gamma : [a, b] \to \mathscr{R}^3_{xyz}$ be a path described by giving r, θ, z as functions of t, that is,

$$c(t) = (r(t), \theta(t), z(t)).$$

Deduce from (a) that

$$s(\gamma) = \int_a^b \left[\left(\frac{dr}{dt} \right)^2 + r^2 \left(\frac{d\theta}{dt} \right)^2 + \left(\frac{dz}{dt} \right)^2 \right]^{1/2} dt.$$

1.10 If $\mathbf{F} : \mathscr{R}^n \to \mathscr{R}^n$ is a $\mathscr{C}^1$ force field and $\gamma : [a, b] \to \mathscr{R}^n$ a $\mathscr{C}^1$ path, use the fact that $\mathbf{F}(\gamma(t)) = m\gamma''(t)$ to show that the work W done by this force field in moving a particle of mass m along the path γ from $\gamma(a)$ to $\gamma(b)$ is

$$W = \tfrac{1}{2} m v(b)^2 - \tfrac{1}{2} m v(a)^2,$$

where $v(t) = |\gamma'(t)|$. Thus the work done by the field equals the increase in the kinetic energy of the particle.

1.11 For each $(x, y) \in \mathscr{R}^2$, let $\mathbf{F}(x, y)$ be a unit vector pointed toward the origin from (x, y). Calculate the work done by the force field $\mathbf{F}$ in moving a particle from $(2a, 0)$ to $(0, 0)$ along the top half of the circle $(x - a)^2 + y^2 = a^2$.

1.12 Let $d\theta = (-y\,dx + x\,dy)/(x^2 + y^2)$. Thinking of polar angles, explain why it seems "obvious" that $\int_\gamma d\theta = 2\pi$, where $\gamma(t) = (a \cos t, b \sin t)$, $t \in [0, 2\pi]$. Accepting this fact, deduce from it that

$$\int_0^{2\pi} \frac{dt}{a^2 \cos^2 t + b^2 \sin^2 t} = \frac{2\pi}{ab}.$$

1.13 Let $d\theta = (-y\,dx + x\,dy)/(x^2 + y^2)$ on $\mathscr{R}^2 - \mathbf{0}$.
(a) If $\gamma : [0, 1] \to \mathscr{R}^2 - \mathbf{0}$ is defined by $\gamma(t) = (\cos 2k\pi t, \sin 2k\pi t)$, where k is an integer, show that $\int_\gamma d\theta = 2k\pi$.

(b) If $\gamma : [0, 1] \to \mathscr{R}^2$ is a closed path (that is, $\gamma(0) = \gamma(1)$) whose image lies in the open first quadrant, conclude from Theorem 1.5 that $\int_\gamma d\theta = 0$.

1.14 Let the differential form ω be defined on the open set $U \subset \mathscr{R}^2$, and suppose there exist continuous functions, $f, g : \mathscr{R} \to \mathscr{R}$ such that $\omega(x, y) = f(x)\,dx + g(y)\,dy$ if $(x, y) \in U$. Apply Theorem 1.5 to show that $\int_\gamma \omega = 0$ if γ is a closed $\mathscr{C}^1$ path in U.

1.15 Prove Theorem 1.7(b).

1.16 Let $\mathbf{F} : U \to \mathscr{R}^n$ be a continuous vector field on $U \subset \mathscr{R}^n$ such that $|\mathbf{F}(\mathbf{x})| \leqq M$ if $\mathbf{x} \in U$. If $\mathscr{C}$ is a curve in U, and γ a parametrization of $\mathscr{C}$, show that

$$\int \mathbf{F} \cdot \mathbf{T}\, ds \leqq Ms(\gamma).$$

1.17 If γ is a parametrization of the oriented curve $\mathscr{C}$ with arclength form ds, show that $ds_{\gamma(t)}(\gamma'(t)) = |\gamma'(t)|$, so

$$\int_\gamma ds = \int_a^b |\gamma'(t)|\, dt = s(\gamma).$$

1.18 Let $\gamma(t) = (\cos t, \sin t)$ for $t \in [0, 2\pi]$ be the standard parametrization of the unit circle in $\mathscr{R}^2$. If

$$\omega = \frac{(x - y)\,dx + (x + y)\,dy}{x^2 + y^2}$$

on $\mathscr{R}^2 - \mathbf{0}$, calculate $\int_\gamma \omega$. Then explain carefully why your answer implies that there is *no* function $f : \mathscr{R}^2 - \mathbf{0} \to \mathscr{R}$ with $df = \omega$.

1.19 Let $\omega = y\,dx + x\,dy + 2z\,dz$ on $\mathscr{R}^3$. Write down by inspection a differentiable function $f : \mathscr{R}^3 \to \mathscr{R}$ such that $df = \omega$. If $\gamma : [a, b] \to \mathscr{R}^3$ is a $\mathscr{C}^1$ path from $(1, 0, 1)$ to $(0, 1, -1)$, what is the value of $\int_\gamma \omega$?

1.20 Let γ be a smooth parametrization of the curve $\mathscr{C}$ in $\mathscr{R}^2$, with unit tangent $\mathbf{T}$ and unit normal $\mathbf{N}$ defined by

$$\mathbf{T}(\gamma(t)) = \frac{\gamma'(t)}{|\gamma'(t)|} = \frac{1}{|\gamma'(t)|}(\gamma_1'(t), \gamma_2'(t))$$

and

$$\mathbf{N}(\gamma(t)) = \frac{1}{|\gamma'(t)|}(\gamma_2'(t), -\gamma_1'(t)).$$

Given a vector field $\mathbf{F} = (F_1, F_2)$ on $\mathscr{R}^2$, let $\omega = -F_2\,dx + F_1\,dy$, and then show that

$$\int_\gamma \mathbf{F} \cdot \mathbf{N}\, ds = \int_\gamma \omega.$$

1.21 Let $\mathscr{C}$ be an oriented curve in $\mathscr{R}^n$, with unit tangent vector $\mathbf{T} = (t_1, \ldots, t_n)$.

(a) Show that the definition [Eq. (16)] of the arclength form ds of C is equivalent to

$$ds = \sum_{i=1}^n t_i\, dx_i.$$

(b) If the vector $\mathbf{v}$ is tangent to C, show that

$$t_i\, ds(\mathbf{v}) = dx_i(\mathbf{v}), \qquad i = 1, \ldots, n.$$

(c) If $\mathbf{F} = (F_1, \ldots, F_n)$ is a vector field, show that the 1-forms

$$\mathbf{F} \cdot \mathbf{T}\, ds \qquad \text{and} \qquad \sum_{i=1}^n F_i\, dx_i$$

agree on vectors that are tangent to C.

2 GREEN'S THEOREM

Let us recall again the fundamental theorem of calculus in the form

$$\int_a^b f'(t)\,dt = f(b) - f(a), \tag{1}$$

where f is a real-valued $\mathscr{C}^1$ function on the interval $[a, b]$. Regarding the interval $I = [a, b]$ as an oriented smooth curve from a to b, we may write $\int_I df$ for the left-hand side of (1). Regarding the right-hand side of (1) as a sort of "0-dimensional integral" of the function f over the boundary ∂I of I (which consists of the two points a and b), and thinking of b as the positive endpoint and a the negative endpoint of I (Fig. 5.9), let us write

$$\int_{\partial I} f = f(b) - f(a).$$

Then Eq. (1) takes the appealing (if artificially contrived) form

$$\int_I df = \int_{\partial I} f. \tag{2}$$

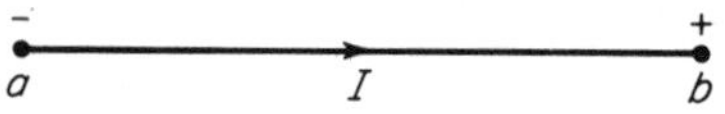

Figure 5.9

Green's theorem is a 2-dimensional generalization of the fundamental theorem of calculus. In order to state it in a form analogous to Eq. (2), we need the notion of the differential of a differential form (to play the role of the differential of a function). Given a $\mathscr{C}^1$ differential form $\omega = P\,dx + Q\,dy$ in two variables, we define its *differential* $d\omega$ by

$$d\omega = \left(\frac{\partial Q}{\partial x} - \frac{\partial P}{\partial y}\right) dx\,dy.$$

Thus $d\omega$ is a *differential 2-form*, that is, an expression of the form $a\,dx\,dy$, where a is a real-valued function of x and y. In a subsequent section we will give a definition of differential 2-forms in terms of bilinear mappings (*à la* the definition of the differential form $a_1\,dx_1 + \cdots + a_n\,dx_n$ in Section 1), but for present purposes it will suffice to simply regard $a\,dx\,dy$ as a formal expression whose role is solely notational.

Given a *continuous* differential 2-form $\alpha = a\,dx\,dy$ (that is, the function a is continuous) and a contented set $D \subset \mathscr{R}^2$, the *integral* of α on D is defined by

$$\int_D \alpha = \iint_D a(x, y)\,dx\,dy,$$

where the right-hand side is an "ordinary" integral.

Now that we have two types of differential forms, we will refer to

$$a_1\,dx_1 + \cdots + a_n\,dx_n \qquad \text{or} \qquad P\,dx + Q\,dy$$

as a *differential* 1-*form*. If moreover we agree to call real-valued functions 0-*forms*, then the differential of a 0-form is a 1-form, while the differential of a 1-form is a 2-form. We will eventually have a full array of differential forms in all dimensions, with the differential of a k-form being a $(k+1)$-form.

We are now ready for a preliminary informal statement of Green's theorem.

Let D be a "nice" region in the plane $\mathscr{R}^2$, whose boundary ∂D consists of a finite number of closed curves, each of which is "positively oriented" with respect to D. If $\omega = P\,dx + Q\,dy$ is a $\mathscr{C}^1$ differential 1-*form defined on D, then*

$$\int_D d\omega = \int_{\partial D} \omega. \tag{3}$$

That is,

$$\iint_D \left(\frac{\partial Q}{\partial x} - \frac{\partial P}{\partial y}\right) dx\,dy = \int_{\partial D} P\,dx + Q\,dy.$$

Note the formal similarity between Eqs. (2) and (3). In each, the left-hand side is the integral over a set of the differential of a form, while the right-hand side is the integral of the form over the boundary of the set. We will later see, in Stokes' theorem, a comprehensive multidimensional generalization of this phenomenon.

The above statement fails in several respects to be (as yet) an actual theorem. We have not yet said what we mean by a "nice" region in the plane, nor what it means for its boundary curves to be "positively oriented." Also, there is a question as to the definition of the integral on the right-hand side of (3), since we have only defined the integral of a 1-form over a $\mathscr{C}^1$ path (rather than a curve, as such).

The last question is the first one we will consider. The continuous path $\gamma : [a, b] \to \mathscr{R}^n$ is called *piecewise smooth* if there is a partition $\mathscr{P} = \{a = a_0 < a_1 < \cdots < a_k = b\}$ of the interval $[a, b]$ such that each restriction γ^i of γ to $[a_{i-1}, a_i]$, defined by

$$\gamma^i(t) = \gamma(t) \qquad \text{for} \quad t \in [a_{i-1}, a_i],$$

is smooth. If ω is a continuous differential 1-form, we then define

$$\int_\gamma \omega = \sum_{i=1}^{k} \int_{\gamma^i} \omega.$$

It is easily verified that this definition of $\int_\gamma \omega$ is independent of the partition $\mathscr{P}$. (Exercise 2.11).

A *piecewise-smooth curve* C in $\mathscr{R}^n$ is the image of a piecewise-smooth path $\gamma : [a, b] \to \mathscr{R}^n$ which is one-to-one on (a, b); C is *closed* if $\gamma(a) = \gamma(b)$. The path γ is a *parametrization* of C, and the pair (C, γ) is called an *oriented* piecewise-smooth curve [although we will ordinarily abbreviate (C, γ) to C].

Now let $\beta : [c, d] \to \mathscr{R}^n$ be a second piecewise-smooth path which is one-to-one on (c, d) and whose image is C. Then we write $(C, \gamma) = (C, \beta)$ [respectively, $(C, \gamma) = -(C, \beta)$], and say that γ and β induce the same orientation [respectively, opposite orientations] of C, provided that their unit tangent vectors are equal [respectively, opposite] at each point where both are defined. It then follows from Exercises 1.4 and 1.5 that, for any continuous differential 1-form ω,

$$\int_\gamma \omega = \int_\beta \omega \tag{4}$$

if γ and β induce the same orientation of C, while

$$\int_\gamma \omega = -\int_\beta \omega \tag{5}$$

if γ and β induce opposite orientations of C (see Exercise 2.12).

Given an *oriented* piecewise-smooth curve C, and a continuous differential 1-form ω defined on C, we may now define the integral of ω over C by

$$\int_C \omega = \int_\gamma \omega,$$

where γ is any parametrization of C. It then follows from (4) and (5) that $\int_C \omega$ is thereby well defined, and that

$$\int_{-C} \omega = -\int_C \omega,$$

where $C = (C, \gamma)$ and $-C = -(C, \gamma) = (C, \beta)$, with γ and β inducing opposite orientations.

Now, finally, the right-hand side of (3) is meaningful, provided that ∂D consists of mutually disjoint oriented piecewise-smooth closed curves $C_1, \ldots, C_r$; then

$$\int_{\partial D} \omega = \sum_{i=1}^{r} \int_{C_i} \omega.$$

A *nice region* in the plane is a connected compact (closed and bounded) set $D \subset \mathscr{R}^2$ whose boundary ∂D is the union of a finite number of mutually disjoint piecewise-smooth closed curves (as above). It follows from Exercise IV.5.1 that every nice region D is contented, so $\int_D \alpha$ exists if α is a continuous 2-form. Figure 5.10 shows a disk, an annular region (or "disk with one hole"), and a "disk with two holes"; each of these is a nice region.

It remains for us to say what is meant by a "positive orientation" of the boundary of the nice region D. The intuitive meaning of the statement that an oriented boundary curve $C = (C, \gamma)$ of a nice region D is positively oriented with respect to D is that the region D stays on one's left as he proceeds around C in the direction given by its parametrization γ. For example, if D is a circular disk, the positive orientation of its boundary circle is its counterclockwise one.

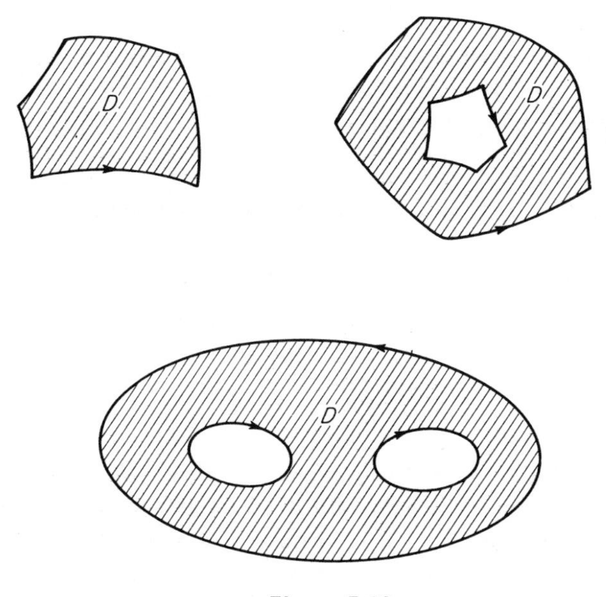

Figure 5.10

According to the Jordan curve theorem, every closed curve C in $\mathscr{R}^2$ separates $\mathscr{R}^2$ into two connected open sets, one bounded and the other unbounded (the interior and exterior components, respectively, of $\mathscr{R}^2 - C$). This is a rather difficult topological theorem whose proof will not be included here. The Jordan curve theorem implies the following fact (whose proof is fairly easy, but will also be omitted): Among all of the boundary curves of the nice region D, there is a unique one whose interior component contains all the other boundary curves (if any) of D. This distinguished boundary curve of D will be called its *outer* boundary curve; the others (if any) will be called *inner* boundary curves of D.

The above "left-hand rule" for positively orienting the boundary is equivalent to the following formulation. The *positive orientation* of the boundary of a nice region D is that for which the outer boundary curve of D is oriented counter-clockwise, and the inner boundary curves of D (if any) are all oriented clockwise.

Although this definition will suffice for our applications, the alert reader will see that there is still a problem—what do the words "clockwise" and "counterclockwise" actually mean? To answer this, suppose, for example, that C is an oriented piecewise-smooth closed curve in $\mathscr{R}^2$ which encloses the origin. Once Green's theorem for nice regions has been proved (in the sense that, given a nice region D, $\int_D d\omega = \int_{\partial D} \omega$ for *some* orientation of ∂D), it will then follow (see Example 3 below) that

$$\int_C d\theta = \pm 2\pi,$$

where $d\theta$ is the usual polar angle form. We can then say that C is oriented *counterclockwise* if $\int_C d\theta = +2\pi$, *clockwise* if $\int_C d\theta = -2\pi$.

Before proving Green's theorem, we give several typical applications.

Example 1 Given a line integral $\int_C \omega$, where C is an oriented closed curve bounding a nice region D, it is sometimes easier to compute $\int_D d\omega$. For example,

$$\int_C 2xy\,dx + x^2\,dy = \iint_D \left[\frac{\partial}{\partial x}(x^2) - \frac{\partial}{\partial y}(2xy)\right] dx\,dy$$

$$= \iint_D (0)\,dx\,dy = 0.$$

Example 2 Conversely, given an integral $\iint_D f(x, y)\,dx\,dy$ which is to be evaluated, it may be easier to think of a differential 1-form ω such that $d\omega = f(x, y)\,dx\,dy$, and then compute $\int_{\partial D} \omega$. For example, if D is a nice region and ∂D is positively oriented, then its area is

$$A = \iint_D 1\,dx\,dy = \iint_D (-\tfrac{1}{2}y\,dx + \tfrac{1}{2}x\,dy)$$

$$= \tfrac{1}{2}\int_{\partial D} -y\,dx + x\,dy.$$

The formulas

$$A = \int_{\partial D} -y\,dx = \int_{\partial D} x\,dy$$

are obtained similarly. For instance, suppose that D is the elliptical disk $x^2/a^2 + y^2/b^2 \leq 1$, whose boundary (the ellipse) is parametrized by $x = a\cos t$, $y = b\sin t$, $t \in [0, 2\pi]$. Then its area is

$$A = \tfrac{1}{2}\int_0^{2\pi} [(-b\sin t)(-a\sin t) + (a\cos t)(b\cos t)]\,dt$$

$$= \pi ab,$$

the familiar formula.

Example 3 Let C be a piecewise-smooth closed curve in $\mathscr{R}^2$ which encloses the origin, and is oriented counterclockwise. Let C_a be a clockwise-oriented circle of radius a centered at $\mathbf{0}$, with a sufficiently small that C_a lies in the interior component of $\mathscr{R}^2 - C$ (Fig. 5.11). Then let D be the annular region bounded by C and C_a. If

$$\omega = d\theta = \frac{-y\,dx + x\,dy}{x^2 + y^2},$$

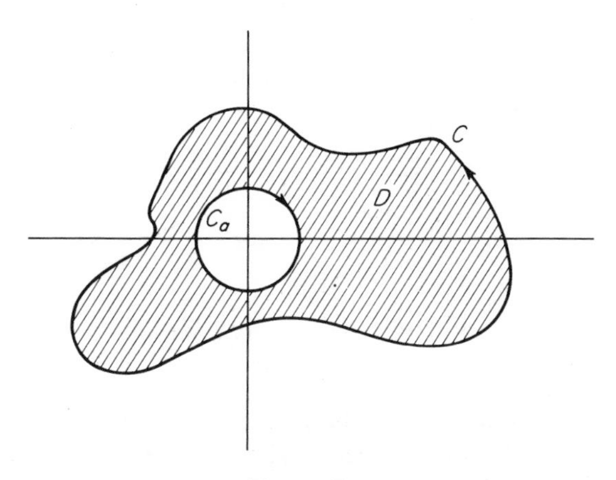

Figure 5.11

then $d\omega = 0$ on D (compute it), so

$$\int_C d\theta + \int_{C_a} d\theta = \int_{\partial D} \omega = \int_D d\omega = 0.$$

Since $\int_{C_a} d\theta = -2\pi$, by essentially the same computation as in Example 3 of Section 1, it follows that

$$\int_C d\theta = -\int_{C_a} d\theta = 2\pi.$$

In particular this is true if C is the ellipse of Exercise 1.12.

On the other hand, if the nice region bounded by C does not contain the origin, then it follows (upon applicaton of Green's theorem to this region) that

$$\int_C d\theta = 0.$$

Equation (3), the conclusion of Green's theorem, can be reformulated in terms of the divergence of a vector field. Given a $\mathscr{C}^1$ vector field $\mathbf{F} : \mathscr{R}^n \to \mathscr{R}^n$, its *divergence* $\operatorname{div} \mathbf{F} : \mathscr{R}^n \to \mathscr{R}$ is the real-valued function defined by

$$\operatorname{div} \mathbf{F} = \frac{\partial F_1}{\partial x_1} + \cdots + \frac{\partial F_n}{\partial x_n},$$

or

$$\operatorname{div} \mathbf{F} = \frac{\partial F_1}{\partial x} + \frac{\partial F_2}{\partial y}$$

in case $n = 2$.

Now let $D \subset \mathscr{R}^2$ be a nice region with ∂D positively oriented, and let $\mathbf{N}$ denote the unit outer normal vector to ∂D, defined as in Exercise 1.20. From that exercise and Green's theorem, we obtain

$$\int_{\partial D} \mathbf{F} \cdot \mathbf{N}\, ds = \int_{\partial D} -F_2\, dx + F_1\, dy$$
$$= \iint_D \left(\frac{\partial F_1}{\partial x} + \frac{\partial F_2}{\partial y} \right) dx\, dy,$$

so

$$\int_{\partial D} \mathbf{F} \cdot \mathbf{N}\, ds = \iint_D \operatorname{div} \mathbf{F}\, dx\, dy \tag{6}$$

for any $\mathscr{C}^1$ vector field $\mathbf{F}$ on D. The number $\int_{\partial D} \mathbf{F} \cdot \mathbf{N}\, ds$ is called the *flux* of the vector field $\mathbf{F}$ across ∂D, and it appears frequently in physical applications. If, for example, $\mathbf{F}$ is the velocity vector field of a moving fluid, then the flux of $\mathbf{F}$ across ∂D measures the rate at which the fluid is leaving the region D.

Example 4 Consider a plane lamina (or thin homogenous plate) in the shape of the open set $U \subset \mathscr{R}^2$, with thermal conductivity k, density ρ, and specific heat c (all constants). Let $u(x, y, t)$ denote the temperature at the point (x, y) at time t. If D is a circular disk in U with boundary circle C, then the heat content of D at time t is

$$h(t) = \iint_D \rho c u(x, y, t)\, dx\, dy.$$

Therefore

$$h'(t) = \iint_D \rho c \frac{\partial u}{\partial t}\, dx\, dy$$

since we can differentiate under the integral sign (see Exercise V.3.5) assuming (as we do) that u is a $\mathscr{C}^2$ function.

On the other hand, since the heat flow vector is $-k\, \nabla u$, the total flux of heat across C (the rate at which heat is leaving D) is

$$\int_C (-k\, \nabla u) \cdot \mathbf{N}\, ds,$$

so

$$h'(t) = \int_C (k \, \nabla u) \cdot \mathbf{N} \, ds$$

$$= \iint_D k \operatorname{div}(\nabla u) \, dx \, dy$$

by Eq. (6).

Equating the above two expressions for $h'(t)$, we find that

$$\iint_D \left[k \operatorname{div}(\nabla u) - \rho c \frac{\partial u}{\partial t} \right] dx \, dy = 0.$$

Since D was an arbitrary disk in U, it follows by continuity that

$$k \operatorname{div}(\nabla u) - \rho c \frac{\partial u}{\partial t} = 0,$$

or

$$\frac{\partial^2 u}{\partial x^2} + \frac{\partial^2 u}{\partial y^2} = a \frac{\partial u}{\partial t} \tag{7}$$

(where $a = \rho c/k$) at each point of U. Equation (7) is the 2-dimensional *heat equation*. Under steady state conditions, $\partial u/\partial t \equiv 0$, it reduces to *Laplace's equation*

$$\nabla^2 u = \frac{\partial^2 u}{\partial x^2} + \frac{\partial^2 u}{\partial y^2} = 0. \tag{8}$$

Having illustrated in the above examples the applicability of Green's theorem, we now embark upon its proof. This proof depends upon an analysis of the geometry or topology of nice regions, and will proceed in several steps. The first step (Lemma 2.1) is a proof of Green's theorem for a single nice region—the *unit square* $I^2 = [0, 1] \times [0, 1] \subset \mathscr{R}^2$; each of the successive steps will employ the previous one to enlarge the class of those nice regions for which we know that Green's theorem holds.

Lemma 2.1 (Green's Theorem for the Unit Square) If $\omega = P\,dx + Q\,dy$ is a $\mathscr{C}^1$ differential 1-form on the unit square I^2, and ∂I^2 is oriented counterclockwise, then

$$\int_{I^2} d\omega = \int_{\partial I^2} \omega. \tag{9}$$

PROOF The proof is by explicit computation. Starting with the left-hand side of (9), and applying in turn Fubini's theorem and the fundamental theorem of calculus, we obtain

$$\begin{aligned}\int_{I^2} d\omega &= \int_0^1 \left(\int_0^1 \frac{\partial Q}{\partial x}\, dx\right) dy - \int_0^1 \left(\int_0^1 \frac{\partial P}{\partial y}\, dy\right) dx \\ &= \int_0^1 Q(1, y)\, dy - \int_0^1 Q(0, y)\, dy - \int_0^1 P(x, 1)\, dx + \int_0^1 P(x, 0)\, dx.\end{aligned}$$

To show that the right-hand side of (9) reduces to the same thing, define the mappings $\gamma_1, \gamma_2, \gamma_3, \gamma_4 : [0, 1] \to \mathscr{R}^2$ by

$$\gamma_1(t) = (t, 0), \qquad \gamma_2(t) = (1, t), \qquad \gamma_3(t) = (1 - t, 1), \qquad \gamma_4(t) = (0, 1 - t);$$

see Fig. 5.12. Then

$$\begin{aligned}\int_{\partial I^2} \omega &= \int_{\gamma_1} \omega + \int_{\gamma_2} \omega + \int_{\gamma_3} \omega + \int_{\gamma_4} \omega \\ &= \int_0^1 P(t, 0)\, dt + \int_0^1 Q(1, t)\, dt + \int_0^1 P(1 - t, 1)\, dt + \int_0^1 Q(0, 1 - t)\, dt \\ &= \int_0^1 P(x, 0)\, dx + \int_0^1 Q(1, y)\, dy - \int_0^1 P(x, 1)\, dx - \int_0^1 Q(0, y)\, dy,\end{aligned}$$

where the last line is obtained from the previous one by means of the substitutions $x = t$, $y = t$, $x = 1 - t$, $y = 1 - t$, respectively. ∎

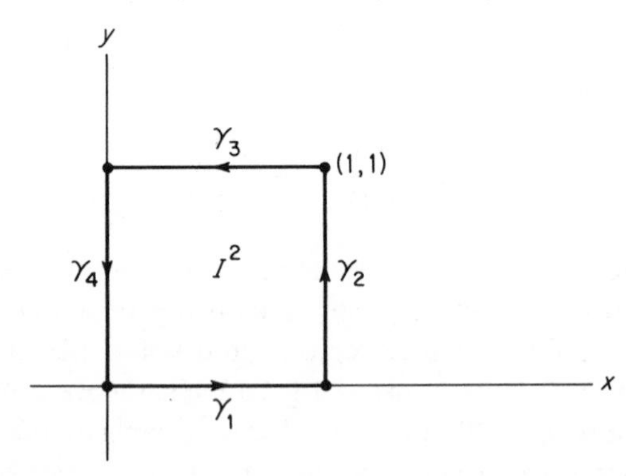

Figure 5.12

Our second step will be to establish Green's theorem for every nice region that is the image of the unit square I^2 under a suitable mapping. The set $D \subset \mathscr{R}^2$ is called an *oriented* (*smooth*) 2-cell if there exists a one-to-one $\mathscr{C}^1$ mapping $F : U \to \mathscr{R}^2$, defined on a neighborhood U of I^2, such that $F(I^2) = D$ and the Jacobian determinant of F is *positive* at each point of U. Notice that the counterclockwise orientation of ∂I^2 induces under F an orientation of ∂D (the positive

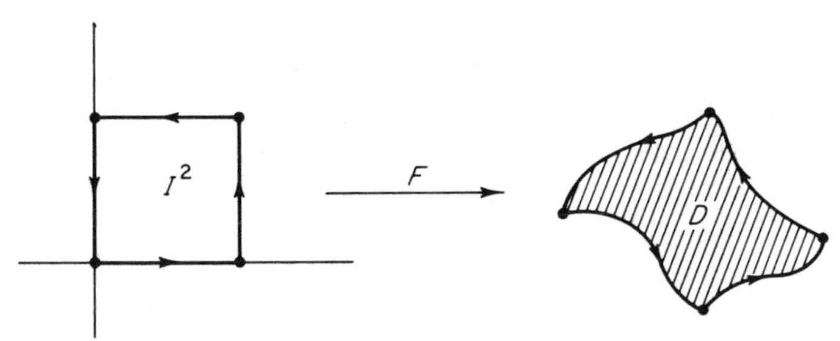

Figure 5.13

or counterclockwise orientation of ∂D). Strictly speaking, an oriented 2-cell should be defined as a pair, consisting of the set D together with this induced orientation of ∂D (Fig. 5.13). Of course it must be verified that this orientation is well defined; that is, if G is a second one-to-one $\mathscr{C}^1$ mapping such that $G(I^2) = D$ and det $G' > 0$, then F and G induce the same orientation of ∂D (Exercise 2.13).

The *vertices* (*edges*) of the oriented 2-cell D are the images under F of the vertices (edges) of the square I^2. Since the differential $dF : \mathscr{R}^2 \to \mathscr{R}^2$ takes straight lines to straight lines (because it is linear), it follows that the *interior angle* (between the tangent lines to the incoming and outgoing edges) at each vertex is $<\pi$. Consequently ∂D is smooth except at the four vertices, but is *not* smooth at the vertices. It follows that neither a circular disk nor a triangular one is an oriented 2-cell, since neither has four "vertices" on its boundary. The other region in Fig. 5.14 fails to be an oriented 2-cell because one of its interior angles is $>\pi$.

If the nice region D is a convex quadrilateral—that is, its single boundary curve contains four straight line edges and each interior angle is $<\pi$—then it can be shown (by explicit construction of the mapping F) that D is an oriented 2-cell (Exercise 2.14). More generally it is true that the nice region D is an oriented 2-cell if its single boundary curve consists of exactly four smooth curves, and the interior angle at each of the four vertices is $<\pi$ (this involves the Jordan curve theorem, and is not so easy).

The idea of the proof of Green's theorem for oriented 2-cells is as follows. Given the differential 1-form ω on the oriented 2-cell D, we will use the mapping $F : I^2 \to D$ to "pull back" ω to a 1-form $F^*\omega$ (defined below) on I^2. We can then apply, to $F^*\omega$, Green's theorem for I^2. We will finally verify that the resulting equation $\int_{I^2} d(F^*\omega) = \int_{\partial I^2}(F^*\omega)$ is equivalent to the equation $\int_D d\omega = \int_{\partial D} \omega$ that we are trying to prove.

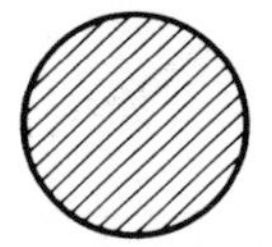
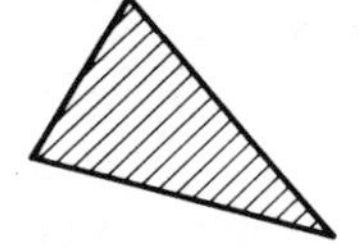
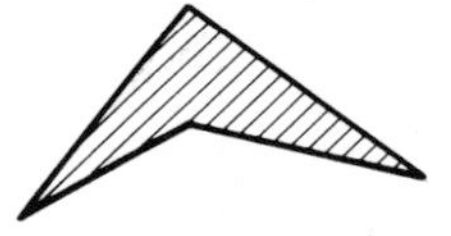

Figure 5.14

Given a $\mathscr{C}^1$ mapping $F: \mathscr{R}^2 \to \mathscr{R}^2$, we must say how differential forms are "pulled back." It will clarify matters to use uv-coordinates in the domain and xy-coordinates in the image, so $F: \mathscr{R}^2_{uv} \to \mathscr{R}^2_{xy}$. A 0-form is just a real valued function $\varphi(x, y)$, and its *pullback* is defined by composition,

$$F^*(\varphi) = \varphi \circ F.$$

That is, $F^*\varphi(u, v) = \varphi(F_1(u, v), F_2(u, v))$, so we are merely using the component functions $x = F_1(u, v)$ and $y = F_2(u, v)$ of F to obtain by substitution a function of u and v (or 0-form on $\mathscr{R}^2_{uv}$) from the given function φ of x and y.

Given a 1-form $\omega = P\,dx + Q\,dy$, we define its *pullback* $F^*\omega = F^*(\omega)$ under F by

$$F^*(\omega) = (F^*P)F^*(dx) + (F^*Q)F^*(dy),$$

where

$$F^*(dx) = \frac{\partial F_1}{\partial u}\,du + \frac{\partial F_1}{\partial v}\,dv \qquad \text{and} \qquad F^*(dy) = \frac{\partial F_2}{\partial u}\,du + \frac{\partial F_2}{\partial v}\,dv.$$

Formally, we obtain $F^*(\omega)$ from ω by simply carrying out the substitution

$$x = F_1(u, v), \qquad y = F_2(u, v),$$

$$dx = \frac{\partial x}{\partial u}\,du + \frac{\partial x}{\partial v}\,dv, \qquad dy = \frac{\partial y}{\partial u}\,du + \frac{\partial y}{\partial v}\,dv.$$

The *pullback* $F^*(\alpha)$ under F of the 2-form $\alpha = g\,dx\,dy$ is defined by

$$F^*(\alpha) = (g \circ F)(\det F')\,du\,dv.$$

The student should verify as an exercise that $F^*(\alpha)$ is the result of making the above formal substitution in $\alpha = g\,dx\,dy$, then multiplying out, making use of the relations $du\,du = dv\,dv = 0$ and $dv\,du = -du\,dv$ which were mentioned at the end of Section IV.5. This phenomenon will be explained once and for all in Section 5 of the present chapter.

Example 5 Let $F: \mathscr{R}^2_{uv} \to \mathscr{R}^2_{xy}$ be given by

$$x = 2u - v, \qquad y = 3u + 2v$$

so $\det F'(u, v) \equiv 7$. If $\varphi(x, y) = x^2 - y^2$, then

$$\begin{aligned} F^*\varphi(u, v) &= (2u - v)^2 - (3u + 2v)^2 \\ &= -5u^2 - 16uv - 3v^2. \end{aligned}$$

If $\omega = -y\,dx + 2x\,dy$, then

$$\begin{aligned} F^*\omega &= -(3u + 2v)(2\,du - dv) + 2(2u - v)(3du + 2\,dv) \\ &= (6u - 10v)\,du + (11u - 2v)\,dv. \end{aligned}$$

If $\alpha = e^{x+y}\, dx\, dy$, then

$$\begin{aligned} F^*\alpha &= e^{(2u-v)+(3u+2v)}(\det F'(u, v))\, du\, dv \\ &= 7e^{5u+v}\, du\, dv. \end{aligned}$$

The following lemma lists the properties of pullbacks that will be needed.

Lemma 2.2 Let $F : U \to \mathscr{R}^2$ be as in the definition of the oriented 2-cell $D = F(I^2)$. Let ω be a $\mathscr{C}^1$ differential 1-form and α a $\mathscr{C}^1$ differential 2-form on D. Then

(a) $\displaystyle\int_{\partial D} \omega = \int_{\partial I^2} F^*\omega,$

(b) $\displaystyle\int_D \alpha = \int_{I^2} F^*\alpha,$

(c) $d(F^*\omega) = F^*(d\omega).$

This lemma will be established in Section 5 in a more general context. However it will be a valuable check on the student's understanding of the definitions to supply the proofs now as an exercise. Part (a) is an immediate consequence of Exercise 2.15. Part (b) follows immediately from the change of variables theorem and the definition of $F^*\alpha$ (in fact this was the motivation for the definition of $F^*\alpha$). Fact (c), which asserts that the differential operation d "commutes" with the pullback operation F^*, follows by direct computation from their definitions.

Lemma 2.3 (Green's Theorem for Oriented 2-Cells) If D is an oriented 2-cell and ω is a $\mathscr{C}^1$ differential 1-form on D, then

$$\int_D d\omega = \int_{\partial D} \omega.$$

PROOF Let $D = F(I^2)$ as in the definition of the oriented 2-cell D. Then

$$\begin{aligned} \int_D d\omega &= \int_{I^2} F^*(d\omega) && \text{by 2.2(b)} \\ &= \int_{I^2} d(F^*\omega) && \text{by 2.2(c)} \\ &= \int_{\partial I^2} F^*\omega && \text{by Green's theorem for } I^2 \\ &= \int_{\partial D} \omega && \text{by 2.2(a).} \end{aligned}$$

∎

The importance of Green's theorem for oriented 2-cells lies not solely in the importance of oriented 2-cells themselves, but in the fact that a more general nice region D may be decomposable into oriented 2-cells in such a way that the truth of Green's theorem for each of these oriented 2-cells implies its truth for D. The following examples illustrate this.

Example 6 Consider an annular region D that is bounded by two concentric circles. D is the union of two oriented 2-cells D_1 and D_2 as indicated in Fig. 5.15.

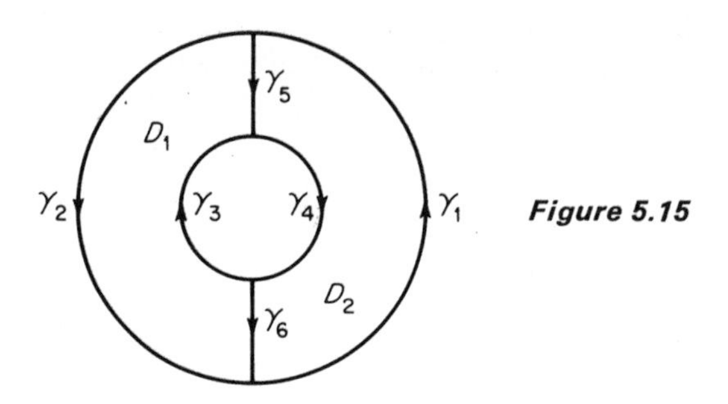

Figure 5.15

In terms of the paths $\gamma_1, \ldots, \gamma_6$ indicated by the arrows, $\partial D_1 = \gamma_2 - \gamma_6 + \gamma_3 - \gamma_5$ and $\partial D_2 = \gamma_1 + \gamma_5 + \gamma_4 + \gamma_6$. Applying Green's theorem to D_1 and D_2, we obtain

$$\int_{D_1} d\omega = \int_{\gamma_2} \omega - \int_{\gamma_6} \omega + \int_{\gamma_3} \omega - \int_{\gamma_5} \omega$$

and

$$\int_{D_2} d\omega = \int_{\gamma_1} \omega + \int_{\gamma_5} \omega + \int_{\gamma_4} \omega + \int_{\gamma_6} \omega.$$

Upon addition of these two equations, the line integrals over γ_5 and γ_6 conveniently cancel to give

$$\begin{aligned} \int_D d\omega &= \int_{D_1} d\omega + \int_{D_2} d\omega \\ &= \left(\int_{\gamma_1} \omega + \int_{\gamma_2} \omega\right) + \left(\int_{\gamma_3} \omega + \int_{\gamma_4} \omega\right) \\ &= \int_{\partial D} \omega, \end{aligned}$$

so Green's theorem holds for the annular region D.

Figure 5.16 shows how to decompose a circular or triangular disk D into oriented 2-cells, so as to apply the method of Example 6 to establish Green's theorem for D. Upon adding the equations obtained by application of Green's theorem to each of the oriented 2-cells, the line integrals over the interior segments will cancel because each is traversed twice in opposite directions, leaving as a result Green's theorem for D.

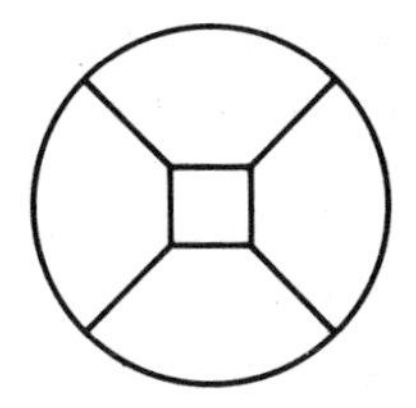

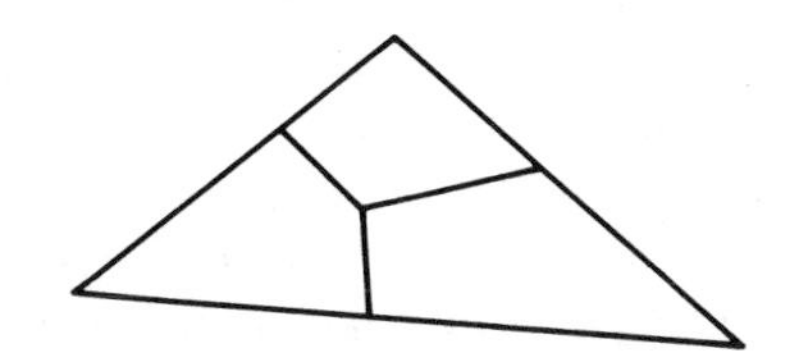

Figure 5.16

Our final version of Green's theorem will result from a formalization of the procedure of these examples. By a (smooth) *cellulation* of the nice region D is meant a (finite) collection $\mathscr{K} = \{D_1, \ldots, D_k\}$ of oriented 2-cells such that

$$D = \bigcup_{i=1}^{k} D_i,$$

with each pair of these oriented 2-cells either being disjoint, or intersecting in a single common vertex, or intersecting in a single common edge on which they induce opposite orientations. Given an edge E of one of these oriented 2-cells D_i, there are two possibilities. If E is also an edge of another oriented 2-cell D_j of $\mathscr{K}$, then E is called an *interior* edge. If D_i is the only oriented 2-cell of $\mathscr{K}$ having E as an edge, then $E \subset \partial D$, and E is a *boundary* edge. Since ∂D is the union of these boundary edges, the cellulation $\mathscr{K}$ induces an orientation of ∂D. This induced orientation of ∂D is unique—any other cellulation of D induces the same orientation of ∂D (Exercise 2.17).

A *cellulated nice region* is a nice region D together with a cellulation $\mathscr{K}$ of D.

Theorem 2.4 (Green's Theorem) If $D \subset \mathscr{R}^2$ is a cellulated nice region and ω is a $\mathscr{C}^1$ differential 1-form on D, then

$$\int_D d\omega = \int_{\partial D} \omega.$$

PROOF We apply Green's theorem for oriented 2-cells to each 2-cell of the cellulation $\mathscr{K} = \{D_1, \ldots, D_k\}$. Then

$$\begin{aligned} \int_D d\omega &= \sum_{i=1}^{k} \int_{D_i} d\omega \\ &= \sum_{i=1}^{k} \int_{\partial D_i} \omega \\ &= \int_{\partial D} \omega \end{aligned}$$

since the line integrals over the interior edges cancel, because each interior edge receives opposite orientations from the two oriented 2-cells of which it is an edge. ∎

Actually Theorem 2.4 suffices to establish Green's theorem for *all* nice regions because every nice region can be cellulated. We refrain from stating this latter result as a formal theorem, because it is always clear in practice how to cellulate a given nice region, so that Theorem 2.4 applies. However an outline of the proof would go as follows. Since we have seen how to cellulate triangles, it suffices to show that every nice region can be "triangulated," that is, decomposed into smooth (curvilinear) triangles that intersect as do the 2-cells of a cellulation. If D has a single boundary curve with three or more vertices, it suffices to pick an interior point $\mathbf{p}$ of D, and then join it to the vertices of ∂D with smooth curves that intersect only in $\mathbf{p}$ (as in Fig. 5.17). Proceeding by

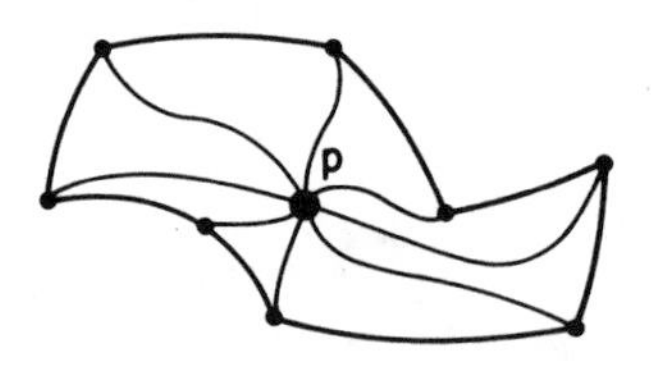

Figure 5.17

induction on the number n of boundary curves of the nice region D, we then notice that, if $n > 1$, D can be expressed as the union of an oriented 2-cell D_1 and a nice region D_2 having $n - 1$ boundary curves (Fig. 5.18). By induction, D_1 and D_2 can then both be triangulated.

This finally brings us to the end of our discussion of the proof of Green's theorem. We mention one more application.

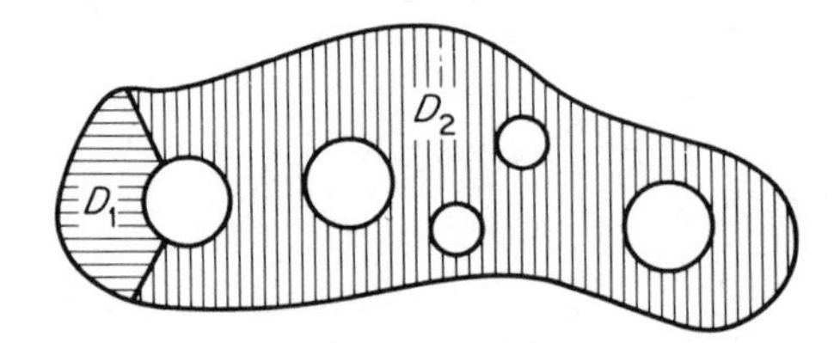

Figure 5.18

Theorem 2.5 If $\omega = P\,dx + Q\,dy$ is a $\mathscr{C}^1$ differential 1-form defined on $\mathscr{R}^2$, then the following three conditions are equivalent:

(a) There exists a function $f : \mathscr{R}^2 \to \mathscr{R}$ such that $df = \omega$.
(b) $\partial Q/\partial x = \partial P/\partial y$ on $\mathscr{R}^2$.
(c) Given points $\mathbf{a}$ and $\mathbf{b}$, the integral $\int_\gamma \omega$ is independent of the piecewise smooth path γ from $\mathbf{a}$ to $\mathbf{b}$.

PROOF We already know that (a) implies both (b) and (c) (Corollary 1.6), so it suffices to show that both (b) and (c) imply (a). Given $(x, y) \in \mathscr{R}^2$, define

$$f(x, y) = \int_{\gamma(x,y)} \omega, \tag{10}$$

where $\gamma(x, y)$ is the straight-line path from $(0, 0)$ to (x, y) (see Fig. 5.19).

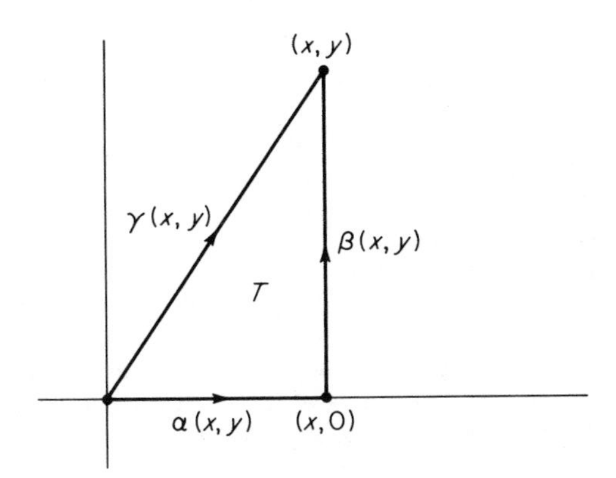

Figure 5.19

Let $\alpha(x, y)$ be the straight-line path from (0, 0) to (x, 0), and $\beta(x, y)$ the straight-line path from (x, 0) to (x, y). Then either immediately from (c), or by application of Green's theorem to the triangle T if we are assuming (b), we see that

$$f(x, y) = \int_{\alpha(x, y)} \omega + \int_{\beta(x, y)} \omega$$
$$= \int_0^x P(t, 0)\, dt + \int_0^y Q(x, t)\, dt.$$

Therefore

$$\frac{\partial f}{\partial y} = \frac{\partial}{\partial y} \int_0^y Q(x, t)\, dt = Q(x, y)$$

by the fundamental theorem of calculus. Similarly $\partial f/\partial x = P$, so $df = \omega$. ∎

The familiar form $d\theta$ shows that the plane $\mathscr{R}^2$ in Theorem 2.5 cannot be replaced by an arbitrary open subset U. However it suffices (as we shall see in Section 8) for U to be star shaped, meaning that U contains the segment $\gamma(x, y)$, for every $(x, y) \in U$.

Example 7 If $\omega = y\, dx + x\, dy$, then the function f defined by (10) is given by

$$f(x_0, y_0) = \int_{\gamma(x_0, y_0)} y\, dx + x\, dy$$
$$= \int_0^1 (y_0 t)(x_0\, dt) + (x_0 t)(y_0\, dt)$$
$$= x_0 y_0 \int_0^1 2t\, dt$$
$$= x_0 y_0,$$

so $f(x, y) = xy$ (as, in the example, could have been seen by inspection).

Exercises

2.1 Let C be the unit circle $x^2 + y^2 = 1$, oriented counterclockwise. Evaluate the following line integrals by using Green's theorem to convert to a double integral over the unit disk D:
(a) $\int_C (3x^2 - y)\,dx + (x + 4y^3)\,dy$,
(b) $\int_C (x^2 + y^2)\,dy$.

2.2 Parametrize the boundary of the ellipse $x^2/a^2 + y^2/b^2 \leqq 1$, and then use the formula $A = \frac{1}{2}\int -y\,dx + x\,dy$ to compute its area.

2.3 Compute the area of the region bounded by the loop of the "folium of Descartes," $x^3 + y^3 = 3xy$. *Hint:* Compute the area of the shaded half D (Fig. 5.20). Set $y = tx$ to discover a parametrization of the curved part of ∂D.

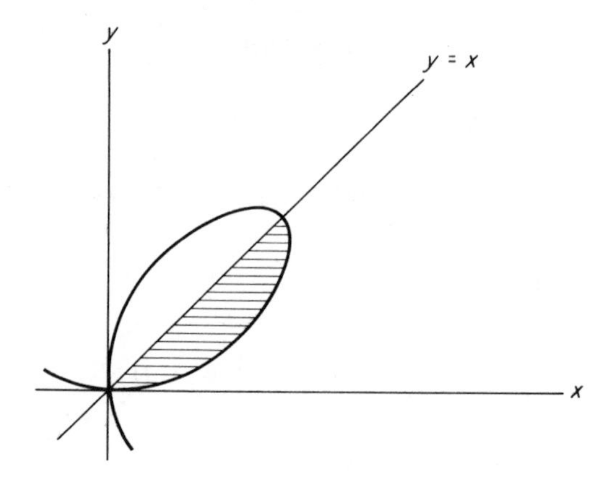

Figure 5.20

2.4 Apply formula (10) in the proof of Theorem 2.5 to find a potential function φ (such that $\nabla\varphi = \mathbf{F}$) for the vector field $\mathbf{F} : \mathscr{R}^2 \to \mathscr{R}^2$, if
(a) $\mathbf{F}(x, y) = (2xy^3, 3x^2y^2)$,
(b) $\mathbf{F}(x, y) = (\sin 2x \cos^2 y, -\sin^2 x \sin 2y)$.

2.5 Let $f, g : [a, b] \to \mathscr{R}$ be two $\mathscr{C}^1$ functions such that $f(x) > g(x) > 0$ on $[a, b]$. Let D be the nice region that is bounded by the graphs $y = f(x)$, $y = g(x)$ and the vertical lines $x = a$, $x = b$. Then the volume, of the solid obtained by revolving D about the x-axis, is

$$V = \pi \int_a^b [f(x)]^2\,dx - \pi \int_a^b [g(x)]^2\,dx$$

by Cavalieri's principle (Theorem IV.4.2). Show first that

$$V = -\pi \int_{\partial D} y^2\,dx.$$

Then apply Green's theorem to conclude that

$$V = \iint_D 2\pi y\,dx\,dy = 2\pi A\bar{y},$$

where A is the area of the region D, and $\bar{y}$ is the y-coordinate of its centroid (see Exercise IV.5.16).

2.6 Let f and g be $\mathscr{C}^2$ functions on an open set containing the nice region D, and denote by $\mathbf{N}$ the unit outer normal vector to ∂D (see Exercise 1.20). The *Laplacian* $\nabla^2 f$ and the *normal derivative* $\partial f/\partial n$ are defined by

$$\nabla^2 f = \operatorname{div}(\nabla f) = \frac{\partial^2 f}{\partial x^2} + \frac{\partial^2 f}{\partial y^2} \qquad \text{and} \qquad \frac{\partial f}{\partial n} = \nabla f \cdot \mathbf{N}.$$

Prove *Green's formulas*:

(a) $$\iint_D (f\nabla^2 g + \nabla f \cdot \nabla g)\,dx\,dy = \int_{\partial D} f\frac{\partial g}{\partial n}\,ds,$$

(b) $$\iint_D (f\nabla^2 g - g\,\nabla^2 f)\,dx\,dy = \int_{\partial D}\left(f\frac{\partial g}{\partial n} - g\frac{\partial f}{\partial n}\right)ds.$$

Hint: For (a) apply Green's theorem with $P = -f\,\partial g/\partial y$ and $Q = f\,\partial g/\partial x$.

2.7 If f is *harmonic* on D, that is, $\nabla^2 f \equiv 0$, set $f = g$ in Green's first formula to obtain

$$\iint_D |\nabla f|^2\,dx\,dy = \int_{\partial D} f\frac{\partial f}{\partial n}\,ds.$$

If $f \equiv 0$ on ∂D, conclude that $f \equiv 0$ on D.

2.8 Suppose that f and g both satisfy *Poisson's equation* on D, that is,

$$\nabla^2 f = \nabla^2 g = \varphi,$$

where $\varphi(x, y)$ is given on D. If $f = g$ on ∂D, apply the previous exercise to $f - g$, to conclude that $f = g$ on D. This is the uniqueness theorem for solutions of Poisson's equation.

2.9 Given $\mathbf{p} \in \mathscr{R}^2$, let $\mathbf{F} : \mathscr{R}^2 - \mathbf{p} \to \mathscr{R}^2$ be the vector field defined by

$$\mathbf{F}(\mathbf{x}) = \nabla \log|\mathbf{x} - \mathbf{p}|^2.$$

If C is a counterclockwise circle centered at $\mathbf{p}$, show by direct computation that

$$\int_C \mathbf{F}\cdot\mathbf{N}\,ds = 4\pi.$$

2.10 Let C be a counterclockwise-oriented smooth closed curve enclosing the points $\mathbf{p}_1, \ldots, \mathbf{p}_k$. If $\mathbf{F}$ is the vector field defined for $\mathbf{x} \neq \mathbf{p}_i$ by

$$\mathbf{F}(\mathbf{x}) = \sum_{i=1}^{k} q_i \nabla \log|\mathbf{x} - \mathbf{p}_i|^2,$$

where $q_1, \ldots, q_k$ are constants, deduce from the previous exercise that

$$\int_C \mathbf{F}\cdot\mathbf{N}\,ds = 4\pi(q_1 + \cdots + q_k).$$

Hint: Apply Green's theorem on the region D which is bounded by C and small circles centered at the points $\mathbf{p}_1, \ldots, \mathbf{p}_k$.

2.11 Show that the integral $\int_\alpha \omega$, of the differential form ω over the piecewise-smooth path γ is independent of the partition $\mathscr{P}$ used in its definition.

2.12 If α and β induce opposite orientations of the piecewise-smooth curve, show that $\int_\alpha \omega = -\int_\beta \omega$.

2.13 Prove that the orientation of the boundary of an oriented 2-cell is well-defined.

2.14 Show that every convex quadrilateral is a smooth oriented 2-cell.

2.15 If the mapping $\gamma : [a, b] \to \mathscr{R}^2$ is $\mathscr{C}^1$ and ω is a continuous 1-form, show that

$$\int_\gamma F^*\omega = \int_{F\circ\gamma} \omega.$$

2.16 Prove Lemma 2.2.
2.17 Prove that the orientation of the boundary of a cellulated nice region is well-defined.

3 MULTILINEAR FUNCTIONS AND THE AREA OF A PARALLELEPIPED

In the first two sections of the chapter we have discussed 1-forms, which are objects that can be integrated over curves (or 1-dimensional manifolds) in $\mathscr{R}^n$. Our eventual goal in this chapter is a similar discussion of differential k-forms, which are objects that can be integrated over k-dimensional manifolds in $\mathscr{R}^n$. The definition of 1-forms involved linear functions on $\mathscr{R}^n$; the definition of k-forms will involve multilinear functions on $\mathscr{R}^n$.

In this section we develop the elementary theory of multilinear functions on $\mathscr{R}^n$, and then apply it to the problem of calculating the area of a k-dimensional parallelepiped in $\mathscr{R}^n$. This computation will be used in our study in Section 4 of k-dimensional surface area in $\mathscr{R}^n$.

Let $(\mathscr{R}^n)^k = \mathscr{R}^n \times \cdots \times \mathscr{R}^n$ (k factors), and consider a function

$$M: (\mathscr{R}^n)^k \to \mathscr{R}.$$

Then M is a function on k-tuples of vectors in $\mathscr{R}^n$; $M(\mathbf{a}^1, \ldots, \mathbf{a}^k) \in \mathscr{R}$ if $\mathbf{a}^1, \ldots, \mathbf{a}^k$ are vectors in $\mathscr{R}^n$. Thus we can regard M as a function of k vector variables. The function M is called *k-multilinear* (or just *multilinear*) if it is linear in each of these variables separately, that is,

$$M(\mathbf{a}^1, \ldots, x\mathbf{a} + y\mathbf{b}, \ldots, \mathbf{a}^k) = xM(\mathbf{a}^1, \ldots, \mathbf{a}, \ldots, \mathbf{a}^k) + yM(\mathbf{a}^1, \ldots, \mathbf{b}, \ldots, \mathbf{a}^k).$$

We will often call M a k-multilinear function on $\mathscr{R}^n$, despite the fact that its domain of definition is actually $(\mathscr{R}^n)^k$. Note that a 1-multilinear function on $\mathscr{R}^n$ is just an (ordinary) linear function on $\mathscr{R}^n$.

We have seen that every linear function on $\mathscr{R}^n$ is a linear combination of certain special ones, namely the projection functions $dx_1, \ldots, dx_n$. Recall that, regarding $\mathbf{a} \in \mathscr{R}^n$ as a column vector, dx_i is the function that picks out the ith row of this vector, $dx_i(\mathbf{a}) = a_i$.

We would like to have a similar description of multilinear functions on $\mathscr{R}^n$. Given a k-tuple $\mathbf{i} = (i_1, \ldots, i_k)$ of integers (not necessarily distinct) between 1 and n, $1 \leqq i_r \leqq n$, we define the function

$$d\mathbf{x}_\mathbf{i} : (\mathscr{R}^n)^k \to \mathscr{R}$$

by

$$dx_{\mathbf{i}}(\mathbf{a}^1, \ldots, \mathbf{a}^k) = \begin{vmatrix} a_{i_1}^1 & a_{i_1}^2 & \cdots & a_{i_1}^k \\ \vdots & \vdots & & \vdots \\ a_{i_k}^1 & a_{i_k}^2 & \cdots & a_{i_k}^k \end{vmatrix}.$$

That is, if A is the $n \times k$ matrix whose column vectors are $\mathbf{a}^1, \ldots, \mathbf{a}^k$,

$$A = (\mathbf{a}^1 \cdots \mathbf{a}^k)$$

and $A_{\mathbf{i}}$ denotes the $k \times k$ matrix whose rth row is the i_rth row of A, then

$$dx_{\mathbf{i}}(\mathbf{a}^1, \ldots, \mathbf{a}^k) = \det A_{\mathbf{i}}.$$

Note that $A_{\mathbf{i}}$ is a $k \times k$ submatrix of A if $\mathbf{i}$ is an *increasing* k-tuple, that is, if $1 \leqq i_1 < i_2 < \cdots < i_k \leqq n$.

It follows immediately from the properties of determinants that $dx_{\mathbf{i}}$ is a k-multilinear function on $\mathscr{R}^n$ and, moreover, is alternating. The k-multilinear function M on $\mathscr{R}^n$ is called *alternating* if

$$M(\mathbf{a}^1, \ldots, \mathbf{a}^k) = 0$$

whenever some pair of the vectors $\mathbf{a}^1, \ldots, \mathbf{a}^k$ are equal, say $\mathbf{a}^r = \mathbf{a}^s$ $(r \neq s)$. The fact that $dx_{\mathbf{i}}$ is alternating then follows from the fact that a determinant vanishes if some pair of its columns are equal.

It is easily proved (Exercise 3.1) that the k-multilinear function M on $\mathscr{R}^n$ is alternating if and only if it changes signs upon the interchange of two of the vectors $\mathbf{a}^1, \ldots, \mathbf{a}^k$, that is,

$$M(\mathbf{a}^1, \ldots, \mathbf{a}^r, \ldots, \mathbf{a}^s, \ldots, \mathbf{a}^k) = -M(\mathbf{a}^1, \ldots, \mathbf{a}^s, \ldots, \mathbf{a}^r, \ldots, \mathbf{a}^k).$$

The notation $dx_{\mathbf{i}}$ generalizes the notation dx_i for the ith projection function on $\mathscr{R}^n$, and we will prove (Theorem 3.4) that every alternating k-multilinear function M on $\mathscr{R}^n$ is a linear combination of the $dx_{\mathbf{i}}$. That is, there exist real numbers $a_{\mathbf{i}}$ such that

$$M = \sum_{\mathbf{i}} a_{\mathbf{i}}\, dx_{\mathbf{i}},$$

the summation being over all k-tuples $\mathbf{i}$ of integers from 1 to n. This generalizes the fact that every linear function on $\mathscr{R}^n$ is of the form $\sum a_i\, dx_i$. Notice that every linear function on $\mathscr{R}^n$ is automatically alternating.

Proposition 3.1 If M is an alternating k-multilinear function on $\mathscr{R}^n$, then

$$M(\mathbf{a}^1, \ldots, \mathbf{a}^k) = 0$$

if the vectors $\mathbf{a}^1, \ldots, \mathbf{a}^k$ are linearly dependent.

This follows immediately from the definitions and the fact that some one of the vectors $\mathbf{a}^1, \ldots, \mathbf{a}^k$ must be a linear combination of the others (Exercise 3.2).

Corollary 3.2 If $k > n$, then the only alternating k-multilinear function on $\mathscr{R}^n$ is the trivial one whose only value is zero.

The following theorem describes the nature of an arbitrary k-multilinear function on $\mathscr{R}^n$ (not necessarily alternating).

Theorem 3.3 Let M be a k-multilinear function on $\mathscr{R}^n$. If $\mathbf{a}^1, \ldots, \mathbf{a}^k$ are vectors in $\mathscr{R}^n$, $\mathbf{a}^j = (a_1{}^j, \ldots, a_n{}^j)$, then

$$M(\mathbf{a}^1, \ldots, \mathbf{a}^k) = \sum_{i_1, \ldots, i_k = 1}^{n} \alpha_{i_1 \cdots i_k} a_{i_1}^1 a_{i_2}^2 \cdots a_{i_k}^k,$$

where

$$\alpha_{i_1 \cdots i_k} = M(\mathbf{e}^{i_1}, \ldots, \mathbf{e}^{i_k}).$$

Here $\mathbf{e}^i$ denotes the ith standard unit basis vector in $\mathscr{R}^n$.

PROOF The proof is by induction on k, the case $k = 1$ being clear by linearity.

For each $r = 1, \ldots, n$, let N_r denote the $(k-1)$-multilinear function on $\mathscr{R}^n$ defined by

$$N_r(\mathbf{a}^1, \ldots, \mathbf{a}^{k-1}) = M(\mathbf{a}^1, \ldots, \mathbf{a}^{k-1}, \mathbf{e}^r).$$

Then

$$\begin{aligned} M(\mathbf{a}^1, \ldots, \mathbf{a}^k) &= \sum_{r=1}^{n} a_r{}^k M(\mathbf{a}^1, \ldots, \mathbf{a}^{k-1}, \mathbf{e}^r) \\ &= \sum_{r=1}^{n} a_r{}^k N_r(\mathbf{a}^1, \ldots, \mathbf{a}^{k-1}) \\ &= \sum_{r=1}^{n} a_r{}^k \left(\sum_{i_j=1}^{n} N_r(\mathbf{e}^{i_1}, \ldots, \mathbf{e}^{i_{k-1}}) a_{i_1}^1 \cdots a_{i_{k-1}}^{k-1} \right) \\ &= \sum_{r=1}^{n} a_r{}^k \left(\sum_{i_j=1}^{n} M(\mathbf{e}^{i_1}, \ldots, \mathbf{e}^{i_{k-1}}, \mathbf{e}^r) a_{i_1}^1 \cdots a_{i_{k-1}}^{k-1} \right) \\ &= \sum_{i_j=1}^{n} \alpha_{i_1 \cdots i_k} a_{i_1}^1 \cdots a_{i_k}^k \end{aligned}$$

as desired. ∎

We can now describe the structure of M in terms of the $d\mathbf{x}_\mathbf{I}$, under the additional hypothesis that M is alternating.

Theorem 3.4 If M is an alternating k-multilinear function on $\mathscr{R}^n$, then

$$M = \sum_{[\mathbf{i}]} \alpha_{\mathbf{i}}\, d\mathbf{x}_{\mathbf{i}},$$

where $\alpha_{\mathbf{i}} = M(\mathbf{e}^{i_1}, \ldots, \mathbf{e}^{i_k})$. Here the notation $[\mathbf{i}]$ signifies summation over all *increasing* k-tuples $\mathbf{i} = (i_1, \ldots, i_k)$.

For the proof we will need the following.

Lemma 3.5 Let $\mathbf{i} = (i_1, \ldots, i_k)$ and $\mathbf{j} = (j_1, \ldots, j_k)$ be increasing k-tuples of integers from 1 to n. Then

$$d\mathbf{x}_{\mathbf{i}}(\mathbf{e}^{j_1}, \ldots, \mathbf{e}^{j_k}) = \begin{cases} 1 & \text{if } \mathbf{i} = \mathbf{j}, \\ 0 & \text{if } \mathbf{i} \neq \mathbf{j}. \end{cases}$$

This lemma follows from the fact that $d\mathbf{x}_{\mathbf{i}}(\mathbf{e}^{j_1}, \ldots, \mathbf{e}^{j_k})$ is the determinant of the matrix $(\delta_i{}^j)$, where

$$\delta_i{}^j = \begin{cases} 1 & \text{if } i = j, \\ 0 & \text{if } i \neq j. \end{cases}$$

If $\mathbf{i} = \mathbf{j}$, then $(\delta_i{}^j)$ is the $k \times k$ identity matrix; otherwise some row of $(\delta_i{}^j)$ is zero, so its determinant vanishes.

PROOF OF THEOREM 3.4 Define the alternating k-multilinear function $\tilde{M}$ on $\mathscr{R}^n$ by

$$\tilde{M}(\mathbf{a}^1, \ldots, \mathbf{a}^k) = \sum_{[\mathbf{i}]} \alpha_{\mathbf{i}}\, d\mathbf{x}_{\mathbf{i}}(\mathbf{a}^1, \ldots, \mathbf{a}^k),$$

where $\alpha_{\mathbf{i}} = M(\mathbf{e}^{i_1}, \ldots, \mathbf{e}^{i_k})$. We want to prove that $M = \tilde{M}$.

By Theorem 3.3 it suffices (because M and $\tilde{M}$ are both alternating) to show that

$$M(\mathbf{e}^{j_1}, \ldots, \mathbf{e}^{j_k}) = \tilde{M}(\mathbf{e}^{j_1}, \ldots, \mathbf{e}^{j_k})$$

for every increasing k-tuple $\mathbf{j} = (j_1, \ldots, j_k)$. But

$$\begin{aligned} \tilde{M}(\mathbf{e}^{j_1}, \ldots, \mathbf{e}^{j_k}) &= \sum_{[\mathbf{i}]} \alpha_{\mathbf{i}}\, d\mathbf{x}_{\mathbf{i}}(\mathbf{e}^{j_1}, \ldots, \mathbf{e}^{j_k}) \\ &= \alpha_{\mathbf{j}} \\ &= M(\mathbf{e}^{j_1}, \ldots, \mathbf{e}^{j_k}) \end{aligned}$$

immediately by Lemma 3.5. ∎

The determinant function, considered as a function of the column vectors of an $n \times n$ matrix, is an alternating n-multilinear function on $\mathscr{R}^n$. This, together with the fact that the determinant of the identity matrix is 1, uniquely characterizes the determinant.

Corollary 3.6 $D = \det$ is the only alternating n-multilinear function on $\mathscr{R}^n$ such that

$$D(\mathbf{e}^1, \ldots, \mathbf{e}^n) = 1.$$

PROOF Exercise 3.3. ∎

As an application of Theorem 3.4, we will next prove the *Binet–Cauchy product formula*, a generalization of the fact that the determinant of the product of two $n \times n$ matrices is equal to the product of their determinants. Recall that, if A is an $n \times k$ matrix, and $\mathbf{i} = (i_1, \ldots, i_k)$, then $A_\mathbf{i}$ denotes that $k \times k$ matrix whose rth row is the i_rth row of A.

Theorem 3.7 Let A be an $k \times n$ matrix and B an $n \times k$ matrix, where $k \leqq n$. Then

$$\det AB = \sum_{[\mathbf{i}]} (\det A_\mathbf{i}^{\mathrm{t}})(\det B_\mathbf{i}).$$

Here A^{t} denotes the transpose of A and, as in Theorem 3.4, $[\mathbf{i}]$ signifies summation over increasing k-tuples.

Note that the case $k = n$, when A and B are both $n \times n$ matrices, is

$$\det AB = (\det A^{\mathrm{t}})(\det B) = (\det A)(\det B).$$

PROOF Let $\mathbf{a}_1, \ldots, \mathbf{a}_k$ be the row vectors of A, and $\mathbf{b}^1, \ldots, \mathbf{b}^k$ the column vectors of B. Since

$$AB = \begin{pmatrix} \mathbf{a}_1 \cdot \mathbf{b}^1 & \cdots & \mathbf{a}_1 \cdot \mathbf{b}^k \\ \vdots & & \vdots \\ \mathbf{a}_k \cdot \mathbf{b}^1 & \cdots & \mathbf{a}_k \cdot \mathbf{b}^k \end{pmatrix},$$

we see that, by holding fixed the matrix A, we obtain an alternating k-multilinear function M of the vectors $\mathbf{b}^1, \ldots, \mathbf{b}^k$,

$$M(\mathbf{b}^1, \ldots, \mathbf{b}^k) = \det AB.$$

Consequently by Theorem 3.4 there exist numbers $\alpha_\mathbf{i}$ (depending upon the matrix A) such that

$$M = \sum_{[\mathbf{i}]} \alpha_\mathbf{i}\, dx_\mathbf{i}.$$

Then

$$\begin{aligned} M(\mathbf{b}^1, \ldots, \mathbf{b}^k) &= \sum_{[\mathbf{i}]} \alpha_\mathbf{i}\, d\mathbf{x}_\mathbf{i}(\mathbf{b}^1, \ldots, \mathbf{b}^k) \\ &= \sum_{[\mathbf{i}]} \alpha_\mathbf{i}(\det B_\mathbf{i}). \end{aligned}$$

But

$$\begin{aligned}\alpha_{\mathbf{i}} &= M(\mathbf{e}^{i_1}, \ldots, \mathbf{e}^{i_k}) \\ &= \det A(\mathbf{e}^{i_1} \ldots \mathbf{e}^{i_k}) \\ &= \begin{vmatrix} a_{1i_1} & a_{1i_2} & \cdots & a_{1i_k} \\ \vdots & \vdots & & \vdots \\ a_{ki_1} & a_{ki_2} & \cdots & a_{ki_k} \end{vmatrix} \\ &= \det A_{\mathbf{i}}^{\mathrm{t}},\end{aligned}$$

so

$$\begin{aligned}\det AB &= M(\mathbf{b}^1, \ldots, \mathbf{b}^k) \\ &= \sum_{[\mathbf{i}]} (\det A_{\mathbf{i}}^{\mathrm{t}})(\det B_{\mathbf{i}})\end{aligned}$$

as desired. ∎

Taking $A = B^{\mathrm{t}}$, we obtain

Corollary 3.8 If A is an $n \times k$ matrix, $k \leqq n$, then

$$\det A^{\mathrm{t}}A = \sum_{[\mathbf{i}]} (\det A_{\mathbf{i}})^2.$$

Our reason for interest in the Binet–Cauchy product formula stems from its application to the problem of computing the area of an k-dimensional parallelepiped in $\mathscr{R}^n$. Let $\mathbf{a}_1, \ldots, \mathbf{a}_k$ be k vectors in $\mathscr{R}^n$. By the k-dimensional *parallelepiped P* which they span is meant the set

$$P = \{\mathbf{x} \in \mathscr{R}^n : \mathbf{x} = \sum_{i=1}^{n} t_i \mathbf{a}_i \text{ with each } t_i \in [0, 1]\}.$$

This is the natural generalization of a parallelogram spanned by two vectors in the plane. If $k = n$, then P is the image of the unit cube I^n, under the linear mapping $\mathscr{L} : \mathscr{R}^n \to \mathscr{R}^n$ such that $L(\mathbf{e}_i) = \mathbf{a}_i$, $i = 1, \ldots, n$. In this case the volume of P is given by the following theorem.

Theorem 3.9 Let P be the parallelepiped in $\mathscr{R}^n$ that is spanned by the vectors $\mathbf{a}_1, \ldots, \mathbf{a}_n$. Then its volume is

$$v(P) = (\det A^{\mathrm{t}}A)^{1/2},$$

where A is the $n \times n$ matrix whose column vectors are $\mathbf{a}_1, \ldots, \mathbf{a}_n$.

PROOF If the linear mapping $L : \mathscr{R}^n \to \mathscr{R}$ is defined by

$$L(\mathbf{x}) = A\mathbf{x},$$

then $L(\mathbf{e}_i) = \mathbf{a}_i$, $i = 1, \ldots, n$. Hence $P = L(I^n)$. Therefore, by Theorem IV.5.1,

$$\begin{aligned} v(P) &= |\det A| \, v(I^n) \\ &= |\det A| \\ &= [(\det A)^2]^{1/2} \\ &= [\det A^t A]^{1/2} \end{aligned}$$

since

$$\det A^t = \det A. \qquad \blacksquare$$

Now let X be a subset of a k-dimensional subspace V of $\mathscr{R}^n$ $(k < n)$. A choice of an orthonormal basis $\mathbf{v}_1, \ldots, \mathbf{v}_k$ for V (which exists by Theorem I.3.3) determines a 1-1 linear mapping $\varphi : V \to \mathscr{R}^k$, defined by $\varphi(\mathbf{v}_i) = \mathbf{e}_i$. We want to define the *k-dimensional area* $a(X)$ by

$$a(X) = v(\varphi(X)).$$

However we must show that $a(X)$ is independent of the choice of basis for V. If $\mathbf{w}_1, \ldots, \mathbf{w}_k$ is a second orthonormal basis for V, determining the one-to-one linear mapping $\psi : V \to \mathscr{R}^k$ by $\psi(\mathbf{w}_i) = \mathbf{e}_i$, then it is easily verified that

$$\psi \circ \varphi^{-1} : \mathscr{R}^k \to \mathscr{R}^k$$

is an orthogonal mapping (Exercise I.6.10). It therefore preserves volumes, by Corollary IV.5.2. Since $\psi \circ \varphi^{-1}(\varphi(X)) = \psi(X)$, we conclude that $v(\varphi(X)) = v(\psi(X))$, so $a(X)$ is well defined. The following theorem shows how to compute it for a k-dimensional parallelepiped.

Theorem 3.10 If P is a k-dimensional parallelepiped in $\mathscr{R}^n$ $(k < n)$ spanned by the vectors $\mathbf{a}_1, \ldots, \mathbf{a}_k$, then

$$a(P) = [\det A^t A]^{1/2},$$

where A is the $n \times k$ matrix whose column vectors are $\mathbf{a}_1, \ldots, \mathbf{a}_k$.

Thus we have the same formula when $k < n$ as when $k = n$.

PROOF Let V be a k-dimensional subspace of $\mathscr{R}^n$ that contains the vectors $\mathbf{a}_1, \ldots, \mathbf{a}_k$, and let $\mathbf{v}_1, \ldots, \mathbf{v}_n$ be an orthonormal basis for $\mathscr{R}^n$ such that $\mathbf{v}_1, \ldots, \mathbf{v}_k$ generate V. Let $\Phi : \mathscr{R}^n \to \mathscr{R}^n$ be the orthogonal mapping defined by $\Phi(\mathbf{v}_i) = \mathbf{e}_i$, $i = 1, \ldots, n$. If $\mathbf{b}_i = \Phi(\mathbf{a}_i)$, $i = 1, \ldots, k$, then $\Phi(P)$ is the k-dimensional parallelepiped in $\mathscr{R}^k$ that is spanned by $\mathbf{b}_1, \ldots, \mathbf{b}_k$. Consequently, using Theorem 3.9

and the fact that the orthogonal mapping Φ preserves inner products (Exercise I.6.10), we have

$$\begin{aligned} a(P) &= v(\Phi(P)) \\ &= [\det B^t B]^{1/2} \qquad (\text{where } B = (\mathbf{b}_1 \cdots \mathbf{b}_k)) \\ &= [\det (\mathbf{b}_i \cdot \mathbf{b}_j)]^{1/2} \\ &= [\det (\mathbf{a}_i \cdot \mathbf{a}_j)]^{1/2} \\ &= [\det A^t A]^{1/2}. \end{aligned}$$

∎

The following formula for $a(P)$ now follows immediately from Theorem 3.10 and Corollary 3.8.

Theorem 3.11 If P and A are as in Theorem 3.10, then

$$a(P) = \left[\sum_{[\mathbf{i}]} (\det A_{\mathbf{i}})^2\right]^{1/2}$$

This result can be interpreted as a *general Pythagorean theorem*. To see this, let $\mathscr{R}_{\mathbf{i}}^k$ denote the k-dimensional coordinate plane in $\mathscr{R}^n$ that is spanned by the unit basis vectors $\mathbf{e}_{i_1}, \ldots, \mathbf{e}_{i_k}$, where $\mathbf{i} = (i_1, \ldots, i_k)$. If $\pi_{\mathbf{i}} : \mathscr{R}^n \to \mathscr{R}_{\mathbf{i}}^k$ is the natural projection mapping, then

$$(\det A_{\mathbf{i}})^2 = \det A_{\mathbf{i}}^t A_{\mathbf{i}}$$

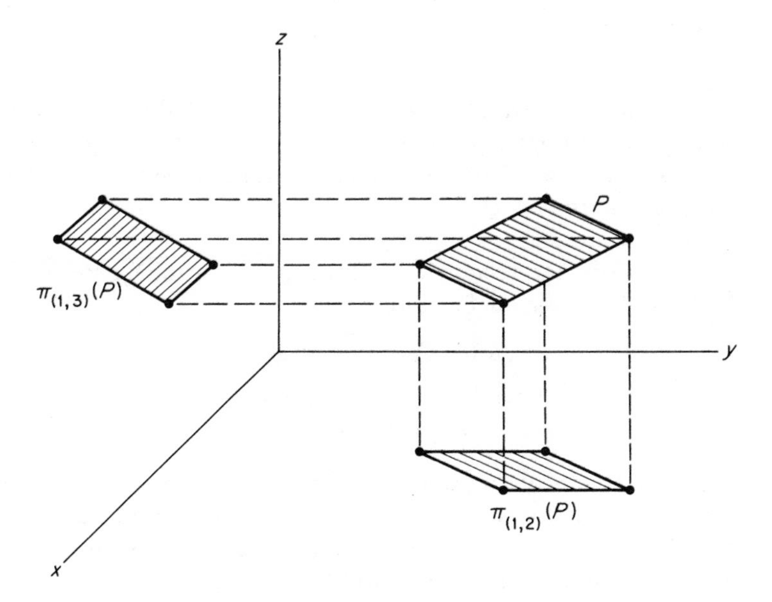

Figure 5.21

is the *square* of the k-dimensional area of the projection $\pi_i(P)$ of P into $\mathscr{R}_i^k$ (see Exercise 3.4). Thus Theorem 3.11 asserts that *the area of the k-dimensional parallelepiped P is equal to the square root of the sum of the squares of the areas of all projections of P into k-dimensional coordinate planes of* $\mathscr{R}^n$ (see Fig. 5.21 for the case $k = 2$, $n = 3$). For $k = 1$, this is just the statement that the length of a vector is equal to the square root of the sum of the squares of its components.

Exercises

3.1 Prove that the k-multilinear function M on $\mathscr{R}^n$ is alternating if and only if the value $M(\mathbf{a}^1, \dots, \mathbf{a}^k)$ changes sign when $\mathbf{a}^r$ and $\mathbf{a}^s$ are interchanged. *Hint:* Consider $M(\mathbf{a}^1, \dots, \mathbf{a}^r + \mathbf{a}^s, \dots, \mathbf{a}^r + \mathbf{a}^s, \dots, \mathbf{a}^k)$.

3.2 Prove Proposition 3.1.

3.3 Prove Corollary 3.6.

3.4 Verify the assertion in the last paragraph of this section, that det $A_i{}^t A_i$ is the square of the k-dimensional area of the projection $\pi_i(P)$, of the parallelepiped spanned by the column vectors of the $n \times k$ matrix A, into $\mathscr{R}_i^k$.

3.5 Let P be the 2-dimensional parallelogram in $\mathscr{R}^3$ spanned by the vectors $\mathbf{a}$ and $\mathbf{b}$. Deduce from Theorem 3.11 the fact that its area is $a(P) = |\mathbf{a} \times \mathbf{b}|$.

3.6 Let P be the 3-dimensional parallelepiped in $\mathscr{R}^3$ that is spanned by the vectors $\mathbf{a}$, $\mathbf{b}$, $\mathbf{c}$. Deduce from Theorem 3.9 that its volume is $v(P) = |\mathbf{a} \cdot \mathbf{b} \times \mathbf{c}|$.

4 SURFACE AREA

In this section we investigate the concept of area for k-dimensional surfaces in $\mathscr{R}^n$. A k-dimensional *surface patch* in $\mathscr{R}^n$ is a $\mathscr{C}^1$ mapping $F: Q \to \mathscr{R}^n$, where Q is a k-dimensional interval (or rectangular parallelepiped) in $\mathscr{R}^k$, which is one-to-one on the interior of Q.

Example 1 Let Q be the rectangle $[0, 2\pi] \times [0, \pi]$ in the $\theta\varphi$-plane $\mathscr{R}^2$, and $F: Q \to \mathscr{R}^3$ the surface patch defined by

$$\begin{aligned} x &= F_1(\theta, \varphi) = a \sin \varphi \cos \theta, \\ y &= F_2(\theta, \varphi) = a \sin \varphi \sin \theta, \\ z &= F_3(\theta, \varphi) = a \cos \varphi, \end{aligned}$$

so a, θ, φ are the spherical coordinates of the point $(x, y, z) = F(\theta, \varphi) \in \mathscr{R}^3$. Then the image of F is the sphere $x^2 + y^2 + z^2 = a^2$ of radius a (Fig. 5.22). Notice that F is one-to-one on the interior of Q, but not on its boundary. The top edge of Q maps to the point $(0, 0, -a)$, the bottom edge to $(0, 0, a)$, and the image of each of the vertical edges is the semicircle $x^2 + z^2 = a^2$, $x \geqq 0$, $y = 0$.

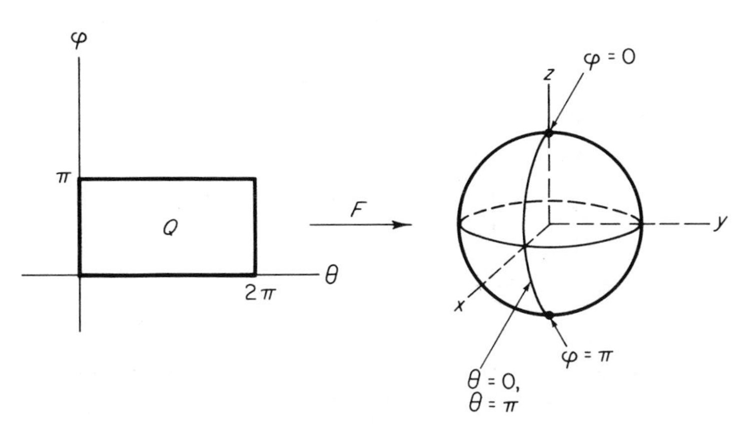

Figure 5.22

Example 2 Let Q be the square $[0, 2\pi] \times [0, 2\pi]$ in the $\theta\varphi$-plane $\mathscr{R}^2$, and $F: Q \to \mathscr{R}^4$ the surface patch defined by

$$x_1 = \cos\theta, \qquad x_3 = \cos\varphi,$$
$$x_2 = \sin\theta, \qquad x_4 = \sin\varphi.$$

The image of F is the so-called "flat torus" $S^1 \times S^1 \subset \mathscr{R}^2 \times \mathscr{R}^2$ in $\mathscr{R}^4$. Note that F is one-to-one on the interior of Q (why?), but not on its boundary [for instance, the four vertices of Q all map to the point $(1, 0, 1, 0) \in \mathscr{R}^4$].

One might expect that the definition of surface area would parallel that of pathlength, so that the area of a surface would be defined as a limit of areas of appropriate inscribed polyhedral surfaces. These inscribed polyhedral surfaces for the surface patch $F: Q \to \mathscr{R}^n$ could be obtained as follows (for $k = 2$; the process could be easily generalized to $k > 2$). Let $\mathscr{P} = \{\sigma_1, \ldots, \sigma_p\}$ be a partition of the rectangle Q into small triangles (Fig. 5.23). Then, for each $i = 1, \ldots, p$, denote by τ_i that triangle in $\mathscr{R}^n$ whose three vertices are the images under F of the three vertices of σ_i. Our inscribed polyhedral approximation is then

$$K_{\mathscr{P}} = \bigcup_{i=1}^{p} \tau_i,$$

and its area is $a(K_{\mathscr{P}}) = \sum_{i=1}^{p} a(\tau_i)$. We would then define the area of the surface patch F by

$$a(F) = \lim_{|\mathscr{P}| \to 0} a(K_{\mathscr{P}}), \tag{1}$$

provided that this limit exists in the sense that, given $\varepsilon > 0$, there is a $\delta > 0$ such that

$$|\mathscr{P}| < \delta \Rightarrow |a(F) - a(K_{\mathscr{P}})| < \varepsilon.$$

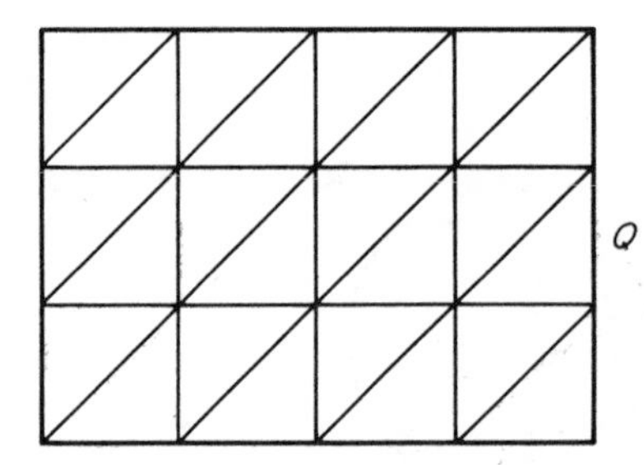

Figure 5.23

However it is a remarkable fact that the limit in (1) almost never exists if $k > 1$! The reason is that the triangles τ_i do not necessarily approximate the surface in the same manner that inscribed line segments approximate a smooth curve. For a partition of small mesh, the inscribed line segments to a curve have slopes approximately equal to those of the corresponding segments of the curve. However, in the case of a surface, the partition $\mathscr{P}$ with small mesh can usually be chosen in such a way that each of the triangles τ_i is very nearly perpendicular to the surface. It turns out that, as a consequence, one can obtain inscribed polyhedra $K_{\mathscr{P}}$ with $|\mathscr{P}|$ arbitrarily small and $a(K_{\mathscr{P}})$ arbitrarily large!

We must therefore adopt a slightly different approach to the definition of surface area. The basic idea is that the difficulties indicated in the previous paragraph can be avoided by the use of circumscribed polyhedral surfaces in place of inscribed ones. A circumscribed polyhedral surface is one consisting of triangles, each of which is tangent to the given surface at some point. Thus circumscribed triangles are automatically free of the fault (indicated above) of inscribed ones (which may be more nearly perpendicular to, than tangent to, the given surface).

Because circumscribed polyhedral surfaces are rather inconvenient to work with, we alter this latter approach somewhat, by replacing triangles with parallelepipeds, and by not requiring that they fit together so nicely as to actually form a continuous surface.

We start with a partition $\mathscr{P} = \{Q_1, \ldots, Q_p\}$ of the k-dimensional interval Q into small subintervals (Fig. 5.24). It will simplify notational matters to assume that Q is a cube in $\mathscr{R}^k$, so we may suppose that each subinterval Q_i is a cube with edgelength h; its volume is then $v(Q_i) = h^k$.

Assuming that we already had an adequate definition of surface area, it would certainly be true that

$$a(F) = \sum_{i=1}^{p} a(F \mid Q_i),$$

where $F \mid Q_i$ denotes the restriction of F to the subinterval Q_i. We now approximate the sum on the right, replacing $a(F \mid Q_i)$ by $a(dF_{\mathbf{q}_i}(Q_i))$, the area of the

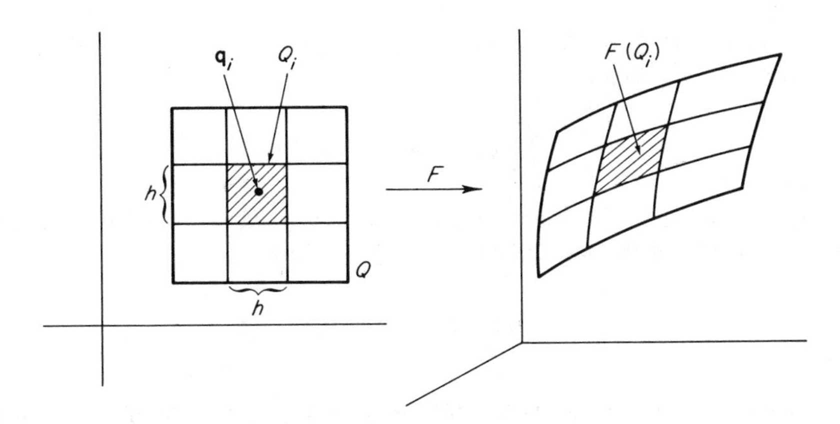

Figure 5.24

k-dimensional parallelepiped onto which Q_i maps under the linear approximation $dF_{\mathbf{q}_i}$ to F, at an arbitrarily selected point $\mathbf{q}_i \in Q_i$ (Fig. 5.25). Then

$$a(F) \approx \sum_{i=1}^{p} a(P_i), \tag{2}$$

where $P_i = dF_{\mathbf{q}_i}(Q_i)$.

Now the parallelepiped P_i is spanned by the vectors

$$h\frac{\partial F}{\partial u_1}, \ldots, h\frac{\partial F}{\partial u_k},$$

evaluated at the point $\mathbf{q}_i \in Q_i$. If A_i is the $n \times k$ matrix having these vectors as its column vectors, then

$$a(P_i) = [\det A_i{}^t A_i]^{1/2}$$

by Theorem 3.10. But it is easily verified that

$$[\det A_i{}^t A_i]^{1/2} = [\det F'(\mathbf{q}_i)^t F'(\mathbf{q}_i)]^{1/2} h^k,$$

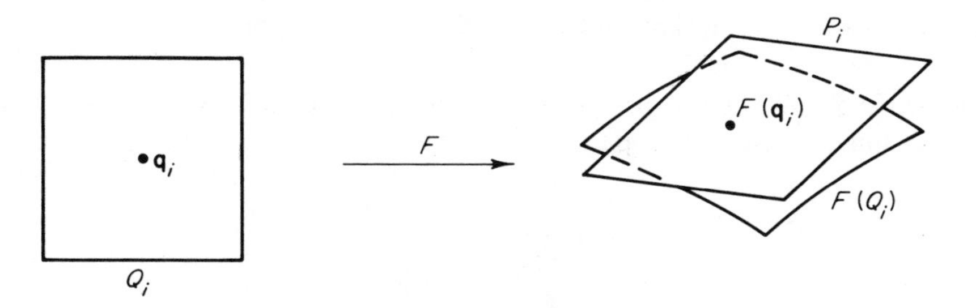

Figure 5.25

since the matrix F' has column vectors $\partial F/\partial u_1, \ldots, \partial F/\partial u_k$. Upon substitution of this into (2), we obtain

$$a(F) \approx \sum_{i=1}^{p} [\det F'(\mathbf{q}_i)^t F'(\mathbf{q}_i)]^{1/2} v(Q_i).$$

Recognizing this as a Riemann sum, we finally *define* the (k-dimensional) *surface area* $a(F)$ of the surface patch $F: Q \to \mathscr{R}^n$ by

$$a(F) = \int_Q [\det F'(\mathbf{u})^t F'(\mathbf{u})]^{1/2} \, d\mathbf{u}. \tag{3}$$

Example 3 If F is the spherical coordinates mapping of Example 1, mapping the rectangle $[0, 2\pi] \times [0, \pi] \subset \mathscr{R}^2_{\theta\varphi}$ onto the sphere of radius a in $\mathscr{R}^3$, then a routine computation yields

$$[\det F'(\theta, \varphi)^t F'(\theta, \varphi)]^{1/2} = a^2 \sin \varphi,$$

so (3) gives

$$a(F) = \int_0^{\pi} \int_0^{2\pi} a^2 \sin \varphi \, d\theta \, d\varphi = 4\pi a^2$$

as expected.

Example 4 If $F: [a, b] \to \mathscr{R}^n$ is a $\mathscr{C}^1$ path, then

$$(F')^t F' = \begin{pmatrix} \dfrac{\partial F_1}{\partial t} & \cdots & \dfrac{\partial F_n}{\partial t} \end{pmatrix} \begin{pmatrix} \dfrac{\partial F_1}{\partial t} \\ \vdots \\ \dfrac{\partial F_n}{\partial t} \end{pmatrix}$$

$$= \sum_{i=1}^{n} \left(\frac{\partial F_i}{\partial t} \right)^2,$$

so we recover from (3) the familiar formula for pathlength.

Example 5 We now consider 2-dimensional surfaces in $\mathscr{R}^n$. Let $\varphi: Q \to \mathscr{R}^n$ be a 2-dimensional surface patch, where Q is a rectangle in the uv-plane. Then

$$(\varphi')^t \varphi' = \begin{pmatrix} \dfrac{\partial \varphi_1}{\partial u} & \cdots & \dfrac{\partial \varphi_n}{\partial u} \\ \dfrac{\partial \varphi_1}{\partial v} & \cdots & \dfrac{\partial \varphi_n}{\partial v} \end{pmatrix} \begin{pmatrix} \dfrac{\partial \varphi_1}{\partial u} & \dfrac{\partial \varphi_1}{\partial v} \\ \vdots & \vdots \\ \dfrac{\partial \varphi_n}{\partial u} & \dfrac{\partial \varphi_n}{\partial v} \end{pmatrix}$$

(equation continues)

$$= \begin{pmatrix} \dfrac{\partial\varphi}{\partial u}\cdot\dfrac{\partial\varphi}{\partial u} & \dfrac{\partial\varphi}{\partial u}\cdot\dfrac{\partial\varphi}{\partial v} \\ \dfrac{\partial\varphi}{\partial v}\cdot\dfrac{\partial\varphi}{\partial u} & \dfrac{\partial\varphi}{\partial v}\cdot\dfrac{\partial\varphi}{\partial v} \end{pmatrix}.$$

Equation (3) therefore yields

$$a(\varphi) = \iint_Q (EG - F^2)^{1/2}\, du\, dv, \tag{4}$$

where

$$E = \frac{\partial\varphi}{\partial u}\cdot\frac{\partial\varphi}{\partial u}, \qquad F = \frac{\partial\varphi}{\partial u}\cdot\frac{\partial\varphi}{\partial v}, \qquad G = \frac{\partial\varphi}{\partial v}\cdot\frac{\partial\varphi}{\partial v}.$$

If we apply Corollary 3.8 to Eq. (3), we obtain

$$a(F) = \int_Q \left[\sum_{[\mathbf{i}]} (\det F_{\mathbf{i}}')^2\right]^{1/2}, \tag{5}$$

where the notation $[\mathbf{i}]$ signifies (as usual) summation over all *increasing* k-tuples $\mathbf{i} = (i_1, \ldots, i_k)$, and $F_{\mathbf{i}}'$ is the $k \times k$ matrix whose rth row is the i_rth row of the derivative matrix $F' = (\partial F_i/\partial u_j)$. With the Jacobian determinant notation

$$\det F_{\mathbf{i}}' = \frac{\partial(F_{i_1}, \ldots, F_{i_k})}{\partial(u_1, \ldots, u_k)},$$

Eq. (5) becomes

$$a(F) = \int_Q \left[\sum_{[\mathbf{i}]} \left(\frac{\partial(F_{i_1}, \ldots, F_{i_k})}{\partial(u_1, \ldots, u_k)}\right)^2\right]^{1/2} du_1 \cdots du_k. \tag{6}$$

We may alternatively explain the $[\mathbf{i}]$ notation with the statement that the summation is over all $k \times k$ submatrices of the $n \times k$ derivative matrix F'.

In the case $k = 2$, $n = 3$ of a 2-dimensional surface patch F in $\mathscr{R}^3$, given by

$$x = F_1(u, v), \qquad y = F_2(u, v), \qquad z = F_3(u, v),$$

(6) reduces to

$$a(F) = \iint_Q \left[\left(\frac{\partial(x, y)}{\partial(u, v)}\right)^2 + \left(\frac{\partial(x, z)}{\partial(u, v)}\right)^2 + \left(\frac{\partial(y, z)}{\partial(u, v)}\right)^2\right]^{1/2} du\, dv. \tag{7}$$

Example 6 We consider the graph in $\mathscr{R}^n$ of a $\mathscr{C}^1$ function $f\colon Q \to \mathscr{R}$, where Q is an interval in $\mathscr{R}^{n-1}$. This graph is the image in $\mathscr{R}^n$ of the surface patch $F\colon Q \to \mathscr{R}^n$ defined by

$$F(x_1, \ldots, x_{n-1}) = (x_1, \ldots, x_{n-1}, f(x_1, \ldots, x_{n-1})) \in \mathscr{R}^n.$$

The derivative matrix of F is

$$F' = \begin{pmatrix} 1 & 0 & \cdots & 0 \\ 0 & 1 & \cdots & 0 \\ \vdots & & & \\ 0 & 0 & \cdots & 1 \\ \dfrac{\partial f}{\partial x_1} & \dfrac{\partial f}{\partial x_2} & \cdots & \dfrac{\partial f}{\partial x_{n-1}} \end{pmatrix}.$$

We want to apply (6) to compute $a(F)$. First note that

$$\frac{\partial(F_1, \ldots, F_{n-1})}{\partial(x_1, \ldots, x_{n-1})} = 1. \tag{8}$$

If $i < n$, then

$$\frac{\partial(F_1, \ldots, \hat{F}_i, \ldots, F_n)}{\partial(x_1, \ldots, x_{n-1})}$$

is the determinant of the $(n-1) \times (n-1)$ matrix A_i which is obtained from F' upon deletion of its ith row. The elements of the ith column of A_i are all zero, except for the last one, which is $\partial f/\partial x_i$. Since the cofactor of $\partial f/\partial x_i$ in A_i is the $(n-2) \times (n-2)$ identity matrix, expansion by minors along the ith column of A_i gives

$$\frac{\partial(F_1, \ldots, \hat{F}_i, \ldots, F_n)}{\partial(x_1, \ldots, x_{n-1})} = \det A_i = (-1)^{n-1+i} \frac{\partial f}{\partial x_i} \tag{9}$$

for $i < n$. Substitution of (8) and (9) into (6) gives

$$\begin{aligned} a(F) &= \int_Q \left[1 + \left(\frac{\partial f}{\partial x_1}\right)^2 + \cdots + \left(\frac{\partial f}{\partial x_{n-1}}\right)^2\right]^{1/2} \qquad (10) \\ &= \int_Q [1 + |\nabla f|^2]^{1/2}. \end{aligned}$$

In the case $n = 3$, where we are considering a graph $z = f(x, y)$ in $\mathscr{R}^3$ over $Q \subset \mathscr{R}^2$, and $F(x, y) = (x, y, f(x, y))$, formula (10) reduces to

$$a(F) = \iint_Q \left[1 + \left(\frac{\partial z}{\partial x}\right)^2 + \left(\frac{\partial z}{\partial y}\right)^2\right]^{1/2} dx\, dy.$$

This, together with (4) and (7) above, are the formulas for surface area in $\mathscr{R}^3$ that are often seen in introductory texts.

Example 7 We now apply formula (10) to calculate the $(n-1)$-dimensional area of the sphere S_r^{n-1} of radius r, centered at the origin in $\mathscr{R}^n$. The upper hemisphere of S_r^{n-1} is the graph in $\mathscr{R}^n$ of the function

$$f(x_1, \ldots, x_{n-1}) = [r^2 - x_1{}^2 - \cdots - x_{n-1}^2]^{1/2}$$
$$= [r^2 - |\mathbf{x}|^2]^{1/2},$$

defined for $\mathbf{x} \in B_r^{n-1} \subset \mathscr{R}^{n-1}$. f is continuous on B_r^{n-1}, and $\mathscr{C}^1$ on int B_r^{n-1}, where

$$\frac{\partial f}{\partial x_i} = \frac{-x_i}{[r^2 - |\mathbf{x}|^2]^{1/2}},$$

so

$$|\nabla f(x)|^2 = \frac{|\mathbf{x}|^2}{r^2 - |\mathbf{x}|^2}$$

and therefore

$$[1 + |\nabla f|^2]^{1/2} = \frac{r}{[r^2 - |\mathbf{x}|^2]^{1/2}}.$$

Hence we are motivated by (10) to *define* $a(S_r^{n-1})$ by

$$a(S_r^{n-1}) = 2\int_{B_r^{n-1}} \frac{r\, dx_1 \cdots dx_{n-1}}{[r^2 - |\mathbf{x}|^2]^{1/2}}.$$

This is an improper integral, but we proceed with impunity to compute it. In order to "legalize" our calculation, one would have to replace integrals over B_r^{n-1} with integrals over $B_{r-\varepsilon}^{n-1}$, and then take limits as $\varepsilon \to 0$.

If $T: \mathscr{R}_{\mathbf{u}}^{n-1} \to \mathscr{R}_{\mathbf{x}}^{n-1}$ is defined by $T(\mathbf{u}) = r\mathbf{u}$, then $T(B^{n-1}) = B_r^{n-1}$ and $\det T' = r^{n-1}$, so the change of variables theorem gives

$$a(S_r^{n-1}) = 2\int_{B_r^{n-1}} \frac{r\, dx_1 \cdots dx_{n-1}}{[r^2 - |\mathbf{x}|^2]^{1/2}}$$
$$= 2\int_{B^{n-1}} \frac{r \cdot r^{n-1}\, du_1 \cdots du_{n-1}}{[r^2 - r^2|\mathbf{u}|^2]^{1/2}}$$
$$= 2r^{n-1}\int_{B^{n-1}} \frac{du_1 \cdots du_{n-1}}{[1 - |\mathbf{u}|^2]^{1/2}},$$
$$a(S_r^{n-1}) = r^{n-1}a(S^{n-1}).$$

Thus the area of an $(n-1)$-dimensional sphere is proportional to the $(n-1)$th power of its radius (as one would expect).

Henceforth we write

$$\omega_n = a(S^{n-1})$$

for the area of the unit $(n-1)$-sphere $S^{n-1} \subset \mathscr{R}^n$. By an application of Fubini's theorem, we obtain (see Fig. 5.26)

$$\begin{aligned}
\omega_n &= 2 \int_{B^{n-1}} \frac{dx_1 \cdots dx_{n-1}}{[1 - |\mathbf{x}|^2]^{1/2}} \\
&= 2 \int_{-1}^{1} \left(\int_{B_r^{n-2}} \frac{dx_1 \cdots dx_{n-2}}{[r^2 - x_1{}^2 \cdots - x_{n-2}^2]^{1/2}} \right) dx_{n-1}
\end{aligned}$$

(where $r = [1 - x_{n-1}^2]^{1/2}$)

$$\begin{aligned}
&= \int_{-1}^{1} \frac{1}{r} a(S_r^{n-2})\, dx_{n-1} \\
&= \int_{-1}^{1} r^{n-3} a(S^{n-2})\, dx_{n-1},
\end{aligned}$$

$$\omega_n = \omega_{n-1} \int_{-1}^{1} (1 - x_{n-1}^2)^{(n-3)/2}\, dx_{n-1}.$$

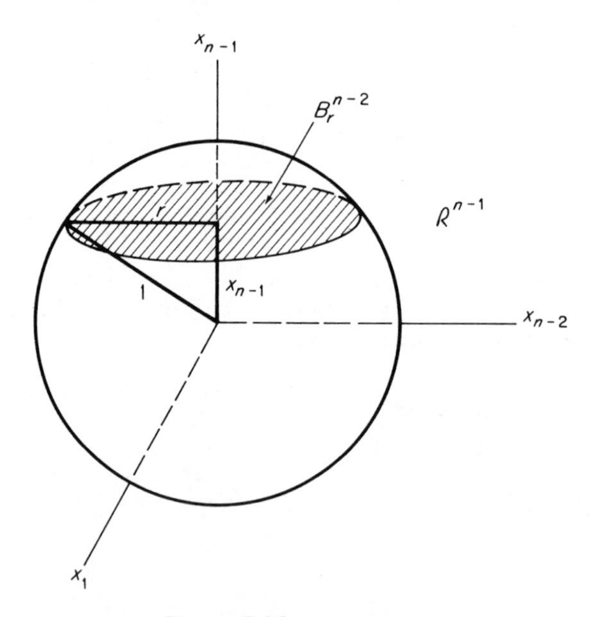

Figure 5.26

With the substitution $x_{n-1} = \cos\theta$ we finally obtain

$$\omega_n = 2I_{n-2}\,\omega_{n-1} \qquad (= 4I_{n-2}\,I_{n-3}\,\omega_{n-2}),$$

where

$$I_k = \int_0^{\pi/2} \sin^k \theta \, d\theta.$$

Since $I_k I_{k-1} = \pi/2k$ by Exercise IV.5.17, it follows that

$$\omega_n = \frac{2\pi}{n-2}\,\omega_{n-2}$$

if $n \geqq 4$. In Exercise 4.10 this recursion relation is used to show that

$$\omega_n = \frac{2\pi^{n/2}}{\Gamma(n/2)}$$

for all $n \geqq 2$.

In the previous example we defined $a(S^{n-1})$ in terms of a particular parametrization of the upper hemisphere of S^{n-1}. There is a valid objection to this procedure—the area of a manifold in $\mathscr{R}^n$ should be defined in an intrinsic manner which is independent of any particular parametrization of it (or of part of it). We now proceed to give such a definition.

First we need to discuss k-cells in $\mathscr{R}^n$. The set $A \subset \mathscr{R}^n$ is called a (smooth) *k-cell* if it is the image of the unit cube $I^k \subset \mathscr{R}^k$ under a one-to-one $\mathscr{C}^1$ mapping $\varphi : U \to \mathscr{R}^n$ which is defined on a neighborhood U of I^k, with the derivative matrix $\varphi'(\mathbf{u})$ having maximal rank k for each $\mathbf{u} \in I^k$ (Fig. 5.27). The mapping

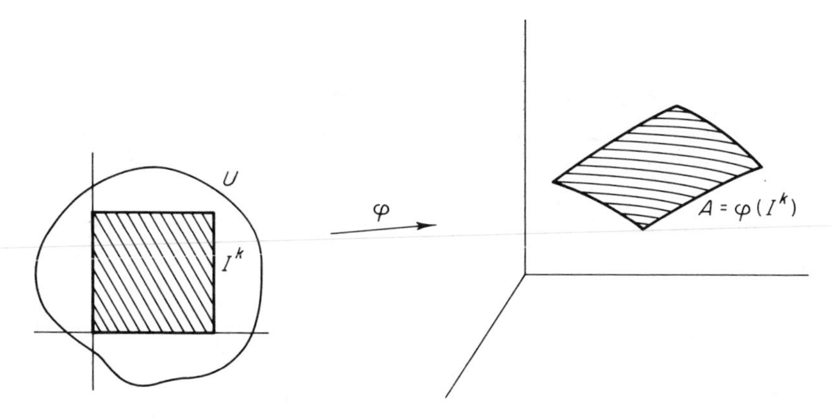

Figure 5.27

φ (restricted to I^k) is then called a *parametrization* of A. The following theorem shows that the area of a k-cell $A \subset \mathscr{R}^n$ is well defined. That is, if φ and ψ are two parametrizations of the k-cell A, then $a(\varphi) = a(\psi)$, so their common value may be denoted by $a(A)$.

Theorem 4.1 If φ and ψ are two parametrizations of the k-cell $A \subset \mathscr{R}^n$, then

$$\int_{I^k} [\det(\varphi')^t\varphi']^{1/2} = \int_{I^k} [\det(\psi')^t\psi']^{1/2}.$$

PROOF Since $\varphi'(\mathbf{u})$ has rank k, the differential $d\varphi_{\mathbf{u}}$ maps $\mathscr{R}^k$ one-to-one onto the tangent k-plane to A at $\varphi(\mathbf{u})$, and the same is true of $d\psi_{\mathbf{u}}$. If

$$T : I^k \to I^k$$

is defined by $T = \psi^{-1} \circ \varphi$, it follows that T is a $\mathscr{C}^1$ one-to-one mapping with $\det T'(\mathbf{u}) \neq 0$ for all $\mathbf{u} \in I^k$ so, by the inverse function theorem, T is $\mathscr{C}^1$-invertible.

Since $\varphi = T \circ \psi$, the chain rule gives

$$\begin{aligned}(\varphi')^t(\varphi') &= ((\psi \circ T)')^t(\psi \circ T)' \\ &= (T')^t(\psi')^t(\psi')(T'),\end{aligned}$$

so

$$[\det(\varphi')^t\varphi']^{1/2} = [\det(\psi')^t\psi']^{1/2}\,|\det T'|.$$

Consequently the change of variables theorem gives

$$\begin{aligned}\int_{I^k} [\det(\varphi')^t\varphi']^{1/2} &= \int_{I^k} [\det(\psi'(T(\mathbf{u})))^t\psi'(T(\mathbf{u}))]^{1/2}\,|\det T'(\mathbf{u})|\,d\mathbf{u} \\ &= \int_{I^k} [\det(\psi')^t\psi']^{1/2}\end{aligned}$$

as desired. ∎

If M is a compact smooth k-manifold in $\mathscr{R}^n$, then M is the union of a finite number of nonoverlapping k-cells. That is, there exist k-cells $A_1, \ldots, A_r$ in $\mathscr{R}^n$ such that $M = \bigcup_{i=1}^r A_i$ and $\operatorname{int} A_i \cap \operatorname{int} A_j = \varnothing$ if $i \neq j$. Such a collection $\{A_1, \ldots, A_r\}$ is called a *paving* of M. The proof that every compact smooth manifold in $\mathscr{R}^n$ is *pavable* (that is, has a paving) is very difficult, and will not be included here. However, if we assume this fact, we can define the (k-dimensional) *area* $a(M)$ of M by

$$a(M) = \sum_{i=1}^{r} a(A_i).$$

Of course it must be verified that $a(M)$ is thereby well defined. That is, if $\{B_1, \ldots, B_s\}$ is a second paving of M, then

$$\sum_{i=1}^{r} a(A_i) = \sum_{j=1}^{s} a(B_j).$$

The proof is outlined in Exercise 4.13.

Example 8 We now employ the above definition, of the area of a smooth manifold in $\mathscr{R}^n$, to give a second computation of $\omega_n = a(S^{n-1})$. Let $\{A_1, \ldots, A_k\}$ be a paving of S^{n-1}. Let $\varphi : I^{n-1} \to A_i$ be a parametrization of A_i. Then, given $\varepsilon > 0$, define $\Phi : I^{n-1} \times [\varepsilon, 1] \to \mathscr{R}^n$ by

$$\Phi(\mathbf{u}, r) = r\varphi(\mathbf{u}) \qquad \text{for} \quad \mathbf{u} \in I^{n-1}, \quad r \in [\varepsilon, 1].$$

Denote by B_i the image of Φ (see Fig. 5.28). Then

$$v(B_i) = \int_{B_i} 1 = \int_{I^{n-1} \times [\varepsilon, 1]} |\det \Phi'| \tag{11}$$

by the change of variables theorem. Now $|\det \Phi'(\mathbf{u}, r)|$ is the volume of the n-dimensional parallelepiped P which is spanned by the vectors

$$\frac{\partial \Phi}{\partial u_1}, \quad \ldots, \quad \frac{\partial \Phi}{\partial u_{n-1}}, \quad \frac{\partial \Phi}{\partial r}$$

[evaluated at $(\mathbf{u}, r)$]. But

$$\frac{\partial \Phi}{\partial r}(\mathbf{u}, r) = \varphi(\mathbf{u}) \in S^{n-1}$$

is a unit vector which is orthogonal to each of the vectors

$$\frac{\partial \Phi}{\partial u_1} = r \frac{\partial \varphi}{\partial u_1}, \ldots, \frac{\partial \Phi}{\partial u_{n-1}} = r \frac{\partial \varphi}{\partial u_{n-1}},$$

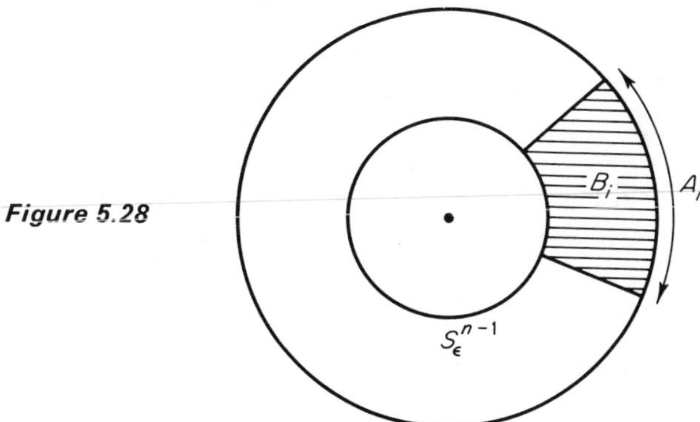

Figure 5.28

since the vectors $r\,\partial\varphi/\partial u_1, \ldots, r\,\partial\varphi/\partial u_{n-1}$ are tangent to S^{n-1}. If Q is the $(n-1)$-dimensional parallelepiped which is spanned by these n − 1 vectors, it follows that $v(P) = a(Q)$. Consequently we have

$$\begin{aligned}
|\det \Phi'(\mathbf{u}, r)| &= v(P) \\
&= a(Q) \\
&= [\det (r\varphi'(\mathbf{u}))^t (r\varphi'(\mathbf{u}))]^{1/2} \\
&= r^{n-1}[\det (\varphi'(\mathbf{u}))^t (\varphi'(\mathbf{u}))]^{1/2}.
\end{aligned}$$

Substituting this into (11), and applying Fubini's theorem, we obtain

$$v(B_i) = \int_\varepsilon^1 r^{n-1}\left(\int_{I^{n-1}} [\det(\varphi'(\mathbf{u}))^t \varphi'(\mathbf{u})]\, d\mathbf{u}\right) dr$$

$$= \int_\varepsilon^1 r^{n-1} a(A_i)\, dr$$

$$v(B_i) = \frac{a(A_i)}{n}(1 - \varepsilon^n).$$

Summing from $i = 1$ to $i = k$, we then obtain

$$v(B^n - \text{int}\, B_\varepsilon^n) = \sum_{i=1}^{k} v(B_i)$$

$$= \sum_{i=1}^{n} \frac{a(A_i)}{n}(1 - \varepsilon^n)$$

$$= \frac{1}{n} a(S^{n-1})(1 - \varepsilon^n).$$

Finally taking the limit as $\varepsilon \to 0$, we obtain

$$v(B^n) = \frac{1}{n} a(S^{n-1}),$$

or

$$a(S^{n-1}) = nv(B^n).$$

Consulting Exercise IV.6.3 for the value of $v(B^n)$, this gives the value for $a(S^{n-1})$ listed at the end of Example 7.

Exercises

4.1 Consider the cylinder $x^2 + y^2 = a^2$, $0 \leqq z \leqq h$ in $\mathscr{R}^3$. Regarding it as the image of a surface patch defined on the rectangle $[0, 2\pi] \times [0, h]$ in the θz-plane, apply formula (3) or (4) to show that its area is $2\pi ah$.

4.2 Calculate the area of the cone $z = \sqrt{x^2 + y^2}$, $z \leqq 1$, noting that it is the image of a surface patch defined on the rectangle $[0, 1] \times [0, 2\pi]$ in the $r\theta$-plane.

4.3 Let T be the torus obtained by rotating the circle $z^2 + (y - a)^2 = b^2$ about the z-axis. Then T is the image of the square $0 \leqq \theta$, $\varphi \leqq 2\pi$ in the $\varphi\theta$-plane under the mapping $F : \mathscr{R}^2 \to \mathscr{R}^3$ defined by

$$x = (a + b\cos\varphi)\cos\theta,$$
$$y = (a + b\cos\varphi)\sin\theta,$$
$$z = b\sin\varphi.$$

Calculate its area (*answer* $= 4\pi^2 ab$).

4.4 Show that the area of the "flat torus" $S^1 \times S^1 \subset \mathscr{R}^4$, of Example 2, is $(2\pi)^2$. Generalize this computation to show that the area, of the "n-dimensional torus" $S^1 \times \cdots \times S^1 \subset \mathscr{R}^2 \times \cdots \times \mathscr{R}^2 = \mathscr{R}^{2n}$, is $(2\pi)^n$.

4.5 Consider the generalized torus $S^2 \times S^2 \subset \mathscr{R}^3 \times \mathscr{R}^3 = \mathscr{R}^6$. Use spherical coordinates in each factor to define a surface patch whose image is $S^2 \times S^2$. Then compute its area.

4.6 Let C be a curve in the yz-plane described by $y = \alpha(t)$, $z = \beta(t)$ for $t \in [a, b]$. The surface swept out by revolving C around the z-axis is the image of the surface patch $\varphi : [0, 2\pi] \times [a, b] \to \mathscr{R}^3$ defined by

$$\begin{aligned} x &= \alpha(t) \cos \theta, \\ y &= \alpha(t) \sin \theta, \\ z &= \beta(t). \end{aligned}$$

Show that $a(\varphi)$ equals the length of C times the distance traveled by its centroid, that is,

$$a(\varphi) = (2\pi\bar{y})L \text{ where } \bar{y} = (1/L)\int_C y\, ds.$$

4.7 Let $\gamma(s) = (\alpha(s), \beta(s))$ be a *unit-speed* curve in the xy-plane, so $(\alpha')^2 + (\beta')^2 = 1$. The "tube surface" with radius r and center-line γ is the image of $\varphi : [0, L] \times [0, 2\pi] \to \mathscr{R}^3$, where φ is given by

$$\begin{aligned} x &= \alpha(s) + r\beta'(s) \cos \theta, \\ y &= \beta(s) - r\alpha'(s) \cos \theta, \\ z &= r \sin \theta. \end{aligned}$$

Show that $a(\varphi) = 2\pi rL$.

4.8 This exercise gives an n-dimensional generalization of "Pappus' theorem for surface area" (Exercise 4.6). Let $\varphi : Q \to \mathscr{R}^{n-1}$ be a k-dimensional surface patch in $\mathscr{R}^{n-1}$, such that $\varphi_{n-1}(\mathbf{u}) \geqq 0$ for all $\mathbf{u} \in Q$. The $(k+1)$-dimensional surface patch

$$\Phi : Q \times [0, 2\pi] \to \mathscr{R}^n,$$

obtained by "revolving φ about the coordinate hyperplane $\mathscr{R}^{n-2}$," is defined by

$$\Phi(\mathbf{u}, \theta) = (\varphi_1(\mathbf{u}), \ldots, \varphi_{n-2}(\mathbf{u}), \varphi_{n-1}(\mathbf{u}) \cos \theta, \varphi_{n-1}(\mathbf{u}) \sin \theta)$$

for $\mathbf{u} \in Q$, $\theta \in [0, 2\pi]$. Write down the matrix Q′, and conclude from inspection of it that

$$(\Phi')^t\Phi' = \left(\begin{array}{c|c} (\varphi')^t\varphi' & \begin{matrix} 0 \\ \vdots \\ 0 \end{matrix} \\ \hline 0 \cdots 0 & \varphi_{n-1} \end{array}\right),$$

so

$$[\det (\Phi')^t\Phi']^{1/2} = \varphi_{n-1}[\det (\varphi')^t\varphi']^{1/2}.$$

Deduce from this that

$$a(\Phi) = 2\pi\bar{x}_{n-1}a(\varphi),$$

where $\bar{x}_{n-1}$ is the $(n-1)$th coordinate of the centroid of φ, defined by

$$\bar{x}_{n-1} = \frac{1}{a(\varphi)} \int_Q \varphi_{n-1} \, [\det (\varphi')^t\varphi']^{1/2}.$$

4.9 Let $T : \mathscr{R}^n \to \mathscr{R}^n$ be the linear mapping defined by $T(\mathbf{x}) = b\mathbf{x}$. If $\varphi : Q \to \mathscr{R}^n$ is a k-dimensional surface patch in $\mathscr{R}^n$, prove that $a(T \circ \varphi) = b^k a(\varphi)$.

4.10 Starting with $\omega_2 = 2\pi$ and $\omega_3 = 4\pi$, use the recursion formula

$$\omega_n = \frac{2\pi}{n-2}\,\omega_{n-2},$$

of Example 7, to establish, by separate inductions on m, the formulas

$$\omega_{2m} = \frac{2\pi^m}{(m-1)!}$$

and

$$\omega_{2m+1} = \frac{2^{m+1}\pi^m}{1 \cdot 3 \cdot 5 \cdot \cdots \cdot (2m-1)}$$

for $\omega_n = a(S^{n-1})$. Consulting Exercise IV.6.2, deduce that

$$\omega_n = \frac{2\pi^{n/2}}{\Gamma(n/2)}$$

for all $n \geqq 2$.

4.11 Use the n-dimensional spherical coordinates of Exercise IV.5.19 to define an $(n-1)$-dimensional surface patch whose image is S^{n-1}. Then show that its area is given by the formula of the previous exercise.

4.12 This exercise is a generalization of Example 8. Let M be a smooth compact $(n-1)$-manifold in $\mathscr{R}^n$, and $\Phi : M \times [a, b] \to \mathscr{R}^n$ a one-to-one $\mathscr{C}^1$ mapping. Write $M_t = \Phi(M \times \{t\})$. Suppose that, for each $\mathbf{x} \in M$, the curve $\gamma_{\mathbf{x}}(t) = \Phi(\mathbf{x}, t)$ is a unit-speed curve which is orthogonal to each of the manifolds M_t. If $A = \Phi(M \times [a, b])$, prove that

$$v(A) = \int_a^b a(M_t)\,dt.$$

4.13 Let M be a smooth k-manifold in $\mathscr{R}^n$, and suppose $\{A_1, \ldots, A_r\}$ and $\{B_1, \ldots, B_s\}$ are two collections of nonoverlapping k-cells such that

$$M = \bigcup_{i=1}^{r} A_i = \bigcup_{j=1}^{s} B_j.$$

Then prove that

$$\sum_{i=1}^{r} a(A_i) = \sum_{j=1}^{s} a(B_j).$$

Hint: Let φ_i and ψ_j be parametrizations of A_i and B_j, respectively. Define

$$Q_{ij} = \varphi_i^{-1}(A_i \cap B_j) \qquad \text{and} \qquad R_{ij} = \psi_j^{-1}(A_i \cap B_j).$$

Then

$$\sum_{i=1}^{r} a(A_i) = \sum_{i=1}^{r}\left(\sum_{j=1}^{s} \int_{Q_{ij}} [\det (\varphi_i')^t \varphi_i']^{1/2}\right),$$

while

$$\sum_{j=1}^{s} a(B_j) = \sum_{j=1}^{s}\left(\sum_{i=1}^{r} \int_{R_{ij}} [\det (\psi_j')^t \psi_j']^{1/2}\right).$$

Show by the method of proof of Theorem 4.1, using Exercises IV.5.3 and IV.5.4, that corresponding terms in the right-hand sums are equal.

5 DIFFERENTIAL FORMS

We have seen that a differential 1-form ω on $\mathscr{R}^n$ is a mapping which associates with each point $\mathbf{x} \in \mathscr{R}^n$ a linear function $\omega(\mathbf{x}) : \mathscr{R}^n \to \mathscr{R}$, and that each linear function on $\mathscr{R}^n$ is a linear combination of the differentials $dx_1, \ldots, dx_n$, so

$$\omega(\mathbf{x}) = a_1(\mathbf{x})\,dx_1 + \cdots + a_n(\mathbf{x})\,dx_n,$$

where $a_1, \ldots, a_n$ are real-valued functions on $\mathscr{R}^n$.

A *differential k-form* α, defined on the set $U \subset \mathscr{R}^n$, is a mapping which associates with each point $\mathbf{x} \in U$ an alternating k-multilinear function $\alpha(\mathbf{x}) = \alpha_{\mathbf{x}}$ on $\mathscr{R}^n$. That is,

$$\alpha : U \to \Lambda^k(\mathscr{R}^n),$$

where $\Lambda^k(\mathscr{R}^n)$ denotes the set of all alternating k-multilinear functions on $\mathscr{R}^n$. Since we have seen in Theorem 3.4 that every alternating k-multilinear function on $\mathscr{R}^n$ is a (unique) linear combination of the "multidifferentials" $d\mathbf{x}_{\mathbf{i}}$, it follows that $\alpha(\mathbf{x})$ can be written in the form

$$\alpha(\mathbf{x}) = \sum_{[\mathbf{i}]} a_{\mathbf{i}}(\mathbf{x})\,d\mathbf{x}_{\mathbf{i}}, \tag{1}$$

where as usual $[\mathbf{i}]$ denotes summation over all increasing k-tuples $\mathbf{i} = (i_1, \ldots, i_k)$ with $1 \leqq i_r \leqq n$, and each $a_{\mathbf{i}}$ is a real-valued function on U. The differential k-form α is called *continuous* (or $\mathscr{C}^1$, etc.) provided that each of the coefficient functions $a_{\mathbf{i}}$ is continuous (or $\mathscr{C}^1$, etc.).

For example, every differential 2-form α on $\mathscr{R}^3$ is of the form

$$\alpha = a_{(1,2)}\,d\mathbf{x}_{(1,2)} + a_{(1,3)}\,d\mathbf{x}_{(1,3)} + a_{(2,3)}\,d\mathbf{x}_{(2,3)},$$

while every differential 3-form β on $\mathscr{R}^3$ is a scalar (function) multiple of the single multidifferential $d\mathbf{x}_{(1,2,3)}$,

$$\beta(\mathbf{x}) = b_{(1,2,3)}(\mathbf{x})\,d\mathbf{x}_{(1,2,3)}.$$

Similarly, every differential 2-form on $\mathscr{R}^n$ is of the form

$$\alpha = \sum_{1 \leqq i < j \leqq n} a_{(i,j)}\,d\mathbf{x}_{(i,j)}.$$

A standard and useful alternative notation for multidifferentials is

$$d\mathbf{x}_{\mathbf{i}} = dx_{i_1} \wedge dx_{i_2} \wedge \cdots \wedge dx_{i_k} \tag{2}$$

if $\mathbf{i} = (i_1, \ldots, i_k)$; we think of the multidifferential $d\mathbf{x}_{\mathbf{i}}$ as a product of the differentials $dx_{i_1}, \ldots, dx_{i_k}$. Recall that, if A is the $n \times k$ matrix whose column vectors are $\mathbf{a}^1, \ldots, \mathbf{a}^k$, then

$$d\mathbf{x}_{\mathbf{i}}(\mathbf{a}^1, \ldots, \mathbf{a}^k) = \det A_{\mathbf{i}},$$

where $A_\mathbf{i}$ denotes the $k \times k$ matrix whose rth row is the i_rth row of A. If $i_r = i_s$, then the rth and sth rows of $A_\mathbf{i}$ are equal, so

$$dx_\mathbf{i}(\mathbf{a}^1, \ldots, \mathbf{a}^k) = 0.$$

In terms of the product notation of (2), it follows that

$$dx_{i_1} \wedge \cdots \wedge dx_{i_k} = 0$$

unless the integers $i_1, \ldots, i_k$ are distinct. In particular,

$$dx_i \wedge dx_i = 0 \tag{3}$$

for each $i = 1, \ldots, n$. Similarly,

$$dx_i \wedge dx_j = -dx_j \wedge dx_i, \tag{4}$$

since the sign of a determinant is changed when two of its rows are interchanged.

The multiplication of differentials extends in a natural way to a multiplication of differential forms. First we define

$$d\mathbf{x_i} \wedge d\mathbf{x_j} = dx_{i_1} \wedge \cdots \wedge dx_{i_k} \wedge dx_{j_1} \wedge \cdots \wedge dx_{j_l} \tag{5}$$

if $\mathbf{i} = (i_1, \ldots, i_k)$ and $\mathbf{j} = (j_1, \ldots, j_l)$. Then, given a differential k-form

$$\alpha = \sum_{[\mathbf{i}]} a_\mathbf{i} \, d\mathbf{x_i}$$

and a differential l-form

$$\beta = \sum_{[\mathbf{j}]} b_\mathbf{j} \, d\mathbf{x_j},$$

their *product* $\alpha \wedge \beta$ (sometimes called *exterior* product) is the differential $(k + l)$-form defined by

$$\alpha \wedge \beta = \sum_{[\mathbf{i}], [\mathbf{j}]} a_\mathbf{i} b_\mathbf{j} \, d\mathbf{x_i} \wedge d\mathbf{x_j}. \tag{6}$$

This means simply that the differential forms α and β are multiplied together in a formal term-by-term way, using (5) and distributivity of multiplication over addition. Strictly speaking, the result of this process, the right-hand side of (6), is not quite a differential form as defined in (1), because the typical $(k + l)$-tuple $(\mathbf{i}, \mathbf{j}) = (i_1, \ldots, i_k, j_1, \ldots, j_l)$ appearing in (6) is not necessarily increasing. However it is clear that, by use of rules (3) and (4), we can rewrite the result in the form

$$\alpha \wedge \beta = \sum_{[\mathbf{i}]} c_\mathbf{i} \, d\mathbf{x_i}$$

with the summation being over all increasing $(k + l)$-tuples. Note that $\alpha \wedge \beta = 0$ if $k + l > n$.

It will be an instructive exercise for the student to deduce from this definition and the anticommutative property (4) that, if α is a k-form and β is an l-form, then

$$\beta \wedge \alpha = (-1)^{kl} \alpha \wedge \beta. \tag{7}$$

Example 1 Let $\alpha = a_1\,dx_1 + a_2\,dx_2 + a_3\,dx_3$ and $\beta = b_1\,dx_1 + b_2\,dx_2 + b_3\,dx_3$ be two 1-forms on $\mathscr{R}^3$. Then

$$\begin{aligned}
\alpha \wedge \beta &= (a_1\,dx_1 + a_2\,dx_2 + a_3\,dx_3) \wedge (b_1\,dx_1 + b_2\,dx_2 + b_3\,dx_3) \\
&= a_1 b_1\,dx_1 \wedge dx_1 + a_1 b_2\,dx_1 \wedge dx_2 + a_1 b_3\,dx_1 \wedge dx_3 \\
&\quad + a_2 b_1\,dx_2 \wedge dx_1 + a_2 b_2\,dx_2 \wedge dx_2 + a_2 b_3\,dx_2 \wedge dx_3 \\
&\quad + a_3 b_1\,dx_3 \wedge dx_1 + a_3 b_2\,dx_3 \wedge dx_2 + a_3 b_3\,dx_3 \wedge dx_3 \\
&= a_1 b_2\,dx_1 \wedge dx_2 + a_2 b_1\,dx_2 \wedge dx_1 + a_1 b_3\,dx_1 \wedge dx_3 \\
&\quad + a_3 b_1\,dx_3 \wedge dx_1 + a_2 b_3\,dx_2 \wedge dx_3 + a_3 b_2\,dx_3 \wedge dx_2, \\
\alpha \wedge \beta &= (a_1 b_2 - a_2 b_1)\,dx_1 \wedge dx_2 + (a_1 b_3 - a_3 b_1)\,dx_1 \wedge dx_3 \\
&\quad + (a_2 b_3 - a_3 b_2)\,dx_2 \wedge dx_3,
\end{aligned}$$

using (3) and (4), respectively in the last two steps. Similarly, consider a 1-form $\omega = P\,dx + Q\,dy + R\,dz$ and 2-form $\alpha = A\,dx \wedge dy + B\,dx \wedge dz + C\,dy \wedge dz$. Applying (3) to delete immediately each multidifferential that contains twice a single differential dx or dy or dz, we obtain

$$\begin{aligned}
\omega \wedge \alpha &= PC\,dx \wedge dy \wedge dz + QB\,dy \wedge dx \wedge dz + RA\,dz \wedge dx \wedge dy \\
&= (PC - QB + RA)\,dx \wedge dy \wedge dz.
\end{aligned}$$

We next define an operation of differentiation for differential k-forms, extending our previous definitions in the case of 0-forms (or functions) on $\mathscr{R}^n$ and 1-forms on $\mathscr{R}^2$. Recall that the differential of the $\mathscr{C}^1$ function $f : \mathscr{R}^n \to \mathscr{R}$ is defined by

$$df = \frac{\partial f}{\partial x_1}\,dx_1 + \cdots + \frac{\partial f}{\partial x_n}\,dx_n .$$

Given a $\mathscr{C}^1$ differential k-form $\alpha = \sum_{[\mathbf{i}]} a_{\mathbf{i}}\,d\mathbf{x}_{\mathbf{i}}$ defined on the open set $U \subset \mathscr{R}^n$, its differential $d\alpha$ is the $(k+1)$-form defined on U by

$$d\alpha = \sum_{[\mathbf{i}]} (da_{\mathbf{i}}) \wedge d\mathbf{x}_{\mathbf{i}} . \tag{8}$$

Note first that the differential operation is clearly additive,

$$d(\alpha + \beta) = d\alpha + d\beta.$$

Example 2 If $\omega = P\,dx + Q\,dy + R\,dz$, then

$$\begin{aligned}
d\omega &= dP \wedge dx + dQ \wedge dy + dR \wedge dz \\
&= \left(\frac{\partial P}{\partial x}\,dx + \frac{\partial P}{\partial y}\,dy + \frac{\partial P}{\partial z}\,dz\right) \wedge dx \\
&\quad + \left(\frac{\partial Q}{\partial x}\,dx + \frac{\partial Q}{\partial y}\,dy + \frac{\partial Q}{\partial z}\,dz\right) \wedge dy \\
&\quad + \left(\frac{\partial R}{\partial x}\,dx + \frac{\partial R}{\partial y}\,dy + \frac{\partial R}{\partial z}\,dz\right) \wedge dz,
\end{aligned}$$

$$d\omega = \left(\frac{\partial R}{\partial y} - \frac{\partial Q}{\partial z}\right) dy \wedge dz + \left(\frac{\partial P}{\partial z} - \frac{\partial R}{\partial x}\right) dz \wedge dx + \left(\frac{\partial Q}{\partial x} - \frac{\partial P}{\partial y}\right) dx \wedge dy.$$

If $\alpha = A\,dy \wedge dz + B\,dz \wedge dx + C\,dx \wedge dy$, then

$$\begin{aligned}
d\alpha &= \left(\frac{\partial A}{\partial x}\,dx + \frac{\partial A}{\partial y}\,dy + \frac{\partial A}{\partial z}\,dz\right) \wedge dy \wedge dz \\
&\quad + \left(\frac{\partial B}{\partial x}\,dx + \frac{\partial B}{\partial y}\,dy + \frac{\partial B}{\partial z}\,dz\right) \wedge dz \wedge dx \\
&\quad + \left(\frac{\partial C}{\partial x}\,dx + \frac{\partial C}{\partial y}\,dy + \frac{\partial C}{\partial z}\,dz\right) \wedge dx \wedge dy,
\end{aligned}$$

$$d\alpha = \left(\frac{\partial A}{\partial x} + \frac{\partial B}{\partial y} + \frac{\partial C}{\partial z}\right) dx \wedge dy \wedge dz.$$

If ω is of class $\mathscr{C}^2$ then, setting $\alpha = d\omega$, we obtain

$$\begin{aligned}
d(d\omega) &= \left[\frac{\partial}{\partial x}\left(\frac{\partial R}{\partial y} - \frac{\partial Q}{\partial z}\right) + \frac{\partial}{\partial y}\left(\frac{\partial P}{\partial z} - \frac{\partial R}{\partial x}\right) + \frac{\partial}{\partial z}\left(\frac{\partial Q}{\partial x} - \frac{\partial P}{\partial y}\right)\right] dx \wedge dy \wedge dz \\
&= 0,
\end{aligned}$$

by the equality, under interchange of order of differentiation, of mixed second order partial derivatives of $\mathscr{C}^2$ functions. The fact, that $d(d\omega) = 0$ if ω is a $\mathscr{C}^2$ differential 1-form in $\mathscr{R}^3$, is an instance of a quite general phenomenon.

Proposition 5.1 If α is a $\mathscr{C}^2$ differential k-form on an open subset of $\mathscr{R}^n$, then $d(d\alpha) = 0$.

PROOF Since $d(\beta + \gamma) = d\beta + d\gamma$, it suffices to verify that $d(d\alpha) = 0$ if

$$\alpha = f\,dx_{i_1} \wedge \cdots \wedge dx_{i_k}.$$

Then

$$d\alpha = \left(\sum_{r=1}^{n} \frac{\partial f}{\partial x_r} dx_r\right) \wedge dx_{i_1} \wedge \cdots \wedge dx_{i_k}$$

$$= \sum_{r=1}^{n} \frac{\partial f}{\partial x_r} dx_r \wedge dx_{i_1} \wedge \cdots \wedge dx_{i_k},$$

so

$$d(d\alpha) = \sum_{r=1}^{n} \left(\sum_{s=1}^{n} \frac{\partial^2 f}{\partial x_s\, \partial x_r} dx_s\right) \wedge dx_r \wedge dx_{i_1} \wedge \cdots \wedge dx_{i_k}$$

$$= \sum_{r,s=1}^{n} \frac{\partial^2 f}{\partial x_s\, \partial x_r} dx_s \wedge dx_r \wedge dx_{i_1} \wedge \cdots \wedge dx_{i_k}.$$

But since $dx_r \wedge dx_s = -dx_s \wedge dx_r$, the terms in this latter sum cancel in pairs, just as in the special case considered in Example 2. ∎

There is a Leibniz-type formula for the differential of a product, but with an interesting twist which results from the anticommutativity of the product operation for forms.

Proposition 5.2 If α is a differential k-form and β a differential l-form, both of class $\mathscr{C}^1$, then

$$d(\alpha \wedge \beta) = (d\alpha) \wedge \beta + (-1)^k \alpha \wedge (d\beta). \tag{9}$$

PROOF By the additivity of the differential operation, it suffices to consider the special case

$$\alpha = a\, dx_{i_1} \wedge \cdots \wedge dx_{i_k}, \qquad \beta = b\, dx_{j_1} \wedge \cdots \wedge dx_{j_l},$$

where a and b are $\mathscr{C}^1$ functions. Then

$$\begin{aligned} d(\alpha \wedge \beta) &= d(ab) \wedge dx_{i_1} \wedge \cdots \wedge dx_{i_k} \wedge dx_{j_1} \wedge \cdots \wedge dx_{j_l} \\ &= (da \wedge b + a \wedge db) \wedge dx_{i_1} \wedge \cdots \wedge dx_{i_k} \wedge \cdots \wedge dx_{j_l} \\ &= (da \wedge dx_{i_1} \wedge \cdots \wedge dx_{i_k}) \wedge (b\, dx_{j_1} \wedge \cdots \wedge dx_{j_l}) \\ &\quad + (-1)^k (a\, dx_{i_1} \wedge \cdots \wedge dx_{i_k}) \wedge (db \wedge dx_{j_1} \wedge \cdots \wedge dx_{j_l}), \end{aligned}$$

$$d(\alpha \wedge \beta) = (d\alpha) \wedge \beta + (-1)^k \alpha \wedge (d\beta),$$

the $(-1)^k$ coming from the application of formula (7) to interchange the 1-form db and the k-form α in the second term. ∎

Recall (from Section 1) that, if ω is a $\mathscr{C}^1$ differential 1-form on $\mathscr{R}^n$, and $\gamma : [a, b] \to \mathscr{R}^n$ is a $\mathscr{C}^1$ path, then the integral of ω over γ is defined by

$$\int_\gamma \omega = \int_a^b \omega_{\gamma(t)}(\gamma'(t))\, dt.$$

We now generalize this definition as follows. If α is a $\mathscr{C}^1$ differential k-form on $\mathscr{R}^n$, and $\varphi : Q \to \mathscr{R}^n$ is a $\mathscr{C}^1$ k-dimensional surface patch, then the *integral of* α *over* φ is defined by

$$\int_\varphi \alpha = \int_Q \alpha_{\varphi(\mathbf{u})}(D_1\varphi(\mathbf{u}), \ldots, D_k\varphi(\mathbf{u}))\, du_1 \cdots du_k, \tag{10}$$

Note that, since the partial derivatives $D_1\varphi, \ldots, D_k\varphi$ are vectors in $\mathscr{R}^n$, the right-hand side of (10) is the "ordinary" integral of a continuous real-valued function on the k-dimensional interval $Q \subset \mathscr{R}^k$.

In the special case $k = n$, the following notational convention is useful. If $\alpha = f\, dx_1 \wedge \cdots \wedge dx_k$ is a $\mathscr{C}^1$ differential k-form on $\mathscr{R}^k$, we write

$$\int_Q \alpha = \int_Q f \tag{11}$$

("ordinary" integral on the right). In other words, $\int_Q \alpha$ is by definition equal to the integral of α over the identity (or inclusion) surface patch $Q \subset \mathscr{R}^k$ (see Exercise 5.7).

The definition in (10) is simply a concise formalization of the result of the following simple and natural procedure. To evaluate the integral

$$\int_Q \left(\sum_{[\mathbf{i}]} a_\mathbf{i}\, dx_{i_1} \wedge \cdots \wedge dx_{i_k}\right),$$

first make the substitutions $x_i = \varphi_i(\mathbf{u})$, $dx_i = \sum_{j=1}^k (\partial\varphi_i/\partial u_j)\, du_j$ throughout. After multiplying out and collecting coefficients, the final result is a differential k-form $\beta = g\, du_1 \wedge \cdots \wedge du_k$ on Q. Then

$$\int_\varphi \left(\sum_{[\mathbf{i}]} a_\mathbf{i}\, dx_{i_1} \wedge \cdots \wedge dx_{i_k}\right) = \int_Q \beta = \int_Q g. \tag{12}$$

Before proving this in general, let us consider the special case in which $\alpha = f\, dy \wedge dz$ is a 2-form on $\mathscr{R}^3$, and $\varphi : Q \to \mathscr{R}^3$ is a 2-dimensional surface patch. Using uv-coordinates in $\mathscr{R}^2$, we obtain

$$\begin{aligned}
\int_Q \alpha &= \int_Q f(\varphi(u, v))\, dy \wedge dz\left(\frac{\partial\varphi}{\partial u}, \frac{\partial\varphi}{\partial v}\right) du\, dv \\
&= \int_Q f(\varphi(u, v))\left(\frac{\partial\varphi_2}{\partial u}\frac{\partial\varphi_3}{\partial v} - \frac{\partial\varphi_3}{\partial u}\frac{\partial\varphi_2}{\partial v}\right) du \wedge dv \\
&= \int_Q f(\varphi(u, v))\left(\frac{\partial\varphi_2}{\partial u}\, du + \frac{\partial\varphi_2}{\partial v}\, dv\right)\left(\frac{\partial\varphi_3}{\partial u}\, du + \frac{\partial\varphi_3}{\partial v}\, dv\right),
\end{aligned}$$

thus verifying (in this special case) the assertion of the preceding paragraph.

In order to formulate precisely (and then prove) the general assertion, we must define the notion of the *pullback* $\varphi^*(\alpha) = \varphi^*\alpha$, of the k-form α on $\mathscr{R}^n$, under a $\mathscr{C}^1$ mapping $\varphi : \mathscr{R}^m_{\mathbf{u}} \to \mathscr{R}^n_{\mathbf{x}}$. This will be a generalization of the pullback defined in Section 2 for differential forms on $\mathscr{R}^2$. We start by defining the pullback of a 0-form (or function) f or differential dx_i by

$$\begin{aligned} \varphi^*(f) &= f \circ \varphi, \\ \varphi^*(dx_i) &= \sum_{j=1}^{m} \frac{\partial \varphi_i}{\partial u_j}\, du_j = d\varphi_i . \end{aligned} \tag{13}$$

We can then extend the definition to arbitrary k-forms on $\mathscr{R}^n$ by requiring that

$$\begin{aligned} \varphi^*(f\alpha) &= (f \circ \varphi)\varphi^*\alpha, \\ \varphi^*(\alpha \wedge \beta) &= \varphi^*\alpha \wedge \varphi^*\beta, \\ &\text{and} \\ \varphi^*(\alpha + \beta) &= \varphi^*\alpha + \varphi^*\beta. \end{aligned} \tag{14}$$

Exercise 5.8 gives an important interpretation of this definition of the pullback operation.

Example 3 Let φ be a $\mathscr{C}^1$ mapping from $\mathscr{R}^2_{uv}$ to $\mathscr{R}^3_{xyz}$. If $\omega = P\,dx + Q\,dy + R\,dz$, then

$$\begin{aligned} \varphi^*\omega &= (P \circ \varphi)\varphi^*(dx) + (Q \circ \varphi)\varphi^*(dy) + (R \circ \varphi)\varphi^*(dz) \\ &= (P \circ \varphi)\left[\frac{\partial \varphi_1}{\partial u}\, du + \frac{\partial \varphi_1}{\partial v}\, dv\right] + (Q \circ \varphi)\left[\frac{\partial \varphi_2}{\partial u}\, du + \frac{\partial \varphi_2}{\partial v}\, dv\right] \\ &\quad + (R \circ \varphi)\left[\frac{\partial \varphi_3}{\partial u}\, du + \frac{\partial \varphi_3}{\partial v}\right] dv, \end{aligned}$$

$$\begin{aligned} \varphi^*\omega &= \left[(P \circ \varphi)\frac{\partial \varphi_1}{\partial u} + (Q \circ \varphi)\frac{\partial \varphi_2}{\partial u} + (R \circ \varphi)\frac{\partial \varphi_3}{\partial u}\right] du \\ &= \left[(P \circ \varphi)\frac{\partial \varphi_1}{\partial v} + (Q \circ \varphi)\frac{\partial \varphi_2}{\partial v} + (R \circ \varphi)\frac{\partial \varphi_3}{\partial v}\right] dv. \end{aligned}$$

If $\alpha = A\,dy \wedge dz$, then

$$\begin{aligned} \varphi^*\alpha &= (A \circ \varphi)\varphi^*(dy) \wedge \varphi^*(dz) \\ &= (A \circ \varphi)\left(\frac{\partial \varphi_2}{\partial u}\, du + \frac{\partial \varphi_2}{\partial v}\, dv\right) \wedge \left(\frac{\partial \varphi_3}{\partial u}\, du + \frac{\partial \varphi_3}{\partial v}\, dv\right), \end{aligned}$$

$$\varphi^*\alpha = (A \circ \varphi)\left(\frac{\partial \varphi_2}{\partial u}\frac{\partial \varphi_3}{\partial v} - \frac{\partial \varphi_3}{\partial u}\frac{\partial \varphi_2}{\partial v}\right) du \wedge dv.$$

In terms of the pullback operation, what we want to prove is that

$$\int_\varphi \alpha = \int_Q \varphi^*\alpha, \tag{15}$$

this being the more precise formulation of Eq. (12). We will need the following lemma.

Lemma 5.3 Let $\omega_1, \ldots, \omega_k$ be k differential 1-forms on $\mathscr{R}^k$, with

$$\omega_i = \sum_{j=1}^{k} a_{ij}\, du_j$$

in **u**-coordinates. Then

$$\omega_1 \wedge \omega_2 \wedge \cdots \wedge \omega_k = (\det A)\, du_1 \wedge \cdots \wedge du_k,$$

where A is the $k \times k$ matrix (a_{ij}).

PROOF Upon multiplying out, we obtain

$$\omega_1 \wedge \omega_2 \wedge \cdots \wedge \omega_k = \sum_{\{\mathbf{j}\}} a_{1j_1} a_{2j_2} \cdots a_{kj_k}\, du_{j_1} \wedge \cdots \wedge du_{j_k},$$

where the notation $\{\mathbf{j}\}$ signifies summation over all permutations $\mathbf{j} = (j_1, \ldots, j_k)$ of $(1, \ldots, k)$. If $\sigma(\mathbf{j})$ denotes the sign of the permutation $\mathbf{j}$, then

$$du_{j_1} \wedge \cdots \wedge du_{j_k} = \sigma(\mathbf{j})\, du_1 \wedge \cdots \wedge du_k,$$

so we have

$$\begin{aligned}\omega_1 \wedge \cdots \wedge \omega_k &= \left(\sum_{\{\mathbf{j}\}} \sigma(\mathbf{j}) a_{1j_1} \cdots a_{kj_k}\right) du_1 \wedge \cdots \wedge du_k \\ &= (\det A)\, du_1 \wedge \cdots \wedge du_k\end{aligned}$$

by the standard definition of det A. ∎

Theorem 5.4 If $\varphi : Q \to \mathscr{R}^n$ is a k-dimensional $\mathscr{C}^1$ surface patch, and α is a differential k-form on $\mathscr{R}^n$, then

$$\int_\varphi \alpha = \int_Q \varphi^*\alpha.$$

PROOF By the additive property of the pullback, it suffices to consider

$$\alpha = a\, dx_{i_1} \wedge \cdots \wedge dx_{i_k}.$$

Then

$$\begin{aligned}\varphi^*\alpha &= (a \circ \varphi)\varphi^*(dx_{i_1}) \wedge \cdots \wedge \varphi^*(dx_{i_k}) \\ &= (a \circ \varphi)\left[\det\left(\frac{\partial \varphi_{i_r}}{\partial u_j}\right)\right] du_1 \wedge \cdots \wedge du_k\end{aligned}$$

by Lemma 5.3 (here $\partial\varphi_{i_r}/\partial u_j$ is the element in the rth row and jth column). Therefore, applying the definitions, we obtain

$$\begin{aligned}\int_\varphi \alpha &= \int_Q a(\varphi(\mathbf{u}))\, dx_{i_1} \wedge \cdots \wedge dx_{i_k}\left(\frac{\partial\varphi}{\partial u_1}, \ldots, \frac{\partial\varphi}{\partial u_k}\right) d\mathbf{u} \\ &= \int_Q a(\varphi(\mathbf{u}))\left[\det\left(\frac{\partial\varphi_{i_r}}{\partial u_j}\right)\right] d\mathbf{u} \\ &= \int_Q \varphi^*\alpha\end{aligned}$$

as desired. ∎

Example 4 Let $Q = [0, 1] \times [0, 1] \subset \mathscr{R}^2$, and suppose $\varphi : Q \to \mathscr{R}^3$ is defined by the equations

$$x = u + v, \qquad y = u - v, \qquad z = uv.$$

We compute the surface integral

$$\int_\varphi x\, dy \wedge dz + y\, dx \wedge dz = \int_\varphi \alpha$$

in two different ways. First we apply the definition in Eq. (10). Since

$$\varphi'(u, v) = \begin{pmatrix} 1 & 1 \\ 1 & -1 \\ v & u \end{pmatrix},$$

we see that

$$dy \wedge dz\left(\frac{\partial\varphi}{\partial u}, \frac{\partial\varphi}{\partial v}\right) = u + v \qquad \text{and} \qquad dx \wedge dz\left(\frac{\partial\varphi}{\partial u}, \frac{\partial\varphi}{\partial v}\right) = u - v.$$

Therefore

$$\begin{aligned}\int_\varphi \alpha &= \int_Q \alpha_{\varphi(u,v)}\left(\frac{\partial\varphi}{\partial u}, \frac{\partial\varphi}{\partial v}\right) du\, dv \\ &= \int_Q [(u+v)(u+v) + (u-v)(u-v)]\, du\, dv \\ &= \int_0^1 \int_0^1 2(u^2 + v^2)\, du\, dv = \tfrac{4}{3}.\end{aligned}$$

Second, we apply Theorem 5.4. The pullback $\varphi^*\alpha$ is simply the result of substituting

$$\begin{aligned} x &= u + v, \qquad & dx &= du + dv, \\ y &= u - v, \qquad & dy &= du - dv, \\ z &= uv, \qquad & dz &= v\, du + u\, dv, \end{aligned}$$

into α. So

$$\begin{aligned}\varphi^*\alpha &= (u+v)(du-dv)\wedge(v\,du+u\,dv)+(u-v)(du+dv)\wedge(v\,du+u\,dv)\\ &=2(u^2+v^2)\,du\wedge dv.\end{aligned}$$

Therefore Theorem 5.4 gives

$$\int_\varphi \alpha = \int_Q \varphi^*\alpha = \int_0^1\int_0^1 2(u^2+v^2)\,du\,dv = \tfrac{4}{3}.$$

Of course the final computation is the same in either case. The point is that Theorem 5.4 enables us to proceed by formal substitution, making use of the equations which define the mapping φ, instead of referring to the original definition of $\int_\varphi \alpha$.

Theorem 5.4 is the k-dimensional generalization of Lemma 2.2(b), which played an important role in the proof of Green's theorem in Section 2. Theorem 5.4 will play a similar role in the proof of Stokes' theorem in Section 6. We will also need the k-dimensional generalization of part (c) of Lemma 2.2—the fact that the differential operation d commutes with pullbacks.

Theorem 5.5 If $\varphi : \mathscr{R}^m \to \mathscr{R}^n$ is a $\mathscr{C}^1$ mapping and α is a $\mathscr{C}^1$ differential k-form on $\mathscr{R}^n$, then

$$d(\varphi^*\alpha) = \varphi^*(d\alpha).$$

PROOF The proof will be by induction on k. When $k=0$, $\alpha = f$, a $\mathscr{C}^1$ function on $\mathscr{R}^n$, and $\varphi^*f = f\circ\varphi$, so

$$d(\varphi^*f) = \sum_{i=1}^m \frac{\partial(f\circ\varphi)}{\partial u_i}\,du_i = \sum_{i=1}^m \left[\sum_{j=1}^n\left(\frac{\partial f}{\partial x_j}\circ\varphi\right)\frac{\partial\varphi_j}{\partial u_i}\right] du_i.$$

But

$$\begin{aligned}\varphi^*(df) &= \varphi^*\left(\sum_{j=1}^n \frac{\partial f}{dx_j}\,dx_j\right)\\ &= \sum_{j=1}^n\left(\frac{\partial f}{\partial x_j}\circ\varphi\right)\varphi^*(dx_j)\\ &= \sum_{j=1}^n\left(\frac{\partial f}{\partial x_j}\circ\varphi\right)\left(\sum_{i=1}^m\frac{\partial\varphi_j}{\partial u_i}\,du_i\right),\end{aligned}$$

which is the same thing.

Supposing inductively that the result holds for $(k-1)$-forms, consider the k-form

$$\alpha = a\,dx_{i_1}\wedge\cdots\wedge dx_{i_k} = dx_{i_1}\wedge\beta,$$

where $\beta = a\, dx_{i_2} \wedge \cdots \wedge dx_{i_k}$. Then

$$\begin{aligned} d\alpha &= d(dx_{i_1}) \wedge \beta - dx_{i_1} \wedge d\beta \\ &= -dx_{i_1} \wedge d\beta \end{aligned}$$

by Proposition 5.2. Therefore

$$\begin{aligned} \varphi^*(d\alpha) &= \varphi^*(-dx_{i_1} \wedge d\beta) \\ &= -\varphi^*(dx_{i_1}) \wedge \varphi^*(d\beta) \\ &= -\varphi^*(dx_{i_1}) \wedge d(\varphi^*\beta), \end{aligned}$$

since $\varphi^*(d\beta) = d(\varphi^*\beta)$ by the inductive assumption. Since $\varphi^*(dx_{i_1}) = d\varphi_{i_1}$ by (13), we now have

$$\begin{aligned} d(\varphi^*\alpha) &= d(\varphi^*(dx_{i_1}) \wedge \varphi^*\beta) \\ &= d(d\varphi_{i_1}) \wedge \varphi^*\beta - \varphi^*(dx_{i_1}) \wedge d(\varphi^*\beta) \\ &= \varphi^*(d\alpha) \end{aligned}$$

using Propositions 5.1 and 5.2. ∎

Our treatment of differential forms in this section has thus far been rather abstract and algebraic. As an antidote to this absence of geometry, the remainder of the section is devoted to a discussion of the "surface area form" of an oriented smooth ($\mathscr{C}^1$) k-manifold in $\mathscr{R}^n$. This will provide an example of an important differential form that appears in a natural geometric setting.

First we recall the basic definitions from Section 4 of Chapter III. A *coordinate patch* on the smooth k-manifold $M \subset \mathscr{R}^n$ is a one-to-one $\mathscr{C}^1$ mapping $\varphi : U \to M$, where U is an open subset of $\mathscr{R}^k$, such that $d\varphi_{\mathbf{u}}$ has rank k for each $\mathbf{u} \in U$: this implies that $[\det(\varphi'(\mathbf{u})^t\varphi'(\mathbf{u})]^{1/2} \neq 0$. An *atlas* for M is a collection $\{\psi_i\}$ of coordinate patches, the union of whose images covers M. An *orientation* for M is an atlas $\{\psi_i\}$ such that the "change of coordinates" mapping, corresponding to any two of these coordinate patches ψ_i and ψ_j whose images $\psi_i(U_i)$ and $\psi_j(U_j)$ overlap, has a *positive* Jacobian determinant. That is, if

$$T_{ij} = \psi_j^{-1} \circ \psi_i : \psi_i^{-1}(\psi_i(U_i) \cap \psi_j(U_j)) \to \psi_j^{-1}(\psi_i(U_i) \cap \psi_j(U_j)),$$

then $\det T'_{ij} > 0$ wherever T_{ij} is defined (see Fig. 5.29). The pair $(M, \{\psi_i\}$, is then called an *oriented* manifold. Finally, the coordinate patch $\varphi : U \to M$ is called *orientation-preserving* if it overlaps positively (in the above sense) with each of the ψ_i, and *orientation-reversing* if it overlaps negatively with each of the ψ_i (that is, the appropriate Jacobian determinants are negative at each point).

The *surface area form* of the oriented k-dimensional manifold $M \subset \mathscr{R}^n$ is the differential k-form

$$dA = \sum_{[\mathbf{i}]} n_{\mathbf{i}}\, d\mathbf{x}_{\mathbf{i}}, \tag{16}$$

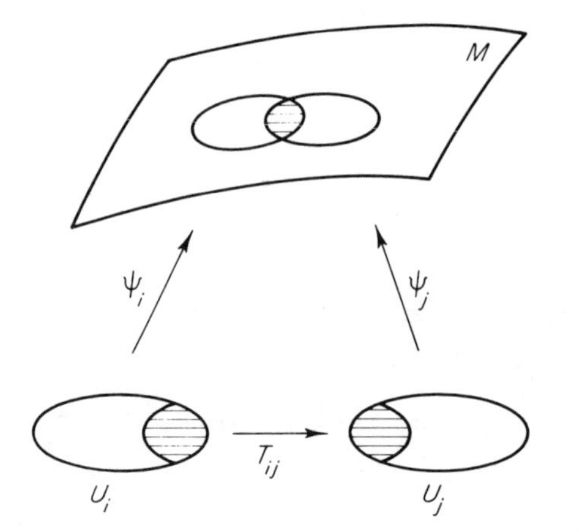

Figure 5.29

whose coefficient functions $n_{\mathbf{i}}$ are defined on M as follows. Given $\mathbf{i} = (i_1, \ldots, i_k)$ and $\mathbf{x} \in M$, choose an orientation-preserving coordinate patch $\varphi : U \to M$ such that $\mathbf{x} = \varphi(\mathbf{u}) \in \varphi(U)$. Then

$$n_{\mathbf{i}}(\mathbf{x}) = \frac{1}{D} \frac{\partial(\varphi_{i_1}, \ldots, \varphi_{i_k})}{\partial(u_1, \ldots, u_k)} = \frac{1}{D} \det \varphi_{\mathbf{i}}'(\mathbf{u}),$$

where

$$D = [\det(\varphi'(\mathbf{u}))^t \varphi'(\mathbf{u})]^{1/2} = [\sum_{[\mathbf{j}]} (\det \varphi_{\mathbf{j}}'(\mathbf{u}))^2]^{1/2}.$$

Example 5 Let $M = S^2$, the unit sphere in $\mathscr{R}^3$. We use the spherical coordinates surface patch defined as usual by

$$x = \sin \varphi \cos \theta, \qquad y = \sin \varphi \sin \theta, \qquad z = \cos \varphi.$$

Here $D = \sin \varphi$ (see Example 3 in Section 4). Hence

$$n_{(1,2)} = \frac{1}{\sin \varphi} \frac{\partial(x, y)}{\partial(\varphi, \theta)} = \cos \varphi = z,$$

$$n_{(1,3)} = \frac{1}{\sin \varphi} \frac{\partial(x, z)}{\partial(\varphi, \theta)} = -\sin \varphi \sin \theta = -y,$$

$$n_{(2,3)} = \frac{1}{\sin \varphi} \frac{\partial(y, z)}{\partial(\varphi, \theta)} = \sin \varphi \cos \theta = x.$$

Thus the area form of S^2 is

$$\begin{aligned} dA &= z\, dx \wedge dy - y\, dx \wedge dz + x\, dy \wedge dz \\ &= x\, dy \wedge dz + y\, dz \wedge dx + z\, dx \wedge dy. \end{aligned}$$

Of course we must prove that $n_{\mathbf{i}}$ is well-defined. So let $\psi : V \to M$ be a second orientation-preserving coordinate patch with $\mathbf{x} = \psi(\mathbf{v}) \in \psi(V)$. If

$$T = \psi^{-1} \circ \varphi : \varphi^{-1}(\varphi(U) \cap \psi(V)) \to \psi^{-1}(\varphi(U) \cap \psi(V)),$$

then $\varphi = \psi \circ T$ on $\varphi^{-1}(\varphi(U) \cap \psi(V))$, so an application of the chain rule gives

$$\begin{aligned}
\det \varphi_{\mathbf{j}}'(\mathbf{u}) &= \frac{\partial(\varphi_{j_1}, \ldots, \varphi_{j_k})}{\partial(u_1, \ldots, u_k)}(\mathbf{u}) \\
&= \det (\psi \circ T)_{\mathbf{j}}'(\mathbf{u}) \\
&= \det \psi_{\mathbf{j}}'(T(\mathbf{u})) \det T'(\mathbf{u}) \\
&= \det \psi_{\mathbf{j}}'(\mathbf{v}) \det T'(\mathbf{u}).
\end{aligned}$$

Therefore

$$\begin{aligned}
\frac{\det \varphi_{\mathbf{i}}'(\mathbf{u})}{\left[\sum_{[\mathbf{j}]} (\det \varphi_{\mathbf{j}}'(\mathbf{u}))^2\right]^{1/2}} &= \frac{\det \psi_{\mathbf{i}}'(\mathbf{v}) \det T'(\mathbf{u})}{\left[\sum_{[\mathbf{j}]} (\det \psi_{\mathbf{j}}'(\mathbf{v}))^2 (\det T'(\mathbf{u}))^2\right]^{1/2}} \\
&= \frac{\det \psi_{\mathbf{i}}'(\mathbf{v})}{\left[\sum_{[\mathbf{j}]} (\det \psi_{\mathbf{j}}'(\mathbf{v}))^2\right]^{1/2}}
\end{aligned}$$

because $\det T'(\mathbf{u}) > 0$. Thus the two orientation-preserving coordinate patches φ and ψ provide the same definition of $n_{\mathbf{i}}(\mathbf{x})$.

The following theorem tells why dA is called the "surface area form" of M.

Theorem 5.6 Let M be an oriented k-manifold in $\mathscr{R}^n$ with surface area form dA. If $\varphi : Q \to M$ is the restriction, to the k-dimensional interval $Q \subset \mathscr{R}^k$, of an orientation-preserving coordinate patch, then

$$a(\varphi) = \int_\varphi dA.$$

PROOF The proof is simply a computation. Using the definition of dA, of the area $a(\varphi)$, and of the integral of a differential form, we obtain

$$\begin{aligned}
a(\varphi) &= \int_Q [\det (\varphi'(\mathbf{u}))^t \varphi'(\mathbf{u})]^{1/2} \, du_1 \cdots du_k \\
&= \int_Q \frac{\det \varphi'(\mathbf{u})^t \varphi'(\mathbf{u})}{D} \, du_1 \cdots du_k \\
&= \int_Q (1/D) \left[\sum_{[\mathbf{j}]} (\det \varphi_{\mathbf{i}}'(\mathbf{u}))^2\right] du_1 \cdots du_k \\
&= \int_Q \left(\sum_{[\mathbf{i}]} n_{\mathbf{i}}(\varphi(\mathbf{u})) \det \varphi_{\mathbf{i}}'(\mathbf{u})\right) du_1 \cdots du_k \\
&= \int_Q \left(\sum_{[\mathbf{i}]} n_{\mathbf{i}}(\varphi(\mathbf{u})) \, d\mathbf{x}_{\mathbf{i}}\left(\frac{\partial \varphi}{\partial u_1}, \ldots, \frac{\partial \varphi}{\partial u_k}\right)\right) du_1 \cdots du_k \\
&= \int_Q dA\left(\frac{\partial \varphi}{\partial u_1}, \ldots, \frac{\partial \varphi}{\partial u_k}\right) du_1 \cdots du_k \\
&= \int_\varphi dA.
\end{aligned}$$

∎

Recall that a paving of the compact smooth k-manifold M is a finite collection $\mathscr{A} = \{A_1, \ldots, A_r\}$ of nonoverlapping k-cells such that $M = \bigcup_{i=1}^{r} A_i$. If M is oriented, then the paving $\mathscr{A}$ is called *oriented* provided that each of the k-cells A_i has a parametrization $\varphi_i : Q_i \to A_i$ which extends to an orientation-preserving coordinate patch for M (defined on a neighborhood of $Q_i \subset \mathscr{R}^k$). Since the k-dimensional area of M is defined by

$$a(M) = \sum_{i=1}^{r} a(A_i) = \sum_{i=1}^{r} a(\varphi_i),$$

we see that Theorem 5.6 gives

$$a(M) = \sum_{i=1}^{r} \int_{\varphi_i} dA. \tag{17}$$

Given a continuous differential k-form α whose domain of definition contains the oriented compact smooth k-manifold M, the *integral of* α on M is defined by

$$\int_M \alpha = \sum_{i=1}^{r} \int_{\varphi_i} \alpha, \tag{18}$$

where $\varphi_1, \ldots, \varphi_r$ are parametrizations of the k-cells of an oriented paving $\mathscr{A}$ of M (as above). So Eq. (17) becomes the pleasant formula

$$a(M) = \int_M dA.$$

The proof that the integral $\int_M \alpha$ is well defined is similar to the proof in Section 4 that $a(M)$ is well defined. The following lemma will play the role here that Theorem 4.1 played there.

Lemma 5.7 Let M be an oriented compact smooth k-manifold in $\mathscr{R}^n$ and α a continuous differential k-form defined on M. Let $\varphi : U \to M$ and $\psi : V \to M$ be two coordinate patches on M with $\varphi(U) = \psi(V)$, and write $T = \psi^{-1} \circ \varphi : U \to V$. Suppose X and Y are contented subsets of U and V, respectively, with $T(X) = Y$. Finally let $\bar{\varphi} = \varphi | X$ and $\bar{\psi} = \psi | Y$. Then

$$\int_{\bar{\varphi}} \alpha = \int_{\bar{\psi}} \alpha$$

if φ and ψ are either both orientation-preserving or both orientation-reversing, while

$$\int_{\bar{\varphi}} \alpha = -\int_{\bar{\psi}} \alpha$$

if one is orientation-preserving and the other is orientation-reversing.

PROOF By additivity we may assume that $\alpha = a\, d\mathbf{x}_{\mathbf{i}}$. Since $\varphi = \psi \circ T$ (Fig. 5.30), an application of the chain rule gives

$$\det \varphi_{\mathbf{i}}'(\mathbf{u}) = \det \psi_{\mathbf{i}}'(T(\mathbf{u})) \det T'(\mathbf{u}).$$

Therefore

$$\begin{aligned}\int_{\bar{\varphi}} \alpha &= \int_X a(\varphi(\mathbf{u})) \det \varphi_{\mathbf{i}}'(\mathbf{u})\, du_1 \cdots du_k \\ &= \int_X a(\psi(T(\mathbf{u}))) \det \psi_{\mathbf{i}}'(T(\mathbf{u})) \det T'(\mathbf{u})\, du_1 \cdots du_k. \qquad (19)\end{aligned}$$

On the other hand, the change of variables theorem gives

$$\begin{aligned}\int_{\bar{\psi}} \alpha &= \int_Y a(\psi(\mathbf{v})) \det \psi_{\mathbf{i}}'(\mathbf{v})\, dv_1 \cdots dv_k \\ &= \int_X a(\psi(T(\mathbf{u}))) \det \psi_{\mathbf{i}}'(T(\mathbf{u})) \,|\det T'(\mathbf{u})|\, du_1 \cdots du_k. \qquad (20)\end{aligned}$$

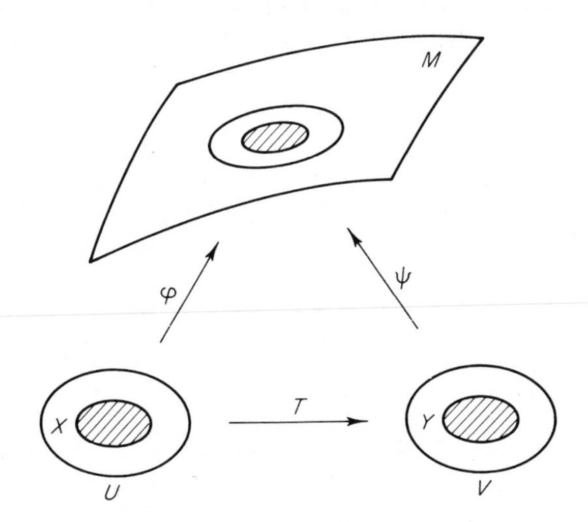

Figure 5.30

Since $\det T' > 0$ either if φ and ψ are both orientation-preserving or if both are orientation-reversing, while $\det T' > 0$ otherwise, the conclusion of the lemma follows immediately from a comparison of formulas (19) and (20). ∎

Now let $\mathscr{A} = \{A_1, \ldots, A_r\}$ and $\mathscr{B} = \{B_1, \ldots, B_s\}$ be two oriented pavings of M. Let φ_i and ψ_j be orientation-preserving parametrizations of A_i and B_j, respectively. Let

$$X_{ij} = \varphi_i^{-1}(A_i \cap B_j) \qquad \text{and} \qquad Y_{ij} = \psi_j^{-1}(A_i \cap B_j).$$

If $\varphi_{ij} = \varphi_i | X_{ij}$ and $\psi_{ij} = \psi_j | Y_{ij}$, then it follows from Lemma 5.7 that

$$\int_{\varphi_{ij}} \alpha = \int_{\psi_{ij}} \alpha$$

for any k-form α defined on M. Therefore

$$\begin{aligned} \sum_{i=1}^{r} \int_{\varphi_i} \alpha &= \sum_{i=1}^{r} \left(\sum_{j=1}^{s} \int_{\varphi_{ij}} \alpha \right) \\ &= \sum_{j=1}^{s} \left(\sum_{i=1}^{r} \int_{\psi_{ij}} \alpha \right) \\ &= \sum_{j=1}^{s} \int_{\psi_j} \alpha, \end{aligned}$$

so the integral $\int_M \alpha$ is indeed well defined by (18).

Integrals of differential forms on manifolds have a number of physical applications. For example, if the 2-dimensional manifold $M \subset \mathscr{R}^3$ is thought of as a lamina with density function ρ, then its mass is given by the integral

$$\int_M \rho \, dA.$$

If M is a closed surface in $\mathscr{R}^3$ with unit outer normal vector field $\mathbf{N}$, and $\mathbf{F}$ is the velocity vector field of a moving fluid in $\mathscr{R}^3$, then the "flux" integral

$$\int_M \mathbf{F} \cdot \mathbf{N} \, dA$$

measures the rate at which the fluid is leaving the region bounded by M. We will discuss such applications as these in Section 7.

Example 6 Let T be the "flat torus" $S^1 \times S^1 \subset \mathscr{R}^2 \times \mathscr{R}^2$, which is the image in $\mathscr{R}^4$ of the surface patch $F: Q = [0, 2\pi]^2 \to \mathscr{R}^4$ defined by

$$\begin{aligned} x_1 &= \cos\theta, & x_3 &= \cos\varphi, \\ x_2 &= \sin\theta, & x_4 &= \sin\varphi. \end{aligned}$$

The surface area form of T is

$$dA = x_2 x_4 \, dx_1 \wedge dx_3 + x_2 x_3 \, dx_4 \wedge dx_1 + x_1 x_4 \, dx_3 \wedge dx_2 + x_1 x_3 \, dx_2 \wedge dx_4$$

(see Exercise 5.11). If Q is subdivided into squares Q_1, Q_2, Q_3, Q_4 as indicated in Fig. 5.31, and $A_i = \varphi(Q_i)$, then $\{A_1, A_2, A_3, A_4\}$ is a paving of T, so

$$a(T) = \int_T dA = \sum_{i=1}^{4} \int_{\varphi | Q_i} dA = \int_{\varphi} dA.$$

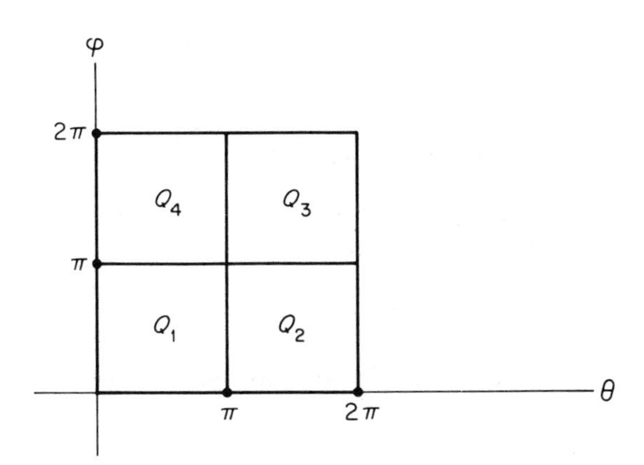

Figure 5.31

Now

$$F' = \begin{pmatrix} -\sin\theta & 0 \\ \cos\theta & 0 \\ 0 & -\sin\varphi \\ 0 & \cos\varphi \end{pmatrix},$$

so

$$dx_1 \wedge dx_3\left(\frac{\partial F}{\partial\theta}, \frac{\partial F}{\partial\varphi}\right) = \sin\theta\sin\varphi,$$

$$dx_4 \wedge dx_1\left(\frac{\partial F}{\partial\theta}, \frac{\partial F}{\partial\varphi}\right) = \sin\theta\cos\varphi,$$

$$dx_3 \wedge dx_2\left(\frac{\partial F}{\partial\theta}, \frac{\partial F}{\partial\varphi}\right) = \cos\theta\sin\varphi,$$

$$dx_2 \wedge dx_4\left(\frac{\partial F}{\partial\theta}, \frac{\partial F}{\partial\varphi}\right) = \cos\theta\cos\varphi.$$

Therefore

$$dA\left(\frac{\partial F}{\partial\theta}, \frac{\partial F}{\partial\varphi}\right) = \sin^2\theta\sin^2\varphi + \sin^2\theta\cos^2\varphi + \cos^2\theta\sin^2\varphi + \cos^2\theta\cos^2\varphi,$$

$$dA\left(\frac{\partial F}{\partial\theta}, \frac{\partial F}{\partial\varphi}\right) = 1.$$

Consequently

$$\begin{aligned} a(T) &= \int_F dA \\ &= \int_0^{2\pi}\int_0^{2\pi} dA\left(\frac{\partial F}{\partial\theta}, \frac{\partial F}{\partial\varphi}\right) d\theta\, d\varphi \\ &= 4\pi^2. \end{aligned}$$

Exercises

5.1 Compute the differentials of the following differential forms.
(a) $\alpha = \sum_{i=1}^{n} (-1)^{i-1} x_i \, dx_1 \wedge \cdots \wedge dx_{i-1} \wedge dx_{i+1} \wedge \cdots \wedge dx_n$.
(b) $r^{-n}\alpha$, where $r = [x_1^2 + \cdots + x_n^2]^{1/2}$.
(c) $\sum_{i=1}^{n} y_i \, dx_i$, where $(x_1, \ldots, x_n, y_1, \ldots, y_n)$ are coordinates in $\mathscr{R}^{2n}$.

5.2 If $F : \mathscr{R}^n \to \mathscr{R}^n$ is a $\mathscr{C}^1$ mapping, show that

$$dF_1 \wedge \cdots \wedge dF_n = \frac{\partial(F_1, \ldots, F_n)}{\partial(x_1, \ldots, x_n)} dx_1 \wedge \cdots \wedge dx_n.$$

5.3 If $\alpha = \sum_{i<j}^{n} a_{ij} \, dx_i \wedge dx_j$ is a differential 2-form on $\mathscr{R}^n$, show that

$$d\alpha = \sum_{i<j<k} \left(\frac{\partial a_{ij}}{\partial x_k} - \frac{\partial a_{ik}}{\partial x_j} + \frac{\partial a_{jk}}{\partial x_i} \right) dx_i \wedge dx_j \wedge dx_k.$$

5.4 The function f is called an *integrating factor* for the 1-form ω if $f(\mathbf{x}) \neq 0$ for all $\mathbf{x}$ and $d(f\omega) = 0$. If the 1-form ω has an integrating factor, show that $\omega \wedge d\omega = 0$.

5.5 (a) If $d\alpha = 0$ and $d\beta = 0$, show that $d(\alpha \wedge \beta) = 0$.
(b) The differential form β is called *exact* if there exists a differential form γ such that $d\gamma = \beta$. If $d\alpha = 0$ and β is exact, prove that $\alpha \wedge \beta$ is exact.

5.6 Verify formula (7) in this section.

5.7 If $\varphi : Q \to \mathscr{R}^k$ is the identity (or inclusion) mapping on the k-dimensional interval $Q \subset \mathscr{R}^k$, and $\alpha = f \, dx_1 \wedge \cdots \wedge dx_k$, show that

$$\int_\varphi \alpha = \int_\varphi f.$$

5.8 Let $\varphi : \mathscr{R}^m \to \mathscr{R}^n$ be a $\mathscr{C}^1$ mapping. If α is a k-form on $\mathscr{R}^n$, prove that

$$(\varphi^*\alpha)_{\mathbf{u}}(\mathbf{v}_1, \ldots, \mathbf{v}_k) = \alpha_{\varphi(\mathbf{u})}(d\varphi_{\mathbf{u}}(\mathbf{v}_1), \ldots, d\varphi_{\mathbf{u}}(\mathbf{v}_k)).$$

This fact, that the value of $\varphi^*\alpha$ on the vectors $\mathbf{v}_1, \ldots, \mathbf{v}_k$ is equal to the value of α on their images under the induced linear mapping $d\varphi$, is often taken as the definition of the pull-back $\varphi^*\alpha$.

5.9 Let C be a smooth curve (or 1-manifold) in $\mathscr{R}^n$, with pathlength form ds. If $\varphi : U \to C$ is a coordinate patch defined for $t \in U \subset \mathscr{R}^1$, show that

$$\varphi^*(ds) = [(\varphi_1'(t))^2 + \cdots + (\varphi_n'(t))^2]^{1/2} \, dt.$$

5.10 Let M be a smooth 2-manifold in $\mathscr{R}^n$, with surface area form dA. If $\varphi : U \to M$ is a coordinate patch defined on the open set $U \subset \mathscr{R}^2_{uv}$, show that

$$\varphi^*(dA) = (EG - F^2)^{1/2} \, du \, dv$$

with E, G, F defined as in Example 5 of Section 4.

5.11 Deduce, from the definition of the surface area form, the area form of the flat torus used in Example 6.

5.12 Let $M_1 \subset \mathscr{R}^m$ be an oriented smooth k-manifold with area form dA_1, and $M_2 \subset \mathscr{R}^n$ an oriented smooth l-manifold with area form dA_2. Regarding dA_1 as a form on $\mathscr{R}^{m+n}$ which involves only the variables $x_1, \ldots, x_m$, and dA_2 as a form in the variables $x_{m+1}, \ldots, x_{m+n}$, show that

$$dA = dA_1 \wedge dA_2$$

is the surface area form of the oriented $(k+l)$-manifold $M_1 \times M_2 \subset \mathscr{R}^{m+n}$. Use this result to obtain the area form of the flat torus in $\mathscr{R}^4$.

5.13 Let M be an oriented smooth $(n-1)$-dimensional manifold in $\mathscr{R}^n$. Define a unit vector field $\mathbf{N} : M \to \mathscr{R}^n$ on M as follows. Given $\mathbf{x} \in M$, choose an orientation-preserving coordinate patch $\varphi : U \to M$ with $\mathbf{x} = \varphi(\mathbf{u})$. Then let the ith component $n_i(\mathbf{x})$ of $\mathbf{N}(\mathbf{x})$ be given by

$$n_i(\mathbf{x}) = \frac{(-1)^{i-1}}{D} \frac{\partial(\varphi_1, \ldots, \hat{\varphi}_i, \ldots, \varphi_n)}{\partial(u_1, \ldots, u_{n-1})}(\mathbf{u}).$$

(a) Show that $\mathbf{N}$ is orthogonal to each of the vectors $\partial\varphi/\partial u_1, \ldots, \partial\varphi/\partial u_{n-1}$, so $\mathbf{N}$ is a unit *normal* vector field on M.
(b) Conclude that the surface area form of M is

$$dA = \sum_{i=1}^{n} (-1)^{i-1} n_i \, dx_1 \wedge \cdots \wedge dx_{i-1} \wedge dx_{i+1} \wedge \cdots dx_n.$$

(c) In particular, conclude (without the use of any coordinate system) that the surface area form of the unit sphere $S^{n-1} \subset \mathscr{R}^n$ is

$$dA = \sum_{i=1}^{n} (-1)^{i-1} x_i \, dx_1 \wedge \cdots \wedge dx_{i-1} \wedge dx_{i+1} \wedge \cdots \wedge dx_n.$$

5.14 (a) If $F : \mathscr{R}^l \to \mathscr{R}^m$ and $G : \mathscr{R}^m \to \mathscr{R}^n$ are $\mathscr{C}^1$ mappings, show that

$$(G \circ F)^* = F^* \circ G^*.$$

That is, if α is a k-form on $\mathscr{R}^n$ and $H = G \circ F$ then $H^*\alpha = F^*(G^*\alpha)$.
(b) Use Theorem 5.4 to deduce from (a) that, if φ is a k-dimensional surface patch in $\mathscr{R}^m$, $F : \mathscr{R}^m \to \mathscr{R}^n$ is a $\mathscr{C}^1$ mapping, and α is a k-form on $\mathscr{R}^n$, then

$$\int_{F \circ \varphi} \alpha = \int_{\varphi} F^*\alpha.$$

5.15 Let α be a differential k-form on $\mathscr{R}^n$. If

$$\mathbf{w}_j = \sum_{i=1}^{k} a_{ij} \mathbf{v}_i, \qquad i = 1, \ldots, k,$$

prove that

$$\alpha(\mathbf{w}_1, \ldots, \mathbf{w}_k) = (\det A)\alpha(\mathbf{v}_1, \ldots, \mathbf{v}_k),$$

where $A = (a_{ij})$.

6 STOKES' THEOREM

Recall that Green's theorem asserts that, if ω is a $\mathscr{C}^1$ differential 1-form on the cellulated region $D \subset \mathscr{R}^2$, then

$$\int_D d\omega = \int_{\partial D} \omega.$$

Stokes' theorem is a far-reaching generalization which says the same thing for a $\mathscr{C}^1$ differential $(k-1)$-form α which is defined on a neighborhood of the cellulated k-dimensional region $R \subset M$, where M is a smooth k-manifold in $\mathscr{R}^n$. That is,

$$\int_R d\alpha = \int_{\partial R} \alpha.$$

Of course we must say what is meant by a cellulated region in a smooth manifold, and also what the above integrals mean, since we thus far have defined integrals of differential forms only on surface patches, cells, and manifolds. Roughly speaking, a cellulation of R will be a collection $\{A_1, \ldots, A_r\}$ of nonoverlapping k-cells which fit together nicely (like the 2-cells of a cellulation of a planar region), and the integral of a k-form ω on R will be defined by

$$\int_R \omega = \sum_{i=1}^{r} \int_{A_i} \omega.$$

Green's theorem is simply the case $n = k = 2$ of Stokes' theorem, while the fundamental theorem of (single-variable) calculus is the case $n = k = 1$ (in the sense of the discussion at the beginning of Section 2). We will see that other cases as well have important physical and geometric applications. Therefore Stokes' theorem may justly be called the "fundamental theorem of multivariable calculus."

The proof in this section of Stokes' theorem will follow the same pattern as the proof in Section 2 of Green's theorem. That is, we will first prove Stokes' theorem for the unit cube $I^k \subset \mathscr{R}^k$. This simplest case will next be used to establish Stokes' theorem for a k-cell on a smooth k-manifold in $\mathscr{R}^n$, using the formal properties of the differential and pullback operations. Finally the general result, for an appropriate region R in a smooth k-manifold, will be obtained by application of Stokes' theorem to the cells of a cellulation of R.

So we start with the unit cube

$$I^k = \{(x_1, \ldots, x_k) \in \mathscr{R}^k : \text{each } x_i \in [0, 1]\}$$

in R^k. Its boundary ∂I^k is the set of all those points $(x_1, \ldots, x_k) \in I^k$ for which one of the k coordinates is either 0 or 1; that is, ∂I^k is the union of the 2^k different $(k-1)$-dimensional *faces* of I^k. These faces are the sets $I_{i,\varepsilon}^{k-1}$ defined by

$$I_{i,\varepsilon}^{k-1} = \{(x_1, \ldots, x_k) \in I^k : x_i = \varepsilon\}$$

for each $i = 1, \ldots, k$ and $\varepsilon = 0$ or $\varepsilon = 1$ (see Fig. 5.32). The (i, ε)*th face* $I_{i,\varepsilon}^{k-1}$ of I^k is the image of the unit cube $I^{k-1} \subset \mathscr{R}^{k-1}$ under the mapping $\imath_{i,\varepsilon} : I^{k-1} \to \mathscr{R}^k$ defined by

$$\imath_{i,\varepsilon}(x_1, \ldots, x_{k-1}) = (x_1, \ldots, x_{i-1}, \varepsilon, x_i, \ldots, x_{k-1}).$$

The mapping $\imath_{i,\varepsilon}$ serves as an orientation for the face $I_{i,\varepsilon}^{k-1}$. The orientations which the faces (edges) of I^2 receive from these mappings are indicated by the

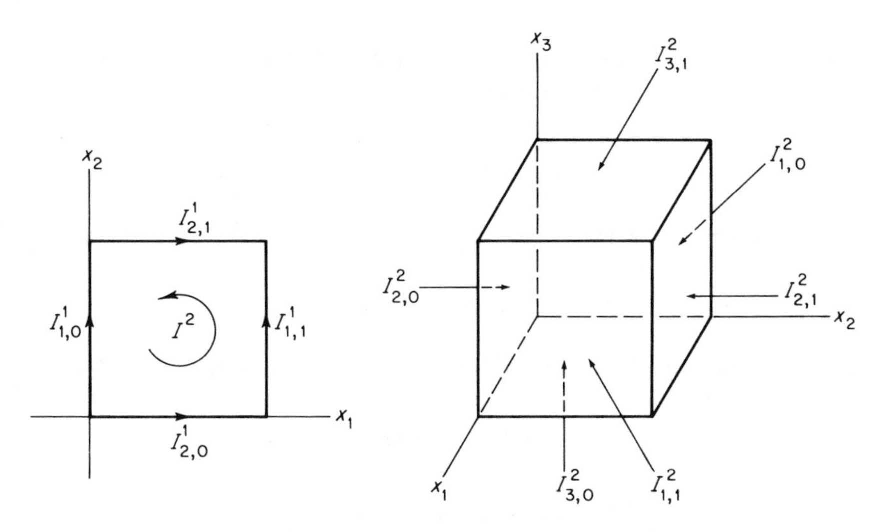

Figure 5.32

arrows in Fig. 5.32. We see that the positive (counterclockwise) orientation of ∂I^2 is given by

$$\partial I^2 = -I^1_{1,0} + I^1_{1,1} - I^2_{2,1} + I^2_{2,0}.$$

If ω is a continuous 1-form, it follows that the integral of ω over the oriented piecewise $\mathscr{C}^1$ closed curve ∂I^2 is given by

$$\begin{aligned}\int_{\partial I^2} \omega &= -\int_{\iota_{1,0}} \omega + \int_{\iota_{1,1}} \omega - \int_{\iota_{2,1}} \omega + \int_{\iota_{2,0}} \omega \\ &= \sum_{i=1}^{2} \sum_{\varepsilon=0,1} (-1)^{i+\varepsilon} \int_{\iota_{i,\varepsilon}} \omega.\end{aligned}$$

The integral of a $(k-1)$-form α over ∂I^k is defined by analogy. That is, we *define*

$$\int_{\partial I^k} \alpha = \sum_{i=1}^{k} \sum_{\varepsilon=0,1} (-1)^{i+\varepsilon} \int_{\iota_{i,\varepsilon}} \alpha. \tag{1}$$

As in Section 5, the integral over I^k of the k-form $f\,dx_1 \wedge \cdots \wedge dx_k$ is defined by

$$\int_{I^k} f\,dx_1 \wedge \cdots \wedge dx_k = \int_{I^k} f.$$

We are now ready to state and prove *Stokes' theorem for the unit cube*. Its proof, like that of Green's theorem for the unit square (Lemma 2.1), will be an explicit computation.

Theorem 6.1 If α is a $\mathscr{C}^1$ differential $(k-1)$-form that is defined on an open set containing the unit cube $I^k \subset \mathscr{R}^k$, then

$$\int_{I^k} d\alpha = \int_{\partial I^k} \alpha. \tag{2}$$

PROOF Let α be given by

$$\alpha = \sum_{i=1}^{k} a_i\, dx_1 \wedge \cdots \wedge \widehat{dx_i} \wedge \cdots dx_k,$$

where $a_1, \ldots, a_k$ are $\mathscr{C}^1$ real-valued functions on a neighborhood of I^k. Then

$$\begin{aligned} d\alpha &= \sum_{i=1}^{k} da_i \wedge dx_1 \wedge \cdots \wedge \widehat{dx_i} \wedge \cdots \wedge dx_k \\ &= \sum_{i=1}^{k} \sum_{j=1}^{k} \frac{\partial a_i}{\partial x_j}\, dx_j \wedge dx_1 \wedge \cdots \wedge \widehat{dx_i} \wedge \cdots \wedge dx_k, \\ d\alpha &= \sum_{i=1}^{k} (-1)^{i-1} \frac{\partial a_i}{\partial x_i}\, dx_1 \wedge \cdots \wedge dx_k, \end{aligned}$$

since

$$dx_j \wedge dx_1 \wedge \cdots \wedge \widehat{dx_i} \wedge \cdots \wedge dx_k = \begin{cases} (-1)^{i-1}\, dx_1 \wedge \cdots \wedge dx_k & \text{if } i = j, \\ 0 & \text{if } i \neq j. \end{cases}$$

We compute both sides of Eq. (2). First

$$\int_{\partial I^k} \alpha = \sum_{i=1}^{k} \sum_{\varepsilon=0,1} (-1)^{i+\varepsilon} \int_{\iota_{i,\varepsilon}} \alpha,$$

where

$$\begin{aligned} \int_{\iota_{i,\varepsilon}} \alpha &= \int_{\iota_{i,\varepsilon}} \left(\sum_{j=1}^{k} a_j\, dx_1 \wedge \cdots \wedge \widehat{dx_j} \wedge \cdots \wedge dx_k \right) \\ &= \int_{I^{k-1}} \sum_{j=1}^{k} (a_j \circ \iota_{i,\varepsilon})\, \iota_{i,\varepsilon}^*(dx_1 \wedge \cdots \wedge \widehat{dx_j} \wedge \cdots \wedge dx_k) \\ &\qquad \text{(by Theorem 5.4)} \\ &= \int_{I^{k-1}} (a_i \circ \iota_{i,\varepsilon})\, \iota_{i,\varepsilon}^*(dx_1 \wedge \cdots \wedge \widehat{dx_i} \wedge \cdots \wedge dx_k), \end{aligned}$$

$$\int_{\iota_{i,\varepsilon}} \alpha = \int_{\iota_{i,\varepsilon}} a_i\, dx_1 \wedge \cdots \wedge \widehat{dx_i} \wedge \cdots \wedge dx_k, \tag{3}$$

because

$$\iota_{i,\varepsilon}^*(dx_1 \wedge \cdots \wedge \widehat{dx_j} \wedge \cdots \wedge dx_k) = \begin{cases} dx_1 \wedge \cdots \wedge \widehat{dx_i} \wedge \cdots \wedge dx_k & \text{if } i = j, \\ 0 & \text{if } i \neq j. \end{cases}$$

To compute the left-hand side of (2), we first apply Fubini's theorem and the (ordinary) fundamental theorem of calculus to obtain

$$\begin{aligned}\int_{I^k} \frac{\partial a_i}{\partial x_i}\, dx_1 \wedge \cdots \wedge dx_k &= \int \left(\int_0^1 \frac{\partial a_i}{\partial x_i}\, dx_i \right) dx_1 \cdots \widehat{dx_i} \cdots dx_k \\ &= \int_0^1 \cdots \int_0^1 [a_i(x_1, \ldots, x_{i-1}, 1, x_{i+1}, \ldots, x_k) \\ &\qquad - a_i(x_1, \ldots, 0, \ldots, x_k)]\, dx_1 \cdots \widehat{dx_i} \cdots dx_k . \\ &= \int_{\iota_{i,1}} a_i\, dx_1 \wedge \cdots \wedge \widehat{dx_i} \wedge \cdots \wedge dx_k \\ &\qquad - \int_{\iota_{i,0}} a_i\, dx_1 \wedge \cdots \wedge \widehat{dx_i} \wedge \cdots \wedge dx_k \\ &= \int_{\iota_{i,1}} \alpha - \int_{\iota_{i,0}} \alpha,\end{aligned}$$

using (3) in the last step. Therefore

$$\begin{aligned}\int_{I^k} d\alpha &= \sum_{i=1}^{k} (-1)^{i-1} \int_{I^k} \frac{\partial a_i}{\partial x_i}\, dx_1 \wedge \cdots \wedge dx_k \\ &= \sum_{i=1}^{k} (-1)^{i+1} \left[\int_{\iota_{i,1}} \alpha - \int_{\iota_{i,0}} \alpha \right] \\ &= \sum_{i=1}^{k} \sum_{\varepsilon=0,1} (-1)^{i+\varepsilon} \int_{\iota_{i,\varepsilon}} \alpha \\ &= \int_{\partial I^k} \alpha\end{aligned}$$

as desired. ∎

Notice that the above proof is a direct generalization of the computation in the proof of Green's theorem for I^2, using only the relevant definitions, Fubini's theorem, and the single-variable fundamental theorem of calculus.

The second step in our program is the proof of Stokes's theorem for an oriented k-cell in a smooth k-manifold in $\mathscr{R}^n$. Recall that a (smooth) k-cell in the smooth k-manifold $M \subset \mathscr{R}^n$ is the image A of a mapping $\varphi : I^k \to M$ which extends to a coordinate patch for M (defined on some open set in $\mathscr{R}^k$ which contains I^k). If M is oriented, then we will say that the k-cell A is *oriented* (positively with respect to the orientation of M) if this coordinate patch is orientation-preserving. To emphasize the importance of the orientation for our purpose here, let us make the following definition. An *oriented k-cell* in the oriented smooth k-manifold $M \subset \mathscr{R}^n$ is a pair (A, φ), where $\varphi : I^k \to M$ is a $\mathscr{C}^1$ mapping which extends to an orientation-preserving coordinate patch, and $A = \varphi(I^k)$ (see Fig. 5.33).

The (i, ε)th *face* $A_{i,\varepsilon}$ of A is the image under φ of the (i, ε)th face of I^k, $A_{i,\varepsilon} = \varphi(I^{k-1}_{i,\varepsilon})$. Thus $A_{i,\varepsilon}$ is the image of I^{k-1} under the mapping

$$\varphi_{i,\varepsilon} = \varphi \circ \iota_{i,\varepsilon} : I^{k-1} \to \mathscr{R}^n.$$

If α is a $(k-1)$-form which is defined on A, we define the integral of α over the oriented boundary ∂A by analogy with Eq. (1),

$$\int_{\partial A} \alpha = \sum_{i=1}^{k} \sum_{\varepsilon=0,1} (-1)^{i+\varepsilon} \int_{\varphi_{i,\varepsilon}} \alpha. \tag{4}$$

Also, if β is a k-form defined on A, we write

$$\int_A \beta = \int_\varphi \beta.$$

With this notation, Stokes' theorem for an oriented k-cell takes the following form.

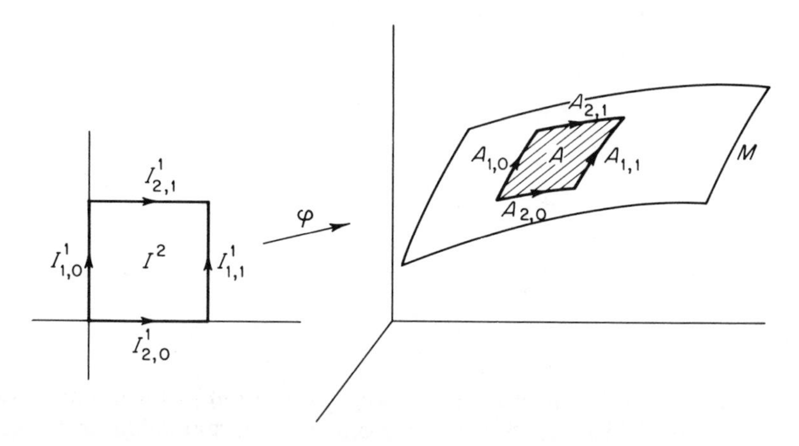

Figure 5.33

Theorem 6.2 Let (A, φ) be an oriented k-cell in an oriented k-manifold in $\mathscr{R}^n$, and let α be a $\mathscr{C}^1$ differential $(k-1)$-form that is defined on an open subset of $\mathscr{R}^n$ which contains A. Then

$$\int_A d\alpha = \int_{\partial A} \alpha. \tag{5}$$

PROOF The proof is a computation which is a direct generalization of the proof of Green's theorem for oriented 2-cells (Lemma 2.3). Applying Stokes' theorem for I^k, and the formal properties of the pullback and differential operations, given in Theorems 5.4 and 5.5, respectively, we obtain

$$
\begin{aligned}
\int_A d\alpha &= \int_\varphi d\alpha && \text{(definition)} \\
&= \int_{I^k} \varphi^*(d\alpha) && \text{(Theorem 5.4)} \\
&= \int_{I^k} d(\varphi^*\alpha) && \text{(Theorem 5.5)} \\
&= \int_{\partial I^k} \varphi^*\alpha && \text{(Theorem 6.1)} \\
&= \sum_{i=1}^{k} \sum_{\varepsilon=0,1} (-1)^{i+\varepsilon} \int_{\iota_{i,\varepsilon}} \varphi^*\alpha && \text{(definition)} \\
&= \sum_{i=1}^{k} \sum_{\varepsilon=0,1} (-1)^{i+\varepsilon} \int_{\varphi \circ \iota_{i,\varepsilon}} \alpha && \text{(Exercise 5.14)} \\
&= \sum_{i=1}^{k} \sum_{\varepsilon=0,1} (-1)^{i+\varepsilon} \int_{\varphi_{i,\varepsilon}} \alpha && \text{(definition of } \varphi_{i,\varepsilon}) \\
&= \int_{\partial A} \alpha && \text{(definition)}
\end{aligned}
$$

as desired. ∎

Our final step will be to extend Stokes' theorem to regions that can be obtained by appropriately piecing together oriented k-cells. Let R be a compact region (the closure of an open subset) in the smooth k-manifold $M \subset \mathscr{R}^n$. By an *oriented cellulation* of R is meant a collection $\mathscr{K} = \{A_1, \ldots, A_p\}$ of oriented k-cells (Fig. 5.34) on M satisfying the following conditions:

(a) $R = \bigcup_{r=1}^{p} A_r$.
(b) For each r and s, the intersection $A_r \cap A_s$ is either empty or is the union of one or more common faces of A_r and A_s.

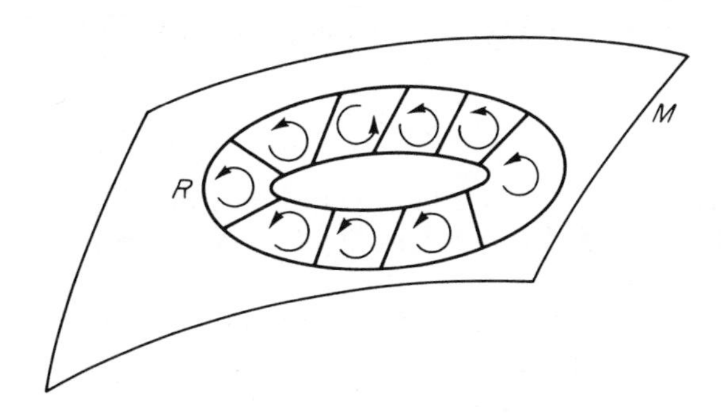

Figure 5.34

(c) If B is a $(k-1)$-dimensional face of A_r, then either A_r is the only k-cell of $\mathscr{K}$ having B as a face, or there is exactly one other k-cell A_s also having B as a face. In the former case B is called a *boundary* face; in the latter B is an *interior* face.

(d) If B is an interior face, with A_r and A_s the two k-cells of $\mathscr{K}$ having B as a face, then A_r and A_s induce opposite orientations on B.

(e) The boundary ∂R of R (as a subset of M) is the union of all the boundary faces of k-cells of $\mathscr{K}$.

The pair $(R, \mathscr{K})$ is called an *oriented cellulated region* in M.

Although an oriented cellulation is conceptually simple—it is simply a collection of nonoverlapping oriented k-cells that fit together in the nicest possible way—the above definition is rather lengthy. This is due to conditions (c), (d), and (e), which are actually redundant—they follow from (a) and (b) and the fact that the k-cells of $\mathscr{K}$ are oriented (positively with respect to the orientation of M). However, instead of proving this, it will be more convenient for us to include these redundant conditions as hypotheses.

Condition (d) means the following. Suppose that B is the (i, α)th face of A_r, and is also the (j, β)th face of A_s. If

$$\varphi : I^k \to A_r \qquad \text{and} \qquad \psi : I^k \to A_s$$

are orientation-preserving parametrizations of A_r and A_s respectively, then

$$B = \varphi_{i,\alpha}(I^{k-1}) = \psi_{j,\beta}(I^{k-1}).$$

What we are assuming in condition (d) is that the mapping

$$\psi_{j,\beta}^{-1} \circ \varphi_{i,\alpha} : I^{k-1} \to I^{k-1}$$

has a negative Jacobian determinant if $(-1)^{i+\alpha} = (-1)^{j+\beta}$, and a positive one if $(-1)^{i+\alpha} = -(-1)^{j+\beta}$. By the proof of Lemma 5.7, this implies that

$$(-1)^{i+\alpha} \int_{\varphi_{i,\alpha}} \omega + (-1)^{j+\beta} \int_{\psi_{j,\beta}} \omega = 0 \tag{6}$$

for any continuous $(k-1)$-form ω. Consequently, if we form the sum

$$\sum_{r=1}^{p} \int_{\partial A_r} \omega,$$

then the integrals over all *interior* faces cancel in pairs. This conclusion is precisely what will be needed for the proof of Stokes' theorem for an oriented cellulated region.

To see the "visual" significance of condition (d) note that, in Fig. 5.35, the arrows, indicating the orientations of two adjacent 2-cells, point in opposite directions along their common edge (or face). This condition is not (and cannot be) satisfied by the "cellulation" of the Möbius strip indicated in Fig. 5.36.

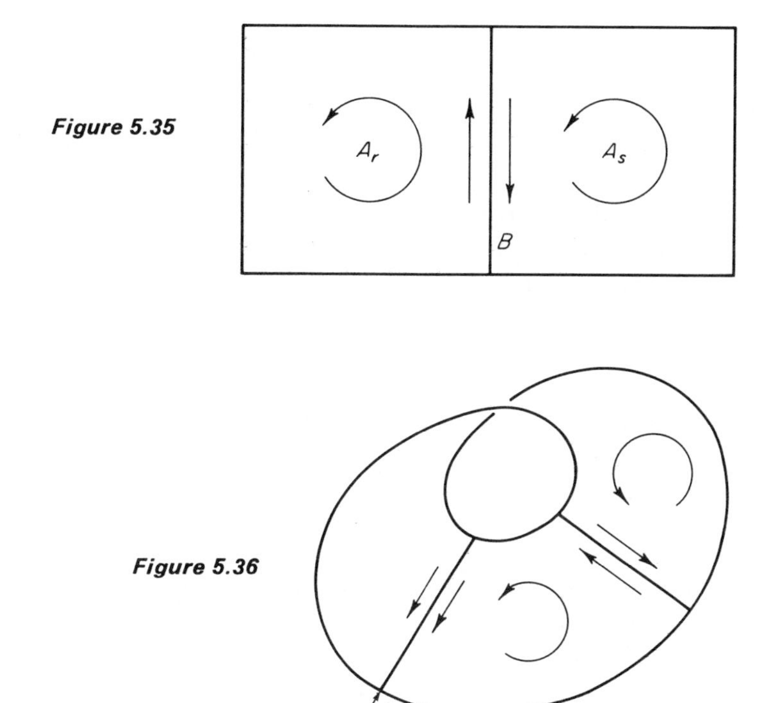

Figure 5.35

Figure 5.36

If $(R, \mathscr{K})$ is an oriented cellulated region in the smooth k-manifold $M \subset \mathscr{R}^n$, and ω is a continuous differential k-form defined on R, we define the integral of ω on R by

$$\int_R \omega = \sum_{r=1}^{p} \int_{A_r} \omega, \tag{7}$$

where $A_1, \ldots, A_p$ are the oriented k-cells of $\mathscr{K}$. Note that this might conceivably depend upon the oriented cellulation $\mathscr{K}$; a more complete notation (which, however, we will not use) would be $\int_{(R, \mathscr{K})} \omega$.

Since the boundary ∂R of R is, by condition (e), the union of all the boundary faces $B_1, \ldots, B_q$ of the k-cells of $\mathscr{K}$, we want to define the integral, of a $(k-1)$-form α on ∂R, as a sum of integrals of α over the $(k-1)$-cells $B_1, \ldots, B_q$. However we must be careful to choose the proper orientation for these boundary faces, as prescribed by the orientation-preserving parametrizations $\varphi^1, \ldots, \varphi^p$ of the k-cells $A_1, \ldots, A_p$ of K. If B_s is the (i_s, ε_s)th face of A_{r_s}, then

$$B_s = \varphi^{r_s}_{i_s, \varepsilon_s}(I^{k-1}).$$

For brevity, let us write

$$\psi_s = \varphi^{r_s}_{i_s, \varepsilon_s} : I^{k-1} \to B_s$$

and

$$\delta_s = i_s + \varepsilon_s .$$

Then $(-1)^{\delta_s} \int_{\psi_s} \alpha$ is the integral corresponding to the face B_s of A_{r_s}, which appears in the sum which defines $\int_{\partial A_{r_s}} \alpha$. We therefore define the integral of the $(k-1)$-form α on ∂R by

$$\int_{\partial R} \alpha = \sum_{s=1}^{q} (-1)^{\delta_s} \int_{\psi_s} \alpha. \tag{8}$$

Although the notation here is quite tedious, the idea is simple enough. We are simply saying that

$$\int_{\partial R} \alpha = \sum_{s=1}^{q} \int_{B_s} \alpha,$$

with the understanding that each boundary face B_s has the orientation that is prescribed for it in the oriented boundary ∂A_{r_s} of the oriented k-cell A_{r_s}.

This is our final definition of an integral of a differential form. Recall that we originally defined the integral of the k-form α on the k-dimensional surface patch φ by formula (10) in Section 5. We then defined the integral of α on a k-cell A in an oriented smooth k-manifold M by

$$\int_A \alpha = \int_\varphi \alpha,$$

where $\varphi : I^k \to A$ is a parametrization that extends to an orientation-preserving coordinate patch for M. The definitions for oriented cellular regions, given above, generalize the definitions for oriented cells, given previously in this section. That is, if $(R, \mathscr{K})$ is an oriented cellular region in which the oriented cellulation $\mathscr{K}$ consists of the single oriented k-cell $R = A$, then the integrals

$$\int_R \omega \quad \text{and} \quad \int_{\partial R} \alpha,$$

as defined above, reduce to the integrals

$$\int_A \omega \quad \text{and} \quad \int_{\partial A} \alpha,$$

as defined previously.

With all this preparation, the proof of Stokes' theorem for oriented cellulated regions is now a triviality.

Theorem 6.3 Let $(R, \mathscr{K})$ be an oriented cellulated region in an oriented smooth k-manifold in $\mathscr{R}^n$. If α is a $\mathscr{C}^1$ differential $(k-1)$-form defined on an open set that contains R, then

$$\int_R d\alpha = \int_{\partial R} \alpha. \tag{9}$$

PROOF Let $A_1, \ldots, A_p$ be the oriented k-cells of $\mathscr{K}$. Then

$$\begin{aligned}\int_R d\alpha &= \sum_{r=1}^{p} \int_{A_r} d\alpha \\ &= \sum_{r=1}^{p} \int_{\partial A_r} \alpha \\ &= \sum_{r=1}^{p} \sum_{i=1}^{k} \sum_{\varepsilon=0,1} (-1)^{i+\varepsilon} \int_{\varphi^r{}_{i,\varepsilon}} \alpha\end{aligned}$$

by Theorem 6.2, $\varphi^1, \ldots, \varphi^p$ being the parametrizations of $A_1, \ldots, A_p$. But this last sum is equal to

$$\int_{\partial R} \alpha,$$

since by Eq. (6) the integrals on interior faces cancel in pairs, while the integral on each boundary face appears once with the "correct" sign—the one given in Eq. (8). ∎

The most common applications of Stokes' theorem are not to oriented cellulated regions as such, but rather to oriented manifolds-with-boundary. A compact oriented smooth *k-manifold-with-boundary* is a compact region V in an oriented k-manifold $M \subset \mathscr{R}^n$, such that its boundary ∂V is a smooth (compact) $(k-1)$-manifold. For example, the unit ball B^n is a compact n-manifold-with-boundary. The ellipsoidal ball

$$E = \left\{(x, y, z) \in \mathscr{R}^3 : \frac{x^2}{a^2} + \frac{y^2}{b^2} + \frac{z^2}{c^2} \leqq 1\right\}$$

is a compact 3-manifold-with-boundary, as is the solid torus obtained by revolving about the z-axis the disk $(y-a)^2 + z^2 \leqq b^2$ in the yz-plane (Fig. 5.37). The hemisphere $x^2 + y^2 + z^2 = 1$, $z \geqq 0$ is a compact 2-manifold-with-boundary, as is the annulus $a^2 \leqq x^2 + y^2 \leqq b^2$.

If V is a compact oriented smooth k-manifold-with-boundary contained in the oriented smooth k-manifold $M \subset \mathscr{R}^n$, then its boundary ∂V is an orientable $(k-1)$-manifold; the *positive* orientation of ∂V is defined as follows. Given $\mathbf{p} \in \partial V$, there exists a coordinate patch $\Phi : U \to M$ such that

(i) $\mathbf{p} \in \Phi(U)$,
(ii) $\Phi^{-1}(\partial V) \subset \mathscr{R}^{k-1}$, and
(iii) $\Phi^{-1}(\text{int } V)$ is contained in the open half-space $x_k > 0$ of $\mathscr{R}^k$.

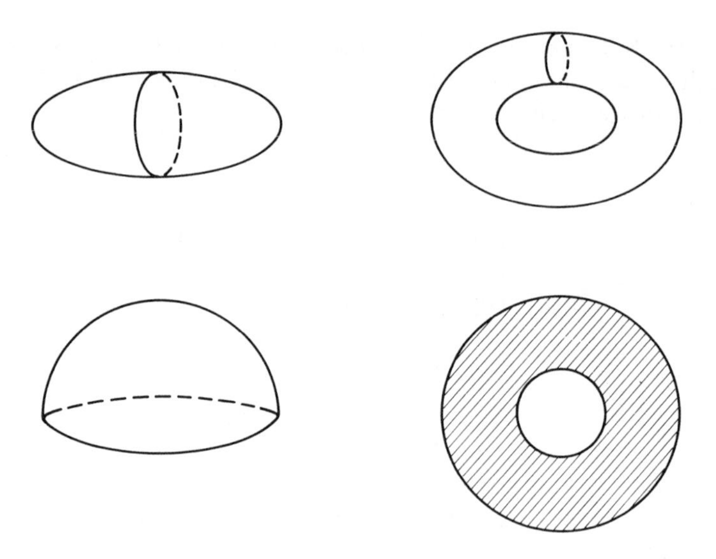

Figure 5.37

We choose Φ to be an orientation-preserving coordinate patch for M if k is even, and an orientation-reversing coordinate patch for M if k is odd. For the existence of such a coordinate patch, see Exercises 4.1 and 4.4 of Chapter III. The reason for the difference between the case k even and the case k odd will appear in the proof of Theorem 6.4 below.

If Φ is such a coordinate patch for M, then $\varphi = \Phi | U \cap \mathscr{R}^{k-1}$ is a coordinate patch for the $(k-1)$-manifold ∂V (see Fig. 5.38). If $\Phi_1, \ldots, \Phi_m$ is a collection of such coordinate patches for M, whose images cover ∂V, then their restrictions $\varphi_1, \ldots, \varphi_m$ to $\mathscr{R}^{k-1}$ form an atlas for ∂V. Since the fact that Φ_i and Φ_j overlap positively (because either both are orientation-preserving or both are orientation-reversing) implies that φ_i and φ_j overlap positively, this atlas $\{\varphi_i\}$ is an orientation for ∂V. It is (by definition) the *positive orientation* of ∂V.

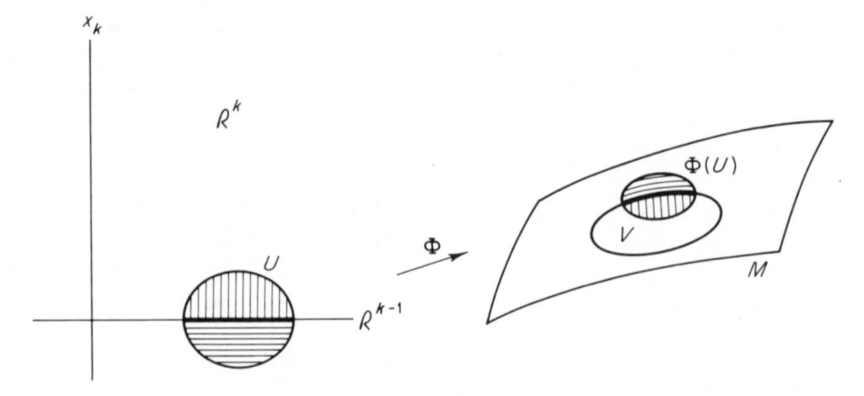

Figure 5.38

As a useful exercise, the reader should check that this construction yields the counterclockwise orientation for the unit circle S^1, considered as the boundary of the disk B^2 in the oriented 2-manifold $\mathscr{R}^2$, and, more generally, that it yields the positive orientation (of Section 2) for any compact 2-manifold-with-boundary in $\mathscr{R}^2$.

A cellulation is a special kind of paving, and we have previously stated that every compact smooth manifold has a paving. It is, in fact, true that every oriented compact smooth manifold-with-boundary possesses an oriented cellulation. If we accept without proof this difficult and deep theorem, we can establish Stokes' theorem for manifolds-with-boundary.

Theorem 6.4 Let V be an oriented compact smooth k-manifold-with-boundary in the oriented smooth k-manifold $M \subset \mathscr{R}^n$. If ∂V has the positive orientation, and α is a $\mathscr{C}^1$ differential $(k-1)$-form on an open set containing V, then

$$\int_V d\alpha = \int_{\partial V} \alpha. \tag{10}$$

PROOF Let $\mathscr{K} = \{A^1, \ldots, A^p\}$ be an oriented cellulation of V. We may assume without loss of generality that the $(k, 0)$th faces $A^1_{k,0}, \ldots, A^q_{k,0}$ of the first q of these oriented k-cells are the boundary faces of the cellulation. Let $\varphi^i : I^k \to A^i$ be the given orientation-preserving parametrization of A^i. Then

$$\int_V d\alpha = \sum_{i=1}^{q} (-1)^k \int_{\varphi^i{}_{k,0}} \alpha \tag{11}$$

by Theorem 6.3.

By the definition in Section 5, of the integral of a differential form over an oriented manifold,

$$\int_{\partial V} \alpha = \sum_{i=1}^{s} \int_{B_i} \alpha,$$

where $B_1, \ldots, B_s$ are the oriented $(k-1)$-cells of any *oriented* paving of the oriented $(k-1)$-manifold ∂V. That is,

$$\int_{\partial V} \alpha = \sum_{i=1}^{s} \int_{\psi_i} \alpha$$

if ψ_i is an orientation-preserving parametrization of B_i, $i = 1, \ldots, s$.

Now the boundary faces $A^1_{k,0}, \ldots, A^q_{k,0}$ constitute a paving of ∂V; the question here is whether or not their parametrizations $\varphi^i_{k,0}$ are orientation-preserving. But from the definition of the positive orientation for ∂V, we see that $\varphi^i_{k,0}$ is orientation-preserving or orientation-reversing according as k is even or odd respectively, since $\varphi^i : I^k \to M$ is orientation-preserving in either case (Fig. 5.39). Therefore

$$\int_{\partial V} \alpha = (-1)^k \sum_{i=1}^{q} \int_{\varphi^i{}_{k,0}} \alpha = \int_V d\alpha$$

by Eq. (11). ∎

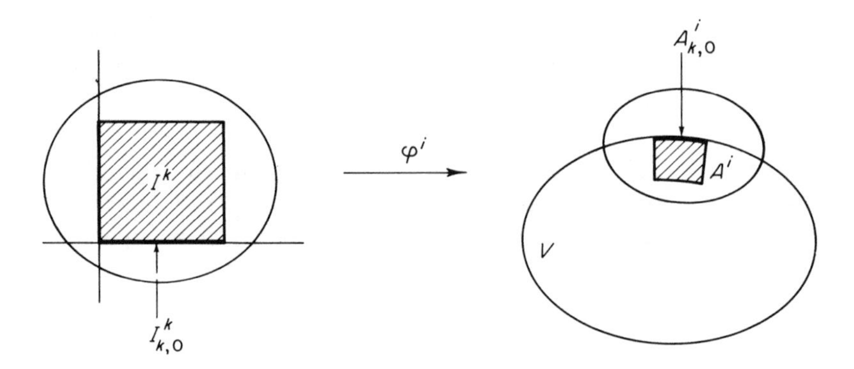

Figure 5.39

There is an alternative proof of Theorem 6.4 which does not make use of oriented cellulations. Instead of splitting the manifold-with-boundary V into oriented cells, this alternative proof employs the device of "partitions of unity" to split the given differential form α into a sum

$$\alpha = \sum_{i=1}^{s} \alpha_i ,$$

where each of the differential forms α_i vanishes outside of some k-cell. Although the partitions of unity approach is not so conceptual as the oriented cellulations approach, it is theoretically preferable because the proof of the existence of partitions of unity is much easier than the proof of the existence of oriented cellulations. For the partitions of unity approach to Stokes' theorem, we refer the reader to Spivak, "Calculus on Manifolds" [18].

Example Let D be a compact smooth n-manifold-with-boundary in $\mathscr{R}^n$. If

$$\alpha = \sum_{i=1}^{n} (-1)^{i-1} x_i \, dx_1 \wedge \cdots \wedge \widehat{dx_i} \wedge \cdots \wedge dx_n ,$$

then

$$\begin{aligned} d\alpha &= \sum_{i=1}^{n} (-1)^{i-1} \, dx_i \wedge dx_1 \wedge \cdots \wedge \widehat{dx_i} \wedge \cdots \wedge dx_n \\ &= n \, dx_1 \wedge \cdots \wedge dx_n . \end{aligned}$$

Theorem 6.4 therefore gives

$$\int_{\partial D} \sum_{i=1}^{n} (-1)^{i-1} x_i \, dx_1 \wedge \cdots \wedge \widehat{dx_i} \wedge \cdots \wedge dx_n = n \int_D dx_1 \wedge \cdots \wedge dx_n,$$

so

$$v(D) = n^{-1} \int_{\partial D} \sum_{i=1}^{n} (-1)^{i-1} x_i \, dx_1 \wedge \cdots \wedge \widehat{dx_i} \wedge \cdots \wedge dx_n \tag{12}$$

if ∂D is positively oriented. Formula (12) is the n-dimensional generalization of the formula

$$A = \tfrac{1}{2} \int_{\partial D} -y \, dx + x \, dy$$

for the area of a region $D \subset \mathscr{R}^2$ (see Example 2 of Section 2).

According to Exercise 5.13, the differential $(n-1)$-form

$$\alpha = \sum_{i=1}^{n} (-1)^{i-1} x_i \, dx_1 \wedge \cdots \wedge \widehat{dx_i} \wedge \cdots \wedge dx_n$$

is the surface area form of the unit sphere $S^{n-1} \subset \mathscr{R}^n$. With $D = B^n$, formula (12) therefore gives

$$v(B^n) = \frac{1}{n} \, a(S^{n-1}),$$

the relationship that we have previously seen in our explicit (and separate) computations of $v(B^n)$ and $a(S^{n-1})$.

The applications of Stokes' theorem are numerous, diverse, and significant. A number of them will be treated in the following exercises, and in subsequent sections.

Exercises

6.1 Let V be a compact oriented smooth $(k+l+1)$-dimensional manifold-with-boundary in $\mathscr{R}^n$. If α is a k-form and β an l-form, each $\mathscr{C}^1$ in a neighborhood of V, use Stokes' theorem to prove the "integration by parts" formula

$$\int_V (d\alpha) \wedge \beta = \int_{\partial V} \alpha \wedge \beta - (-1)^k \int_V \alpha \wedge d\beta.$$

6.2 If ω is a $\mathscr{C}^1$ differential k-form on $\mathscr{R}^n$ such that

$$\int_M \omega = 0$$

for every compact oriented smooth k-manifold $M \subset \mathscr{R}^n$, use Stokes' theorem to show that ω is closed, that is, $d\omega = 0$.

6.3 Let α be a $\mathscr{C}^1$ differential $(k-1)$-form defined in a neighborhood of the oriented compact smooth k-manifold $M \subset \mathscr{R}^n$. Then prove that

$$\int_M d\alpha = 0.$$

Hint: If B is a smooth k-dimensional ball in M (see Fig. 5.40), and $V = M - \text{int } B$, then

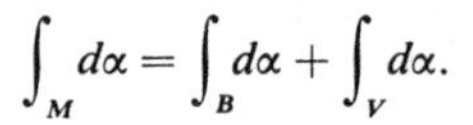

$$\int_M d\alpha = \int_B d\alpha + \int_V d\alpha.$$

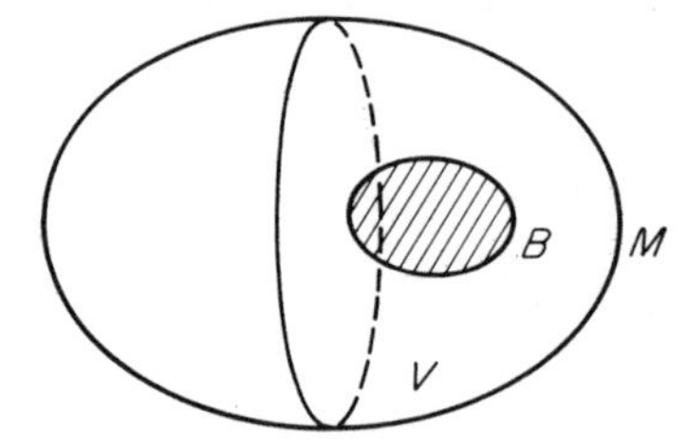

Figure 5.40

6.4 Let V_1 and V_2 be two compact n-manifolds-with-boundary in $\mathscr{R}^n$ such that $V_2 \subset \text{int } V_1$. If α is a differential $(n-1)$-form such that $d\alpha = 0$ on $W = V_1 - \text{int } V_2$, show that

$$\int_{\partial V_1} \alpha = \int_{\partial V_2} \alpha$$

if ∂V_1 and ∂V_2 are both positively oriented.

6.5 We consider in this exercise the differential $(n-1)$-form $d\Theta$ defined on $\mathscr{R}^n - \mathbf{0}$ by

$$d\Theta = \frac{1}{\rho^n} \sum_{i=1}^{n} (-1)^{i-1} x_i \, dx_1 \wedge \cdots \wedge \widehat{dx_i} \wedge \cdots \wedge dx_n,$$

where $\rho^2 = x_1{}^2 + x_2{}^2 + \cdots + x_n{}^2$. For reasons that will appear in the following exercise, this differential form is called the *solid angle form* on $\mathscr{R}^n$, and this is the reason for the notation $d\Theta$. Note that $d\Theta$ reduces to the familiar

$$d\theta = \frac{-y\,dx + x\,dy}{x^2 + y^2}$$

in the case $n = 2$.

(a) Show that $d\Theta$ is closed, $d(d\Theta) = 0$.

(b) Note that, on the unit sphere S^{n-1}, $d\Theta$ equals the surface area form of S^{n-1}. Conclude from Exercise 6.3 that $d\Theta$ is not exact, that is, there does not exist a differential $(n-2)$-form α on $\mathscr{R}^n - \mathbf{0}$ such that $d\alpha = d\Theta$.

(c) If M is an oriented compact smooth $(n-1)$-manifold in $\mathscr{R}^n$ which does not enclose the origin, show that

$$\int_M d\Theta = 0.$$

(d) If M is an oriented compact smooth $(n-1)$-manifold in $\mathscr{R}^n$ which does enclose the origin, use Exercise 6.4 to show that

$$\int_M d\Theta = a(S^{n-1}).$$

6.6 Let $M \subset \mathscr{R}^n - \mathbf{0}$ be an oriented compact $(n-1)$-dimensional manifold-with-boundary such that every ray through $\mathbf{0}$ intersects M in at most one point. The union of those rays through $\mathbf{0}$ which do intersect M is a "solid cone" $C(M)$. We assume that the intersection $N_\rho = C(M) \cap S_\rho^{n-1}$, of $C(M)$ and the sphere S_ρ^{n-1} of radius ρ (Fig. 5.41), is an oriented compact $(n-1)$-manifold-with-boundary in S_ρ^{n-1} (with its positive orientation). The *solid angle* $\Theta(M)$ subtended by M is defined by

$$\Theta(M) = a(N_1) = a(C(M) \cap S^{n-1}).$$

(a) Show that $\Theta(M) = 1/\rho^{n-1} a(N_\rho)$.
(b) Prove that $\Theta(M) = \int_M d\Theta$.
Hints: Choose $\rho > 0$ sufficiently small that N_ρ is "nearer to $\mathbf{0}$" than M. Let V be the compact region in $\mathscr{R}^n$ that is bounded by $M \cup N_\rho \cup R$, where R denotes the portion of the boundary of $C(M)$ which lies between M and N_ρ. Assuming that V has an oriented cellulation, apply Stokes' theorem. To show that

$$\int_R d\Theta = 0,$$

let $\{A_1, \ldots, A_m\}$ be an oriented paving of ∂N_ρ. Then $\{B_1, \ldots, B_n\}$ is an oriented paving of R, where $B_i = R \cap C(A_i)$. Denote by $l(\mathbf{x})$ the length of the segment between N_ρ and M, of the ray through $\mathbf{x} \in \partial N_\rho$. If

$$\varphi_i : I^{n-1} \to A_i$$

is a parametrization of A_i, and

$$\Phi_i : I^{n-2} \times I \to A_i$$

is defined by $\Phi_i(\mathbf{x}, t) = \varphi_i(\mathbf{x})$, finally show that

$$\int_{B_i} d\Theta = \int_{\Phi_i} l\, d\Theta = 0.$$

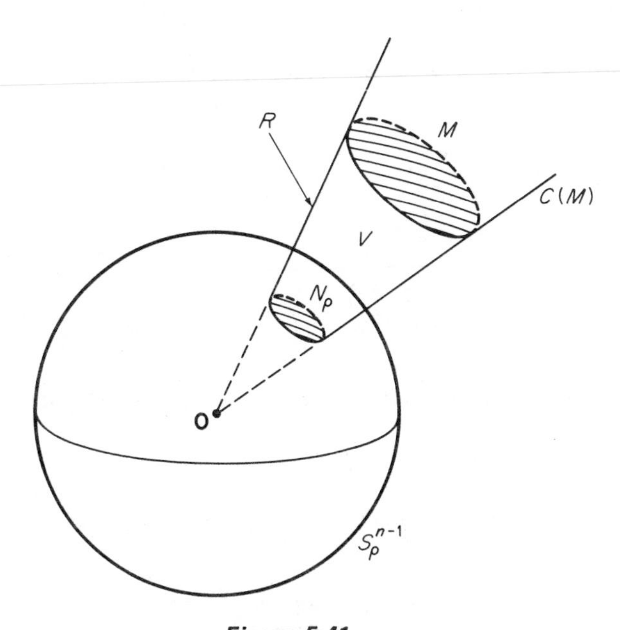

Figure 5.41

6.7 If V is a compact smooth n-dimensional manifold-with-boundary in $\mathscr{R}^n$, then there are two unit normal vector fields on ∂V—the *outer* normal, which at each point of ∂V points out of V, and the *inner* normal, which points into V. On the other hand, if M is an oriented smooth $(n-1)$-manifold in $\mathscr{R}^n$, then we have seen in Exercise 5.13 that the orientation of M uniquely determines a normal vector field $\mathbf{N}$ on M. If $M = \partial V$ is positively oriented (as the boundary of V), show that $\mathbf{N}$ is the *outer* normal vector field. Hence we can conclude from Exercise 5.13 that the surface area form of ∂V is

$$dA = \sum_{i=1}^{n} (-1)^{i-1} n_i \, dx_1 \wedge \cdots \wedge \widehat{dx_i} \wedge \cdots \wedge dx_n,$$

where the n_i are the components of the *outer* normal vector field on ∂V.

7 THE CLASSICAL THEOREMS OF VECTOR ANALYSIS

In this section we give the classical vector formulations of certain important special cases of the general Stokes' theorem of Section 6. The case $n = 3$, $k = 1$ yields the original (classical) form of Stokes' theorem, while the case $n = 3$, $k = 2$ yields the "divergence theorem," or Gauss' theorem.

If $\mathbf{F} : \mathscr{R}^n \to \mathscr{R}^n$ is a $\mathscr{C}^1$ vector field with component functions $F_1, \ldots, F_n$, the *divergence* of $\mathbf{F}$ is the real-valued function $\operatorname{div} \mathbf{F} : \mathscr{R}^n \to \mathscr{R}$ defined by

$$\operatorname{div} \mathbf{F} = \sum_{i=1}^{n} \frac{\partial F_i}{\partial x_i}. \tag{1}$$

The reason for this terminology will be seen in Exercise 7.10. The following result is a restatement of the case $k = n - 1$ of Stokes' theorem.

Theorem 7.1 Let $\mathbf{F}$ be a $\mathscr{C}^1$ vector field defined on a neighborhood of the compact oriented smooth n-manifold-with-boundary $V \subset \mathscr{R}^n$. Then

$$\int_V \operatorname{div} \mathbf{F} = \int_{\partial V} \sum_{i=1}^{n} (-1)^{i-1} F_i \, dx_1 \wedge \cdots \wedge \widehat{dx_i} \wedge \cdots \wedge dx_n, \tag{2}$$

PROOF If $\alpha = \sum_{i=1}^{n} (-1)^{i-1} F_i \, dx_1 \wedge \cdots \wedge \widehat{dx_i} \wedge \cdots \wedge dx_n$, then

$$\begin{aligned} d\alpha &= \sum_{i=1}^{n} (-1)^{i-1} \left(\sum_{j=1}^{n} \frac{\partial F_i}{\partial x_j} dx_j \right) \wedge dx_1 \wedge \cdots \wedge \widehat{dx_i} \wedge \cdots \wedge dx_n \\ &= \sum_{i=1}^{n} (-1)^{i-1} \frac{\partial F_i}{dx_i} dx_i \wedge dx_1 \wedge \cdots \wedge \widehat{dx_i} \wedge \cdots \wedge dx_n \\ &= (\operatorname{div} \mathbf{F}) \, dx_1 \wedge \cdots \wedge dx_n. \end{aligned}$$

∎

The *divergence theorem* in $\mathscr{R}^n$ is the statement that, if $\mathbf{F}$ and V are as in Theorem 7.1, then

$$\int_V \operatorname{div} \mathbf{F} = \int_{\partial V} \mathbf{F} \cdot \mathbf{N} \, dA, \tag{3}$$

where $\mathbf{N}$ is the unit *outer* normal vector field on ∂V, and ∂V is positively oriented (as the boundary of $V \subset \mathscr{R}^n$). The following result will enable us to show that the right-hand side of (2) is equal to the right-hand side of (3).

Theorem 7.2 Let M be an oriented smooth $(n-1)$-manifold in $\mathscr{R}^n$ with surface area form

$$dA = \sum_{i=1}^{n} (-1)^{i-1} n_i \, dx_1 \wedge \cdots \wedge \widehat{dx_i} \wedge \cdots \wedge dx_n.$$

Then, for each $i = 1, \ldots, n$, the restrictions to each tangent plane of M, of the differential $(n-1)$-forms

$$n_i \, dA \qquad \text{and} \qquad (-1)^{i-1} \, dx_1 \wedge \cdots \wedge \widehat{dx_i} \wedge \cdots \wedge dx_n,$$

are equal.

That is, if the vectors $\mathbf{v}_1, \ldots, \mathbf{v}_{n-1}$ are tangent to M at some point $\mathbf{x}$, then

$$n_i(\mathbf{x}) \, dA_{\mathbf{x}}(\mathbf{v}_1, \ldots, \mathbf{v}_{n-1}) = (-1)^{i-1} \, dx_1 \wedge \cdots \wedge \widehat{dx_i} \wedge \cdots \wedge dx_n(\mathbf{v}_1, \ldots, \mathbf{v}_{n-1}).$$

For brevity we write

$$n_i \, dA = (-1)^{i-1} \, dx_1 \wedge \cdots \wedge \widehat{dx_i} \wedge \cdots \wedge dx_n, \tag{4}$$

remembering that we are only asserting the equality of the values of these two differential $(n-1)$-forms on $(n-1)$-tuples of *tangent vectors* to M.

PROOF Let $\mathbf{v}_1, \ldots, \mathbf{v}_{n-1}$ be tangent vectors to M at the point $\mathbf{x} \in M$, and let $\varphi : U \to M$ be an orientation-preserving coordinate patch with $\mathbf{x} = \varphi(\mathbf{u}) \in \varphi(U)$. Since the vectors

$$\frac{\partial \varphi}{\partial u_1}(\mathbf{u}), \ldots, \frac{\partial \varphi}{\partial u_{n-1}}(\mathbf{u})$$

constitute a basis for the tangent plane to M at $\mathbf{x}$, there exists an $(n-1) \times (n-1)$ matrix $A = (a_{ij})$ such that

$$\mathbf{v}_i = \sum_{j=1}^{n} a_{ij} \frac{\partial \varphi}{\partial u_j}, \qquad i = 1, \ldots, n-1.$$

If α is any $(n-1)$-multilinear function on $\mathscr{R}^n$, then

$$\alpha(\mathbf{v}_1, \ldots, \mathbf{v}_{n-1}) = (\det A)\alpha\left(\frac{\partial \varphi}{\partial u_1}, \ldots, \frac{\partial \varphi}{\partial u_{n-1}}\right)$$

by Exercise 5.15. It therefore suffices to show that

$$n_i \, dA\left(\frac{\partial \varphi}{\partial u_1}, \ldots, \frac{\partial \varphi}{\partial u_{n-1}}\right)$$
$$= (-1)^{i-1} \, dx_1 \wedge \cdots \wedge \widehat{dx_i} \wedge \cdots \wedge dx_n\left(\frac{\partial \varphi}{\partial u_1}, \ldots, \frac{\partial \varphi}{\partial u_{n-1}}\right).$$

We have seen in Exercise 5.13 that

$$n_i(\mathbf{x}) = \frac{(-1)^{i-1}}{D} \frac{\partial(\varphi_1, \ldots, \widehat{\varphi_i}, \ldots, \varphi_n)}{\partial(u_1, \ldots, u_{n-1})}(\mathbf{u}),$$

where

$$D = \left[\sum_{j=1}^{n} \left(\frac{\partial(\varphi_1, \ldots, \widehat{\varphi_j}, \ldots, \varphi_n)}{\partial(u_1, \ldots, u_{n-1})}(\mathbf{u})\right)^2\right]^{1/2}.$$

Consequently

$$\begin{aligned}
n_i\, dA\left(\frac{\partial \varphi}{\partial u_1}, \ldots, \frac{\partial \varphi}{\partial u_{n-1}}\right)
&= n_i \sum_{j=1}^{n} (-1)^{j-1} n_j\, dx_1 \wedge \cdots \wedge \widehat{dx_j} \wedge \cdots \wedge dx_n \left(\frac{\partial \varphi}{\partial u_1}, \ldots, \frac{\partial \varphi}{\partial u_{n-1}}\right) \\
&= n_i \sum_{j=1}^{n} (-1)^{j-1} \frac{(-1)^{j-1}}{D} \left(\frac{\partial(\varphi_1, \ldots, \widehat{\varphi_j}, \ldots, \varphi_n)}{\partial(u_1, \ldots, u_{n-1})}\right)^2 \\
&= \frac{(-1)^{i-1}}{D^2} \frac{\partial(\varphi_1, \ldots, \widehat{\varphi_i}, \ldots, \varphi_n)}{\partial(u_1, \ldots, u_{n-1})} \sum_{j=1}^{n} \left(\frac{\partial(\varphi_1, \ldots, \widehat{\varphi_j}, \ldots, \varphi_n)}{\partial(u_1, \ldots, u_{n-1})}\right)^2 \\
&= (-1)^{i-1} \frac{\partial(\varphi_1, \ldots, \widehat{\varphi_i}, \ldots, \varphi_n)}{\partial(u_1, \ldots, u_{n-1})} \\
&= (-1)^{i-1} dx_1 \wedge \cdots \wedge \widehat{dx_i} \wedge \cdots \wedge dx_n \left(\frac{\partial \varphi}{\partial u_1}, \ldots, \frac{\partial \varphi}{\partial u_{n-1}}\right)
\end{aligned}$$

as desired. ∎

In our applications of this theorem, the following interpretation, of the coefficient functions n_i of the surface area form

$$dA = \sum_{i=1}^{n} (-1)^{i-1} n_i\, dx_1 \wedge \cdots \wedge \widehat{dx_i} \wedge \cdots \wedge dx_n$$

of the oriented $(n-1)$-manifold M, will be important. *If M is the positively-oriented boundary of the compact oriented n-manifold-with-boundary $V \subset \mathscr{R}^n$, then the n_i are simply the components of the unit outer normal vector field* $\mathbf{N}$ *on* $M = \partial V$ (see Exercise 6.7).

Example 1 We consider the case $n = 3$. If V is a compact 3-manifold-with-boundary in $\mathscr{R}^3$, and $\mathbf{N} = (n_1, n_2, n_3)$ is the unit outer normal vector field on ∂V, then the surface area form of ∂V is

$$dA = n_1\, dy \wedge dz + n_2\, dz \wedge dx + n_3\, dx \wedge dy, \tag{5}$$

and Eq. (4) yields

$$\begin{aligned} n_1\, dA &= dy \wedge dz, \\ n_2\, dA &= dz \wedge dx, \\ n_3\, dA &= dx \wedge dy. \end{aligned} \tag{6}$$

Now let $\mathbf{F} = (F_1, F_2, F_3)$ be a $\mathscr{C}^1$ vector field in a neighborhood of V. Then Eqs. (6) give

$$F_1\, dy \wedge dz + F_2\, dz \wedge dx + F_3\, dx \wedge dy = F_1 n_1\, dA + F_2 n_2\, dA + F_3 n_3\, dA,$$

so

$$F_1\, dy \wedge dz + F_2\, dz \wedge dx + F_3\, dx \wedge dy = \mathbf{F} \cdot \mathbf{N}\, dA. \tag{7}$$

Upon combining this equation with Theorem 7.1, we obtain

$$\begin{aligned} \int_V \operatorname{div} \mathbf{F} &= \int_{\partial V} F_1\, dy \wedge dz + F_2\, dz \wedge dx + F_3\, dx \wedge dy \\ &= \int_{\partial V} \mathbf{F} \cdot \mathbf{N}\, dA, \end{aligned}$$

the divergence theorem in dimension 3. Equation (7), which in essence expresses an arbitrary 2-form on ∂V as a multiple of ∂A, is frequently used in applications of the divergence theorem.

The divergence theorem is often applied to compute a given surface integral by "transforming" it into the corresponding volume integral. The point is that volume integrals are usually easier to compute than surface integrals. The following two examples illustrate this.

Example 2 To compute $\int_{\partial V} x\, dy \wedge dz - y\, dz \wedge dx$, we take $\mathbf{F} = (x, -y, 0)$. Then $\operatorname{div} \mathbf{F} \equiv 0$, so

$$\begin{aligned} \int_{\partial V} x\, dy \wedge dz - y\, dz \wedge dx &= \int_{\partial V} \mathbf{F} \cdot \mathbf{N}\, dA \\ &= \int_V \operatorname{div} \mathbf{F} = 0. \end{aligned}$$

Example 3 To compute $\int_{S^2} y\, dz \wedge dx + xz\, dx \wedge dy$, take $\mathbf{F} = (0, y, xz)$. Then $\operatorname{div} \mathbf{F} = 1 + x$, so

$$\begin{aligned} \int_{S^2} y\, dz \wedge dx + xz\, dx \wedge dy &= \int_{B^3} (1 + x)\, dx\, dy\, dz \\ &= v(B^3) + \int_{B^3} x\, dx\, dy\, dz \\ &= \frac{4\pi}{3}, \end{aligned}$$

since the last integral is obviously zero by symmetry.

Example 4 Let S_1 and S_2 be two spheres centered at the origin in $\mathscr{R}^3$ with S_1 interior to S_2, and denote by R the region between them. Suppose that $\mathbf{F}$ is a vector field such that $\operatorname{div} \mathbf{F} = 0$ at each point of R. If $\mathbf{N}_0$ is the outer normal vector field on ∂R (pointing outward at each point of S_2, and inward at each point of S_1, as in Fig. 5.42), then the divergence theorem gives

$$\int_{\partial R} \mathbf{F} \cdot \mathbf{N}\, dA = \int_R \operatorname{div} \mathbf{F} = 0.$$

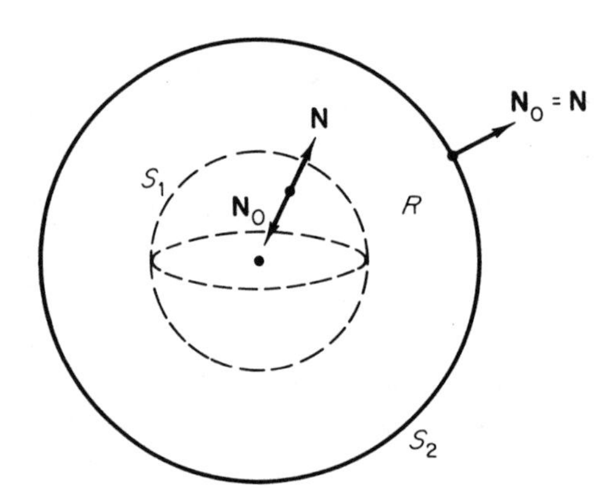

Figure 5.42

If $\mathbf{N}$ denotes the outer normal on both spheres (considering each as the positively oriented boundary of a ball), then $\mathbf{N} = \mathbf{N}_0$ on S_2, while $\mathbf{N} = -\mathbf{N}_0$ on S_1. Therefore

$$\int_{\partial R} \mathbf{F} \cdot \mathbf{N}_0\, dA = \int_{S_2} \mathbf{F} \cdot \mathbf{N}\, dA - \int_{S_1} \mathbf{F} \cdot \mathbf{N}\, dA,$$

so we conclude that

$$\int_{S_1} \mathbf{F} \cdot \mathbf{N}\, dA = \int_{S_2} \mathbf{F} \cdot \mathbf{N}\, dA.$$

The proof of the n-dimensional divergence theorem is simply a direct generalization of the proof in Example 1 of the 3-dimensional divergence theorem.

Theorem 7.3 If $\mathbf{F}$ is a $\mathscr{C}^1$ vector field defined on a neighborhood of the compact n-manifold-with-boundary $V \subset \mathscr{R}^n$, then

$$\int_V \operatorname{div} \mathbf{F} = \int_{\partial V} \mathbf{F} \cdot \mathbf{N}\, dA, \tag{3}$$

where $\mathbf{N}$ and dA are the outer normal and surface area form of the positively-oriented boundary ∂V.

PROOF In computing the right-hand integral, the $(n-1)$-form $\mathbf{F} \cdot \mathbf{N}\, dA$ is applied only to $(n-1)$-tuples of tangent vectors to ∂V. Therefore we can apply Theorem 7.2 to obtain

$$\begin{aligned}\mathbf{F}\cdot\mathbf{N}\,dA &= \sum_{i=1}^{n} F_i n_i\,dA \\ &= \sum_{i=1}^{n} (-1)^{i-1} F_i\,dx_1 \wedge \cdots \wedge \widehat{dx_i} \wedge \cdots \wedge dx_n.\end{aligned}$$

Thus Eq. (3) follows immediately from Theorem 7.1. ∎

The integral $\int_{\partial V} \mathbf{F}\cdot\mathbf{N}\,dA$ is sometimes called the *flux* of the vector field $\mathbf{F}$ across the surface ∂V. This terminology is motivated by the following physical interpretation. Suppose that $\mathbf{F} = \rho\mathbf{v}$, where $\mathbf{v}$ is the velocity vector field of a fluid which is flowing throughout an open set containing V, and ρ is its density distribution function (both functions of x, y, z, and t). We want to show that $\int_{\partial V} \mathbf{F}\cdot\mathbf{N}\,dA$ is the rate at which mass is leaving the region V, that is, that

$$\int_{\partial V} \mathbf{F}\cdot\mathbf{N}\,dA = -M'(t),$$

where

$$M(t) = \int_V \rho(x, y, z, t)\,dx\,dy\,dz,$$

is the total mass of the fluid in V at time T. Let $\{A_1, \ldots, A_k\}$ be an oriented paving of the oriented $(n-1)$-manifold ∂V, with the $(n-1)$-cells $A_1, \ldots, A_k$ so small that each is approximately an $(n-1)$-dimensional parallelepiped (Fig. 5.43). Let $\mathbf{N}_i$, $\mathbf{v}_i$, ρ_i, and $\mathbf{F}_i = \rho_i\mathbf{v}_i$ be the values of $\mathbf{N}$, $\mathbf{v}$, ρ, and $\mathbf{F}$ at a selected point of A_i. Then the volume of that fluid which leaves the region V through the cell A_i, during the short time interval Δt, is approximately equal to the volume of an n-dimensional parallelepiped with base A_i and height $(\mathbf{v}_i\,\Delta t)\cdot\mathbf{N}_i$, and the density of this portion of the fluid is approximately ρ_i. Therefore, if ΔM denotes the change in the mass of the fluid within V during the time interval Δt, then

$$\begin{aligned}\Delta M &\approx -\sum_{i=1}^{k} \rho_i(\mathbf{v}_i\,\Delta t)\cdot\mathbf{N}_i\,a(A_i) \\ &= -\Delta t\sum_{i=1}^{k} \mathbf{F}_i\cdot\mathbf{N}_i\,a(A_i) \\ &= -\Delta t\sum_{i=1}^{k}\int_{A_i} \mathbf{F}_i\cdot\mathbf{N}_i\,dA.\end{aligned}$$

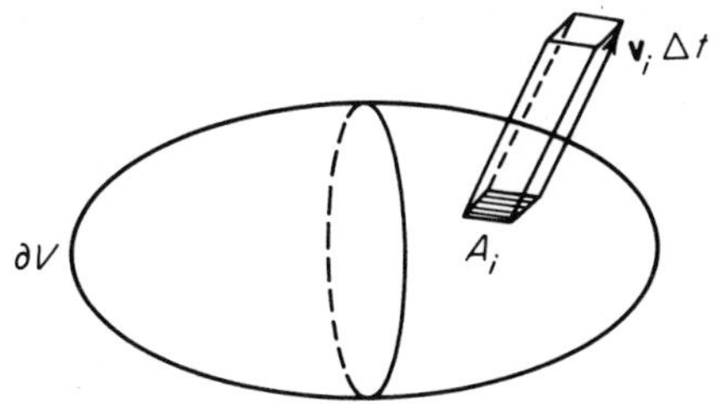

Figure 5.43

Taking the limit as $\Delta t \to 0$, we conclude that

$$M'(t) = -\int_{\partial V} \mathbf{F} \cdot \mathbf{N}\, dA$$

as desired.

We can now apply the divergence theorem to obtain

$$M'(t) = -\int_V \operatorname{div} \mathbf{F}.$$

On the other hand, if $\partial\rho/\partial t$ is continuous, then by differentiation under the integral sign we obtain

$$M'(t) = \int_V \frac{\partial \rho}{\partial t}.$$

Consequently

$$\int_V \left(\operatorname{div} \mathbf{F} + \frac{\partial \rho}{\partial t}\right) = 0.$$

Since this must be true for any region V within the fluid flow, we conclude by the usual continuity argument that

$$\operatorname{div}(\rho \mathbf{V}) + \frac{\partial \rho}{\partial t} = 0.$$

This is the *equation of continuity* for fluid flow. Notice that, for an *incompressible* fluid (one for which $\rho =$ constant), it reduces to

$$\operatorname{div} \mathbf{v} = 0.$$

We now turn our attention to the special case $n = 3$, $k = 1$ which, as we shall see, leads to the following classical formulation of Stokes' theorem in $\mathscr{R}^3$:

$$\int_D (\operatorname{curl} \mathbf{F}) \cdot \mathbf{N}\, dA = \int_{\partial D} \mathbf{F} \cdot \mathbf{T}\, ds, \tag{8}$$

where D is an oriented compact 2-manifold-with-boundary in $\mathscr{R}^3$, $\mathbf{N}$ and $\mathbf{T}$ are the appropriate unit normal and unit tangent vector fields on D and ∂D, respectively, $\mathbf{F} = (F_1, F_2, F_3)$ is a $\mathscr{C}^1$ vector field, and curl $\mathbf{F}$ is the vector field defined by

$$\operatorname{curl} \mathbf{F} = \left(\frac{\partial F_3}{\partial y} - \frac{\partial F_2}{\partial z}, \frac{\partial F_1}{\partial z} - \frac{\partial F_3}{\partial x}, \frac{\partial F_2}{\partial x} - \frac{\partial F_1}{\partial y}\right).$$

This definition of curl $\mathbf{F}$ can be remembered in the form

$$\operatorname{curl} \mathbf{F} = \nabla \times \mathbf{F} = \begin{vmatrix} \mathbf{e}_1 & \mathbf{e}_2 & \mathbf{e}_3 \\ \dfrac{\partial}{\partial x} & \dfrac{\partial}{\partial y} & \dfrac{\partial}{\partial z} \\ F_1 & F_2 & F_3 \end{vmatrix}.$$

That is, curl **F** is the result of formal expansion, along its first row, of the 3×3 determinant obtained by regarding curl **F** as the cross product of the gradient operator $\Delta = (\partial/\partial x, \partial/\partial y, \partial/\partial z)$ and the vector **F**.

If $\omega = F_1\,dx + F_2\,dy + F_3\,dz$ is the differential 1-form whose coefficient functions are the components of **F**, then

$$d\omega = \left(\frac{\partial F_3}{\partial y} - \frac{\partial F_2}{\partial z}\right) dy \wedge dz + \left(\frac{\partial F_1}{\partial z} - \frac{\partial F_3}{\partial x}\right) dz \wedge dx + \left(\frac{\partial F_2}{\partial x} - \frac{\partial F_1}{\partial y}\right) dx \wedge dy$$

by Example 2 of Section 5. Notice that the coefficient functions of the 2-form $d\omega$ are the components of the vector curl **F**. This correspondence, between ω and **F** on the one hand, and between $d\omega$ and curl **F** on the other, is the key to the vector interpretation of Stokes' theorem.

The unit normal **N** and unit tangent **T**, in formula (8) above, are defined as follows. The oriented compact 2-manifold-with-boundary D is (by definition) a subset of an oriented smooth 2-manifold $M \subset \mathscr{R}^3$. The orientation of M prescribes a unit normal vector field **N** on M as in Exercise 5.13. Specifically,

$$\mathbf{N} = \frac{\dfrac{\partial \varphi}{\partial u} \times \dfrac{\partial \varphi}{\partial v}}{\left|\dfrac{\partial \varphi}{\partial u} \times \dfrac{\partial \varphi}{\partial v}\right|}, \tag{9}$$

where φ is an orientation-preserving coordinate patch on M. If **n** is the outer normal vector field on ∂D—that is, at each point of ∂D, **n** is the tangent vector to M which points out of D (Fig. 5.44)—then we define the unit tangent on ∂D by

$$\mathbf{T} = \mathbf{N} \times \mathbf{n}. \tag{10}$$

The reader may check "visually" that this definition of **T** yields the same orientation of ∂D as in the statement of Green's theorem, in case D is a plane region. That is, if one proceeds around ∂D in the direction prescribed by **T**, remaining upright (regarding the direction of **N** as "up"), then the region D remains on his left.

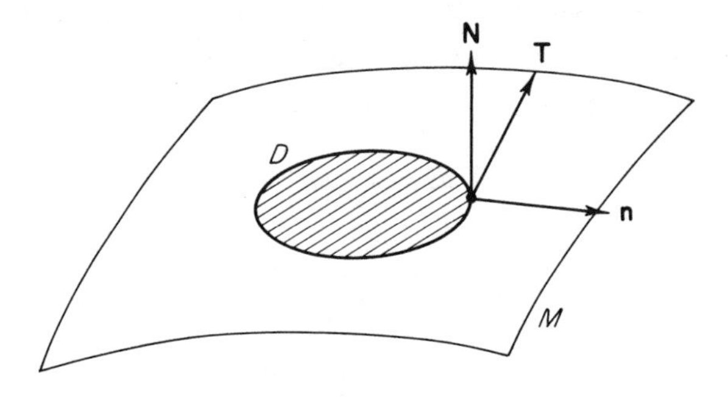

Figure 5.44

Theorem 7.4 Let D be an oriented compact 2-manifold-with-boundary in $\mathscr{R}^3$, and let $\mathbf{N}$ and $\mathbf{T}$ be the unit normal and unit tangent vector fields, on D and ∂D, respectively, defined above. If $\mathbf{F}$ is a $\mathscr{C}^1$ vector field on an open set containing D, then

$$\int_D (\text{curl } \mathbf{F}) \cdot \mathbf{N} \, dA = \int_{\partial D} \mathbf{F} \cdot \mathbf{T} \, ds. \tag{8}$$

PROOF The orientation of ∂D prescribed by (10) is the positive orientation of ∂D, as defined in Section 6. If

$$\omega = F_1 \, dx + F_2 \, dy + F_3 \, dz,$$

it follows that

$$\int_{\partial D} \omega = \int_{\partial D} \mathbf{F} \cdot \mathbf{T} \, ds. \tag{11}$$

See Exercise 1.21 or the last paragraph of Section 1.

Applying Theorem 7.2 in the form of Eq. (6), we obtain

$$\begin{aligned} d\omega &= \left(\frac{\partial F_3}{\partial y} - \frac{\partial F_2}{\partial z}\right) dy \wedge dz + \left(\frac{\partial F_1}{\partial z} - \frac{\partial F_3}{\partial x}\right) dz \wedge dy + \left(\frac{\partial F_2}{\partial x} - \frac{\partial F_1}{\partial y}\right) dx \wedge dy \\ &= \left(\frac{\partial F_3}{\partial y} - \frac{\partial F_2}{\partial z}\right) n_1 \, dA + \left(\frac{\partial F_1}{\partial z} - \frac{\partial F_3}{\partial x}\right) n_2 \, dA + \left(\frac{\partial F_2}{\partial x} - \frac{\partial F_1}{\partial y}\right) n_3 \, dA \\ &= (\text{curl } \mathbf{F}) \cdot \mathbf{N} \, dA, \end{aligned}$$

so

$$\int_D d\omega = \int_D (\text{curl } \mathbf{F}) \cdot \mathbf{N} \, dA. \tag{12}$$

Since $\int_D d\omega = \int_{\partial D} \omega$ by the general Stokes' theorem, Eqs. (11) and (12) imply Eq. (8). ∎

Stokes' Theorem is frequently applied to evaluate a given line integral, by "transforming" it to a surface integral whose computation is simpler.

Example 5 Let $\mathbf{F}(x, y, z) = (x, x + y, x + y + z)$, and denote by C the ellipse in which the plane $z = y$ intersects the cylinder $x^2 + y^2 = 1$, oriented counterclockwise around the cylinder. We wish to compute $\int_C \mathbf{F} \cdot \mathbf{T} \, ds$. Let D be the elliptical disk bounded by C. The semiaxes of D are 1 and $\sqrt{2}$, so $a(D) = \pi\sqrt{2}$. Its unit normal is $\mathbf{N} = (0, -1/\sqrt{2}, 1/\sqrt{2})$, and curl $\mathbf{F} = (1, -1, 1)$. Therefore

$$\begin{aligned} \int_C \mathbf{F} \cdot \mathbf{T} \, ds &= \int_D (\text{curl } \mathbf{F}) \cdot \mathbf{N} \, dA \\ &= \int_D \frac{2}{\sqrt{2}} \, dA \\ &= \frac{2}{\sqrt{2}} \, a(D) \\ &= 2\pi. \end{aligned}$$

Another typical application of Stokes' theorem is to the computation of a given surface integral by "replacing" it with a simpler surface integral. The following example illustrates this.

Example 6 Let $\mathbf{F}$ be the vector field of Example 5, and let D be the upper hemisphere of the unit sphere S^2. We wish to compute $\int_D (\text{curl } \mathbf{F}) \cdot \mathbf{N} \, dA$. Let B be the unit disk in the xy-plane, and $C = \partial D = \partial B$, oriented counterclockwise. Then two applications of Stokes' theorem give

$$\begin{aligned}\int_D (\text{curl } \mathbf{F}) \cdot \mathbf{N} \, dA &= \int_C \mathbf{F} \cdot \mathbf{T} \, ds \\ &= \int_B (\text{curl } \mathbf{F}) \cdot \mathbf{N} \, dA \\ &= \int_B (1, -1, 1) \cdot (0, 0, 1) \, dA \\ &= a(B) \\ &= \pi.\end{aligned}$$

We have interpreted the divergence vector in terms of fluid flow; the curl vector also admits such an interpretation. Let $\mathbf{F}$ be the velocity vector field of an incompressible fluid flow. Then the integral

$$\int_C \mathbf{F} \cdot \mathbf{T} \, ds$$

is called the *circulation* of the field $\mathbf{F}$ around the oriented closed curve C. If C_r is the boundary of a small disk D_r of radius r, centered at the point $\mathbf{p}$ and normal to the unit vector $\mathbf{b}$ (Fig. 5.45), then

$$\int_{C_r} \mathbf{F} \cdot \mathbf{T} \, ds = \int_D (\text{curl } \mathbf{F}) \cdot \mathbf{b} \, dA \approx \pi r^2 (\text{curl } F(\mathbf{p})) \cdot \mathbf{b}.$$

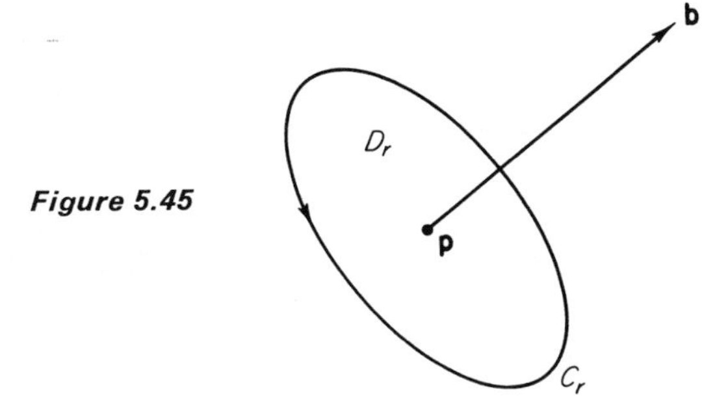

Figure 5.45

Taking the limit as $r \to 0$, we see that

$$(\text{curl } F(\mathbf{p})) \cdot \mathbf{b} = \lim_{r \to 0} \frac{1}{\pi r^2} \int_{C_r} \mathbf{F} \cdot \mathbf{T} \, ds.$$

Thus the **b**-component of curl **F** measures the circulation (or rotation) of the fluid around the vector **b**. For this reason the fluid flow is called *irrotational* if curl $\mathbf{F} = \mathbf{0}$.

For convenience we have restricted our attention in this section to *smooth* manifolds-with-boundary. However the divergence theorem and the classical Stokes' theorem hold for more general regions. For example, if V is an oriented cellulated n-dimensional region in $\mathscr{R}^n$, and $\mathbf{F}$ is a $\mathscr{C}^1$ vector field, then

$$\int_V \operatorname{div} \mathbf{F} = \int_{\partial V} \mathbf{F} \cdot \mathbf{N} \, dA$$

(just as in Theorem 7.3), with the following definition of the surface integral on the right. Let $\mathscr{K}$ be an oriented cellulation of V, and let $A_1, \ldots, A_p$ be the boundary $(n-1)$-cells of $\mathscr{K}$, so $\partial V = \bigcup_{i=1}^p A_i$. Let A_i be oriented in such a way that the procedure of Exercise 5.13 gives the *outer* normal vector field on A_i, and denote by $\mathbf{N}_i$ this outer normal on A_i. Then we define

$$\int_{\partial V} \mathbf{F} \cdot \mathbf{N} \, dA = \sum_{i=1}^{p} \int_{A_i} \mathbf{F} \cdot \mathbf{N}_i \, dA.$$

We will omit the details, but this more general divergence theorem could be established by the method of proof of the general Stokes' theorem in Section 6 —first prove the divergence theorem for an oriented n-cell in $\mathscr{R}^n$, and then piece together the n-cells of an oriented cellulation.

Example 7 Let V be the solid cylinder $x^2 + y^2 \leqq 1$, $0 \leqq z \leqq 1$. Denote by D_0 and D_1 the bottom and top disks respectively, and by R the cylindrical surface (Fig. 5.46). Then the outer normal is given by

$$\mathbf{N} = \begin{cases} (0, 0, -1) & \text{on} \quad D_0, \\ (0, 0, 1) & \text{on} \quad D_1, \\ (x, y, 0) & \text{at} \quad (x, y, z) \in R. \end{cases}$$

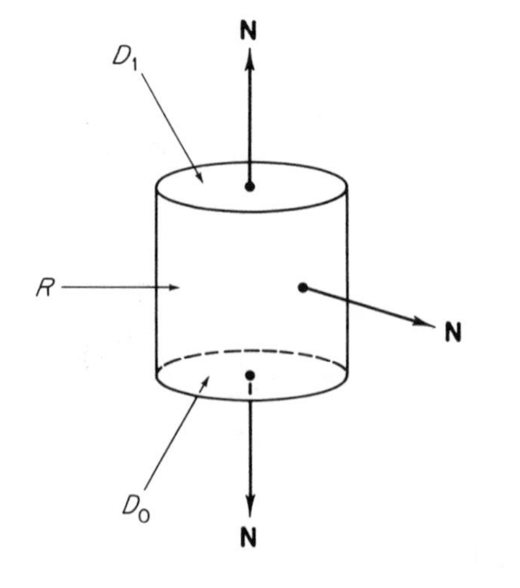

Figure 5.46

If $\mathbf{F} = (x, y, z)$, then

$$\begin{aligned}\int_{\partial V} \mathbf{F} \cdot \mathbf{N}\, dA &= \int_{D_0} \mathbf{F} \cdot \mathbf{N}\, dA + \int_{D_1} \mathbf{F} \cdot \mathbf{N}\, dA + \int_{R} \mathbf{F} \cdot \mathbf{N}\, dA \\ &= \int_{D_0} 0\, dA + \int_{D_1} 1\, dA + \int_{R} 1\, dA \\ &= a(D_1) + a(R) = \pi + 2\pi \\ &= 3\pi.\end{aligned}$$

Alternatively, we can apply the divergence theorem to obtain

$$\int_{\partial V} \mathbf{F} \cdot \mathbf{N}\, dA = \int_{V} \operatorname{div} \mathbf{F} = 3v(V) = 3\pi.$$

Similarly, Stokes' theorem holds for *piecewise smooth* compact oriented surfaces in $\mathscr{R}^3$. A piecewise smooth compact oriented surface S in $\mathscr{R}^3$ is the union of a collection $\{A_1, \ldots, A_p\}$ of oriented 2-cells in $\mathscr{R}^3$, satisfying conditions (b), (c), (d) in the definition of an oriented cellulation (Section 6), with the union ∂S of the boundary edges of these oriented 2-cells being a finite number of oriented closed curves. If $\mathbf{F}$ is a $\mathscr{C}^1$ vector field, then

$$\int_{S} (\operatorname{curl} \mathbf{F}) \cdot \mathbf{N}\, dA = \int_{\partial S} \mathbf{F} \cdot \mathbf{T}\, ds$$

(just as in Theorem 7.4), with the obvious definition of these integrals (as in the above discussion of the divergence theorem for cellulated regions). For example, Stokes' theorem holds for a compact oriented *polyhedral surface* (one which consists of a nonoverlapping triangles).

Exercises

7.1 The moment of inertia I of the region $V \subset \mathscr{R}^3$ about the z-axis is defined by

$$I = \int_V (x^2 + y^2).$$

Show that

$$I = \tfrac{1}{4} \textstyle\int_{\partial V} (x^3 + xy^2)\, dy \wedge dz + (x^2 y + y^3)\, dz \wedge dx.$$

7.2 Let $\rho = (x^2 + y^2 + z^2)^{1/2}$.

(a) If $\mathbf{F}(x, y, z) = \rho \cdot (x, y, z)$, show that $\operatorname{div} \mathbf{F} = 4\rho$. Use this fact and the divergence theorem to show that

$$\int_{B_a{}^3} \rho = \pi a^4$$

($B_a{}^3$ being the ball of radius a) by converting to a surface integral that can be evaluated by inspection.

(b) If $\mathbf{F} = \rho^2 \cdot (x, y, z)$, compute the integral $\int_{B_a{}^3} \operatorname{div} \mathbf{F}$ in a similar manner.

7.3 Find a function $g(\rho)$ such that, if $\mathbf{F}(x, y, z) = g(\rho)(x, y, z)$, then $\operatorname{div} \mathbf{F} = \rho^m$ $[m \geqq 0$, $\rho = (x^2 + y^2 + z^2)^{1/2}]$. Use it to prove that

$$\int_V \rho^m = \frac{1}{m+3} \int_{\partial V} \rho^m (x, y, z) \cdot \mathbf{N} \, dA$$

if V is a compact 3-manifold-with-boundary in $\mathscr{R}^3$. For example, show that

$$\int_{B^3} \rho^5 = \frac{\pi}{2}.$$

7.4 In each of the following, let C be the curve of intersection of the cylinder $x^2 + y^2 = 1$ and the given surface $z = f(x, y)$, oriented counterclockwise around the cylinder. Use Stokes' theorem to compute the line integral by first converting it to a surface integral.

(a) $\int_C y \, dx + z \, dy + x \, dz, \quad z = xy.$

(b) $\int_C z(x-1) \, dy + y(x+1) \, dz, \quad z = xy + 1.$

(c) $\int_C z \, dx - x \, dz, \quad z = 1 - y.$

7.5 Let the 2-form α be defined on $\mathscr{R}^3 - \mathbf{p}$, $\mathbf{p} = (a, b, c)$, by

$$\alpha = \frac{(x-a) \, dy \wedge dz + (y-b) \, dz \wedge dx + (z-c) \, dx \wedge dy}{[(x-a)^2 + (y-b)^2 + (z-c)^2]^{3/2}}.$$

(a) Show that $d\alpha = 0$.
(b) Conclude that $\int_M \alpha = 0$ if M is a compact smooth 2-manifold not enclosing the point $\mathbf{p}$.
(c) Show that $\int_M \alpha = 4\pi$ if M is a sphere centered at $\mathbf{p}$.
(d) Show that $\int_M \alpha = 4\pi$ if M is any compact positively oriented smooth 2-manifold enclosing the point $\mathbf{p}$.

7.6 The *potential* $\varphi(x, y, z)$ at $\mathbf{x} = (x, y, z)$ due to a collection of charges $q_1, \ldots, q_m$ at the points $\mathbf{p}_1, \ldots, \mathbf{p}_m$ is

$$\varphi(x, y, z) = \sum_{i=1}^{m} \frac{q_i}{r_i},$$

where $r_i = |\mathbf{x} - \mathbf{p}_i|$. If $\mathbf{E} = -\nabla\varphi$, the *electric field* vector, apply the previous problem to show that

$$\int_M \mathbf{E} \cdot \mathbf{N} \, dA = 4\pi(q_1 + \cdots + q_m)$$

if M is a smooth 2-manifold enclosing these charges. This is *Gauss' law*.

7.7 Let f and g be $\mathscr{C}^1$ functions on an open set containing the compact n-manifold-with-boundary $V \subset \mathscr{R}^n$, and let $\mathbf{N}$ be the unit outer normal on ∂V. The *Laplacian* $\nabla^2 f$ and the *normal* derivative $\partial f/\partial n$ are defined by

$$\nabla^2 f = \operatorname{div}(\nabla f) \qquad \text{and} \qquad \frac{\partial f}{\partial n} = \nabla f \cdot \mathbf{N}.$$

Prove *Green's formulas* in $\mathscr{R}^n$:

(a) $$\int_V (f\nabla^2 g + \nabla f \cdot \nabla g) = \int_{\partial V} f\frac{\partial g}{\partial n}\,dA,$$

(b) $$\int_V (f\nabla^2 g - g\,\nabla^2 f) = \int_{\partial F} \left(f\frac{\partial g}{\partial n} - g\frac{\partial f}{\partial n}\right) dA.$$

Hint: For (a), apply the divergence theorem with $\mathbf{F} = f\nabla g$.

7.8 Use Green's formulas in $\mathscr{R}^n$ to generalize Exercises 2.7 and 2.8 to $\mathscr{R}^n$. In particular, if f and g are both harmonic on $V \subset \mathscr{R}^n$, $\nabla^2 f = \nabla^2 g = 0$, and $f = g$ on ∂V, prove that $f = g$ throughout V.

7.9 Let f be a harmonic function on the open set $U \subset \mathscr{R}^n$ ($n > 2$). If B is an n-dimensional ball in U with center $\mathbf{p}$ and radius a, and $S = \partial B$, prove that

$$f(\mathbf{p}) = \frac{1}{a(S)} \int_S f\,dA.$$

That is, the value of f at the center of the ball B is the average of its values on the boundary S of B. This is the *average value property* for harmonic functions. *Outline*: Without loss of generality, we may assume that $\mathbf{p}$ is the origin. Define the function $g : \mathscr{R}^n - \mathbf{0} \to \mathscr{R}$ by

$$g(\mathbf{x}) = \frac{1}{r^{n-2}}, \qquad r = |\mathbf{x}|.$$

Denote by S_ε a small sphere of radius $\varepsilon > 0$ centered at $\mathbf{0}$, and by V the region between S_ε and S (Fig. 5.47).

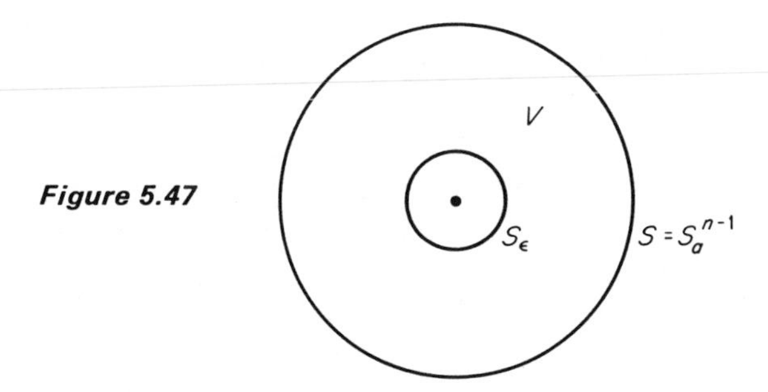

Figure 5.47

(a) Show that g is harmonic, and then apply the second Green's formula (Exercise 7.7) to obtain the formula

$$\int_S \left(f\frac{\partial g}{\partial n} - g\frac{\partial f}{\partial n}\right) dA = \int_{S_\varepsilon} \left(f\frac{\partial g}{\partial n} - g\frac{\partial f}{\partial n}\right) dA. \qquad (*)$$

(b) Notice that

$$\int_S \left(f\frac{\partial g}{\partial n} - g\frac{\partial f}{\partial n}\right) dA = \frac{2-n}{a^{n-1}} \int_S f\,dA - \frac{1}{a^{n-2}} \int_S \frac{\partial f}{\partial n}\,dA,$$

and similarly for the right-hand side of (∗), with a replaced by ε, because $\partial g/\partial n = (2-n)/r^{n-1}$. Use the divergence theorem to show that

$$\int_S \frac{\partial f}{\partial n}\, dA = \int_{S_\varepsilon} \frac{\partial f}{\partial n}\, dA = 0,$$

so formula (∗) reduces to

$$\frac{2-n}{a^{n-1}} \int_S f\, dA = \frac{2-n}{\varepsilon^{n-1}} \int_{S_\varepsilon} f\, dA. \qquad (**)$$

(c) Now obtain the average value property for f by taking the limit in (∗∗) as $\varepsilon \to 0$.

7.10 Let $\mathbf{F}$ be a $\mathscr{C}^1$ vector field in $\mathscr{R}^n$. Denote by B_ε the ball of radius ε centered at $\mathbf{p}$, and $S_\varepsilon = \partial B_\varepsilon$. Use the divergence theorem to show that

$$\operatorname{div} \mathbf{F}(\mathbf{p}) = \lim_{\varepsilon \to 0} \frac{1}{v(B_\varepsilon)} \int_{S_\varepsilon} \mathbf{F} \cdot \mathbf{N}\, dA.$$

If we think of $\mathbf{F}$ as the velocity vector field of a fluid flow, then this shows that $\operatorname{div} \mathbf{F}(\mathbf{p})$ is the rate (per unit volume) at which the fluid is "diverging" away from the point $\mathbf{p}$.

7.11 Let V be a compact 3-manifold-with-boundary in the lower half-space $z < 0$ of $\mathscr{R}^3$. Think of V as an object submerged in a fluid of uniform density ρ, its surface at $z = 0$ (Fig. 5.48). The *buoyant force* $\mathbf{B}$ on V, due to the fluid, is defined by

$$\mathbf{B} = -\int_{\partial V} \mathbf{F} \cdot \mathbf{N}\, dA,$$

where $\mathbf{F} = (0, 0, \rho z)$. Use the divergence theorem to show that

$$\mathbf{B} = \int_V \rho.$$

Thus the buoyant force on the object is equal to the weight of the fluid that it displaces (Archimedes).

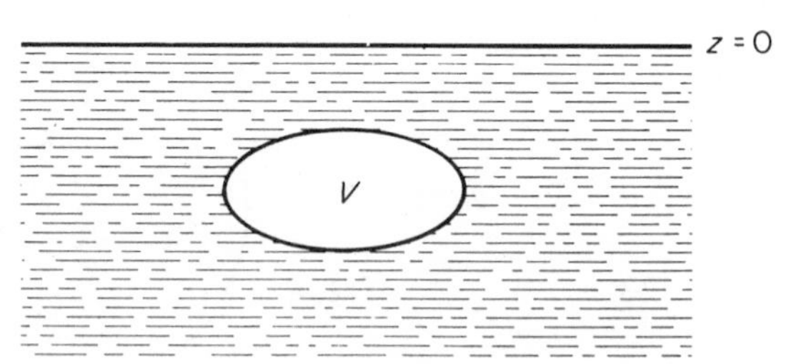

Figure 5.48

7.12 Let U be an open subset of $\mathscr{R}^3$ that is *star-shaped* with respect to the origin. This means that, if $\mathbf{p} \in U$, then U contains the line segment from $\mathbf{0}$ to $\mathbf{p}$. Let $\mathbf{F}$ be a $\mathscr{C}^1$ vector field on U. Then use Stokes' theorem to prove that $\operatorname{curl} \mathbf{F} = \mathbf{0}$ on U if and only if

$$\int_C \mathbf{F} \cdot \mathbf{T} = 0$$

for every polygonal closed curve C in U. *Hint:* To show that $\operatorname{curl} \mathbf{F} \equiv 0$ implies that the integral vanishes, let C consist of the line segments $L_1, \ldots, L_p$. For each $i = 1, \ldots, p$, denote by T_i the triangle whose vertices are the origin and the endpoints of L_i. Apply Stokes' theorem to each of these triangles, and then add up the results.

8 CLOSED AND EXACT FORMS

Let ω be a $\mathscr{C}^1$ differential k-form defined on the open set U in $\mathscr{R}^n$. Then ω is called *closed* (on U) if $d\omega = 0$, and *exact* (on U) if there exists a $(k-1)$-form α on U such that $d\alpha = \omega$.

Since $d(d\omega) = 0$ by Proposition 5.1, we see immediately that every exact form is closed. Our object in this section is to discuss the extent to which the converse is true.

According to Theorem 2.5, every closed 1-form, which is defined on all of $\mathscr{R}^2$, is exact. However, we have seen in Section 2 that the 1-form

$$\omega = \frac{-y\,dx + x\,dy}{x^2 + y^2}$$

is closed on $\mathscr{R}^2 - \mathbf{0}$, but is not exact; that is, there is no function $f: \mathscr{R}^2 - \mathbf{0} \to \mathscr{R}$ such that $df = \omega$. These facts suggest that the question as to whether the closed 1-form ω on U is exact, depends upon the geometry of the set U.

The "Poincaré lemma" (Theorem 8.1 below) asserts that every closed form defined on a star-shaped region is exact. The open set $U \subset \mathscr{R}^n$ is called *star-shaped* with respect to the point $\mathbf{a}$ provided that, given $\mathbf{x} \in U$, the line segment from $\mathbf{a}$ to $\mathbf{x}$ is contained in U. The open set U is *star-shaped* if there exists $\mathbf{a} \in U$ such that U is star-shaped with respect to $\mathbf{a}$. For example, $\mathscr{R}^n$ itself is star-shaped (with respect to any point), as is every open ball or open n-dimensional interval in $\mathscr{R}^n$.

Theorem 8.1 Every closed $\mathscr{C}^1$ differential k-form defined on a star-shaped open subset U of $\mathscr{R}^n$ is exact.

PROOF We may assume (after a translation if necessary) that U is star-shaped with respect to the origin $\mathbf{0}$. We want to define, for each positive integer k, a certain function I, from k-forms on U to $(k-1)$-forms, such that $I(0) = 0$.

First we need the following notation. Given a k-tuple $\mathbf{i} = (i_1, i_2, \ldots, i_k)$, write

$$\mu_{\mathbf{i}} = \sum_{j=1}^{k} (-1)^{j-1} x_{i_j}\, dx_{i_1} \wedge \cdots \widehat{dx_{i_j}} \cdots \wedge dx_{i_k}.$$

Note that

$$d\mu_{\mathbf{i}} = k\, d\mathbf{x}_{\mathbf{i}} = k\, dx_{i_1} \wedge \cdots \wedge dx_{i_k} \tag{1}$$

and, if $j\mathbf{i} = (j, i_1, \ldots, i_k)$, then

$$\mu_{j\mathbf{i}} = x_j\, d\mathbf{x}_{\mathbf{i}} - dx_j \wedge \mu_{\mathbf{i}} \tag{2}$$

(Exercise 8.4). Now, given a k-form

$$\omega = \sum_{[\mathbf{i}]} a_{\mathbf{i}}\, d\mathbf{x}_{\mathbf{i}}$$

on U, we define the $(k-1)$-form $I\omega$ on U by

$$I\omega(\mathbf{x}) = \sum_{[\mathbf{i}]} \left(\int_0^1 t^{k-1} a_{\mathbf{i}}(t\mathbf{x})\, dt \right) \mu_{\mathbf{i}}. \tag{3}$$

Note that this makes sense because U is star-shaped with respect to $\mathbf{0}$, so $t\mathbf{x} \in U$ if $\mathbf{x} \in U$ and $t \in [0, 1]$. Clearly $I(0) = 0$, and

$$\begin{aligned} d(I\omega) &= \sum_{[\mathbf{i}]} \left[d\left(\int_0^1 t^{k-1} a_{\mathbf{i}}(t\mathbf{x})\, dt \right) \wedge \mu_{\mathbf{i}} + \left(\int_0^1 t^{k-1} a_{\mathbf{i}}(t\mathbf{x})\, dt \right) d\mu_{\mathbf{i}} \right] \\ &= \sum_{[\mathbf{i}]} \left[\left(\sum_{j=1}^n \frac{\partial}{\partial x_j} \left(\int_0^1 t^{k-1} a_{\mathbf{i}}(t\mathbf{x})\, dt \right) dx_j \right) \wedge \mu_{\mathbf{i}} + k\left(\int_0^1 t^{k-1} a_{\mathbf{i}}(t\mathbf{x})\, dt \right) d\mathbf{x}_{\mathbf{i}} \right], \end{aligned}$$

$$d(I\omega) = \sum_{[\mathbf{i}]} \left[\sum_{j=1}^n \left(\int_0^1 t^k \frac{\partial a_{\mathbf{i}}}{\partial x_j}(t\mathbf{x})\, dt \right) dx_j \wedge \mu_{\mathbf{i}} + k\left(\int_0^1 t^{k-1} a_{\mathbf{i}}(t\mathbf{x})\, dt \right) d\mathbf{x}_{\mathbf{i}} \right]. \tag{4}$$

If ω is closed on U, then

$$\begin{aligned} 0 &= I(d\omega) \\ &= I\left(\sum_{[\mathbf{i}]} \sum_{j=1}^n \frac{\partial a_{\mathbf{i}}}{\partial x_j}\, dx_j \wedge d\mathbf{x}_{\mathbf{i}} \right) \\ &= \sum_{[\mathbf{i}]} \sum_{j=1}^n \left(\int_0^1 t^k \frac{\partial a_{\mathbf{i}}}{\partial x_j}(t\mathbf{x})\, dt \right) \mu_{j\mathbf{i}}, \end{aligned}$$

$$0 = \sum_{[\mathbf{i}]} \sum_{j=1}^n \left(\int_0^1 t^k \frac{\partial a_{\mathbf{i}}}{\partial x_j}(t\mathbf{x})\, dt \right) (x_j\, d\mathbf{x}_{\mathbf{i}} - dx_j \wedge \mu_{\mathbf{i}}). \tag{5}$$

Upon addition of Eqs. (4) and (5), we obtain

$$\begin{aligned} d(I\omega) &= \sum_{[\mathbf{i}]} \left[\sum_{j=1}^n \left(\int_0^1 t^k \frac{\partial a_{\mathbf{i}}}{\partial x_j}(t\mathbf{x}) x_j\, dt \right) + k\left(\int_0^1 t^{k-1} a_{\mathbf{i}}(t\mathbf{x})\, dt \right) \right] d\mathbf{x}_{\mathbf{i}} \\ &= \sum \left[\int_0^1 \left(k t^{k-1} a_{\mathbf{i}}(t\mathbf{x}) + t^k \frac{d}{dt}(a_{\mathbf{i}}(t\mathbf{x})) \right) dt \right] d\mathbf{x}_{\mathbf{i}} \\ &= \sum \left(\int_0^1 \frac{d}{dt}[t^k a_{\mathbf{i}}(t\mathbf{x})]\, dt \right) d\mathbf{x}_{\mathbf{i}} \\ &= \sum_{[\mathbf{i}]} [t^k a_{\mathbf{i}}(t\mathbf{x})]_{t=0}^1\, d\mathbf{x}_{\mathbf{i}} \\ &= \sum_{[\mathbf{i}]} a_{\mathbf{i}}(\mathbf{x})\, d\mathbf{x}_{\mathbf{i}}, \end{aligned}$$

$$d(I\omega) = \omega.$$

Thus we have found a $(k-1)$-form whose differential is ω, as desired. ∎

Example 1 Consider the closed 2-form

$$\omega = xy\,dy \wedge dz + y\,dz \wedge dx - (z + yz)\,dx \wedge dy$$

on $\mathscr{R}^2$. In the notation of the proof of Theorem 8.1,

$$dy \wedge dz = d\mathbf{x}_{(2,3)}, \qquad dz \wedge dx = d\mathbf{x}_{(3,1)}, \qquad dx \wedge dy = d\mathbf{x}_{(1,2)},$$

and

$$\mu_{(2,3)} = y\,dz - z\,dy, \qquad \mu_{(3,1)} = z\,dx - x\,dz, \qquad \mu_{(1,2)} = x\,dy - y\,dx.$$

Equation (3) therefore gives

$$\begin{aligned} I\omega &= \left(\int_0^1 t(t^2xy)\,dt\right)(y\,dz - z\,dy) + \left(\int_0^1 t(ty)\,dt\right)(z\,dx - x\,dz) \\ &\quad + \left(\int_0^1 t(-tz - t^2yz)\,dt\right)(x\,dy - y\,dx) \\ &= \frac{xy}{4}(y\,dz - z\,dy) + \frac{y}{3}(z\,dx - x\,dz) \\ &\quad + \left(-\frac{z}{3} - \frac{yz}{4}\right)(x\,dy - y\,dx), \end{aligned}$$

$$I\omega = \left(\frac{2yz}{3} + \frac{y^2z}{4}\right)dx - \left(\frac{yxz}{2} + \frac{xz}{3}\right)dy + \left(\frac{xy^2}{4} - \frac{xy}{3}\right)dz,$$

a 1-form whose differential is ω.

Theorem 8.1 has as special cases two important facts of vector analysis. Recall that, if $\mathbf{F} = (P, Q, R)$ is a $\mathscr{C}^1$ vector field on an open set U in $\mathscr{R}^3$, then its divergence and curl are defined by

$$\operatorname{div} \mathbf{F} = \frac{\partial P}{\partial x} + \frac{\partial Q}{\partial y} + \frac{\partial R}{\partial z},$$

$$\operatorname{curl} \mathbf{F} = \left(\frac{\partial R}{\partial y} - \frac{\partial Q}{\partial z}, \frac{\partial P}{\partial z} - \frac{\partial R}{\partial x}, \frac{\partial Q}{\partial x} - \frac{\partial P}{\partial y}\right).$$

It follows easily that $\operatorname{div} \mathbf{F} = 0$ if $\mathbf{F} = \operatorname{curl} \mathbf{G}$ for some vector field $\mathbf{G}$ on U, while $\operatorname{curl} \mathbf{F} = \mathbf{0}$ if $\mathbf{F} = \operatorname{grad} f = \nabla f$ for some scalar function f. The following theorem asserts that the converses are true if U is star-shaped.

Theorem 8.2 Let $\mathbf{F}$ be a $\mathscr{C}^1$ vector field on the star-shaped open set $U \subset \mathscr{R}^3$. Then

(a) $\operatorname{curl} \mathbf{F} = \mathbf{0}$ if and only if there exists $f: U \to \mathscr{R}$ such that $\mathbf{F} = \operatorname{grad} f$,

(b) $\operatorname{div} \mathbf{F} = 0$ if and only if there exists $\mathbf{G}: U \to \mathscr{R}^3$ such that $\mathbf{F} = \operatorname{curl} \mathbf{G}$.

PROOF Given a vector field $\mathbf{F} = (P, Q, R)$, we define a 1-form $\alpha_{\mathbf{F}}$, a 2-form $\beta_{\mathbf{F}}$, and a 3-form $\gamma_{\mathbf{F}}$ by

$$\begin{aligned}
\alpha_{\mathbf{F}} &= P\,dx + Q\,dy + R\,dz, \\
\beta_{\mathbf{F}} &= P\,dy \wedge dz + Q\,dz \wedge dx + R\,dx \wedge dy, \\
\gamma_{\mathbf{F}} &= (P + Q + R)\,dx \wedge dy \wedge dz.
\end{aligned}$$

Then routine computations yield the formulas

$$df = \alpha_{\text{grad } f}, \tag{6}$$

$$d\alpha_{\mathbf{F}} = \beta_{\text{curl } \mathbf{F}}, \tag{7}$$

$$d\beta_{\mathbf{F}} = \gamma_{\text{div } \mathbf{F}}. \tag{8}$$

To prove that curl(grad f) = $\mathbf{0}$, note that

$$\beta_{\text{curl(grad } f)} = d(\alpha_{\text{grad } f}) = d(df) = 0$$

by (6), (7), and Proposition 5.1. To prove that div(curl $\mathbf{G}$) = 0, note that

$$\gamma_{\text{div(curl } \mathbf{G})} = d(\beta_{\text{curl } \mathbf{G}}) = d(d\alpha_{\mathbf{G}}) = 0$$

by (7), (8), and Proposition 5.1.

If curl $\mathbf{F} = \mathbf{0}$ on U, then $d\alpha_{\mathbf{F}} = \beta_{\text{curl } \mathbf{F}} = 0$ by (7), so Theorem 8.1 gives a function $f: U \to \mathscr{R}$ such that $df = \alpha_{\mathbf{F}}$, that is

$$D_1 f\,dx + D_2 f\,dy + D_3 f\,dz = P\,dx + Q\,dy + R\,dz,$$

so $\nabla f = \mathbf{F}$. This proves (a).

If div $\mathbf{F} = 0$ on U, then $d\beta_{\mathbf{F}} = \gamma_{\text{div } \mathbf{F}} = 0$ by (8), so Theorem 8.1 gives a 1-form

$$\omega = G_1\,dx + G_2\,dy + G_3\,dz$$

such that $d\omega = \beta_{\mathbf{F}}$. But if $\mathbf{G} = (G_1, G_2, G_3)$, then

$$d\omega = d\alpha_{\mathbf{G}} = \beta_{\text{curl } \mathbf{G}}$$

by (7), so it follows that curl $\mathbf{G} = \mathbf{F}$. This proves (b). ∎

Example 2 As a typical physical application of the Poincaré lemma, we describe the reduction of Maxwell's electromagnetic field equations to the inhomogeneous wave equation. Maxwell's equations relate the electric field vector $\mathbf{E}(x, y, z, t)$, the magnetic field vector $\mathbf{H}(x, y, z, t)$, the charge density $\rho(x, y, z)$, and the current density $\mathbf{J}(x, y, z)$; we assume these are all defined on $\mathscr{R}^3$ (for each t). With the standard notation div $\mathbf{F} = \nabla \cdot \mathbf{F}$ and curl $\mathbf{F} = \nabla \times \mathbf{F}$, these equations are

$$\nabla \cdot \mathbf{E} = \rho, \tag{9}$$

$$\nabla \cdot \mathbf{H} = 0, \tag{10}$$

$$\nabla \times \mathbf{E} + \frac{\partial \mathbf{H}}{\partial t} = \mathbf{0}, \tag{11}$$

$$\nabla \times \mathbf{H} - \frac{\partial \mathbf{E}}{\partial t} = \mathbf{J}. \tag{12}$$

Equation (10) asserts that there are no magnetic sources, while (9), (11), and (12) are, respectively, Gauss' law, Faraday's law, and Ampere's law.

Let us introduce the following differential forms on $\mathscr{R}^4$:

$$\begin{aligned}
E &= E_1\,dx + E_2\,dy + E_3\,dz,\\
*E &= E_1\,dy \wedge dz + E_2\,dz \wedge dx + E_3\,dx \wedge dy,\\
H &= H_1\,dy \wedge dz + H_2\,dz \wedge dx + H_3\,dx \wedge dy,\\
*H &= H_1\,dx + H_2\,dy + H_3\,dz,\\
J &= J_1\,dy \wedge dz + J_2\,dz \wedge dx + J_3\,dx \wedge dy,
\end{aligned}$$

where $\mathbf{E} = (E_1, E_2, E_3)$, $\mathbf{H} = (H_1, H_2, H_3)$, and $\mathbf{J} = (J_1, J_2, J_3)$. Then Eqs. (10) and (11) are equivalent to the equation

$$d(E \wedge dt + H) = 0, \tag{13}$$

while Eqs. (9) and (12) are equivalent to

$$d(*H \wedge dt - *E) = J \wedge dt - \rho\,dx \wedge dy \wedge dz. \tag{14}$$

Thus (13) and (14) are Maxwell's equations in the notation of differential forms.

Because of (13), the Poincaré lemma implies the existence of a 1-form

$$\alpha = A_1\,dx + A_2\,dy + A_3\,dz - \varphi\,dt$$

such that

$$d\alpha = E \wedge dt + H. \tag{15}$$

Of course the 1-form α which satisfies (15) is not unique; so does $\alpha + df$ for any differentiable function $f: \mathscr{R}^4 \to \mathscr{R}$. In particular, if f is a solution of the inhomogeneous wave equation

$$\nabla^2 f - \frac{\partial^2 f}{\partial t^2} = -\left(\frac{\partial A_1}{\partial x} + \frac{\partial A_2}{\partial y} + \frac{\partial A_3}{\partial z} + \frac{\partial \varphi}{\partial t}\right), \tag{16}$$

then the new 1-form

$$\beta = \alpha + df = G_1\,dx + G_2\,dy + G_3\,dz - g\,dt$$

satisfies both the equation

$$d\beta = E \wedge dt + H \tag{17}$$

and the condition

$$\frac{\partial G_1}{\partial x} + \frac{\partial G_2}{\partial y} + \frac{\partial G_3}{\partial z} + \frac{\partial g}{\partial t} = 0. \tag{18}$$

Computing $d\beta$, we find that (17) implies that

$$H_1 = \frac{\partial G_3}{\partial y} - \frac{\partial G_2}{\partial z}, \qquad H_2 = \frac{\partial G_1}{\partial z} - \frac{\partial G_3}{\partial x}, \qquad H_3 = \frac{\partial G_2}{\partial x} - \frac{\partial G_1}{\partial y},$$
$$E_1 = -\frac{\partial G_1}{\partial t} - \frac{\partial g}{\partial x}, \qquad E_2 = -\frac{\partial G_2}{\partial t} - \frac{\partial g}{\partial y}, \qquad E_3 = -\frac{\partial G_3}{dt} - \frac{\partial g}{\partial z}. \tag{19}$$

Substituting these expressions for the components of E and H, and making use of (18), a straightforward computation gives

$$\begin{aligned} d(*H \wedge dt - *E) = {} & \left(\frac{\partial^2 G_1}{\partial t^2} - \nabla^2 G_1\right) dy \wedge dz \wedge dt \\ & + \left(\frac{\partial^2 G_2}{\partial t^2} - \nabla^2 G_2\right) dz \wedge dx \wedge dt \\ & + \left(\frac{\partial^2 G_3}{\partial t^2} - \nabla^2 G_3\right) dx \wedge dy \wedge dt \\ & + \left(\nabla^2 g - \frac{\partial^2 g}{\partial t^2}\right) dx \wedge dy \wedge dz. \end{aligned}$$

Comparing this result with Eq. (14), we conclude that the vector field $\mathbf{G} = (G_1, G_2, G_3)$ on $\mathscr{R}^3$ and the function g satisfy the inhomogeneous wave equations

$$\nabla^2 \mathbf{G} - \frac{\partial^2 \mathbf{G}}{\partial t^2} = -\mathbf{J} \qquad \text{and} \qquad \nabla^2 g - \frac{\partial^2 g}{\partial t^2} = -\rho. \tag{20}$$

Thus the solution of Maxwell's equations reduces to the problem of solving the inhomogeneous wave equation. In particular, if $\mathbf{G}$ and g satisfy Eq. (18) and (20), then the vector fields

$$\mathbf{E} = -\nabla g - \frac{\partial \mathbf{G}}{\partial t} \qquad \text{and} \qquad \mathbf{H} = \nabla \times \mathbf{G},$$

defined by Eqs. (19), satisfy Maxwell's equations (9)–(12).

Exercises

8.1 For each of the following forms ω, find α such that $d\alpha = \omega$.
(a) $\omega = (3x^2y^2 + 8xy^3)\,dx + (2x^3y + 12x^2y^2 + 4y)\,dy$ on $\mathscr{R}^2$.
(b) $\omega = (y^2 + 2xz^2)\,dx + (2xy + 3y^2z^3)\,dy + (2x^2z + 3y^3z^2)\,dz$ on $\mathscr{R}^3$.
(c) $\omega = (2y - 4)\,dy \wedge dz + (y^2 - 2x)\,dz \wedge dx + (3 - x - 2yz)\,dx \wedge dy$.
(d) $\omega = (xy^2 + yz^2 + zx^2)\,dx \wedge dy \wedge dz$.

8.2 Let ω be a closed 1-form on $\mathscr{R}^2 - \mathbf{0}$ such that $\int_{S^1} \omega = 0$. Prove that ω is exact on $\mathscr{R}^2 - \mathbf{0}$.
Hint: Given $(x, y) \in \mathscr{R}^2 - \mathbf{0}$, define $f(x, y)$ to be the integral of ω along the path $\gamma_{(x,y)}$

which follows an arc of the unit circle from (1, 0) to $(x/|x|, y|y|)$, and then follows the radial straight line segment to (x, y). Then show that $df = \omega$.

8.3 If ω is closed 1-form on $\mathscr{R}^3 - \mathbf{0}$, prove that ω is exact on $\mathscr{R}^3 - 0$. *Hint*: Given $\mathbf{p} \in \mathscr{R}^3 - \mathbf{0}$, define $f(\mathbf{p})$ to be the integral of ω along the path $\gamma_{\mathbf{p}}$ which follows a great-circle arc on S^2 from (1, 0, 0) to $\mathbf{p}/|\mathbf{p}| \in S^2$, and then follows the radial straight line segment to $\mathbf{p}$. Apply Stokes' theorem to show that $f(\mathbf{p})$ is independent of the chosen great circle. Then show that $df = \omega$.

8.4 Verify formulas (1) and (2).

8.5 Verify formulas (6), (7), and (8).

8.6 Let $\omega = dx + z\,dy$ on $\mathscr{R}^3$. Given $f(x, y, z)$, compute $d(f\omega)$. Conclude that ω does not have an integrating factor (see Exercise 5.4).

VI

The Calculus of Variations

The calculus of variations deals with a certain class of maximum–minimum problems, which have in common the fact that each of them is associated with a particular sort of integral expression. The simplest example of such a problem is the following. Let $f: \mathscr{R}^3 \to \mathscr{R}$ be a given $\mathscr{C}^2$ function, and denote by $\mathscr{C}^1[a, b]$ the set of all real-valued $\mathscr{C}^1$ functions defined on the interval $[a, b]$. Given $\psi \in \mathscr{C}^1[a, b]$, define $F(\psi) \in \mathscr{R}$ by

$$F(\psi) = \int_a^b f(\psi(t), \psi'(t), t)\, dt. \qquad (*)$$

Then F is a real-valued function on $\mathscr{C}^1[a, b]$. The problem is then to find that element $\varphi \in \mathscr{C}^1[a, b]$, if any, at which the function $F : \mathscr{C}^1[a, b] \to \mathscr{R}$ attains its maximum (or minimum) value, subject to the condition that φ has given preassigned values $\varphi(a) = \alpha$ and $\varphi(b) = \beta$ at the endpoints of $[a, b]$.

For example, if we want to find the $\mathscr{C}^1$ function $\varphi : [a, b] \to \mathscr{R}$ whose graph $x = \varphi(t)$ (in the tx-plane, Fig. 6.1) joins the points (a, α) and (b, β) and has minimal length, then we want to minimize the function $F : \mathscr{C}^1[a, b] \to \mathscr{R}$ defined by

$$F(\psi) = \int_a^b [1 + (\psi'(t))^2]^{1/2}\, dt.$$

Here the function $f: \mathscr{R}^3 \to \mathscr{R}$ is defined by $f(x, y, t) = \sqrt{1 + y^2}$.

Note that in the above problem the "unknown" is a *function* $\varphi \in \mathscr{C}^1[a, b]$, rather than a *point* in $\mathscr{R}^n$ (as in the maximum–minimum problems which we considered in Chapter II). The "function space" $\mathscr{C}^1[a, b]$, is, like $\mathscr{R}^n$, a vector space, albeit an infinite-dimensional one. The purpose of this chapter is to appropriately generalize the finite-dimensional methods of Chapter II so as to treat some of the standard problems of the calculus of variations.

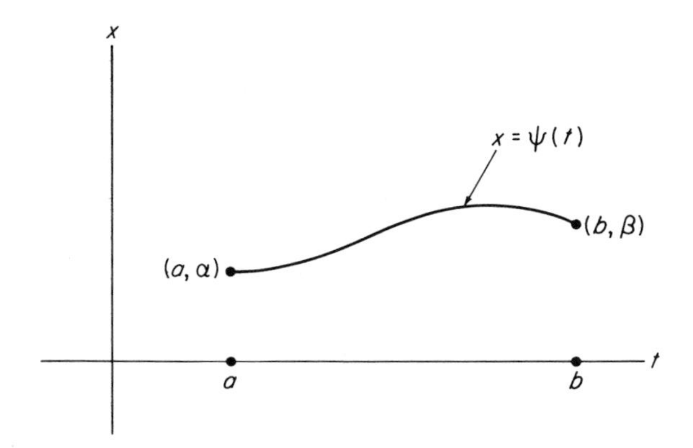

Figure 6.1

In Section 3 we will show that, if $\varphi \in \mathscr{C}^1[a, b]$ maximizes (or minimizes) the function $F : \mathscr{C}^1[a, b] \to \mathscr{R}$ defined by (*), subject to the conditions $\varphi(a) = \alpha$ and $\varphi(b) = \beta$, then φ satisfies the Euler–Lagrange equation

$$\frac{\partial f}{\partial x}(\varphi(t), \varphi'(t), t) - \frac{d}{dt}\frac{\partial f}{\partial y}(\varphi(t), \varphi'(t), t) = 0.$$

The proof of this will involve a generalized version of the familiar technique of "setting the derivative equal to zero, and then solving for the unknown."

In Section 4 we discuss the so-called "isoperimetric problem," in which it is desired to maximize or minimize the function

$$F(\psi) = \int_a^b f(\psi(t), \psi'(t), t)\, dt,$$

subject to the conditions $\psi(a) = \alpha$, $\psi(b) = \beta$ and the "constraint"

$$G(\psi) = \int_a^b g(\psi(t), \psi'(t), t)\, dt = c.$$

Here f and g are given $\mathscr{C}^2$ functions on $\mathscr{R}^3$. We will see that, if $\varphi \in \mathscr{C}^1[a, b]$ is a solution for this problem, then there exists a number $\lambda \in \mathscr{R}$ such that φ satisfies the Euler–Lagrange equation

$$\frac{\partial h}{\partial x}(\varphi(t), \varphi'(t), t) - \frac{d}{dt}\frac{\partial h}{\partial y}(\varphi(t), \varphi'(t), t) = 0$$

for the function $h : \mathscr{R}^3 \to \mathscr{R}$ defined by

$$h(x, y, t) = f(x, y, t) - \lambda g(x, y, t).$$

This assertion is reminiscent of the constrained maximum–minimum problems of Section II.5, and its proof will involve a generalization of the Lagrange multiplier method that was employed there.

As preparation for these general maximum–minimum methods, we need to develop the rudiments of "calculus in normed vector spaces." This will be done in Sections 1 and 2.

1 NORMED VECTOR SPACES AND UNIFORM CONVERGENCE

The introductory remarks above indicate that certain typical calculus of variations problems lead to a consideration of "function spaces" such as the vector space $\mathscr{C}^1[a, b]$ of all continuously differentiable functions defined on the interval $[a, b]$. It will be important for our purpose to define a norm on the vector space $\mathscr{C}^1[a, b]$, thereby making it into a *normed* vector space. As we will see in Section 2, this will make it possible to study real-valued functions on $\mathscr{C}^1[a, b]$ by the methods of differential calculus.

Recall (from Section I.3) that a *norm* on the vector space V is a real-valued function $x \to |x|$ on V satisfying the following conditions:

N1 $|x| > 0$ if $x \neq 0$,

N2 $|ax| = |a|\,|x|$,

N3 $|x + y| \leqq |x| + |y|$,

for all $x, y \in V$ and $a \in \mathscr{R}$. The norm $|x|$ of the vector $x \in V$ may be thought of as its length or "size." Also, given $x, y \in V$, the norm $|x - y|$ of $x - y$ may be thought of as the "distance" from x to y.

Example 1 We have seen in Section I.3 that each of the following definitions gives a norm on $\mathscr{R}^n$:

$$|\mathbf{x}|_0 = \max(|x_1|, \ldots, |x_n|) \qquad \text{(the sup norm)},$$

$$|\mathbf{x}|_1 = |x_1| + \cdots + |x_n| \qquad \text{(the 1-norm)},$$

$$|\mathbf{x}|_2 = (x_1{}^2 + \cdots + x_n{}^2)^{1/2} \qquad \text{(the Euclidean norm)},$$

where $\mathbf{x} = (x_1, \ldots, x_n) \in \mathscr{R}^n$. It was (and is) immediate that $|\ |_0$ and $|\ |_1$ are norms on $\mathscr{R}^n$, while the verification that $|\ |_2$ satisfies the triangle inequality (Condition N3) required the Cauchy–Schwarz inequality (Theorem I.3.1).

Any two norms on $\mathscr{R}^n$ are equivalent in the following sense. The two norms $|\ |_1$ and $|\ |_2$ on the vector space V are said to be *equivalent* if there exist positive numbers a and b such that

$$a|x|_1 \leqq |x|_2 \leqq b|x|_1$$

for all $x \in V$. We leave it as an easy exercise for the reader to check that this relation between norms on V is an equivalence relation (in particular, if the norms $|\ |_1$ and $|\ |_2$ on V are equivalent, and also $|\ |_2$ and $|\ |_3$ are equivalent, then it follows that $|\ |_1$ and $|\ |_3$ are equivalent norms on V).

We will not include here the full proof that any two norms on $\mathscr{R}^n$ are equivalent. However let us show that any *continuous* norm $|\ |$ on $\mathscr{R}^n$ (that is, $|\mathbf{x}|$ is a continuous function of $\mathbf{x}$) is equivalent to the Euclidean norm $|\ |_2$. Since $|\ |$ is continuous on the unit sphere

$$S^{n-1} = \{\mathbf{x} \in \mathscr{R}^n : |\mathbf{x}|_2 = 1\},$$

there exist positive numbers m and M such that

$$m|\mathbf{y}|_2 = m \leqq |\mathbf{y}| \leqq M = M|\mathbf{y}|_2$$

for all $\mathbf{y} \in S^{n-1}$. Given $\mathbf{x} \in \mathscr{R}^n$, choose $a > 0$ such that $\mathbf{x} = a\mathbf{y}$ with $\mathbf{y} \in S^{n-1}$. Then the above inequality gives

$$ma|\mathbf{y}|_2 \leqq a|\mathbf{y}| \leqq Ma|\mathbf{y}|_2,$$

so

$$m|\mathbf{x}|_2 \leqq |\mathbf{x}| \leqq M|\mathbf{x}|_2$$

as desired. In particular, the sup norm and the 1-norm of Example 1 are both equivalent to the Euclidean norm, since both are obviously continuous. For the proof that *every* norm on $\mathscr{R}^n$ is continuous, see Exercise 1.7.

Equivalent norms on the vector space V give rise to equivalent notions of sequential convergence in V. The sequence $\{x_n\}_1^\infty$ of points of V is said to *converge* to $x \in V$ with respect to the norm $|\ |$ if and only if

$$\lim_{n \to 0} |x_n - x| = 0.$$

Lemma 1.1 Let $|\ |_1$ and $|\ |_2$ be equivalent norms on the vector space V. Then the sequence $\{x_n\}_1^\infty \subset V$ converges to $x \in V$ with respect to the norm $|\ |_1$ if and only if it converges to x with respect to the norm $|\ |_2$.

This follows almost immediately from the definitions; the easy proof is left to the reader.

Example 2 Let V be the vector space of all continuous real-valued functions on the closed interval $[a, b]$, with the vector space operations defined by

$$(\varphi + \psi)(x) = \varphi(x) + \psi(x), \qquad (a\varphi)(x) = a\varphi(x),$$

where $\varphi, \psi \in V$ and $a \in \mathscr{R}$. We can define various norms on V, analogous to the norms on $\mathscr{R}^n$ in Example 1, by

$$\|\varphi\|_0 = \max_{x \in [a, b]} |\varphi(x)| \qquad \text{(the sup norm)},$$

$$\|\varphi\|_1 = \int_a^b |\varphi(x)|\, dx \qquad \text{(the 1-norm)},$$

$$\|\varphi\|_2 = \left(\int_a^b |\varphi(x)|^2\, dx\right)^{1/2} \qquad \text{(the 2-norm)}.$$

Again it is an elementary matter to show that $\| \ \|_0$ and $\| \ \|_1$ are indeed norms on V, while the verification that $\| \ \|_2$ satisfies the triangle inequality requires the Cauchy–Schwarz inequality (see the remarks following the proof of Theorem 3.1 in Chapter I). We will denote by $\mathscr{C}^0[a, b]$ the normed vector space V *with the sup norm* $\| \ \|_0$.

No two of the three norms in Example 2 are equivalent. For example, to show that $\| \ \|_0$ and $\| \ \|_1$ are not equivalent, consider the sequence $\{\varphi_n\}_1^\infty$ of elements of V defined by

$$\varphi_n(x) = \begin{cases} 1 - nx & \text{if } x \in \left[0, \dfrac{1}{n}\right], \\ 0 & \text{if } x \in \left[\dfrac{1}{n}, 1\right], \end{cases}$$

and let $\varphi_0(x) = 0$ for $x \in [0, 1]$ (here the closed interval $[a, b]$ is the unit interval $[0, 1]$). Then it is clear that $\{\varphi_n\}_1^\infty$ converges to φ_0 with respect to the 1-norm $\| \ \|_1$ because

$$\|\varphi_n - \varphi_0\|_1 = \int_0^1 |\varphi_n(x) - \varphi_0(x)| \, dx = \frac{1}{2n} \to 0.$$

However $\{\varphi_n\}_1^\infty$ does *not* converge to φ_0 with respect to the sup norm $\| \ \|_0$ because

$$\|\varphi_n - \varphi_0\|_0 = \|\varphi_n\|_0 = 1$$

for all n. It therefore follows from Lemma 1.1 that the norms $\| \ \|_0$ and $\| \ \|_1$ are not equivalent.

Next we want to prove that $\mathscr{C}^0[a, b]$, the vector space of Example 2 with the sup norm, is complete. The normed vector space V is called *complete* if every Cauchy sequence of elements of V converges to some element of V. Just as in $\mathscr{R}^n$, the sequence $\{x_n\}_1^\infty \subset V$ is called a *Cauchy sequence* if, given $\varepsilon > 0$, there exists a positive integer N such that

$$m, n \geqq N \Rightarrow |x_m - x_n| < \varepsilon.$$

These definitions are closely related to the following properties of sequences of functions. Let $\{f_n\}_1^\infty$ be a sequence of real-valued functions, each defined on the set S. Then $\{f_n\}_1^\infty$ is called a *uniformly Cauchy sequence* (of functions on S) if, given $\varepsilon > 0$, there exists N such that

$$m, n \geqq N \Rightarrow |f_m(x) - f_n(x)| < \varepsilon$$

for every $x \in S$. We say that $\{f_n\}_1^\infty$ converges *uniformly* to the function $f: S \to \mathscr{R}$ if, given $\varepsilon > 0$, there exists N such that

$$n \geqq N \Rightarrow |f_n(x) - f(x)| < \varepsilon$$

for every $x \in S$.

Example 3 Consider the sequence $\{f_n\}_1^\infty$, where $f_n : [0, 1] \to \mathscr{R}$ is the function whose graph is shown in Fig. 6.2. Then $\lim_{n\to\infty} f_n(x) = 0$ for every $x \in [0, 1]$. However the sequence $\{f_n\}_1^\infty$ does not converge uniformly to its pointwise limit $f(x) \equiv 0$, because $\|f_n\| = 1$ for every n (sup norm).

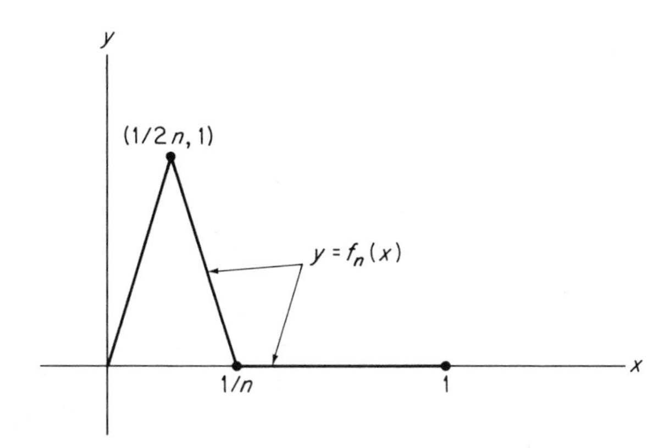

Figure 6.2

Example 4 Let $f_n(x) = x^n$ on $[0, 1]$. Then

$$\lim_{n\to\infty} f_n(x) = f(x) = \begin{cases} 0 & \text{if} \quad x < 1, \\ 1 & \text{if} \quad x = 1, \end{cases}$$

so the pointwise limit function is not continuous. It follows from Theorem 1.2 below that the sequence $\{f_n\}$ does not converge uniformly to f.

Comparing the above definitions, note that a Cauchy sequence in $\mathscr{C}^0[a, b]$ is simply a uniformly Cauchy sequence of continuous functions on $[a, b]$, and that the sequence $\{\varphi_n\}_1^\infty \subset \mathscr{C}^0[a, b]$ converges with respect to the sup norm to $\varphi \in \mathscr{C}^0[a, b]$ if and only if it converges uniformly to φ. Therefore, in order to prove that $\mathscr{C}^0[a, b]$ is complete, we need to show that every uniformly Cauchy sequence of continuous functions on $[a, b]$ converges uniformly to some continuous function on $[a, b]$. The first step is the proof that, if a sequence of continuous functions converges uniformly, then the limit function is continuous.

Theorem 1.2 Let S be a subset of $\mathscr{R}^k$. Let $\{f_n\}_1^\infty$ be a sequence of continuous real-valued functions on S. If $\{f_n\}_1^\infty$ converges uniformly to the function $f: S \to \mathscr{R}$, then f is continuous.

PROOF Given $\varepsilon > 0$, first choose N sufficiently large that $|f_N(x) - f(x)| < \varepsilon/3$ for all $x \in S$ (by the uniform convergence). Then, given $x_0 \in S$, choose $\delta > 0$ such that

$$x \in S, \ |x - x_0| < \delta \Rightarrow |f_N(x) - f_N(x_0)| < \frac{\varepsilon}{3}$$

(by the continuity of f_N at x_0). If $x \in S$ and $|x - x_0| < \delta$, then it follows that

$$\begin{aligned}|f(x) - f(x_0)| &\leqq |f(x) - f_N(x)| + |f_N(x) - f_N(x_0)| + |f_N(x_0) - f(x)| \\ &< \frac{\varepsilon}{3} + \frac{\varepsilon}{3} + \frac{\varepsilon}{3} = \varepsilon,\end{aligned}$$

so f is continuous at x_0. ∎

Corollary 1.3 $\mathscr{C}^0[a, b]$ is complete.

PROOF Let $\{\varphi_n\}_1^\infty$ be a Cauchy sequence of elements of $\mathscr{C}^0[a, b]$, that is, a uniformly Cauchy sequence of continuous real-valued functions on $[a, b]$. Given $\varepsilon > 0$, choose N such that

$$m, n \geqq N \Rightarrow \|\varphi_m - \varphi_n\| < \frac{\varepsilon}{2} \qquad \text{(sup norm)}.$$

Then, in particular, $|\varphi_m(x) - \varphi_n(x)| < \varepsilon/2$ for each $x \in [a, b]$. Therefore $\{\varphi_n(x)\}_1^\infty$ is a Cauchy sequence of real numbers, and hence converges to some real number $\varphi(x)$. It remains to show that the sequence of functions $\{\varphi_n\}$ converges uniformly to φ; if so, Theorem 1.2 will then imply that $\varphi \in \mathscr{C}^0[a, b]$.

We assert that $n \geqq N$ (same N as above, n fixed) implies that

$$|\varphi(x) - \varphi_n(x)| < \varepsilon \qquad \text{for all} \quad x \in [a, b].$$

To see this, choose $m \geqq N$ sufficiently large (depending upon x) that $|\varphi(x) - \varphi_m(x)| < \varepsilon/2$. Then it follows that

$$\begin{aligned}|\varphi(x) - \varphi_n(x)| &\leqq |\varphi(x) - \varphi_m(x)| + |\varphi_m(x) - \varphi_n(x)| \\ &< \frac{\varepsilon}{2} + \|\varphi_m - \varphi_n\| \\ &< \frac{\varepsilon}{2} + \frac{\varepsilon}{2} = \varepsilon.\end{aligned}$$

Since $x \in [a, b]$ was arbitrary, it follows that $\|\varphi_n - \varphi\| < \varepsilon$ as desired. ∎

Example 5 Let $\mathscr{C}^1[a, b]$ denote the vector space of all continuously differentiable real-valued functions on $[a, b]$, with the vector space operations of $\mathscr{C}^0[a, b]$, but with the "$\mathscr{C}^1$-norm" defined by

$$\|\varphi\| = \max_{x \in [a, b]} |\varphi(x)| + \max_{x \in [a, b]} |\varphi'(x)|$$

for $\varphi \in \mathscr{C}^1[a, b]$. We leave it as an exercise for the reader to verify that this does define a norm on $\mathscr{C}^1[a, b]$.

In Section 4 we will need to know that the normed vector space $\mathscr{C}^1[a, b]$ is complete. In order to prove this, we need a result on the termwise-differentiation of a sequence of continuously differentiable functions on $[a, b]$. Suppose that the sequence $\{f_n\}_1^\infty$ converges (pointwise) to f, and that the sequence of derivatives $\{f_n'\}_1^\infty$ converges to g. We would like to know whether $f' = g$. Note that this is the question as to whether

$$f' = D(\lim_{n\to\infty} f_n) \stackrel{?}{=} \lim_{n\to\infty} (Df_n) = g,$$

an "interchange of limits" problem of the sort considered at the end of Section IV.3. The following theorem asserts that this interchange of limits is valid, provided that the convergence of the sequence of derivatives is uniform.

Theorem 1.4 Let $\{f_n\}_1^\infty$ be a sequence of continuously differentiable real-valued functions on $[a, b]$, converging (pointwise) to f. Suppose that the $\{f_n'\}_1^\infty$ converges uniformly to a function g. Then $\{f_n\}_1^\infty$ converges uniformly to f, and f is differentiable, with $f' = g$.

PROOF By the fundamental theorem of calculus, we have

$$f_n(x) = f_n(a) + \int_a^x f_n'$$

for each n and each $x \in [a, b]$. From this and Exercise IV.3.4 (on the termwise-integration of a uniformly convergent sequence of continuous functions) we obtain

$$\begin{aligned} f(x) &= \lim_{n\to\infty} f_n(x) \\ &= \lim_{n\to\infty} f_n(a) + \lim_{n\to\infty} \int_a^x f_n', \end{aligned}$$

$$f(x) = f(a) + \int_a^x g.$$

Another application of the fundamental theorem yields $f' = g$ as desired.

To see that the convergence of $\{f_n\}_1^\infty$ to f is uniform, note that

$$\begin{aligned} |f_n(x) - f(x)| &= \left| \int_a^x f_n' - \int_a^x g \right| + |f_n(a) - f(a)| \\ &\leqq \int_a^x |f_n' - g| + |f_n(a) - f(a)| \\ &\leqq (b-a)\|f_n' - g\|_0 + |f_n(a) - f(a)|. \end{aligned}$$

The uniform convergence of the sequence $\{f_n\}_1^\infty$ therefore follows from that of the sequence $\{f_n'\}_1^\infty$. ∎

Corollary 1.5 $\mathscr{C}^1[a, b]$ is complete.

PROOF Let $\{\varphi_n\}_1^\infty$ be a Cauchy sequence of elements of $\mathscr{C}^1[a, b]$. Since

$$\max_{x\in[a,b]} |\varphi_m(x) - \varphi_n(x)| \leqq \|\varphi_m - \varphi_n\| \qquad (\mathscr{C}^1\text{-norm}),$$

we see that $\{\varphi_n\}_1^\infty$ is a uniformly Cauchy sequence of continuous functions. It follows from Corollary 1.3 that $\{\varphi_n\}_1^\infty$ converges uniformly to a continuous function $\varphi : [a, b] \to \mathscr{R}$. Similarly

$$\max_{x \in [a,b]} |\varphi_m'(x) - \varphi_n'(x)| \leqq \|\varphi_m - \varphi_n\| \qquad (\mathscr{C}^1\text{-norm}),$$

so it follows, in the same way from Corollary 1.3, that $\{\varphi_n'\}_1^\infty$ converges uniformly to a continuous function $\psi : [a, b] \to \mathscr{R}$. Now Theorem 1.4 implies that φ is differentiable with $\varphi' = \psi$, so $\varphi \in \mathscr{C}^1[a, b]$. Since

$$\max_{x \in [a,b]} |\varphi_n(x) - \varphi(x)| = \|\varphi_n - \varphi\|_0$$

and (sup norm),

$$\max_{x \in [a,b]} |\varphi_n'(x) - \varphi'(x)| = \|\varphi_n' - \varphi'\|_0$$

the uniform convergence of the sequences $\{\varphi_n\}_1^\infty$ and $\{\varphi_n'\}_1^\infty$ implies that the sequence $\{\varphi_n\}_1^\infty$ converges to φ with respect to the $\mathscr{C}^1$-norm of $\mathscr{C}^1[a, b]$. Thus every Cauchy sequence in $\mathscr{C}^1[a, b]$ converges. ∎

Example 6 Let $\mathscr{C}^1([a, b]), \mathscr{R}^n)$ denote the vector space of all $\mathscr{C}^1$ paths in $\mathscr{R}^n$ (mappings $\varphi : [a, b] \to \mathscr{R}^n$), with the norm

$$\|\varphi\| = \max_{t \in [a,b]} |\varphi(t)|_0 + \max_{t \in [a,b]} |\varphi'(t)|_0,$$

where $|\ |_0$ denotes the sup norm in $\mathscr{R}^n$. Then it follows, from a coordinatewise application of Corollary 1.5, that the normed vector space $\mathscr{C}^1([a, b], \mathscr{R}^n)$ is complete.

Exercises

1.1 Verify that $\|\ \|_0$, $\|\ \|_1$, and $\|\ \|_2$, as defined in Example 2, are indeed norms on the vector space of all continuous functions defined on $[a, b]$.

1.2 Show that the norms $\|\ \|_1$ and $\|\ \|_2$ of Example 2 are not equivalent. *Hint*: Truncate the function $1/\sqrt{t}$ near 0.

1.3 Let E and F be normed vector spaces with norms $\|\ \|_E$ and $\|\ \|_F$, respectively. Then the the product set $E \times F$ is a vector space, with the vector space operations defined coordinatewise. Prove that the following are equivalent norms on $E \times F$:

$$\|(x, y)\|_0 = \max(\|x\|_E, \|y\|_F),$$
$$\|(x, y)\|_1 = \|x\|_E + \|y\|_F,$$
$$\|(x, y)\|_2 = [(\|x\|_E)^2 + (\|y\|_F)^2]^{1/2}.$$

1.4 If the normed vector spaces E and F are complete, prove that the normed vector space $E \times F$, of the previous exercise, is also complete.

1.5 Show that a closed subspace of a complete normed vector space is complete.

1.6 Denote by $\mathscr{P}_n[a, b]$ the subspace of $\mathscr{C}^0[a, b]$ that consists of all polynomials of degree at most n. Show that $\mathscr{P}_n[a, b]$ is a closed subspace of $\mathscr{C}^0[a, b]$. *Hint*: Associate the polynomial $\varphi(t) = \sum_{k=0}^n a_k t^k$ with the point $\mathbf{a} = (a_0, a_1, \ldots, a_k) \in \mathscr{R}^{k+1}$, and compare $\|\varphi\|_0$ with $|\mathbf{a}|$.

1.7 If $\|\ \|$ is a norm on $\mathscr{R}^n$, prove that the function $\mathbf{x} \to \|\mathbf{x}\|$ is continuous on $\mathscr{R}^n$. *Hint*: If

$$\mathbf{x} = \sum_{i=1}^{n} x_i \mathbf{e}_i,$$

then

$$\|\mathbf{x}\| \leqq \sum_{i=1}^{n} |x_i|\ \|\mathbf{e}_i\|.$$

The Cauchy–Schwarz inequality then gives

$$\|\mathbf{x}\| \leqq M|\mathbf{x}|,$$

where $M = [\sum_{i=1}^{n} \|\mathbf{e}_i\|^2]^{1/2}$ and $|\ |$ is the Euclidean norm on $\mathscr{R}^n$.

2 CONTINUOUS LINEAR MAPPINGS AND DIFFERENTIALS

In this section we discuss the concepts of linearity, continuity, and differentiability for mappings from one normed vector space to another. The definitions here will be simply repetitions of those (in Chapters I and II) for mappings from one Euclidean space to another.

Let E and F be vector spaces. Recall that the mapping $\varphi : E \to F$ is called *linear* if

$$\varphi(ax + by) = a\varphi(x) + b\varphi(y)$$

for all $x, y \in E$ and $a, b \in \mathscr{R}$.

Example 1 The real-valued function $I : \mathscr{C}^0[a, b] \to \mathscr{R}$, defined on the vector space of all continuous functions on $[a, b]$ by

$$I(f) = \int_a^b f,$$

is clearly a linear mapping from $\mathscr{C}^0[a, b]$ to $\mathscr{R}$.

If the vector spaces E and F are normed, then we can talk about limits (of mappings from E to F). Given a mapping $f : E \to F$ and $x_0 \in E$, we say that

$$\lim_{x \to x_0} f(x) = L \in F$$

if, given $\varepsilon > 0$, there exists $\delta > 0$ such that

$$0 < |x - x_0| < \delta \Rightarrow |f(x) - L| < \varepsilon.$$

The mapping f is *continuous* at $x_0 \in E$ if

$$\lim_{x \to x_0} f(x) = f(x_0).$$

We saw in Section I.7 (Example 8) that every linear mapping between Euclidean spaces is continuous (everywhere). However, for mappings between infinite-dimensional normed vector spaces, linearity does not, in general, imply continuity.

Example 2 Let V be the vector space of all continuous real-valued functions on [0, 1], as in Example 2 of Section 1. Let E_1 denote V with the 1-norm $\| \ \|_1$, and let E_0 denote V with the sup norm $\| \ \|_0$. Consider the identity mapping $\lambda : V \to V$ as a mapping from E_1 to E_0,

$$\lambda : E_1 \to E_0 .$$

We inquire as to whether λ is continuous at $0 \in E_1$ (the constant zero function on [0, 1]). If it were, then, given $\varepsilon > 0$, there would exist $\delta > 0$ such that

$$\|\varphi\|_1 < \delta \Rightarrow \|\varphi\|_0 < \varepsilon.$$

However we saw in Section 1 that, given $\delta > 0$, there exists a function $\varphi : [0, 1] \to \mathscr{R}$ such that

$$\|\varphi\|_1 < \delta \qquad \text{while} \qquad \|\varphi\|_0 = 1.$$

It follows that $\lambda : E_1 \to E_0$ is *not* continuous at $0 \in E_1$.

The following theorem provides a useful criterion for continuity of linear mappings.

Theorem 2.1 Let $L : E \to F$ be a linear mapping where E and F are normed vector spaces. Then the following three conditions on L are equivalent:

(a) There exists a number $c > 0$ such that

$$|L(v)| \leqq C|v|$$

for all $v \in E$.

(b) L is continuous (everywhere).

(c) L is continuous at $0 \in E$.

PROOF Suppose first that (a) holds. Then, given $x_0 \in E$ and $\varepsilon > 0$,

$$\begin{aligned} |x - x_0| < \frac{\varepsilon}{c} \Rightarrow |L(x) - L(x_0)| &= |L(x - x_0)| \\ &\leqq c|x - x_0| \\ &< \varepsilon, \end{aligned}$$

so it follows that L is continuous at x_0. Thus (a) implies (b).

To see that (c) implies (a), assume that L is continuous at 0, and choose $\delta > 0$ such that $|x| \leqq \delta$ implies $|L(x)| < 1$. Then, given $v \neq 0 \in E$, it follows that

$$\begin{aligned} |L(v)| &= L\left(\frac{|v|}{\delta} \cdot \frac{\delta}{|v|} v\right) \\ &= \frac{|v|}{\delta} L\left(\frac{\delta}{|v|} v\right) \qquad \left(\text{take } x = \frac{\delta}{|v|} v\right) \\ &< \frac{1}{\delta} |v|, \end{aligned}$$

so we may take $c = 1/\delta$. ∎

Example 3 Let E_0 and E_1 be the normed vector spaces of Example 2, but this time consider the identity mapping on their common underlying vector space V as a linear mapping from E_0 to E_1,

$$\mu : E_0 \to E_1.$$

Since

$$\|\varphi\|_1 = \int_0^1 |\varphi(t)|\, dt \leqq \max_{t \in [0,1]} |\varphi(t)| = \|\varphi\|_0,$$

we see from Theorem 2.1 (with $c = 1$) that μ is continuous.

Thus the inverse of a one-to-one continuous linear mapping of one normed vector space onto another need not be continuous. Let $L : E \to F$ be a linear mapping which is both one-to-one and surjective (that is, $L(E) = F$). Then we will call L an *isomorphism* (of normed vector spaces) if and only if both L and L^{-1} are continuous.

Example 4 Let $A, B : [a, b] \to \mathscr{R}^n$ be continuous paths in $\mathscr{R}^n$, and consider the linear function

$$L : \mathscr{C}^1([a, b], \mathscr{R}^n) \to \mathscr{R}$$

defined by

$$L(\varphi) = \int_a^b [A(t) \cdot \varphi(t) + B(t) \cdot \varphi'(t)]\, dt,$$

where the dot denotes the usual inner product in $\mathscr{R}^n$. Then the Cauchy–Schwarz inequality yields

$$\begin{aligned} |L(\varphi)| &= \left| \int_a^b [A(t) \cdot \varphi(t) + B(t) \cdot \varphi'(t)]\, dt \right| \\ &\leqq \int_a^b (|A(t)|\, |\varphi(t)| + |B(t)|\, |\varphi'(t)|)\, dt \\ &\leqq (b - a)(\|A\|_0 + \|B\|_0)\|\varphi\|_1, \end{aligned}$$

so an application of Theorem 2.1, with $c = (b - a)(\|A\|_0 + \|B\|_0)$, shows that L is continuous.

We are now prepared to discuss differentials of mappings of normed vector spaces. The definition of differentiability, for mappings of normed vector spaces, is the same as its definition for mappings of Euclidean spaces, except that we must explicitly require the approximating linear mapping to be continuous. The mapping $f: E \to F$ is *differentiable* at $x \in E$ if and only if there exists a *continuous* linear mapping $L : E \to F$ such that

$$\lim_{h \to 0} \frac{f(x + h) - f(x) - L(h)}{|h|} = 0. \tag{1}$$

The continuous linear mapping L, if it exists, is unique (Exercise 2.3), and it is easily verified that a linear mapping L satisfying (1) is continuous at 0 if and only if f is continuous at x (Exercise 2.4).

If $f: E \to F$ is differentiable at $x \in E$, then the continuous linear mapping which satisfies (1) is called the *differential* of f at x, and is denoted by

$$df_x : E \to F.$$

In the finite-dimensional case $E = \mathscr{R}^n$, $F = \mathscr{R}^m$ that we are already familiar with, the $m \times n$ matrix of the linear mapping df_x is the derivative $f'(x)$. Here we will be mainly interested in mappings between infinite-dimensional spaces, whose differential linear mappings are not representable by matrices, so derivatives will not be available.

Example 5 If $f: E \to F$ is a continuous linear mapping, then

$$\frac{f(x + h) - f(x) - f(h)}{|h|} = 0$$

for all x, $h \in E$ with $h \neq 0$, so f is differentiable at x, with $df_x = f$. Thus a continuous linear mapping is its own differential (at every point $x \in E$).

Example 6 We now give a less trivial computation of a differential. Let $g : \mathscr{R} \to \mathscr{R}$ be a $\mathscr{C}^2$ function, and define

$$f: \mathscr{C}^0[a, b] \to \mathscr{C}^0[a, b]$$

by

$$f(\varphi) = g \circ \varphi.$$

We want to show that f is differentiable, and to compute its differential.

If f is differentiable at $\varphi \in \mathscr{C}^0[a, b]$, then $df_\varphi(h)$ should be the linear (in $h \in \mathscr{C}^0[a, b]$) part of $f(\varphi + h) - f(\varphi)$. To investigate this difference, we write down the second degree Taylor expansion of g at $\varphi(t)$:

$$g(\varphi(t) + h(t)) - g(\varphi(t)) = g'(\varphi(t))h(t) + R(h)(t),$$

where

$$R(h)(t) = \tfrac{1}{2}g''(\xi(t))(h(t))^2 \tag{2}$$

for some $\xi(t)$ between $\varphi(t)$ and $\varphi(t) + h(t)$. Then

$$f(\varphi + h) - f(\varphi) = L(h) + R(h),$$

where $L(h) \in \mathscr{C}^0[a, b]$ is defined by

$$L(h)(t) = g'(\varphi(t))h(t).$$

It is clear that $L : \mathscr{C}^0[a, b] \to \mathscr{C}^0[a, b]$ is a continuous linear mapping, so in order to prove that f is differentiable at φ with $df_\varphi = L$, it suffices to prove that

$$\lim_{h \to 0} \frac{R(h)}{\|h\|_0} = 0. \tag{3}$$

Note that, since g is a $\mathscr{C}^2$ function, there exists $M > 0$ such that $|g''(\xi(t))| < 2M$ when $\|h\|_0$ is sufficiently small (why?). It then follows from (2) that

$$\frac{|R(h)(t)|}{\|h\|_0} < \frac{M|h(t)|^2}{\|h\|_0} \leqq M\|h\|_0,$$

and this implies (3) as desired. Thus the differential $df_\varphi : \mathscr{C}^0[a, b] \to \mathscr{C}^0[a, b]$ of f at φ is defined by

$$df_\varphi(h)(t) = g'(\varphi(t))h(t).$$

The *chain rule* for mappings of normed vector spaces takes the expected form.

Theorem 2.2 Let U and V be open subsets of the normed vector spaces E and F respectively. If the mappings $f : U \to F$ and $g : V \to G$ (a third normed vector space) are differentiable at $x \in U$ and $f(x) \in V$ respectively, then their composition $h = g \circ f$ is differentiable at x, and

$$dh_x = dg_{f(x)} \circ df_x .$$

The proof is precisely the same as that of the finite-dimensional chain rule (Theorem II.3.1), and will not be repeated.

There is one case in which derivatives (rather than differentials) are important. Let $\varphi : \mathscr{R} \to E$ be a path in the normed vector space E. Then the familiar limit

$$\varphi'(t) = \lim_{h \to 0} \frac{\varphi(t + h) - \varphi(t)}{h} \in E,$$

if it exists, is the *derivative* or *velocity vector* of φ at t. It is easily verified that $\varphi'(t)$ exists if and only if φ is differentiable at t, in which case $d\varphi_t(h) = \varphi'(t)h$ (again, just as in the finite-dimensional case).

In the following sections we will be concerned with the problem of minimizing (or maximizing) a differentiable real-valued function $f: E \to \mathscr{R}$ on a subset M of the normed vector space E. In order to state the result which will play the role that Lemma II.5.1 does in finite-dimensional maximum–minimum problems, we need the concept of tangent sets. Given a subset M of the normed vector space E, the *tangent set* TM_x of M at the point $x \in M$ is the set of all those vectors $v \in E$, for which there exists a differentiable path $\varphi : \mathscr{R} \to M \subset E$ such that $\varphi(0) = x$ and $\varphi'(0) = v$. Thus TM_x is simply the set of all velocity vectors at x of differentiable paths in M which pass through x. Hence the tangent set of an arbitrary subset of a normed vector space is the natural generalization of the tangent space of a submanifold of $\mathscr{R}^n$.

The following theorem gives a necessary condition for local maxima or local minima (we state it for minima).

Theorem 2.3 Let the function $f: E \to \mathscr{R}$ be differentiable at the point x of the subset M of the normed vector space E. If $f(v) \geqq f(x)$ for all $v \in M$ sufficiently close to x, then

$$df_x \mid TM_x = 0.$$

That is, $df_x(v) = 0$ for all $v \in TM_x$.

PROOF Given $v \in TM_x$, let $\varphi : \mathscr{R} \to M$ be a differentiable path in E such that $\varphi(0) = x$ and $\varphi'(0) = v$. Then the composition $g = f \circ \varphi : \mathscr{R} \to \mathscr{R}$ has a local minimum at 0, so $g'(0) = 0$. The chain rule therefore gives

$$\begin{aligned} 0 = g'(0) = dg_0(1) &= df_{\varphi(0)}(d\varphi_0(1)) \\ &= df_x(\varphi'(0)), \\ 0 &= df_x(v) \end{aligned}$$

as desired. ∎

In Section 4 we will need the implicit mapping theorem for mappings of *complete* normed vector spaces. Both its statement and its proof, in this more general context, are essentially the same as those of the finite-dimensional implicit mapping theorem in Chapter III.

For the statement, we need to define the *partial* differentials of a differentiable mapping which is defined on the product of two normed vector spaces E and F. The product set $E \times F$ is made into a vector space by defining

$$(x_1, y_1) + (x_2, y_2) = (x_1 + x_2, y_1 + y_2) \qquad \text{and} \qquad a(x, y) = (ax, ay).$$

If $| \ |_E$ and $| \ |_F$ denote the norms on E and F, respectively, then

$$|(x, y)| = \max(|x|_E, |y|_F)$$

defines a norm on $E \times F$, and $E \times F$ is complete if both E and F are (Exercise 1.4). For instance, if $E = F = \mathscr{R}$ with the ordinary norm (absolute value), then this definition gives the sup norm on the plane $\mathscr{R} \times \mathscr{R} = \mathscr{R}^2$ (see Example 1 of Section 1).

Now let the mapping

$$f: E \times F \to G$$

be differentiable at the point $(a, b) \in E \times F$. It follows easily (Exercise 2.5) that the mappings

$$\varphi : E \to G \qquad \text{and} \qquad \psi : F \to G,$$

defined by

$$\varphi(x) = f(x, b) \qquad \text{and} \qquad \psi(y) = f(a, y), \tag{4}$$

are differentiable at $a \in E$ and $b \in F$, respectively. Then $d_x f_{(a, b)}$ and $d_y f_{(a, b)}$, the *partial differentials* of f at (a, b), with respect to $x \in E$ and $y \in F$, respectively, are defined by

$$d_x f_{(a, b)} = d\varphi_a \qquad \text{and} \qquad d_y f_{(a, b)} = d\psi_b .$$

Thus $d_x f_{(a, b)}$ is the differential of the mapping $E \to G$ obtained from $f: E \times F \to G$ by fixing $y = b$, and $d_y f_{(a, b)}$ is obtained similarly by holding x fixed. This generalizes our definition in Chapter III of the partial differentials of a mapping from $\mathscr{R}^{m+n} = \mathscr{R}^m \times \mathscr{R}^n$ to $\mathscr{R}^k$.

With this notation and terminology, the statement of the implicit mapping theorem is as follows.

Implicit Mapping Theorem *Let $f: E \times F \to G$ be a $\mathscr{C}^1$ mapping, where E, F, and G are complete normed vector spaces. Suppose that $f(a, b) = 0$, and that*

$$d_y f_{(a, b)} : F \to G$$

is an isomorphism. Then there exists a neighborhood U of a in E, a neighborhood W of (a, b) in $E \times F$, and a $\mathscr{C}^1$ mapping $\varphi : U \to F$ such that the following is true: If $(x, y) \in W$ and $x \in U$, then $f(x, y) = 0$ if and only if $y = \varphi(x)$.

This statement involves $\mathscr{C}^1$ mappings from one normed vector space to another, whereas we have defined only differentiable ones. The mapping $g : E \to F$ of normed vector spaces is called *continuously differentiable*, or $\mathscr{C}^1$, if it is differentiable and $dg_x(v)$ is a continuous function of (x, v), that is, the mapping $(x, y) \to dg_x(v)$ from $E \times E$ to F is continuous.

As previously remarked, the proof of the implicit mapping theorem in complete normed vector spaces is essentially the same as its proof in the finite-dimensional case. In particular, it follows from the inverse mapping theorem for complete normed vector spaces in exactly the same way that the finite-dimensional implicit mapping theorem follows from the finite-dimensional inverse mapping theorem (see the proof of Theorem III.3.4). The general inverse mapping theorem is identical to the finite-dimensional case (Theorem III.3.3), except that Euclidean space $\mathscr{R}^n$ is replaced by a complete normed vector space E. Moreover the proof is essentially the same, making use of the contraction mapping theorem. Finally, the only property of $\mathscr{R}^n$ that was used in the proof of contraction mapping theorem, is that it is complete. It would be instructive for the student to reread the proofs of these three theorems in Chapter III, verifying that they generalize to complete normed vector spaces.

Exercises

2.1 Show that the function $I(f) = \int_a^b f$ is continuous on $\mathscr{C}^0[a, b]$.

2.2 If $L : \mathscr{R}^n \to E$ is a linear mapping of $\mathscr{R}^n$ into a normed vector space, show that L is continuous.

2.3 Let $f: E \to F$ be differentiable at $x \in E$, meaning that there exists a continuous linear mapping $L : E \to F$ satisfying Eq. (1) of this section. Show that L is unique.

2.4 Let $f: E \to F$ be a mapping and $L : E \to F$ a linear mapping satisfying Eq. (1). Show that L is continuous if and only if f is continuous at x.

2.5 Let $f: E \times F \to G$ be a differentiable mapping. Show that the restrictions $\varphi : E \to G$ and $\psi : F \to G$ of Eq. (4) are differentiable.

2.6 If the mapping $f: E \times F \to G$ is differentiable at $p \in E \times F$, show that $d_x f_p(r) = df_p(r, 0)$ and $d_y f_p(s) = df_p(0, s)$.

2.7 Let $M \subset E$ be a translate of the closed subspace V of the normed vector space E. That is, given $x \in M$, $M = \{x + y : y \in V\}$. Then prove that $TM_x = V$.

3 THE SIMPLEST VARIATIONAL PROBLEM

We are now prepared to discuss the first problem mentioned in the introduction to this chapter. Given $f: \mathscr{R}^3 \to \mathscr{R}$, we seek to minimize (or maximize) the function $F : \mathscr{C}^1[a, b] \to \mathscr{R}$ defined by

$$F(\psi) = \int_a^b f(\psi(t), \psi'(t), t)\, dt, \tag{1}$$

amongst those functions $\psi \in \mathscr{C}^1[a, b]$ such that $\psi(a) = \alpha$ and $\psi(b) = \beta$ (where α and β are given real numbers).

Let M denote the subset of $\mathscr{C}^1[a, b]$ consisting of those functions ψ that satisfy the endpoint conditions $\psi(a) = \alpha$ and $\psi(b) = \beta$. Then we are interested in the local extrema of F on the subset M. If F is differentiable at $\varphi \in M$, and $F \mid M$ has a local extremum at φ, then Theorem 2.3 implies that

$$dF_\varphi \mid TM_\varphi = 0. \tag{2}$$

We will say that the function $\varphi \in M$ is an *extremal* for F on M if it satisfies the *necessary* condition (2). We will not consider here the difficult matter of finding sufficient conditions under which an extremal actually yields a local extremum.

In order that we be able to ascertain whether a given function $\varphi \in M$ is or is not an extremal for F on M, that is, whether or not $dF_\varphi | TM_\varphi = 0$, we must (under appropriate conditions on f) explicitly compute the differential dF_φ of F at φ, and we must determine just what the tangent set TM_φ of M at φ is.

The latter problem is quite easy. First pick a fixed element $\varphi_0 \in M$. Then, given any $\varphi \in M$, the difference $\varphi - \varphi_0$ is an element of the subspace

$$\mathscr{C}_0{}^1[a, b] = \{\psi \in \mathscr{C}^1[a, b] : \psi(a) = \psi(b) = 0\}$$

of $\mathscr{C}^1[a, b]$, consisting of those $\mathscr{C}^1$ functions on $[a, b]$ that are zero at both endpoints. Conversely, if $\psi \in \mathscr{C}_0{}^1[a, b]$, then clearly $\varphi_0 + \psi \in M$. Thus M is a *hyperplane* in $\mathscr{C}^1[a, b]$, namely the translate by φ_0 of the subspace $\mathscr{C}_0{}^1[a, b]$. But the tangent set of a hyperplane is, at every point, simply the subspace of which it is a translate (see Exercise 2.7). Therefore

$$TM_\varphi = \mathscr{C}_0{}^1[a, b] \tag{3}$$

for every $\varphi \in M$.

The following theorem gives the computation of dF_φ.

Theorem 3.1 Let $F : \mathscr{C}^1[a, b] \to \mathscr{R}$ be defined by (1), with $f : \mathscr{R}^3 \to \mathscr{R}$ being a $\mathscr{C}^2$ function. Then F is differentiable with

$$dF_\varphi(h) = \int_a^b \left[\frac{\partial f}{\partial x}(\varphi(t), \varphi'(t), t)h(t) + \frac{\partial f}{\partial y}(\varphi(t), \varphi'(t), t)h'(t)\right] dt \tag{4}$$

for all $\varphi, h \in \mathscr{C}^1[a, b]$.

In the use of partial derivative notation on the right-hand side of (4), we are thinking of $(x, y, t) \in \mathscr{R}^3$.

PROOF If F is differentiable at $\varphi \in \mathscr{C}^1[a, b]$, then $dF_\varphi(h)$ should be the linear (in h) part of $F(\varphi + h) - F(\varphi)$. To investigate this difference, we write down the second degree Taylor expansion of f at the point $(\varphi(t), \varphi'(t), t) \in \mathscr{R}^3$:

$$\begin{aligned} &f(\varphi(t) + h(t), \varphi'(t) + h'(t), t) - f(\varphi(t), \varphi'(t), t) \\ &\quad = \frac{\partial f}{\partial x}(\varphi(t), \varphi'(t), t)h(t) + \frac{\partial f}{\partial y}(\varphi(t), \varphi'(t), t)h'(t) + r(h(t)), \end{aligned} \tag{5}$$

where

$$r(h(t)) = \frac{1}{2!}\left[\frac{\partial^2 f}{\partial x^2}(\xi(t))(h(t))^2 + 2\frac{\partial^2 f}{\partial x\,\partial y}(\xi(t))h(t)h'(t) + \frac{\partial^2 f}{\partial y^2}(\xi(t))(h'(t))^2\right]$$

for some point $\xi(t)$ of the line segment in $\mathscr{R}^3$ from $(\varphi(t), \varphi'(t), t)$ to $(\varphi(t) + h(t), \varphi'(t) + h'(t), t)$. If B is a large ball in $\mathscr{R}^3$ that contains in its interior the image of the continuous path $t \to (\varphi(t), \varphi'(t), t)$, $t \in [a, b]$, and M is the maximum of the absolute values of the second order partial derivatives of f at points of B, then it follows easily that

$$|r(h(t))| \leqq \frac{M}{2}(|h(t)| + |h'(t)|)^2 \tag{6}$$

for all $t \in [a, b]$, if $\|h\|_0$ is sufficiently small.

From (5) we obtain

$$F(\varphi + h) - F(\varphi) = L(h) + R(h),$$

with

$$L(h) = \int_a^b \left[\frac{\partial f}{\partial x}(\varphi(t), \varphi'(t), t)h(t) + \frac{\partial f}{\partial y}(\varphi(t), \varphi'(t), t)h'(t)\right] dt$$

and

$$R(h) = \int_D^b r(h(t))dt.$$

In order to prove that F is differentiable at φ with $dF_\varphi = L$ as desired, it suffices to note that $L : \mathscr{C}^1[a, b] \to \mathscr{R}$ is a continuous linear function (by Example 4 of Section 2), and then to show that,

$$\lim_{\|h\|_1 \to 0} \frac{|R(h)|}{\|h\|_1} = 0. \tag{7}$$

But it follows immediately from (6) that

$$\begin{aligned} |R(h)| &\leqq \int_a^b |r(h(t))| \, dt \\ &\leqq \int_a^b \frac{M}{2}(2\|h\|_1)^2 \, dt = 2M(b - a)(\|h\|_1)^2, \end{aligned}$$

and this implies (7). ∎

With regard to condition (2), we are interested in the value of $dF_\varphi(h)$ when $h \in TM_\varphi = \mathscr{C}_0^{\ 1}[a, b]$.

Corollary 3.2 Assume, in addition to the hypotheses of Theorem 3.1, that φ is a $\mathscr{C}^2$ function and that $h \in \mathscr{C}_0^{\ 1}[a, b]$. Then

$$dF_\varphi(h) = \int_a^b \left[\frac{\partial f}{\partial x}(\varphi(t), \varphi'(t), t) - \frac{d}{dt}\frac{\partial f}{\partial y}(\varphi(t), \varphi'(t), t)\right] h(t) \, dt. \tag{8}$$

PROOF If φ is a $\mathscr{C}^2$ function, then $\partial f/\partial y(\varphi(t), \varphi'(t), t)$ is a $\mathscr{C}^1$ function of $t \in [a, b]$. A simple integration by parts therefore gives

$$\int_a^b \frac{\partial f}{\partial y}(\varphi(t), \varphi'(t), t)h'(t)\,dt$$

$$= \left[\frac{\partial f}{\partial y}(\varphi(t), \varphi'(t), t)h(t)\right]_a^b - \int_a^b \frac{d}{dt}\left(\frac{df}{\partial y}(\varphi(t), \varphi'(t), t)\right) h(t)\,dt$$

$$= -\int_a^b \frac{d}{dt}\left(\frac{\partial f}{\partial y}(\varphi(t), \varphi'(t), t)\right)h(t)\,dt$$

because $h(a) = h(b) = 0$. Thus formula (8) follows from formula (4) in this case. ∎

This corollary shows that the $\mathscr{C}^2$ function $\varphi \in M$ is an extremal for F on M if and only if

$$\int_a^b \left[\frac{\partial f}{\partial x}(\varphi(t), \varphi'(t), t) - \frac{d}{dt}\frac{\partial f}{\partial y}(\varphi(t), \varphi'(t), t)\right] h(t)\,dt = 0 \tag{9}$$

for every $\mathscr{C}^1$ function h on $[a, b]$ such that $h(a) = h(b) = 0$. The following lemma verifies the natural guess that (9) can hold for *all* such h only if the function within the brackets in (9) vanishes identically on $[a, b]$.

Lemma 3.3 If $\varphi : [a, b] \to \mathscr{R}$ is a continuous function such that

$$\int_a^b \varphi(t)h(t)\,dt = 0$$

for every $h \in \mathscr{C}_0{}^1[a, b]$, then φ is identically zero on $[a, b]$.

PROOF Suppose, to the contrary, that $\varphi(t_0) \neq 0$ for some $t_0 \in [a, b]$. Then, by continuity, φ is nonzero on some interval containing t_0, say, $\varphi(t) > 0$ for $t \in [t_1, t_2] \subset [a, b]$. If h is defined on $[a, b]$ by

$$h(t) = \begin{cases} (t - t_1)^2(t - t_2)^2 & \text{if} \quad t \in [t_1, t_2], \\ 0 & \text{otherwise}, \end{cases}$$

then $h \in \mathscr{C}_0{}^1[a, b]$, and

$$\int_a^b \varphi(t)h(t)\,dt = \int_{t_1}^{t_2} \varphi(t)(t - t_1)^2(t - t_2)^2\,dt > 0,$$

because the integrand is positive except at the endpoints t_1 and t_2 (Fig. 6.3). This contradiction proves that $\varphi \equiv 0$ on $[a, b]$. ∎

The fundamental necessity condition for extremals, the Euler–Lagrange equation, follows immediately from Corollary 3.2 and Lemma 3.3.

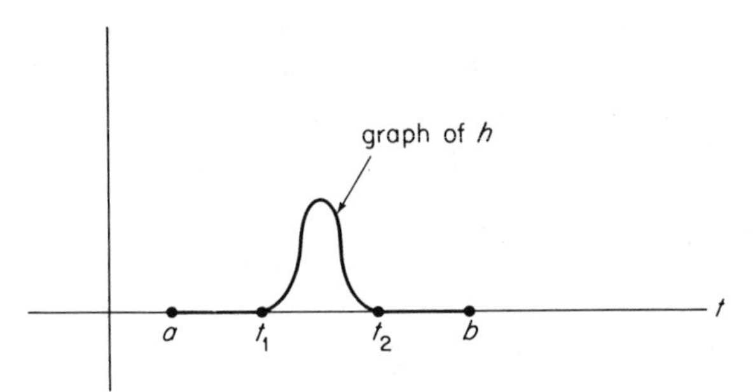

Figure 6.3

Theorem 3.4 Let $F : \mathscr{C}^1[a, b] \to \mathscr{R}$ be defined by (1), with $f : \mathscr{R}^3 \to \mathscr{R}$ being a $\mathscr{C}^2$ function. Then the $\mathscr{C}^2$ function $\varphi \in M$ is an extremal for F on $M = \{\psi \in \mathscr{C}^1[a, b] : \psi(a) = \alpha \text{ and } \psi(b) = \beta\}$ if and only if

$$\frac{\partial f}{\partial x}(\varphi(t), \varphi'(t), t) - \frac{d}{dt}\frac{\partial f}{\partial y}(\varphi(t), \varphi'(t), t) = 0 \tag{10}$$

for all $t \in [a, b]$.

Equation (10) is the *Euler–Lagrange equation* for the extremal φ. Note that it is actually a second order (ordinary) differential equation for φ, since the chain rule gives

$$\frac{d}{dt}\frac{\partial f}{\partial y} = \frac{\partial^2 f}{\partial x\, \partial y}\varphi' + \frac{\partial^2 f}{\partial y^2}\varphi'' + \frac{\partial^2 f}{\partial t\, \partial y},$$

where the partial derivatives of f are evaluated at $(\varphi(t), \varphi'(t), t) \in \mathscr{R}^3$.

REMARKS The hypothesis in Theorem 3.4 that the extremal φ is a $\mathscr{C}^2$ function (rather than merely $\mathscr{C}^1$) is actually unnecessary. First, if φ is an extremal which is only assumed to be $\mathscr{C}^1$ then, by a more careful analysis, it can still be proved that $\partial f/\partial y(\varphi(t), \varphi'(t), t)$ is a differentiable function (of t) satisfying the Euler–Lagrange equation. Second, if φ is a $\mathscr{C}^1$ extremal such that

$$\frac{\partial^2 f}{\partial x\, \partial y}(\varphi(t), \varphi'(t), t) \neq 0$$

for all $t \in [a, b]$, then it can be proved that φ is, in fact, a $\mathscr{C}^2$ function. We will not include these refinements because Theorem 3.4 as stated, with the additional hypothesis that φ is $\mathscr{C}^2$, will suffice for our purposes.

We illustrate the applications of the Euler–Lagrange equation with two standard first examples.

Example 1 We consider a special case of the problem of finding the path of minimal length joining the points (a, α) and (b, β) in the tx-plane. Suppose in particular that $\varphi : [a, b] \to \mathscr{R}$ is a $\mathscr{C}^2$ function with $\varphi(a) = \alpha$ and $\varphi(b) = \beta$ whose graph has minimal length, in comparison with the graphs of all other such functions. Then φ is an extremal (subject to the endpoint conditions) of the function

$$F(\psi) = \int_a^b [1 + (\psi'(t))^2]^{1/2}\, dt,$$

whose integrand function is

$$f(x, y, t) = (1 + y^2)^{1/2}.$$

Since $\partial f/\partial x = 0$ and $\partial f/\partial y = y/(1 + y^2)^{1/2}$, the Euler–Lagrange equation for φ is therefore

$$\frac{d}{dt} \frac{\varphi'}{[1 + (\varphi')^2]^{1/2}} = 0,$$

which upon computation reduces to

$$\frac{\varphi''}{(1 + (\varphi')^2)^{3/2}} = 0.$$

Therefore $\varphi'' = 0$ on $[a, b]$, so φ is a linear function on $[a, b]$, and its graph is (as expected) the straight line segment from (a, α) to (b, β).

Example 2 We want to minimize the area of a surface of revolution. Suppose in particular that $\varphi : [a, b] \to \mathscr{R}$ is a $\mathscr{C}^2$ function with $\varphi(a) = \alpha$ and $\varphi(b) = \beta$, such that the surface obtained by revolving the curve $x = \varphi(t)$ about the t-axis has minimal area, in comparison with all other surfaces of revolution obtained in this way (subject to the endpoint conditions). Then φ is an extremal of the function

$$F(\psi) = \int_a^b 2\pi\psi(t)[1 + (\psi'(t))^2]^{1/2}\, dt,$$

whose integrand function is

$$f(x, y, t) = 2\pi x(1 + y^2)^{1/2}.$$

Here

$$\frac{\partial f}{\partial x} = 2\pi(1 + y^2)^{1/2} \qquad \text{and} \qquad \frac{\partial f}{\partial y} = \frac{2\pi x y}{(1 + y^2)^{1/2}}.$$

Upon substituting $x = \varphi(t)$, $y = \varphi'(t)$ into the Euler–Lagrange equation, and simplifying, we obtain

$$\frac{\partial f}{\partial x} - \frac{d}{dt}\frac{\partial f}{\partial y} = 2\pi \frac{1 + (\varphi')^2 - \varphi\varphi''}{[1 + (\varphi')^2]^{3/2}}.$$

It follows that

$$\frac{\varphi}{[1 + (\varphi')^2]^{1/2}} = c \qquad \text{(constant)}$$

(differentiate the latter equation), or

$$\varphi' = \left(\frac{\varphi^2 - c^2}{c^2}\right)^{1/2}.$$

The general solution of this first order equation is

$$\varphi(t) = c \cosh \frac{t + d}{c}, \tag{11}$$

where d is a second constant. Thus the curve $x = \varphi(t)$ is a *catenary* (Fig. 6.4) passing through the given points (a, α) and (b, β).

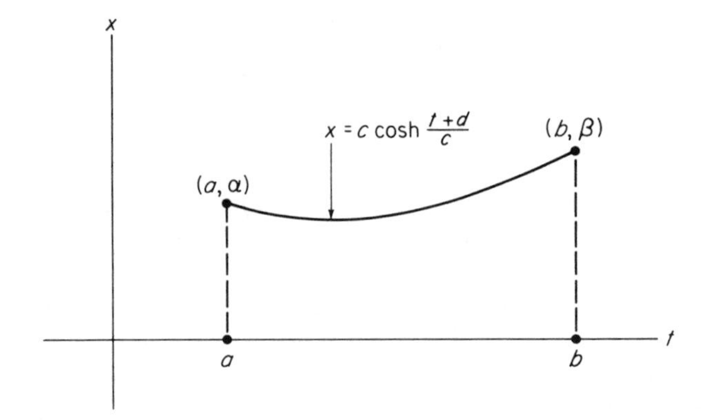

Figure 6.4

It can be shown that, if $b - a$ is sufficiently large compared to α and β, then *no* catenary of the form (11) passes through the given points (a, α) and (b, β), so in this case there will not exist a smooth extremal.

This serves to emphasize the fact that the Euler–Lagrange equation merely provides a necessary condition that a given function φ maximize or minimize the given integral functional F. It may happen either that there exist no extremals (solutions of the Euler–Lagrange equation that satisfy the endpoint conditions), or that a given extremal does not maximize or minimize F (just as a critical point of a function on $\mathscr{R}^n$ need not provide a maximum or minimum).

All of our discussion thus far can be generalized from the real-valued to the vector-valued case, that is obtained by replacing the space $\mathscr{C}^1[a, b]$ of real-valued functions with the space $\mathscr{C}^1([a, b], \mathscr{R}^n)$ of $\mathscr{C}^1$ paths in $\mathscr{R}^n$. The proofs are all essentially the same, aside from the substitution of vector notation for scalar notation, so we shall merely outline the results.

Given a $\mathscr{C}^2$ function $f: \mathscr{R}^n \times \mathscr{R}^n \times \mathscr{R} \to \mathscr{R}$, we are interested in the extrema of the function $F: \mathscr{C}^1([a, b], \mathscr{R}^n) \to \mathscr{R}$ defined by

$$F(\psi) = \int_a^b f(\psi(t), \psi'(t), t)\, dt, \tag{12}$$

amongst those $\mathscr{C}^1$ paths $\psi : [a, b] \to \mathscr{R}^n$ such that $\psi(a) = \alpha$ and $\psi(b) = \beta$, where α and β are given points in $\mathscr{R}^n$.

Denoting by M the subset of $\mathscr{C}^1([a, b], \mathscr{R}^n)$ consisting of those paths that satisfy the endpoint conditions, the path $\varphi \in M$ is an *extremal* for F on M if and only if

$$dF_\varphi \mid TM_\varphi = 0.$$

We find, just as before, that M is a hyperplane. For each $\varphi \in M$,

$$TM_\varphi = \mathscr{C}_0{}^1([a, b], \mathscr{R}^n) = \{\psi \in \mathscr{C}^1([a, b], \mathscr{R}^n) : \psi(a) = \psi(b) = 0\}.$$

With the notation $(\mathbf{x}, \mathbf{y}, t) \in \mathscr{R}^n \times \mathscr{R}^n \times \mathscr{R}$, let us write

$$\frac{\partial f}{\partial \mathbf{x}} = \left(\frac{\partial f}{\partial x_1}, \ldots, \frac{\partial f}{\partial x_n}\right) \qquad \text{and} \qquad \frac{\partial f}{\partial \mathbf{y}} = \left(\frac{\partial f}{\partial y_1}, \ldots, \frac{\partial f}{\partial y_n}\right),$$

so $\partial f/\partial \mathbf{x}$ and $\partial f/\partial \mathbf{y}$ are vectors. If φ is a $\mathscr{C}^2$ path in $\mathscr{R}^n$ and $h \in \mathscr{C}_0{}^1([a, b], \mathscr{R}^n)$, then we find (by generalizing the proofs of Theorem 3.1 and Corollary 3.2) that

$$dF_\varphi(h) = \int_a^b \left[\frac{\partial f}{\partial \mathbf{x}}(\varphi(t), \varphi'(t), t) - \frac{d}{dt}\frac{\partial f}{\partial \mathbf{y}}(\varphi(t), \varphi'(t), t)\right] \cdot h(t)\, dt. \tag{13}$$

Compare this with Eq. (8); here the dot denotes the Euclidean inner product in $\mathscr{R}^n$.

By an n-dimensional version of Lemma 3.3, it follows from (13) that the $\mathscr{C}^2$ path $\varphi \in M$ is an extremal for F on M if and only if

$$\frac{\partial f}{\partial \mathbf{x}}(\varphi(t), \varphi'(t), t) - \frac{d}{dt}\frac{\partial f}{\partial \mathbf{y}}(\varphi(t), \varphi'(t), t) = 0. \tag{14}$$

This is the Euler–Lagrange equation in vector form. Taking components, we obtain the scalar Euler–Lagrange equations

$$\frac{\partial f}{\partial x_i} - \frac{d}{dt}\frac{\partial f}{\partial y_i} = 0, \qquad i = 1, \ldots, n. \tag{15}$$

Example 3 Suppose that $\varphi : [a, b] \to \mathscr{R}^n$ is a minimal-length $\mathscr{C}^2$ path with endpoints $\alpha = \varphi(a)$ and $\beta = \varphi(b)$. Then φ is an extremal for the function $F: \mathscr{C}^1([a, b], \mathscr{R}^n) \to \mathscr{R}$ defined by

$$F(\psi) = \int_a^b [(\psi'(t))^2 + \cdots + (\psi_n'(t))^2]^{1/2}\, dt,$$

whose integrand function is

$$f(\mathbf{x}, \mathbf{y}, t) = (y_1{}^2 + \cdots + y_n{}^2)^{1/2}.$$

Since $\partial f/\partial x_i = 0$ and $\partial f/\partial y_i = y_i/(y_1{}^2 + \cdots + y_n{}^2)^{1/2}$, the Euler–Lagrange equations for φ give

$$\frac{\varphi_i'(t)}{[(\varphi_1'(t))^2 + \cdots + (\varphi_n'(t))^2]^{1/2}} = \text{constant}, \qquad i = 1, \ldots, n.$$

Therefore the unit tangent vector $\varphi'(t)/|\varphi'(t)|$ is constant, so it follows that the image of φ is the straight line segment from α to β.

Exercises

3.1 Suppose that a particle of mass m moves in the force field $F: \mathscr{R}^3 \to \mathscr{R}^3$, where $F(\mathbf{x}) = -\nabla V(\mathbf{x})$ with $V: \mathscr{R}^3 \to \mathscr{R}$ a given potential energy function. According to Hamilton's principle, the path $\varphi: [a, b] \to \mathscr{R}^3$ of the particle is an extremal of the integral of the difference of the kinetic and potential energies of the particle,

$$\int_a^b [\tfrac{1}{2}m|\varphi'(t)|^2 - V(\varphi(t))]\, dt.$$

Show that the Euler–Lagrange equations (15) for this problem reduce to Newton's law of motion

$$F(\varphi(t)) = m\varphi''(t).$$

3.2 If $f(x, y, t)$ is actually independent of t, so $\partial f/\partial t = 0$, and $\varphi: [a, b] \to \mathscr{R}$ satisfies the Euler–Lagrange equation

$$\frac{\partial f}{\partial x} - \frac{d}{dt}\left(\frac{\partial f}{\partial y}\right) = 0,$$

show that $y\, \partial f/\partial y - f$ is constant, that is,

$$\varphi'(t)\frac{\partial f}{\partial y}(\varphi(t), \varphi'(t), t) - f(\varphi(t), \varphi'(t), t) = k$$

for all $t \in [a, b]$.

3.3 (The brachistochrone) Suppose a particle of mass m slides down a frictionless wire connecting two fixed points in a vertical plane (Fig. 6.5). We wish to determine the shape $y = \varphi(x)$ of the wire if the time of descent is minimal. Let us take the origin as the initial point, with the y-axis pointing downward. The velocity v of the particle is determined by the energy equation $\tfrac{1}{2}mv^2 = mgy$, whence $v = \sqrt{2gy}$. The time T of descent from $(0, 0)$ to (x_1, y_1) is therefore given by

$$T = \int_0^{x_1} \frac{ds}{v} = \frac{1}{(2g)^{1/2}} \int_0^{x_1} \frac{1}{y^{1/2}} \left[1 + \left(\frac{dy}{dx}\right)^2\right]^{1/2} dx.$$

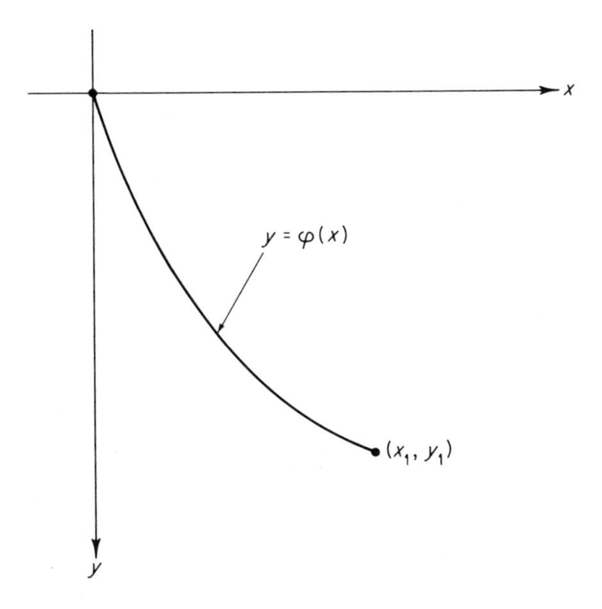

Figure 6.5

Show that the curve of minimal descent time is the *cycloid*

$$x = a(\theta - \sin\theta), \qquad y = a(1 - \cos\theta)$$

generated by the motion of a fixed point on the circumference of a circle of radius a which rolls along the x-axis [the constant a being determined by the condition that it pass through the point (x_1, y_1)]. *Hint*: Noting that

$$f(y, y', x) = \frac{[1 + (y')^2]^{1/2}}{y^{1/2}}$$

is independent of x, apply the result of the previous exercise,

$$y' \frac{\partial f}{\partial y'} - f = k = \frac{1}{(2a)^{1/2}}.$$

Make the substitution $y = 2a \sin^2 \theta/2$ in order to integrate this equation.

Geodesics. In the following five problems we discuss geodesics (shortest paths) on a surface S in $\mathscr{R}^3$. Suppose that S is parametrized by $T : \mathscr{R}^2_{uv} \to \mathscr{R}^3_{xyz}$, and that the curve $\gamma : [a, b] \to S$ is the composition $\gamma = T \circ c$, where $c : [a, b] \to \mathscr{R}^2_{uv}$. Then, by Exercise V.1.8, the length of γ is

$$s(\gamma) = \int_a^b \left[E\left(\frac{du}{dt}\right)^2 + 2F\frac{du}{dt}\frac{dv}{dt} + G\left(\frac{dv}{dt}\right)^2\right]^{1/2} dt,$$

where

$$E = \frac{\partial T}{\partial u} \cdot \frac{\partial T}{\partial u}, \qquad F = \frac{\partial T}{\partial u} \cdot \frac{\partial T}{\partial v}, \qquad G = \frac{\partial T}{\partial v} \cdot \frac{\partial T}{\partial v}.$$

In order for γ to be a minimal-length path on S from $\gamma(a)$ to $\gamma(b)$, it must therefore be an extremal for the integral $s(\gamma)$. We say that γ is a *geodesic* on S if it is an extremal (with endpoints fixed) for the integral

$$\int_a^b \left[E\left(\frac{du}{dt}\right)^2 + 2F\frac{du}{dt}\frac{dv}{dt} + G\left(\frac{du}{dt}\right)^2\right] dt, \tag{$*$}$$

which is somewhat easier to work with.

3.4 (a) Suppose that $f(x_1, x_2, y_1, y_2, t)$ is independent of t, so $\partial f/\partial t = 0$. If $\varphi(t) = (x_1(t), x_2(t))$ is an extremal for

$$\int_a^b f\left(x_1, x_2, \frac{dx_1}{dt}, \frac{dx_2}{dt}, t\right) dt,$$

prove that

$$x_1'(t)\frac{\partial f}{\partial y_1}(\varphi(t), \varphi'(t), t) + x_2'(t)\frac{\partial f}{\partial y_2}(\varphi(t), \varphi'(t), t) - f(\varphi(t), \varphi'(t), t) = c$$

is constant for $t \in [a, b]$. *Hint*: Show that

$$\frac{d}{dt}\left(x_1'\frac{\partial f}{\partial y_1} + x_2'\frac{\partial f}{\partial y_2} - f\right) = -\sum_{i=1}^{2}\left[\frac{\partial f}{\partial x_i} - \frac{d}{dt}\left(\frac{\partial f}{\partial y_i}\right)\right] - \frac{\partial f}{\partial t}.$$

(b) If $f(u, v, u', v') = E(u, v)(u')^2 + 2F(u, v)u'v' + G(u, v)(v')^2$, show that

$$u'\frac{\partial f}{\partial u'} + v'\frac{\partial f}{\partial v'} = 2f.$$

(c) Conclude from (a) and (b) that a geodesic φ on the surface S is a *constant-speed* curve, $|\varphi'(t)| =$ constant.

3.5 Deduce from the previous problem [part (c)] that, if $\gamma : [a, b] \to S$ is a geodesic on the surface S, then γ is an extremal for the pathlength integral $s(\gamma)$. *Hint*: Compare the Euler–Lagrange equations for the two integrals.

3.6 Let S be the vertical cylinder $x^2 + y^2 = r^2$ in $\mathscr{R}^3$, and parametrize S by $T: \mathscr{R}^2_{\theta z} \to \mathscr{R}^3_{xyz}$, where $T(\theta, z) = (r\cos\theta, r\sin\theta, z)$. If $\gamma(t) = T(\theta(t), z(t))$ is a geodesic on S, show that the Euler–Lagrange equations for the integral $(*)$ reduce to

$$\frac{d\theta}{dt} = \text{constant} \qquad \text{and} \qquad \frac{dz}{dt} = \text{constant},$$

so $\theta(t) = at + b$, $z(t) = ct + d$. The case $a = 0$ gives a vertical straight line, the case $c = 0$ gives a horizontal circle, while the case $a \neq 0$, $c \neq 0$ gives a helix on S (see Exercise II.1.12).

3.7 Generalize the preceding problem to the case of a "generalized cylinder" which consists of all vertical straight lines through the smooth curve $C \subset \mathscr{R}^2$.

3.8 Show that the geodesics on a sphere S are the great circles on S.

3.9 Denote by $\mathscr{C}^2[a, b]$ the vector space of twice continuously differentiable functions on $[a, b]$, and by $\mathscr{C}_0{}^2[a, b]$ the subspace consisting of those functions $\psi \in \mathscr{C}^2[a, b]$ such that $\psi(a) = \psi'(a) = \psi(b) = \psi'(b) = 0$.

(a) Show that

$$\|\psi\| = \max_{t \in [a,b]} |\psi(t)| + \max_{t \in [a,b]} |\psi'(t)| + \max_{t \in [a,b]} |\psi''(t)|$$

defines a norm on $\mathscr{C}^2[a, b]$.

(b) Given a $\mathscr{C}^2$ function $f : \mathscr{R}^4 \to \mathscr{R}$, define $F : \mathscr{C}^2[a, b] \to \mathscr{R}$ by

$$F(\psi) = \int_a^b f(\psi(t), \psi'(t), \psi''(t), t)\, dt.$$

Then prove, by the method of proof of Theorem 3.1, that F is differentiable with

$$dF_\varphi(h) = \int_a^b \left(\frac{\partial f}{\partial x} h(t) + \frac{\partial f}{\partial y} h'(t) + \frac{\partial f}{\partial z} h''(t) \right) dt,$$

where the partial derivatives of f are evaluated at $(\varphi(t), \varphi'(t), \varphi''(t), t)$.

(c) Show, by integration by parts as in the proof of Corollary 3.2, that

$$dF_\varphi(h) = \int_a^b \left[\frac{\partial f}{\partial x} - \frac{d}{dt}\left(\frac{\partial f}{\partial y}\right) + \frac{d^2}{dt^2}\left(\frac{\partial f}{\partial z}\right) \right] h(t)\, dt$$

if $h \in \mathscr{C}_0^2[a, b]$.

(d) Conclude that φ satisfies the second order Euler–Lagrange equation

$$\frac{\partial f}{\partial x} - \frac{d}{dt}\left(\frac{\partial f}{\partial y}\right) + \frac{d^2}{dt^2}\left(\frac{\partial f}{\partial z}\right) = 0$$

if φ is an extremal for F, subject to given endpoint conditions on φ and φ' (assuming that φ is of class $\mathscr{C}^4$—note that the above equation is a fourth order ordinary differential equation in φ).

4 THE ISOPERIMETRIC PROBLEM

In this section we treat the so-called isoperimetric problem that was mentioned in the introduction to this chapter. Given functions $f, g : \mathscr{R}^3 \to \mathscr{R}$, we wish to maximize or minimize the function

$$F(\psi) = \int_a^b f(\psi(t), \psi'(t), t)\, dt, \tag{1}$$

subject to the endpoint conditions $\psi(a) = \alpha$, $\psi(b) = \beta$ and the constraint

$$G(\psi) = \int_a^b f(\psi(t), \psi'(t), t)\, dt = c \qquad \text{(constant)}. \tag{2}$$

If, as in Section 3, we denote by M the hyperplane in $\mathscr{C}^1[a, b]$ consisting of all those $\mathscr{C}^1$ functions $\psi : [a, b] \to \mathscr{R}$ such that $\psi(a) = \alpha$ and $\psi(b) = \beta$, then our problem is to locate the local extrema of the function F on the set $M \cap G^{-1}(c)$.

The similarity between this problem and the constrained maximum–minimum problems of Section II.5 should be obvious—the only difference is that here the functions F and G are defined on the infinite-dimensional normed vector space $\mathscr{C}^1[a, b]$, rather than on a finite-dimensional Euclidean space. So our method of attack will be to appropriately generalize the method of Lagrange multipliers so that it will apply in this context.

First let us recall Theorem II.5.5 in the following form. Let $F, G : \mathscr{R}^n \to \mathscr{R}$ be $\mathscr{C}^1$ functions such that $G(\mathbf{0}) = 0$ and $\nabla G(\mathbf{0}) \neq \mathbf{0}$. If F has a local maximum or local minimum at $\mathbf{0}$ subject to the constraint $G(\mathbf{x}) = 0$, then there exists a number λ such that

$$\nabla F(\mathbf{0}) = \lambda \nabla G(\mathbf{0}). \tag{3}$$

Since the differentials $dF_{\mathbf{0}}, dG_{\mathbf{0}} : \mathscr{R}^n \to \mathscr{R}$ are given by

$$dF_{\mathbf{0}}(\mathbf{v}) = \nabla F(\mathbf{0}) \cdot \mathbf{v} \qquad \text{and} \qquad dG_{\mathbf{0}}(\mathbf{v}) = \nabla G(\mathbf{0}) \cdot \mathbf{v},$$

Eq. (3) can be rewritten

$$dF_{\mathbf{0}} = \Lambda \circ dG_{\mathbf{0}}, \tag{4}$$

where $\Lambda : \mathscr{R} \to \mathscr{R}$ is the linear function defined by $\Lambda(t) = \lambda t$.

Equation (4) presents the Lagrange multiplier method in a form which is suitable for generalization to normed vector spaces (where differentials are available, but gradient vectors are not). For the proof we will need the following elementary algebraic lemma.

Lemma 4.1 Let α and β be real-valued linear functions on the vector space E such that

$$\operatorname{Ker} \alpha \supset \operatorname{Ker} \beta \qquad \text{and} \qquad \operatorname{Im} \beta = R.$$

Then there exists a linear function $\Lambda : \mathscr{R} \to \mathscr{R}$ such that $\alpha = \Lambda \circ \beta$. That is, the following diagram of linear functions "commutes."

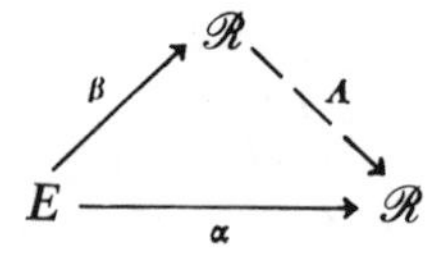

PROOF Given $t \in \mathscr{R}$, pick $x \in E$ such that $\beta(x) = t$, and define

$$\Lambda(t) = \alpha(x).$$

In order to show that Λ is well defined, we must see that, if y is another element of E with $\beta(y) = t$, then $\alpha(x) = \alpha(y)$. But if $\beta(x) = \beta(y) = t$, then $x - y \in \operatorname{Ker} \beta \subset \operatorname{Ker} \alpha$, so $\alpha(x - y) = 0$, which immediately implies that $\alpha(x) = \alpha(y)$.

If $\beta(x) = s$ and $\beta(y) = t$, then

$$\begin{aligned} \Lambda(as + bt) &= \alpha(ax + by) \\ &= a\alpha(x) + b\alpha(y) \\ &= a\Lambda(s) + b\Lambda(t), \end{aligned}$$

so Λ is linear. ∎

The following theorem states the Lagrange multiplier method in the desired generality.

Theorem 4.2 Let F and G be real-valued $\mathscr{C}^1$ functions on the complete normed vector space E, with $G(0) = 0$ and $dG_0 \neq 0$ (so $\operatorname{Im} dG_0 = \mathscr{R}$). If $F : E \to \mathscr{R}$ has a local extremum at 0 subject to the constraint $G(x) = 0$, then there exists a linear function $\Lambda : \mathscr{R} \to \mathscr{R}$ such that

$$dF_0 = \Lambda \circ dG_0. \tag{5}$$

Of course the statement, that "F has a local extremum at 0 subject to $G(x) = 0$," means that the restriction $F | G^{-1}(0)$ has a local extremum at $0 \in G^{-1}(0)$.

PROOF This will follow from Lemma 4.1, with $\alpha = dF_0$ and $\beta = dG_0$, if we can prove that Ker dF_0 contains Ker dG_0. In order to do this, let us first assume the fact (to be established afterward) that, given $v \in \operatorname{Ker} dG_0$, there exists a differentiable path $\gamma : (-\varepsilon, \varepsilon) \to E$ whose image lies in $G^{-1}(0)$, such that $\gamma(0) = 0$ and $\gamma'(0) = v$ (see Fig. 6.6).

Then the composition $h = F \circ \gamma : (-\varepsilon, \varepsilon) \to \mathscr{R}$ has a local extremum at 0, so $h'(0) = 0$. The chain rule therefore gives

$$\begin{aligned} 0 &= h'(0) = dh_0(1) = dF_{\gamma(0)}(d\gamma_0(1)) \cdot \\ &= dF_0(\gamma'(0)), \\ 0 &= dF_0(v) \end{aligned}$$

as desired.

We will use the implicit function theorem to verify the existence of the differentiable path γ used above. If $X = \operatorname{Ker} dG_0$ then, since $dG_0 : E \to \mathscr{R}$ is continuous, X is a closed subspace of E, and is therefore complete (by Exercise 1.5). Choose $w \in E$ such that $dG_0(w) = 1$, and denote by Y the closed subspace of E consisting of all scalar multiples of w; then Y is a "copy" of $\mathscr{R}$.

It is clear that $X \cap Y = 0$. Also, if $e \in E$ and $a = dG_0(e) \in \mathscr{R}$, then

$$dG_0(e - aw) = dG_0(e) - a\,dG_0(w) = 0$$

so $e - aw \in X$. Therefore

$$e = x + y$$

with $x \in X$ and $y = aw \in Y$. Thus E is the algebraic direct sum of the subspaces X and Y. Moreover, it is true (although we omit the proof) that the norm on E is equivalent to the product norm on $X \times Y$, so we may write $E = X \times Y$.

In order to apply the implicit function theorem, we need to know that

$$d_y G_0 : Y \to \mathscr{R}$$

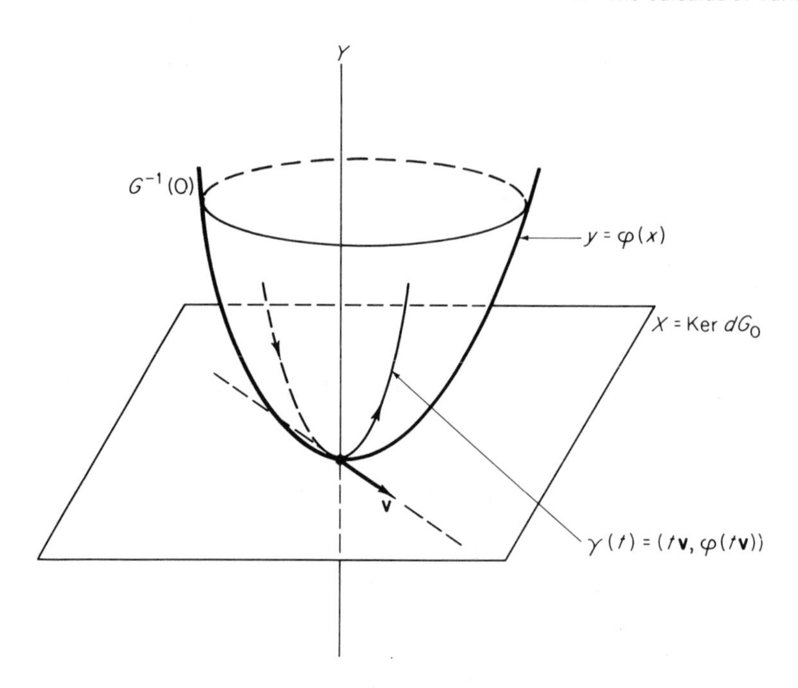

Figure 6.6

is an isomorphism. Since $Y \approx \mathscr{R}$, we must merely show that $d_y G_0 \neq 0$. But, given $(r, s) \in X \times Y = E$, we have

$$\begin{aligned} dG_0(r, s) &= dG_0(r, 0) + dG_0(0, s) \\ &= dG_0(0, s) \\ &= d_y G_0(s) \end{aligned}$$

by Exercise 2.6, so the assumption that $d_y G_0 = 0$ would imply that $dG_0 = 0$, contrary to hypothesis.

Consequently the implicit function theorem provides a $\mathscr{C}^1$ function $\varphi : X \to Y$ whose graph $y = \varphi(x)$ in $X \times Y = E$ coincides with $G^{-1}(0)$, inside some neighborhood of 0. If $H(x) = G(x, \varphi(x))$, then $H(x) = 0$ for x near 0, so

$$\begin{aligned} 0 = dH_0(u) &= d_x G_0(u) + d_y G_0(d\varphi_0(u)) \\ &= d_y G_0(d\varphi_0(u)) \end{aligned}$$

for all $u \in X$. It therefore follows that $d\varphi_0 = 0$, because $d_y G_0$ is an isomorphism.

Finally, given $v = (u, 0) \in \operatorname{Ker} dG_0$, define $\gamma : \mathscr{R} \to E$ by $\gamma(t) = (tu, \varphi(tu))$. Then $\gamma(0) = 0$ and $\gamma(t) \in G^{-1}(0)$ for t sufficiently small, and

$$\gamma'(0) = (u, d\varphi_0(u)) = (u, 0) = v$$

as desired. ∎

We are now prepared to deal with the isoperimetric problem. Let f and g be real-valued $\mathscr{C}^2$ functions on $\mathscr{R}^3$, and define the real-valued functions F and G on $\mathscr{C}^1[a, b]$ by

$$F(\psi) = \int_a^b f(\psi(t), \psi'(t), t)\,dt \tag{6}$$

and

$$G(\psi) = \int_a^b g(\psi(t), \psi'(t), t)\,dt - c, \tag{7}$$

where $c \in \mathscr{R}$. Assume that φ is a $\mathscr{C}^2$ element of $\mathscr{C}^1[a, b]$ at which F has a local extremum on $M \cap G^{-1}(0)$, where M is the usual hyperplane in $\mathscr{C}^1[a, b]$ that is determined by the endpoint conditions $\psi(a) = \alpha$ and $\psi(b) = \beta$.

We have seen (in Section 3) that M is the translate (by any fixed element of M) of the subspace $\mathscr{C}_0{}^1[a, b]$ of $\mathscr{C}^1[a, b]$ consisting of these elements $\psi \in \mathscr{C}^1[a, b]$ such that $\psi(a) = \psi(b) = 0$. Let $T: \mathscr{C}_0{}^1[a, b] \to M$ be the translation defined by

$$T(\psi) = \psi + \varphi,$$

and note that $T(0) = \varphi$, while

$$dT_0 : \mathscr{C}_0{}^1[a, b] \to \mathscr{C}_0{}^1[a, b] = TM_\varphi$$

is the identity mapping.

Now consider the real-valued functions $F \circ T$ and $G \circ T$ on $\mathscr{C}_0{}^1[a, b]$. The fact that F has a local extremum on $M \cap G^{-1}(0)$ at φ implies that $F \circ T$ has a local extremum at 0 subject to the condition $G \circ T(\psi) = 0$.

Let us assume that φ *is not an extremal for G on M*, that is, that

$$dG_\varphi | TM_\varphi \neq 0,$$

so $d(G \circ T)_0 \neq 0$. Then Theorem 4.2 applies to give a linear function $\Lambda : \mathscr{R} \to \mathscr{R}$ such that

$$d(F \circ T)_0 = \Lambda \circ d(G \circ T)_0 .$$

Since dT_0 is the identity mapping on $C_0{}^1[a, b]$, the chain rule gives

$$dF_\varphi = \Lambda \circ dG_\varphi$$

on $\mathscr{C}_0{}^1[a, b]$. Writing $\Lambda(t) = \lambda t$ and applying the computation of Corollary 3.2 for the differentials dF_φ and dG_φ, we conclude that

$$\int_a^b \left[\frac{\partial f}{\partial x}(\varphi(t), \varphi'(t), t) - \frac{d}{dt}\frac{\partial f}{\partial y}(\varphi(t), \varphi'(t), t)\right] u(t)\,dt$$

$$= \lambda \int_a^b \left[\frac{\partial g}{\partial x}(\varphi(t), \varphi'(t), t) - \frac{d}{dt}\frac{\partial g}{\partial y}(\varphi(t), \varphi'(t), t)\right] u(t)\,dt$$

for all $u \in \mathscr{C}_0{}^1[a, b]$.

If $h : \mathscr{R}^3 \to \mathscr{R}$ is defined by

$$h(x, y, t) = f(x, y, t) - \lambda g(x, y, t),$$

it follows that

$$\int_a^b \left[\frac{\partial h}{\partial x}(\varphi(t), \varphi'(t), t) - \frac{d}{dt}\frac{\partial h}{\partial y}(\varphi(t), \varphi'(t), t)\right] u(t)\, dt = 0$$

for all $u \in \mathscr{C}_0^{\,1}[a, b]$. An application of Lemma 3.3 finally completes the proof of the following theorem.

Theorem 4.3 Let F and G be the real-valued functions on $\mathscr{C}^1[a, b]$ defined by (6) and (7), where f and g are $\mathscr{C}^2$ functions on $\mathscr{R}^3$. Let $\varphi \in M$ be a $\mathscr{C}^2$ function which is not an extremal for G. If F has a local extremum at φ subject to the conditions

$$\psi(a) = \alpha, \qquad \psi(b) = \beta, \qquad \text{and} \qquad G(\psi) = 0,$$

then there exists a real number λ such that φ satisfies the Euler–Lagrange equation for the function $h = f - \lambda g$, that is,

$$\frac{\partial h}{\partial x}(\varphi(t), \varphi'(t), t) - \frac{d}{dt}\frac{\partial h}{\partial y}(\varphi(t), \varphi'(t), t) = 0 \tag{8}$$

for all $t \in [a, b]$.

The following application of this theorem is the one which gave such constraint problems in the calculus of variations their customary name—isoperimetric problems.

Example Suppose $\varphi : [a, b] \to \mathscr{R}$ is that nonnegative function (if any) with $\varphi(a) = \varphi(b) = 0$ whose graph $x = \varphi(t)$ has length L, such that the area under its graph is maximal. We want to prove that the graph $x = \varphi(t)$ must be an arc of

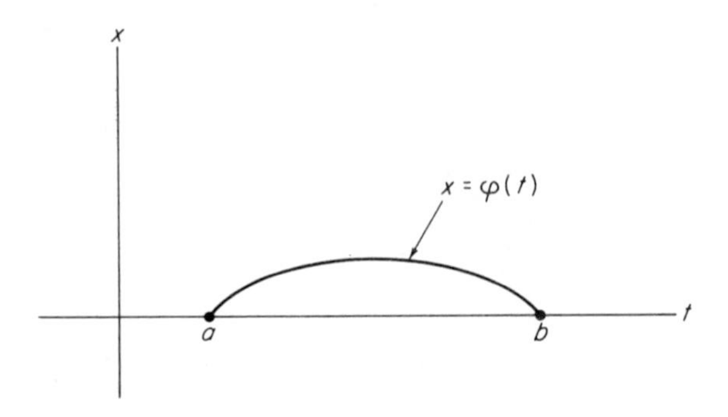

Figure 6.7

a circle (Fig. 6.7). If $f(x, y, t) = x$ and $g(x, y, t) = \sqrt{1 + y^2}$, then φ maximizes the integral

$$\int_a^b f(\varphi(t), \varphi'(t), t)\, dt,$$

subject to the conditions

$$\varphi(a) = \varphi(b) = 0 \qquad \text{and} \qquad \int_a^b g(\varphi(t), \varphi'(t), t)\, dt = L.$$

Since $\partial f/\partial x = 1$, $\partial f/\partial y = \partial g/\partial x = 0$, $\partial g/\partial y = y/\sqrt{1 + y^2}$, the Euler–Lagrange equation (8) is

$$1 - \lambda \frac{d}{dt}\left(\frac{\varphi'(t)}{1 + (\varphi'(t))^2}\right) = 0,$$

or

$$\frac{\varphi''(t)}{[1 + (\varphi'(t))^2]^{3/2}} = \frac{1}{\lambda}.$$

This last equation just says that the curvature of the curve $t \to (t, \varphi(t))$ is the *constant* $1/\lambda$. Its image must therefore be part of a circle.

The above discussion of the isoperimetric problem generalizes in a straightforward manner to the case in which there is more than one constraint. Given $\mathscr{C}^2$ functions $f, g_1, \ldots, g_k : \mathscr{R}^3 \to \mathscr{R}$, we wish to minimize or maximize the function

$$F(\psi) = \int_a^b f(\psi(t), \psi'(t), t)\, dt, \tag{9}$$

subject to the endpoint conditions $\psi(a) = \alpha$, $\psi(b) = \beta$ and the constraints

$$G_i(\psi) = \int_a^b g_i(\psi(t), \psi'(t), t)\, dt - c_i = 0. \tag{10}$$

Our problem then is to locate the local extrema of F on $M \cap G^{-1}(\mathbf{0})$, where M is the usual hyperplane in $\mathscr{C}^1[a, b]$ and

$$G = (G_1, \ldots, G_k) : \mathscr{C}^1[a, b] \to \mathscr{R}^k.$$

The result, analogous to Theorem 4.3, is as follows.

Let $\varphi \in M$ be a $\mathscr{C}^2$ function which is not an extremal for any linear combination of the functions $G_1, \ldots, G_k$. If F has a local extremum at φ subject to the conditions

$$\psi(a) = \alpha, \qquad \psi(b) = \beta, \qquad \text{and} \qquad G(\psi) = \mathbf{0},$$

then there exist numbers $\lambda_1, \ldots, \lambda_k$ such that φ satisfies the Euler–Lagrange equation for the function

$$h = f - \sum_{i=1}^{k} \lambda_i g_i .$$

Inclusion of the complete details of the proof would be repetitious, so we simply outline the necessary alterations in the proof of Theorem 4.3.

First Lemma 4.1 and Theorem 4.2 are slightly generalized as follows. In Lemma 4.1 we take β to be a linear mapping from E to $\mathscr{R}^k$ with $\operatorname{Im} \beta = \mathscr{R}^k$, and in Theorem 4.2 we take G to be a $\mathscr{C}^1$ mapping from E to $\mathscr{R}^k$ such that $G(0) = \mathbf{0}$ and $\operatorname{Im} dG_0 = \mathscr{R}^k$. The only other change is that, in the conclusion of each, Λ becomes a real-valued linear function on $\mathscr{R}^k$. The proofs remain essentially the same.

We then apply the generalized Theorem 4.2 to the mappings $F : \mathscr{C}^1[a, b] \to \mathscr{R}$ and $G : \mathscr{C}^1[a, b] \to \mathscr{R}^k$, defined by (9) and (10), in the same way that the original Theorem 4.2 was applied (in the proof of Theorem 4.3) to the functions defined by (6) and (7). The only additional observation needed is that, if φ is not an extremal for any linear combination of the component functions $G_1, \ldots, G_k$, then it follows easily that dG_φ maps TM_φ *onto* $\mathscr{R}^k$. We then conclude as before that

$$dF_\varphi = \Lambda \circ dG_\varphi \qquad \text{on} \quad \mathscr{C}_0{}^1[a, b]$$

for some linear function $\Lambda : \mathscr{R}^k \to \mathscr{R}$. Writing $\Lambda(x_1, \ldots, x_k) = \sum_{i=1}^k \lambda_i x_i$, we conclude that

$$\int_a^b \left[\frac{\partial f}{\partial x}(\varphi(t), \varphi'(t), t) - \frac{d}{dt}\frac{\partial f}{\partial y}(\varphi(t), \varphi'(t), t)\right] u(t)\, dt$$

$$= \sum_{i=1}^{k} \lambda_i \int_a^b \left[\frac{\partial g_i}{\partial x}(\varphi(t), \varphi'(t), t) - \frac{d}{dt}\frac{\partial g_i}{\partial y}(\varphi(t), \varphi'(t), t)\right] u(t)\, dt$$

for all $u \in \mathscr{C}_0{}^1[a, b]$. An application of Lemma 3.3 then implies that φ satisfies the Euler–Lagrange equation for $h = f - \sum \lambda_i g_i$.

Exercises

4.1 Consulting the discussion at the end of Section 3, generalize the isoperimetric problem to the vector-valued case as follows: Let $f, g : \mathscr{R}^n \times \mathscr{R}^n \times \mathscr{R} \to \mathscr{R}$ be given functions, and suppose $\varphi : [a, b] \to \mathscr{R}^n$ is an extremal for

$$F(\psi) = \int_a^b f(\psi(t), \psi'(t), t)\, dt,$$

subject to the conditions $\psi(a) = \alpha$, $\psi(b) = \beta$ and

$$G(\psi) = \int_a^b g(\psi(t), \psi'(t), t)\,dt = c.$$

Then show under appropriate conditions that, for some number λ, the path φ satisfies the Euler–Lagrange equations

$$\frac{\partial h}{\partial x_i} - \frac{d}{dt}\frac{\partial h}{\partial y_i} = 0, \qquad i = 1, \ldots, n$$

for the function $h = f - \lambda g$.

4.2 Let $\varphi : [a, b] \to \mathscr{R}^2$ be a closed curve in the plane, $\varphi(a) = \varphi(b)$, and write $\varphi(t) = (x(t), y(t))$. Apply the result of the previous problem to show that, if φ maximizes the area integral

$$\tfrac{1}{2}\int_a^b (xy' - yx')\,dt,$$

subject to

$$\int_a^b ((x')^2 + (y')^2)^{1/2}\,dt = L \qquad \text{(constant)},$$

then the image of φ is a circle.

4.3 With the notation and terminology of the previous problem, establish the following reciprocity relationship. The closed path φ is an extremal for the area integral, subject to the arclength integral being constant, if and only if φ is an extremal for the arclength integral subject to the area integral being constant. Conclude that, if φ has minimal length amongst curves enclosing a given area, then the image of φ is a circle.

4.4 Formulate (along the lines of Exercise 3.9) a necessary condition that $\varphi : [a, b] \to \mathscr{R}$ minimize

$$\int_a^b f(\varphi(t), \varphi'(t), \varphi''(t), t)\,dt,$$

subject to

$$\int_a^b g\,(\varphi(t), \varphi'(t), \varphi''(t), t)\,dt = \text{constant}.$$

This is the isoperimetric problem with second derivatives.

4.5 Suppose that $r = f(\theta)$, $\theta \in [0,\pi]$ describes (in polar coordinates) a closed curve of length L that encloses maximal area. Show that it is a circle by maximizing

$$A = \frac{1}{2}\int_0^\pi r^2\,d\theta,$$

subject to the condition

$$\int_0^\pi \left[\left(\frac{dr}{d\theta}\right)^2 - r^2\right]^{1/2} d\theta = L.$$

4.6 A uniform flexible cable of fixed length hangs between two fixed points. If it hangs in such a way as to minimize the height of its center of gravity, show that its shape is that of a catenary (see Example 2 of Section 3). *Hint:* Note that Exercise 3.2 applies.

4.7 If a hanging flexible cable of fixed length supports a *horizontally* uniform load, show that its shape is that of a parabola.

5 MULTIPLE INTEGRAL PROBLEMS

Thus far, we have confined our attention to extremum problems associated with the simple integral $\int_a^b f(\psi(t), \psi'(t), t)\, dt$, where ψ is a function of one variable. In this section we briefly discuss the analogous problems associated with a multiple integral whose integrand involves an "unknown" function of several variables.

Let D be a cellulated n-dimensional region in $\mathscr{R}^n$. Given $f: \mathscr{R}^{2n+1} \to \mathscr{R}$, we seek to maximize or minimize the function F defined by

$$F(\psi) = \int_D f\left(x_1, \ldots, x_n, \psi(x_1, \ldots, x_n), \frac{\partial \psi}{\partial x_1}, \ldots, \frac{\partial \psi}{\partial x_n}\right) d\mathbf{x}, \tag{1}$$

amongst those $\mathscr{C}^1$ functions $\psi : D \to \mathscr{R}$ which agree with a given fixed function $\psi_0 : D \to \mathscr{R}$ on the boundary ∂D of the region D. In terms of the gradient vector

$$\nabla\psi(\mathbf{x}) = (D_1\psi(\mathbf{x}), \ldots, D_n\psi(\mathbf{x})),$$

we may rewrite (1) as

$$F(\psi) = \int_D f(\mathbf{x}, \psi(\mathbf{x}), \nabla\psi(\mathbf{x}))\, d\mathbf{x}. \tag{2}$$

Throughout this section we will denote the first n coordinates in $\mathscr{R}^{2n+1}$ by $x_1, \ldots, x_n$, the $(n+1)$th coordinate by y, and the last n coordinates in $\mathscr{R}^{2n+1}$ by $z_1, \ldots, z_n$. Thus we are thinking of the Cartesian factorization $\mathscr{R}^{2n+1} = \mathscr{R}^n \times \mathscr{R} \times \mathscr{R}^n$, and therefore write $(\mathbf{x}, y, \mathbf{z})$ for the typical point of $\mathscr{R}^{2n+1}$. In terms of this notation, we are interested in the function

$$F(\psi) = \int_D f(\mathbf{x}, y, \mathbf{z})\, d\mathbf{x},$$

where $y = \psi(\mathbf{x})$ and $z = \nabla\psi(\mathbf{x})$.

The function F is defined by (2) on the vector space $\mathscr{C}^1(D)$ that consists of all real-valued $\mathscr{C}^1$ functions on D (with the usual pointwise addition and scalar multiplication). We make $\mathscr{C}^1(D)$ into a normed vector space by defining

$$\|\psi\| = \max_{\mathbf{x} \in D} |\psi(\mathbf{x})| + \max_{\mathbf{x} \in D} |\nabla\psi(\mathbf{x})|. \tag{3}$$

It can then be verified that the normed vector space $\mathscr{C}^1(D)$ is *complete*. The proof of this fact is similar to that of Corollary 1.5 (that $\mathscr{C}^1[a, b]$ is complete), but is somewhat more tedious, and will be omitted (being unnecessary for what follows in this section).

Let M denote the subset of $\mathscr{C}^1(D)$ consisting of those functions ψ that satisfy the "boundary condition"

$$\psi(\mathbf{x}) = \psi_0(\mathbf{x}) \qquad \text{if} \quad \mathbf{x} \in \partial D.$$

Then, given any $\psi \in M$, the difference $\psi - \psi_0$ is an element of the subspace

$$\mathscr{C}_0{}^1(D) = \{\varphi \in \mathscr{C}^1(D) : \varphi(\mathbf{x}) = 0 \quad \text{if} \quad \mathbf{x} \in \partial D\}$$

of $\mathscr{C}^1(D)$, consisting of all those $\mathscr{C}^1$ functions on D that vanish on ∂D. Conversely, if $\varphi \in \mathscr{C}_0{}^1(D)$, then clearly $\psi_0 + \varphi \in M$. Thus M is a hyperplane in $\mathscr{C}^1(D)$, namely, the translate by the fixed element $\psi_0 \in M$ of the subspace $\mathscr{C}_0{}^1(D)$. Consequently

$$TM_\psi = \mathscr{C}_0{}^1(D)$$

for all $\psi \in M$.

If $F : \mathscr{C}^1(D) \to \mathscr{R}$ is differentiable at $\varphi \in M$, and $F|M$ has a local extremum at φ, Theorem 2.3 implies that

$$dF_\varphi(h) = 0 \qquad \text{for all} \quad h \in \mathscr{C}_0{}^1(D). \tag{4}$$

Just as in the single-variable case, we will call the function $\varphi \in M$ an *extremal* for F on M if it satisfies the necessary condition (4).

The following theorem is analogous to Theorem 3.1, and gives the computation of the differential dF_φ when F is defined by (1).

Theorem 5.1 Suppose that D is a compact cellulated n-dimensional region in $\mathscr{R}^n$, and that $f : \mathscr{R}^{2n+1} \to \mathscr{R}$ is a $\mathscr{C}^2$ function. Then the function $F : \mathscr{C}^1(D) \to \mathscr{R}$ defined by (1) is differentiable with

$$dF_\varphi(h) = \int_D \left[\frac{\partial f}{\partial y} \cdot h(\mathbf{x}) + \sum_{i=1}^n \frac{\partial f}{\partial z_i} \cdot \frac{\partial h}{\partial x_i}(\mathbf{x})\right] d\mathbf{x} \tag{5}$$

for all φ, $h \in \mathscr{C}^1(D)$. The partial derivatives

$$\frac{\partial f}{\partial y}, \frac{\partial f}{\partial z_1}, \ldots, \frac{\partial f}{\partial z_n}$$

in (5) are evaluated at the point $(\mathbf{x}, \varphi(\mathbf{x}), \nabla\varphi(\mathbf{x})) \in \mathscr{R}^{2n+1}$.

The method of proof of Theorem 5.1 is the same as that of Theorem 3.1, making use of the second degree Taylor expansion of f. The details will be left to the reader.

In view of condition (4), we are interested in the value of $dF_\varphi(h)$ when $h \in TM_\varphi = \mathscr{C}_0{}^1(D)$. The following theorem is analogous to Corollary 3.2.

Theorem 5.2 Assume, in addition to the hypotheses of Theorem 5.1, that φ is a $\mathscr{C}^2$ function and that $h \in \mathscr{C}_0{}^1(D)$. Then

$$dF_\varphi(h) = \int_D \left[\frac{\partial f}{\partial y} - \sum_{i=1}^n \frac{\partial}{\partial x_i}\left(\frac{\partial f}{\partial z_i}\right)\right] h(\mathbf{x})\, d\mathbf{x}. \tag{6}$$

Here also the partial derivatives of f are evaluated at $(\mathbf{x}, \varphi(\mathbf{x}), \nabla\varphi(\mathbf{x})) \in \mathscr{R}^{2n+1}$.

PROOF Consider the differential $(n-1)$-form defined on D by

$$\omega = \sum_{i=1}^n (-1)^{i+1} \frac{\partial f}{\partial z_i} h\, dx_1 \wedge \cdots \wedge \widehat{dx_i} \wedge \cdots \wedge dx_n.$$

A routine computation gives

$$d\omega = \left[\sum_{i=1}^n \left(\frac{\partial f}{\partial z_i}\frac{\partial h}{\partial x_i} + \frac{\partial}{\partial x_i}\left(\frac{\partial f}{\partial z_i}\right) h\right)\right] d\mathbf{x},$$

where $d\mathbf{x} = dx_1 \wedge \cdots \wedge dx_n$. Hence

$$\left(\sum_{i=1}^n \frac{\partial f}{\partial z_i}\frac{\partial h}{\partial x_i}\right) d\mathbf{x} = d\omega - \left(\sum_{i=1}^n \frac{\partial}{\partial x_i}\left(\frac{\partial f}{\partial z_i}\right) h\right) d\mathbf{x}.$$

Substituting this into Eq. (5), we obtain

$$dF_\varphi(h) = \int_D \left[\frac{\partial f}{\partial y} - \sum_{i=1}^n \frac{\partial}{\partial x_i}\left(\frac{\partial f}{\partial z_i}\right)\right] h\, d\mathbf{x} + \int_D d\omega. \tag{7}$$

But $\int_D d\omega = \int_{\partial D} \omega = 0$ by Stokes' theorem and the fact that $\omega = 0$ on ∂D because $h \in \mathscr{C}_0{}^1(D)$. Thus Eq. (7) reduces to the desired Eq. (6). ∎

Theorem 5.2 shows that the $\mathscr{C}^2$ function $\varphi \in M$ is an extremal for F on M if and only if

$$\int_D \left[\frac{\partial f}{\partial y} - \sum_{i=1}^n \frac{\partial}{\partial x_i}\left(\frac{\partial f}{\partial z_i}\right)\right] h(\mathbf{x})\, d\mathbf{x} = 0$$

for *every* $h \in \mathscr{C}_0{}^1(D)$. From this result and the obvious multivariable analog of Lemma 3.3 we immediately obtain the multivariable Euler–Lagrange equation.

Theorem 5.3 Let $F : \mathscr{C}^1(D) \to \mathscr{R}$ be defined by Eq. (1), with $f : \mathscr{R}^{2n+1} \to \mathscr{R}$ being a $\mathscr{C}^2$ function. Then the $\mathscr{C}^2$ function $\varphi \in M$ is an extremal for F on M if and only if

$$\frac{\partial f}{\partial y}(\mathbf{x}, \varphi(\mathbf{x}), \nabla\varphi(\mathbf{x})) - \sum_{i=1}^n \frac{\partial^2 f}{\partial x_i\, \partial z_i}(\mathbf{x}, \varphi(\mathbf{x}), \nabla\varphi(\mathbf{x})) = 0$$

for all $\mathbf{x} \in D$.

The equation

$$\frac{\partial f}{\partial y} - \sum_{i=1}^{n} \frac{\partial}{\partial x_i}\left(\frac{\partial f}{\partial z_i}\right) = 0, \tag{8}$$

with the partial derivatives of f evaluated at $(\mathbf{x}, \varphi(\mathbf{x}), \nabla\varphi(\mathbf{x}))$, is the *Euler–Lagrange equation* for the extremal φ. We give some examples to illustrate its applications.

Example 1 (minimal surfaces) If D is a disk in the plane, and $\varphi_0 : D \to \mathscr{R}$ a function, then the graph of φ_0 is a disk in $\mathscr{R}^3$. We consider the following question. Under what conditions does the graph (Fig. 6.8) of the function $\varphi : D \to \mathscr{R}$ have minimal surface area, among the graphs of all those functions $\psi : D \to \mathscr{R}$ that agree with φ_0 on the boundary curve ∂D of the disk D?

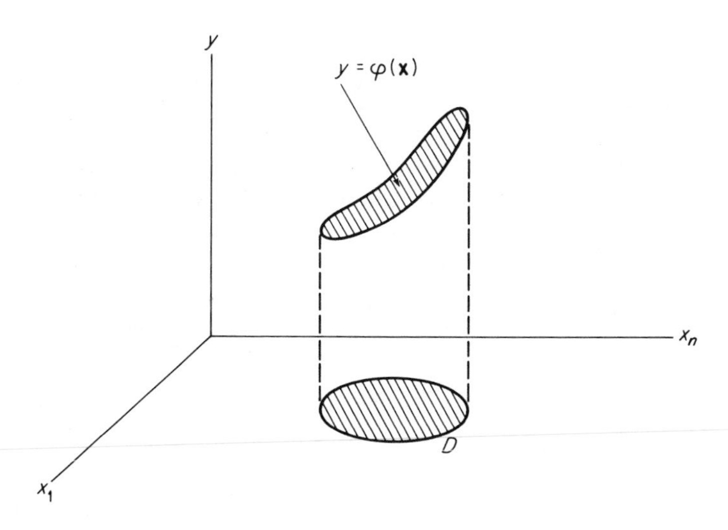

Figure 6.8

We can just as easily discuss the n-dimensional generalization of this question. So we start with a smooth compact n-manifold-with-boundary $D \subset \mathscr{R}^n$, and a $\mathscr{C}^1$ function $\varphi_0 : D \to \mathscr{R}$, whose graph $y = \varphi_0(\mathbf{x})$ is an n-manifold-with-boundary in $\mathscr{R}^{n+1}$.

The area $F(\varphi)$ of the graph of the function $\varphi : D \to \mathscr{R}$ is given by formula (10) of Section V.4,

$$F(\varphi) = \int_D [1 + |\nabla\varphi(\mathbf{x})|^2]^{1/2}\, d\mathbf{x}.$$

We therefore want to minimize the function $F : \mathscr{C}^1(D) \to \mathscr{R}$ defined by (1) with

$$f(\mathbf{x}, y, \mathbf{z}) = (1 + {z_1}^2 + \cdots + {z_n}^2)^{1/2}.$$

Since

$$\frac{\partial f}{\partial y} = 0, \qquad \frac{\partial f}{\partial z_i} = \frac{z_i}{(1 + z_1{}^2 + \cdots + z_k{}^2)^{1/2}},$$

the Euler–Lagrange equation (8) for this problem is

$$\sum_{i=1}^{n} \frac{\partial}{\partial x_i}\left(\frac{\partial \varphi}{\partial x_i}[1 + |\nabla\varphi|^2]^{-1/2}\right) = 0.$$

Upon calculating the indicated partial derivatives and simplifying, we obtain

$$(1 + |\nabla\varphi|^2)\,\nabla^2\varphi = \sum_{i,j=1}^{n} \frac{\partial \varphi}{\partial x_i}\frac{\partial \varphi}{\partial x_j}\frac{\partial^2 \varphi}{\partial x_i\,\partial x_j}, \tag{9}$$

where

$$\nabla^2\varphi = \sum_{i=1}^{n} \frac{\partial^2 \varphi}{\partial x_i{}^2}$$

as usual. Equation (9) therefore gives a necessary condition that the area of the graph of $y = \varphi(\mathbf{x})$ be minimal, among all n-manifolds-with-boundary in $\mathscr{R}^{n+1}$ that have the same boundary.

In the original problem of 2-dimensional minimal surfaces, it is customary to use the notation

$$z = \varphi(x, y), \qquad p = \frac{\partial z}{\partial x}, \qquad q = \frac{\partial z}{\partial y},$$

$$r = \frac{\partial^2 z}{\partial x^2}, \qquad s = \frac{\partial^2 z}{\partial x\,\partial y}, \qquad t = \frac{\partial^2 z}{\partial y^2}.$$

With this notation, Eq. (9) takes the form

$$(1 + q^2)r - 2pqs + (1 + p^2)t = 0.$$

This is of course a second order partial differential equation for the unknown function $z = \varphi(x, y)$.

Example 2 (vibrating membrane) In this example we apply Hamilton's principle to derive the wave equation for the motion of a vibrating n-dimensional "membrane." The cases $n = 1$ and $n = 2$ correspond to a vibrating string and an "actual" membrane, respectively.

We assume that the equilibrium position of the membrane is the compact n-manifold-with-boundary

$$W \subset \mathscr{R}^n = \mathscr{R}^n \times \{0\} \subset \mathscr{R}^{n+1},$$

and that it vibrates with its boundary fixed. Let its motion be described by the function

$$\varphi : W \times \mathscr{R} \to \mathscr{R},$$

in the sense that the graph $y = \varphi(\mathbf{x}, t)$ is the position in $\mathscr{R}^{n+1}$ of the membrane at time t (Fig. 6.9).

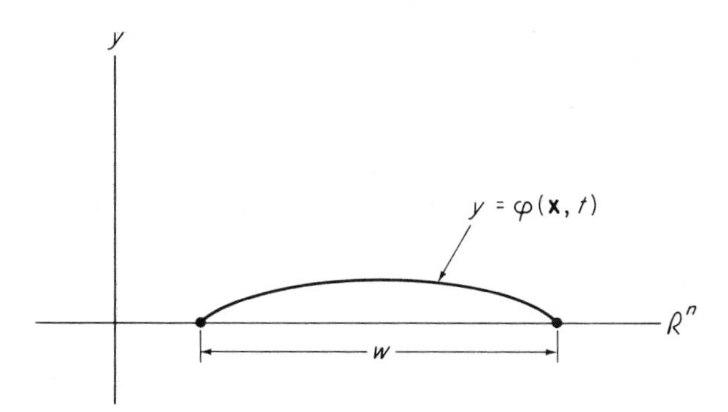

Figure 6.9

If the membrane has constant density σ, then its kinetic energy at time t is

$$T(t) = \frac{\sigma}{2} \int_W [D_2 \varphi(\mathbf{x}, t)]^2 \, d\mathbf{x} = \frac{\sigma}{2} \int_W \left(\frac{\partial \varphi}{\partial t} \right)^2 d\mathbf{x}. \tag{10}$$

We assume initially that the potential energy V of the membrane is proportional to the increase in its surface area, that is,

$$V(t) = \tau(a(t) - a(0)),$$

where $a(t)$ is the area of the membrane at time t. The constant τ is called the "surface tension." By formula (10) of Section V.4 we then have

$$V(t) = \tau \left(\int_W [1 + |\nabla \varphi|^2]^{1/2} \, d\mathbf{x} - \int_W d\mathbf{x} \right)$$

$$= \tau \int_W [(1 + \tfrac{1}{2} |\nabla \varphi|^2 + \cdots) - 1] \, d\mathbf{x}.$$

We now suppose that the deformation of the membrane is so slight that the higher order terms (indicated by the dots) may be neglected. The potential energy of the membrane at time t is then given by

$$V(t) = \frac{\tau}{2} \int_W |\nabla \varphi(\mathbf{x})|^2 \, d\mathbf{x}. \tag{11}$$

According to Hamilton's principle of physics, the motion of the membrane is such that the value of the integral

$$\int_a^b [T(t) - V(t)] \, dt = \int_a^b \left(\int_W \left[\frac{\sigma}{2} \left(\frac{\partial \varphi}{\partial t} \right)^2 - \frac{\tau}{2} |\nabla \varphi|^2 \right] d\mathbf{x} \right) dt$$

is minimal for every time interval $[a, b]$. That is, if $D = W \times [a, b]$, then the actual motion φ is an extremal for the function $F: \mathscr{C}^1(D) \to \mathscr{R}$ defined by

$$F(\psi) = \int_D \left[\frac{\sigma}{2}\left(\frac{\partial\psi}{\partial t}\right)^2 - \frac{\tau}{2}\left(\frac{\partial\psi}{\partial x_1}\right)^2 - \cdots - \frac{\tau}{2}\left(\frac{\partial\psi}{\partial x_n}\right)^2\right], \tag{12}$$

on the hyperplane $M \subset \mathscr{C}^1(D)$ consisting of those functions ψ that agree with φ on ∂D.

If we temporarily write $t = x_{n+1}$ and define f on $\mathscr{R}^{2n+3} = \mathscr{R}^{2n+1} \times \mathscr{R} \times \mathscr{R}^{2n+1}$ by

$$f(\mathbf{x}, y, \mathbf{z}) = -\frac{\tau}{2}(z_1{}^2 + \cdots + z_n{}^2) + \frac{\sigma}{2} z_{n+1}^2,$$

then we may rewrite (12) as

$$F(\psi) = \int_D f(\mathbf{x}, y, \mathbf{z})\, d\mathbf{x},$$

where $y = \psi(\mathbf{x})$ and $\mathbf{z} = \nabla\psi(\mathbf{x})$. Since

$$\frac{\partial f}{\partial y} = 0, \qquad \frac{\partial f}{\partial z_i} = -\tau z_i \quad \text{if} \quad i \leqq n, \quad \text{and} \quad \frac{\partial f}{\partial z_{n+1}} = \sigma z_{n+1},$$

it follows that the Euler–Lagrange equation (8) for this problem is

$$\tau\left(\sum_{i=1}^n \frac{\partial z_i}{\partial x_i}\right) - \sigma\frac{\partial z_{n+1}}{\partial x_{n+1}} = 0, \qquad \mathbf{z} = \nabla\varphi(\mathbf{x}),$$

or

$$\tau\, \nabla^2\varphi = \sigma\frac{\partial^2\varphi}{\partial t^2}. \tag{13}$$

Equation (13) is the n-dimensional wave equation.

APPENDIX

The Completeness of $\mathscr{R}$

In this appendix we give a self-contained treatment of the various consequences of the completeness of the real number system that are used in the text.

We start with the least upper bound axiom, regarding it as the basic completeness property of $\mathscr{R}$. The set S of real numbers is *bounded above* (respectively *below*) if there exists a number c such that $c \geqq x$ (respectively $c \leqq x$) for all $x \in S$. The number c is then called an *upper* (respectively *lower*) *bound* for S. The set S is *bounded* if it is both bounded above and bounded below.

A *least upper* (respectively *greatest lower*) *bound* for S is an upper bound (respectively lower bound) b such that $b \leqq c$ (respectively $b \geqq c$) for every upper (respectively lower) bound c for S. We can now state our axiom.

Least Upper Bound Axiom *If the set S of real numbers is bounded above, then it has a least upper bound.*

By consideration of the set $\{x \in \mathscr{R} : -x \in S\}$, it follows easily that, if S is bounded below, then S has a greatest lower bound.

The sequence $\{x_n\}_1^\infty$ of real numbers is called *nondecreasing* (respectively *nonincreasing*) if $x_n \leqq x_{n+1}$ (respectively $x_n \geqq x_{n+1}$) for each $n \geqq 1$. We call a sequence *monotone* if it is either nonincreasing or nondecreasing. The following theorem gives the "bounded monotone sequence property" of $\mathscr{R}$.

Theorem A.1 Every bounded monotone sequence of real numbers converges.

PROOF If, for example, the sequence $S = \{x_n\}_1^\infty$ is bounded and nondecreasing, and a is the least upper bound for S that is provided by the axiom, then it follows immediately from the definitions that $\lim_{n \to \infty} x_n = a$. ▮

The following theorem gives the "nested interval property" of $\mathscr{R}$.

Theorem A.2 If $\{I_n\}_1^\infty$ is a sequence of closed intervals ($I_n = [a_n, b_n]$) such that

(i) $I_n \supset I_{n+1}$ for each $n \geqq 1$, and

(ii) $\lim_{n\to\infty}(b_n - a_n) = 0$,

then there exists precisely one number c such that $c \in I_n$ for each n, so $c = \bigcap_{n=1}^\infty I_n$.

PROOF It is clear from (ii) that there is at most one such number c. The sequence $\{a_n\}_1^\infty$ is bounded and nondecreasing, while the sequence $\{b_n\}_1^\infty$ is bounded and nonincreasing. Therefore $a = \lim_{n\to\infty} a_n$ and $b = \lim_{n\to\infty} b_n$ exist by Theorem A.1. Since $a_n \leqq b_n$ for each $n \geqq 1$, it follows easily that $a \leqq b$. But then (ii) implies that $a = b$. Clearly this common value is a number belonging to each of the intervals $\{I_n\}_1^\infty$. ∎

We can now prove the intermediate value theorem.

Theorem A.3 If the function $f: [a, b] \to \mathscr{R}$ is continuous and $f(a) < 0 < f(b)$, then there exists $c \in (a, b)$ such that $f(c) = 0$.

PROOF Let $I_1 = [a, b]$. Having defined I_n, let I_{n+1} denote that closed half-interval of I_n such that $f(x)$ is positive at one endpoint of I_{n+1} and negative at the other. Then the sequence $\{I_n\}_1^\infty$ satisfies the hypotheses of Theorem A.2. If $c = \bigcap_{n=1}^\infty I_n$, then the continuity of f implies that $f(c)$ can be neither positive nor negative (why?), so $f(c) = 0$. ∎

Lemma A.4 If the function $f: [a, b] \to \mathscr{R}$ is continuous, then f is bounded on $[a, b]$.

PROOF Supposing to the contrary that f is not bounded on $I_1 = [a, b]$, let I_2 be a closed half-interval of I_1 on which f is not bounded. In general, let I_{n+1} be a closed half-interval of I_n on which f is not bounded.

If $c = \bigcap_{n=1}^\infty I_n$, then, by continuity, there exists a neighborhood U of c such that f is bounded on U (why?). But $I_n \subset U$ if n is sufficiently large. This contradiction proves that f is bounded on $[a, b]$. ∎

We can now prove the maximum value theorem.

Theorem A.5 If $f: [a, b] \to \mathscr{R}$ is continuous, then there exists $c \in [a, b]$ such that $f(x) \leqq f(c)$ for all $x \in [a, b]$.

PROOF The set $f([a, b])$ is bounded by Lemma A.4, so let M be its least upper bound; that is, M is the least upper bound of $f(x)$ on $[a, b]$. Then the least upper bound of $f(x)$ on at least one of the two closed half-intervals of $[a, b]$ is also M; denote it by I_1. Given I_n, let I_{n+1} be a closed half-interval of I_n on which the least upper bound of $f(x)$ is M. If $c = \bigcap_{n=1}^{\infty} I_n$, then it follows easily from the continuity of f that $f(c) = M$. ∎

As a final application of the "method of bisection," we establish the "Bolzano–Weierstrass property" of $\mathscr{R}$.

Theorem A.6 Every bounded, infinite subset S of $\mathscr{R}$ has a limit point.

PROOF If I_0 is a closed interval containing S, denote by I_1 one of the closed half-intervals of I_0 that contains infinitely many points of S. Continuing in this way, we define a nested sequence of intervals $\{I_n\}$, each of which contains infinitely many points of S. If $c = \bigcap_{n=1}^{\infty} I_n$, then it is clear that c is a limit point of S. ∎

We now work toward the proof that a sequence of real numbers converges if and only if it is a Cauchy sequence. The sequence $\{a_n\}_1^{\infty}$ is called a *Cauchy sequence* if, given $\varepsilon > 0$, there exists N such that

$$m, n \geqq N \Rightarrow |a_m - a_n| < \varepsilon.$$

It follows immediately from the triangle inequality that every convergent sequence is a Cauchy sequence.

Lemma A.7 Every bounded sequence of real numbers has a convergent subsequence.

PROOF If the sequence $\{a_n\}_1^{\infty}$ contains only finitely many distinct points, the the conclusion is trivial and obvious. Otherwise we are dealing with a bounded infinite set, to which the Bolzano–Weierstrass theorem applies, giving us a limit point a. If, for each integer $k \geqq 1$, a_{n_k} is a point of the sequence such that $|a_{n_k} - a| < 1/k$, then it is clear that $\{a_{n_k}\}_{k=1}^{\infty}$ is a convergent subsequence. ∎

Theorem A.8 Every Cauchy sequence of real numbers converges.

PROOF Given a Cauchy sequence $\{a_n\}_1^{\infty}$, choose N such that

$$m, n \geqq N \Rightarrow |a_m - a_n| < 1.$$

Then $a_n \in [a_N - 1, a_N + 1]$ if $n \geqq N$, so it follows that the sequence $\{a_n\}_1^{\infty}$ is bounded. By Lemma A.7 it therefore has convergent subsequence $\{a_{n_k}\}_1^{\infty}$.

If $a = \lim_{k\to\infty} a_{n_k}$, we want to prove that $\lim_{n\to\infty} a_n = a$. Given $\varepsilon > 0$, choose M such that

$$m, n \geqq M \Rightarrow |a_m - a_n| < \frac{\varepsilon}{2}.$$

Then choose K such that $n_K \geqq M$ and $|a_{n_K} - a| < \varepsilon/2$. Then

$$n \geqq M \Rightarrow |a_n - a| \leqq |a_n - a_{n_K}| + |a_{n_K} - a| < \varepsilon$$

as desired. ∎

The sequence $\{\mathbf{a}_n\}_1^\infty$ of points in $\mathscr{R}^k$ is called a *Cauchy sequence* if, given $\varepsilon > 0$, there exists N such that

$$m, n \geqq N \Rightarrow |\mathbf{a}_m - \mathbf{a}_n| < \varepsilon$$

(either Euclidean or sup norm). It follows easily, by coordinatewise application of Theorem A.8, that every Cauchy sequence of points in $\mathscr{R}^k$ converges.

Suggested Reading

One goal of this book is to motivate the serious student to go deeper into the topics introduced here. We therefore provide some suggestions for further reading in the references listed below.

For the student who wants to take another look at single-variable calculus, we recommend the excellent introductory texts by Kitchen [8] and Spivak [17]. Also several introductory analysis books, such as those by Lang [9], Rosenlicht [13], and Smith [16], begin with a review of single-variable calculus from a more advanced viewpoint.

Courant's treatment of multivariable calculus [3] is rich in applications, intuitive insight and geometric flavor, and is my favorite among the older advanced calculus books.

Cartan [1], Dieudonné [5], Lang [9], and Loomis and Sternberg [10] all deal with differential calculus in the context of normed vector spaces. Dieudonne's text is a classic in this area; Cartan's treatment of the calculus of normed vector spaces is similar but easier to read. Both are written on a considerably more abstract and advanced level than this book.

Milnor [11] gives a beautiful exposition of the application of the inverse function theorem to establish such results as the fundamental theorem of algebra (every polynomial has a complex root) and the Brouwer fixed point theorem (every continuous mapping of the n-ball into itself has a fixed point).

The method of successive approximations, by which we proved the inverse and implicit function theorems in Chapter III, also provides the best approach to the basic existence and uniqueness theorems for differential equations. For this see the chapters on differential equations in Cartan [1], Lang [9], Loomis and Sternberg [10], and Rosenlicht [13].

Sections 2 and 5 of Chapter IV were influenced by the chapter on multiple

(Riemann) integrals in Lang [9]. Smith [16] gives a very readable undergraduate-level exposition of Lebesgue integration in $\mathscr{R}^n$. In Section IV.6 we stopped just short of substantial applications of improper integrals. As an example we recommend the elegant little book on Fourier series and integrals by Seeley [14].

For an excellent discussion of surface area, see the chapter on this topic in Smith [16]. Cartan [2] and Spivak [18] give more advanced treatments of differential forms; in particular, Spivak's book is a superb exposition of an alternative approach to Stokes' theorem. Flanders [6] discusses a wide range of applications of differential forms to geometry and physics. The best recent book on elementary differential geometry is that of O'Neill [12]; it employs differential forms on about the same level as in this book. Our discussion of closed and exact forms in Section V.8 is a starting point for the algebraic topology of manifolds—see Singer and Thorpe [15] for an introduction.

Our treatment of the calculus of variations was influenced by the chapter on this topic in Cartan [2]. For an excellent summary of the classical applications see the chapters on the calculus of variations in Courant [3] and Courant and Hilbert [4]. For a detailed study of the calculus of variations we recommend Gel'fand and Fomin [7].

REFERENCES

[1] H. Cartan, "Differential Calculus," Houghton Mifflin, Boston, 1971.
[2] H. Cartan, "Differential Forms," Houghton Mifflin, Boston, 1970.
[3] R. Courant, "Differential and Integral Calculus," Vol. II, Wiley (Interscience), New York, 1937.
[4] R. Courant and D. Hilbert, "Methods of Mathematical Physics," Vol. I, Wiley (Interscience), New York, 1953.
[5] J. Dieudonné, "Foundations of Modern Analysis," Academic Press, New York, 1960.
[6] H. Flanders, "Differential Forms: With Applications to the Physical Sciences," Academic Press, New York, 1963.
[7] I. M. Gel'fand and S. V. Fomin, "Calculus of Variations," Prentice-Hall, Englewood Cliffs, New Jersey, 1963.
[8] J. W. Kitchen, Jr., "Calculus of One Variable," Addison-Wesley, Reading, Massachusetts, 1968.
[9] S. Lang, "Analysis I," Addison-Wesley, Reading, Massachusetts, 1968.
[10] L. H. Loomis and S. Sternberg, "Advanced Calculus," Addison-Wesley, Reading, Massachusetts, 1968.
[11] J. Milnor, "Topology from the Differentiable Viewpoint," Univ. Press of Virginia, Charlottesville, Virginia, 1965.
[12] B. O'Neill, "Elementary Differential Geometry," Academic Press, New York, 1966.

[13] M. Rosenlicht, "Introduction to Analysis," Scott-Foresman, Glenview, Illinois, 1968.
[14] R. T. Seeley, "An Introduction to Fourier Series and Integrals," Benjamin, New York, 1966.
[15] I. M. Singer and J. A. Thorpe, "Lecture Notes on Elementary Topology and Geometry," Scott-Foresman, Glenview, Illinois, 1967.
[16] K. T. Smith, "Primer of Modern Analysis," Bogden and Quigley, Tarrytown-on-Hudson, New York, 1971.
[17] M. Spivak, "Calculus," Benjamin, New York, 1967.
[18] M. Spivak, "Calculus on Manifolds," Benjamin, New York, 1965.

Subject Index

N

O

P

Q

A CATALOG OF SELECTED

DOVER BOOKS

IN SCIENCE AND MATHEMATICS

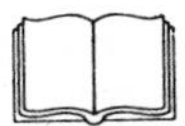

THE HISTORICAL BACKGROUND OF CHEMISTRY, Henry M. Leicester. Evolution of ideas, not individual biography. Concentrates on formulation of a coherent set of chemical laws. 260pp. 5⅜ x 8½. 61053-5

A SHORT HISTORY OF CHEMISTRY, J. R. Partington. Classic exposition explores origins of chemistry, alchemy, early medical chemistry, nature of atmosphere, theory of valency, laws and structure of atomic theory, much more. 428pp. 5⅜ x 8½. (Available in U.S. only.) 65977-1

GENERAL CHEMISTRY, Linus Pauling. Revised 3rd edition of classic first-year text by Nobel laureate. Atomic and molecular structure, quantum mechanics, statistical mechanics, thermodynamics correlated with descriptive chemistry. Problems. 992pp. 5⅜ x 8½. 65622-5

Engineering

DE RE METALLICA, Georgius Agricola. The famous Hoover translation of greatest treatise on technological chemistry, engineering, geology, mining of early modern times (1556). All 289 original woodcuts. 638pp. 6¾ x 11. 60006-8

FUNDAMENTALS OF ASTRODYNAMICS, Roger Bate et al. Modern approach developed by U.S. Air Force Academy. Designed as a first course. Problems, exercises. Numerous illustrations. 455pp. 5⅜ x 8½. 60061-0

DYNAMICS OF FLUIDS IN POROUS MEDIA, Jacob Bear. For advanced students of ground water hydrology, soil mechanics and physics, drainage and irrigation engineering and more. 335 illustrations. Exercises, with answers. 784pp. 6⅛ x 9¼. 65675-6

ANALYTICAL MECHANICS OF GEARS, Earle Buckingham. Indispensable reference for modern gear manufacture covers conjugate gear-tooth action, gear-tooth profiles of various gears, many other topics. 263 figures. 102 tables. 546pp. 5⅜ x 8½. 65712-4

MECHANICS, J. P. Den Hartog. A classic introductory text or refresher. Hundreds of applications and design problems illuminate fundamentals of trusses, loaded beams and cables, etc. 334 answered problems. 462pp. 5⅜ x 8½. 60754-2

MECHANICAL VIBRATIONS, J. P. Den Hartog. Classic textbook offers lucid explanations and illustrative models, applying theories of vibrations to a variety of practical industrial engineering problems. Numerous figures. 233 problems, solutions. Appendix. Index. Preface. 436pp. 5⅜ x 8½. 64785-4

STRENGTH OF MATERIALS, J. P. Den Hartog. Full, clear treatment of basic material (tension, torsion, bending, etc.) plus advanced material on engineering methods, applications. 350 answered problems. 323pp. 5⅜ x 8½. 60755-0

A HISTORY OF MECHANICS, René Dugas. Monumental study of mechanical principles from antiquity to quantum mechanics. Contributions of ancient Greeks, Galileo, Leonardo, Kepler, Lagrange, many others. 671pp. 5⅜ x 8½. 65632-2

Math–Geometry and Topology

ELEMENTARY CONCEPTS OF TOPOLOGY, Paul Alexandroff. Elegant, intuitive approach to topology from set-theoretic topology to Betti groups; how concepts of topology are useful in math and physics. 25 figures. 57pp. $5\frac{3}{8} \times 8\frac{1}{2}$. 60747-X

COMBINATORIAL TOPOLOGY, P. S. Alexandrov. Clearly written, well-organized, three-part text begins by dealing with certain classic problems without using the formal techniques of homology theory and advances to the central concept, the Betti groups. Numerous detailed examples. 654pp. $5\frac{3}{8} \times 8\frac{1}{2}$. 40179-0

EXPERIMENTS IN TOPOLOGY, Stephen Barr. Classic, lively explanation of one of the byways of mathematics. Klein bottles, Moebius strips, projective planes, map coloring, problem of the Koenigsberg bridges, much more, described with clarity and wit. 43 figures. 210pp. $5\frac{3}{8} \times 8\frac{1}{2}$. 25933-1

CONFORMAL MAPPING ON RIEMANN SURFACES, Harvey Cohn. Lucid, insightful book presents ideal coverage of subject. 334 exercises make book perfect for self-study. 55 figures. 352pp. $5\frac{5}{8} \times 8\frac{1}{4}$. 64025-6

THE GEOMETRY OF RENÉ DESCARTES, René Descartes. The great work founded analytical geometry. Original French text, Descartes's own diagrams, together with definitive Smith-Latham translation. 244pp. $5\frac{3}{8} \times 8\frac{1}{2}$. 60068-8

THE THIRTEEN BOOKS OF EUCLID'S ELEMENTS, translated with introduction and commentary by Sir Thomas L. Heath. Definitive edition. Textual and linguistic notes, mathematical analysis. 2,500 years of critical commentary. Unabridged. 1,414pp. $5\frac{3}{8} \times 8\frac{1}{2}$. Three-vol. set.
Vol. I: 60088-2 Vol. II: 60089-0 Vol. III: 60090-4

GEOMETRY OF COMPLEX NUMBERS, Hans Schwerdtfeger. Illuminating, widely praised book on analytic geometry of circles, the Moebius transformation, and two-dimensional non-Euclidean geometries. 200pp. $5\frac{5}{8} \times 8\frac{1}{4}$. 63830-8

DIFFERENTIAL GEOMETRY, Heinrich W. Guggenheimer. Local differential geometry as an application of advanced calculus and linear algebra. Curvature, transformation groups, surfaces, more. Exercises. 62 figures. 378pp. $5\frac{3}{8} \times 8\frac{1}{2}$. 63433-7

CURVATURE AND HOMOLOGY: Enlarged Edition, Samuel I. Goldberg. Revised edition examines topology of differentiable manifolds; curvature, homology of Riemannian manifolds; compact Lie groups; complex manifolds; curvature, homology of Kaehler manifolds. New Preface. Four new appendixes. 416pp. $5\frac{3}{8} \times 8\frac{1}{2}$. 40207-X

TOPOLOGY, John G. Hocking and Gail S. Young. Superb one-year course in classical topology. Topological spaces and functions, point-set topology, much more. Examples and problems. Bibliography. Index. 384pp. $5\frac{5}{8} \times 8\frac{1}{4}$. 65676-4

Physics

OPTICAL RESONANCE AND TWO-LEVEL ATOMS, L. Allen and J. H. Eberly. Clear, comprehensive introduction to basic principles behind all quantum optical resonance phenomena. 53 illustrations. Preface. Index. 256pp. 5⅜ x 8½. 65533-4

ULTRASONIC ABSORPTION: An Introduction to the Theory of Sound Absorption and Dispersion in Gases, Liquids and Solids, A. B. Bhatia. Standard reference in the field provides a clear, systematically organized introductory review of fundamental concepts for advanced graduate students, research workers. Numerous diagrams. Bibliography. 440pp. 5⅜ x 8½. 64917-2

QUANTUM THEORY, David Bohm. This advanced undergraduate-level text presents the quantum theory in terms of qualitative and imaginative concepts, followed by specific applications worked out in mathematical detail. Preface. Index. 655pp. 5⅜ x 8½. 65969-0

ATOMIC PHYSICS (8th edition), Max Born. Nobel laureate's lucid treatment of kinetic theory of gases, elementary particles, nuclear atom, wave-corpuscles, atomic structure and spectral lines, much more. Over 40 appendices, bibliography. 495pp. 5⅜ x 8½. 65984-4

AN INTRODUCTION TO HAMILTONIAN OPTICS, H. A. Buchdahl. Detailed account of the Hamiltonian treatment of aberration theory in geometrical optics. Many classes of optical systems defined in terms of the symmetries they possess. Problems with detailed solutions. 1970 edition. xv + 360pp. 5⅜ x 8½. 67597-1

THIRTY YEARS THAT SHOOK PHYSICS: The Story of Quantum Theory, George Gamow. Lucid, accessible introduction to influential theory of energy and matter. Careful explanations of Dirac's anti-particles, Bohr's model of the atom, much more. 12 plates. Numerous drawings. 240pp. 5⅜ x 8½. 24895-X

ELECTRONIC STRUCTURE AND THE PROPERTIES OF SOLIDS: The Physics of the Chemical Bond, Walter A. Harrison. Innovative text offers basic understanding of the electronic structure of covalent and ionic solids, simple metals, transition metals and their compounds. Problems. 1980 edition. 582pp. 6⅛ x 9¼. 66021-4

HYDRODYNAMIC AND HYDROMAGNETIC STABILITY, S. Chandrasekhar. Lucid examination of the Rayleigh-Benard problem; clear coverage of the theory of instabilities causing convection. 704pp. 5⅜ x 8¼. 64071-X

INVESTIGATIONS ON THE THEORY OF THE BROWNIAN MOVEMENT, Albert Einstein. Five papers (1905–8) investigating dynamics of Brownian motion and evolving elementary theory. Notes by R. Fürth. 122pp. 5⅜ x 8½. 60304-0

THE PHYSICS OF WAVES, William C. Elmore and Mark A. Heald. Unique overview of classical wave theory. Acoustics, optics, electromagnetic radiation, more. Ideal as classroom text or for self-study. Problems. 477pp. 5⅜ x 8½. 64926-1

METHODS OF THERMODYNAMICS, Howard Reiss. Outstanding text focuses on physical technique of thermodynamics, typical problem areas of understanding, and significance and use of thermodynamic potential. 1965 edition. 238pp. 5⅜ x 8½.
69445-3

TENSOR ANALYSIS FOR PHYSICISTS, J. A. Schouten. Concise exposition of the mathematical basis of tensor analysis, integrated with well-chosen physical examples of the theory. Exercises. Index. Bibliography. 289pp. 5⅜ x 8½. 65582-2

RELATIVITY IN ILLUSTRATIONS, Jacob T. Schwartz. Clear nontechnical treatment makes relativity more accessible than ever before. Over 60 drawings illustrate concepts more clearly than text alone. Only high school geometry needed. Bibliography. 128pp. 6⅛ x 9¼. 25965-X

THE ELECTROMAGNETIC FIELD, Albert Shadowitz. Comprehensive undergraduate text covers basics of electric and magnetic fields, builds up to electromagnetic theory. Also related topics, including relativity. Over 900 problems. 768pp. 5⅝ x 8¼. 65660-8

GREAT EXPERIMENTS IN PHYSICS: Firsthand Accounts from Galileo to Einstein, edited by Morris H. Shamos. 25 crucial discoveries: Newton's laws of motion, Chadwick's study of the neutron, Hertz on electromagnetic waves, more. Original accounts clearly annotated. 370pp. 5⅜ x 8½. 25346-5

RELATIVITY, THERMODYNAMICS AND COSMOLOGY, Richard C. Tolman. Landmark study extends thermodynamics to special, general relativity; also applications of relativistic mechanics, thermodynamics to cosmological models. 501pp. 5⅜ x 8½. 65383-8

LIGHT SCATTERING BY SMALL PARTICLES, H. C. van de Hulst. Comprehensive treatment including full range of useful approximation methods for researchers in chemistry, meteorology and astronomy. 44 illustrations. 470pp. 5⅜ x 8½.
64228-3

STATISTICAL PHYSICS, Gregory H. Wannier. Classic text combines thermodynamics, statistical mechanics and kinetic theory in one unified presentation of thermal physics. Problems with solutions. Bibliography. 532pp. 5⅜ x 8½. 65401-X

Paperbound unless otherwise indicated. Available at your book dealer, online at **www.doverpublications.com**, or by writing to Dept. GI, Dover Publications, Inc., 31 East 2nd Street, Mineola, NY 11501. For current price information or for free catalogues (please indicate field of interest), write to Dover Publications or log on to **www.doverpublications.com** and see every Dover book in print. Dover publishes more than 500 books each year on science, elementary and advanced mathematics, biology, music, art, literary history, social sciences, and other areas.